PROCEEDINGS

OF THE

1993 INTERNATIONAL CONFERENCE

ON

PARALLEL PROCESSING

August 16 – 20, 1993

Vol. I Architecture
C.Y. Roger Chen and P. Bruce Berra, Editors
Syracuse University

Sponsored by

THE PENNSYLVANIA STATE UNIVERSITY

CRC Press
Boca Raton Ann Arbor Tokyo London

ISSN 0190-3918
ISBN 0-8493-8983-6 (set)
ISBN 0-8493-8984-4 (vol. I)
ISBN 0-8493-8985-2 (vol. II)
ISBN 0-8493-8986-0 (vol. III)
IEEE Computer Society Order Number 4500-22

Additional copies may be obtained from:

CRC Press, Inc.
2000 Corporate Blvd., N.W.
Boca Raton, Florida 33431

Preface

Interest in parallel processing continues to be high; this year we received 477 papers from 28 different countries, representing a multiplicity of disciplines. In order to accommodate as many presentations as possible and yet maintain the high quality of the conference, we have accepted papers as regular and concise. Regular papers offer well-conceived ideas, good results, and clear presentations. While concise papers also present well-conceived ideas and good results, they do so in fewer pages. The table below summarizes the number of submissions by area:

Area	Submitted	Accepted	
		Regular	Concise
Architecture	166	16 (9.6%)	42 (25.3%)
Software	154	16 (10.4%)	39 (25.3%)
Algorithms/Applications	157	21 (13.4%)	31 (19.7%)
Total	477	53 (11.1%)	112 (23.5%)

This year the program activity was structured slightly differently than in past years since we had a program chair and three co-program chairs. With this organization we worked as a team and assigned papers to the three primary areas in somewhat of a balanced mode. Then each of the co-program chairs was largely responsible for their area. Each submitted paper was reviewed by at least three external reviewers. To avoid conflicts of interest, we did not process the papers from our colleagues at Syracuse University. These were handled by Professor Chita Das of Pennsylvania State University. We are very grateful for his professional and prompt coordination of the reviewing process.

We can truly state that the refereeing process ran much more smoothly than we had originally expected. We attribute this to two main reasons. The first is that we sent one copy of each paper outside of the United States for review; these reviews were thorough and timely. The second reason is that we used e-mail almost exclusively. We would like to express our sincere appreciation to the referees for their part in making the selection process a success. The quality of the conference can be maintained only through such strong support.

This year's keynote speaker is Dr. Ken Kennedy, an internationally known researcher in parallel processing. In addition, we have two super panels chaired by Dr. H.J. Siegel and Dr. Kai Hwang who are also internationally known in parallel processing. We are honored to have Dr.'s Kennedy, Siegel and Hwang sharing their visions on parallel processing with us.

Finally, the success of a conference rests with many people. We would like to thank Dr. Tse-yun Feng for his valuable guidance in the preparation of the program. In addition many graduate and undergraduate students aided us in the process and we gratefully acknowledge their help. We would like to thank Faraz Kohari for his help with the database management functions and Emily Yakawiak for putting the proceedings together. We gratefully acknowledge the New York State Center for Advanced Technology in Computer Applications and Software Engineering (CASE) and the Department of Electrical and Computer Engineering for their support.

P. Bruce Berra, Program Chair
C.Y. Roger Chen, Co-Program Chair-Architecture
Alok Choudhary, Co-Program Chair-Software
Salim Hariri, Co-Program Chair-Algorithms/Applications

The Case Center
Electrical and Computer Engineering
Syracuse University
Syracuse, NY 13244

Abdennadher, N.	Bhagavathi, D.	Christopher, T.	Eswar, K.
Aboelaze, M. A.	Bhuyan, L.	Chu, D.	Etiemble, D.
Abu-Ghazaleh, N.	Bhuyan, L. N.	Chung, S. M.	Evans, J.
AbuAyyash, S.	Bianchini, R.	Clary, J. S.	Exman, I.
Agarwal, A.	Bic, L.	Clifton, C.	Farkas, K.
Agarwal, M.	Bitzaros, S.	Cloonan, T. J.	Felderman, R. E.
Agha, G.	Blackston, D.	Cook, G.	Feng, C.
Agrawal, D. P.	Boku, T.	Copty, N.	Feng, W.
Agrawal, P.	Bordawekar, R.	Cortadella, J.	Fernandes, R.
Agrawal, V.	Boura, Y.	Costicoglou, S.	Field, A. J.
Alijani, G.	Bove Jr., V. M.	Crovella, M.	Fienup, M. A.
Amano, H.	Brent, R. P.	Cukic, B.	Fraser, D.
Andre, F.	Breznay, P.	Cytron, R.	Fu, J.
Annaratone, M.	Brown, J. C.	Dahlgren, F.	Ganger, G.
Antonio, J. K.	Burago, A.	Dandamudi, S.	Garg, S.
Anupindi, K.	Burkhardt, W. H.	Das, A.	Genjiang, Z.
Applebe, B.	Butler, M.	Das, C.	Gessesse, G.
Arif, G.	Cam, H.	Das, C. R.	Gewali, L. P.
Ariyawansa, K. A.	Capel, M.	Das, S.	Ghafoor, A.
Arthur, R. M.	Casavant, T. L.	Das, S. K.	Ghonaimy, M. A. R.
Arunachalam, M.	Casulleras, J.	Datta, A. K.	Ghosh, K.
Atiquzzaman, M.	Cavallaro, J.	Davis, M. H.	Ghozati, S. A.
Audet, D.	Celenk, M.	Debashis, B.	Gibson, G. A.
Avalani, B.	Chalasani, S.	Defu, Z.	Gokhale, M.
Ayani, R.	Chamberlain, R.	Delgado-Frias, J. G.	Goldberg
Ayguade, E.	Chandra, A.	Demuynck, M.	Gong, C.
Babb, J.	Chandy, J.	Deshmukh, R. G.	Gonzalez, M.
Babbar, D.	Chang, M. F.	Dimopoulos, N. J.	Gornish, E.
Bagherzadeh, N.	Chang, Y.	Dimpsey, R.	Granston, E.
Baglietto, P.	Chao, D.	Douglass, B.	Greenwood, G. W.
Baldwin, R.	Chao, J.	Dowd, P. W.	Grimshaw, A. S.
Banerjee, C.	Chao, L.	Drach, N.	Gupta, A. K.
Banerjee, P.	Chase, C.	Du, D. H. C.	Gupta, S.
Banerjee, S.	Chase, C. M.	Duesterwald, E.	Gupta, S. K.
Barada, H.	Chen, C.	Duke, D. W.	Gurla, H.
Barriga, L.	Chen, C. L.	Durand, D.	Gursoy, A.
Basak, D.	Chen, D.	Dutt, N.	Haddad, E.
Baskiyar, S.	Chen, K.	Dutt, S.	Hamdi, M.
Bastani, F.	Chen, S.	Dwarkadas, S.	Hameurlain, A.
Bayoumi, M. A.	Chen, Y.	Edirisooriya, S.	Han, Y.
Beauvais, J.	Cheng, J.	Efe, K.	Hanawa, T.
Beckmann, C.	Cheong, H.	El-Amawy, A.	Hao, Y.
Bellur, U.	Chiang, C.	Elmohamed, S.	Harathi, K.
Benson, G. D.	Chien, C.	Enbody, R. J.	Harper, M.
Berkovich, S.	Chiueh, T.	Erdogan, S.	Hauck, S.
Berson, D.	Choi, J. H.	Erdogan, S. S.	Heiss, H.

Novack, S.	Raatikainen, P.	Sheu, J.	Tout, W.
Nutt, G.	Radia, N.	Sheu, T.	Traff, J. L.
Obeng, M. S.	Rafieymehr, A.	Shi, H.	Trahan, J.
Oehring, S.	Raghavan, R.	Shin, K. G.	Tsai, W.
Oh, H.	Raghavendra, C. S.	Shing, H.	Tseng, W.
Okabayashi, I.	Raghunath, M. T.	Shiokawa, S.	Tseng, Y.
Okawa, Y.	Rajagopalan, U.	Shirazi, B.	Tzeng, N.
Olariu, S.	Ramachandran, U.	Shoari, S.	Ulusoy, O.
Olive, A.	Ramany, S.	Shu, W.	Unrau, R.
Omer, J.	Ravikumar, C.	Sibai, F. N.	Vaidya, N.
Omiecinski, E.	Ray, S.	Siegel, H. J.	Vaidyanathan, R.
Oruc, A. Y.	Reese, D.	Simms, D.	Valero-Garcia, M.
Ouyang, P.	Reichmeyer, F.	Sinclair, J. B.	Varma, G.
Paden, R.	Rigoutsos, I.	Singh, S.	Varma, G. S. D.
Panda, D.	Ripoli, A.	Singh, U.	Varman, P. J.
Panda, D. K.	Robertazzi, T.	Sinha, A.	deVel, O.
Pao, D.	Rokusawa, K.	Sinha, A. B.	Verma, R. M.
Parashar, M.	Rowland, M.	Sinha, B. P.	Vuppala, V.
Park, B. S.	Rowley, R.	Slimani, Y.	Wagh, M.
Park, C.	Roysam, B.	Snelick, R.	Wahab, A.
Park, H.	Ruighaver, A. B.	So, K.	Wakatani, A.
Park, J.	Ryan, C.	Soffa, M.	Wallace, D.
Park, K. H.	Saghi, G.	Son, S.	Wang, C.
Park, S.	Saha, A.	Song, J.	Wang, D.
Park, Y.	Sajeev, A.	Song, Q. W.	Wang, D. T.
Parsons, I.	Salamon, A.	Srimani, P. K.	Wang, H.
Patnaik, L. M.	Salinas, J.	Su, C.	Wang, H. C.
Pears, A. N.	Sarikaya, B.	Suguri, T.	Wang, M.
Pei-Yung-Hsiao	Sass, R.	Sun, X.	Watts, T.
Perkins, S.	Schaeffer, J.	Sundareswaran, P.	Weissman, J. B.
Peterson, G.	Schall, M.	Sunderam, V.	Wen, Z.
Petkov, N.	Schoinas, I.	Sussman, A.	Wills, S.
Pfeiffer, P.	Schwederski, T.	Sy, Y. K.	Wilson, A.
Phanindra, M.	Schwiebert, L.	Szafron, D.	Wilson, D.
Picano, S.	Seigel, H. J.	Szymanski, T.	Wilton, S. F.
Pinto, A. D.	Seo, K.	Takizawa, M.	Wilton, S. J. E.
Pissinou, N.	Sha, E.	Tan, K.	Wittie, L.
Podlubny, I.	Shah, G.	Tandri, S.	Wong, W.
Poulsen, D.	Shang, W.	Taylor, V. E.	Wu, C.
Pourzandi, M.	Shankar, R.	Tayyab, A.	Wu, C. E.
Pradhan, D. K.	Sharp, D.	Temam, O.	Wu, H.
Pramanick, I.	Sheffler, T. J.	Thakur, R.	Wu, J.
Prasanna, V. K.	Sheikh, S.	Thapar, M.	Wu, M.
Pravin, D.	Shekhar	Thayalan, K.	Wu, M. Y.
Puthukattukaran, J. J.	Shen, X.	Thekkath, R.	Xu, C.
Qiao, C.	Sheng, M. J.	Torrellas, J.	Xu, H.

Xu, Z.
Yacoob, Y.
Yalamanchili, S.
Yamashita, H.
Yang, C.
Yang, C. S.
Yang, Y.
Yen, I.
Yeung, D.
Youn, H. Y.
Young, C.
Young, H.
Young, H. C.
Youngseun, K.
Yousif, M.
Yu, C.
Yu, C. S.
Yu, S.
Yuan, S.
Yuan, S. M.
Yum, T. K.
Zaafrani
ZeinElDine, O.
Zeng, N.
Zhang, J.
Zhang, X.
Zhen, S. Q.
Zheng, S. Q.
Zhou, B. B.
Zhu, H.
Ziavras, S. G.
Zievers, W. C.
Ziv, A.
Zubair, M.

Volume I = Architecture
Volume II = Software
Volume III = Algorithms & Applications

TABLE OF CONTENTS
VOLUME I - ARCHITECTURE

xiv

SESSION 1A

CACHE MEMORY (I)

Automatic Partitioning of Parallel Loops for Cache-Coherent Multiprocessors

Anant Agarwal, David Kranz
Laboratory for Computer Science
Massachusetts Institute of Technology
Cambridge, MA 02139

Venkat Natarajan
Motorola Cambridge Research Center
Cambridge, MA 02139

Abstract

This paper presents a theoretical framework for automatically partitioning parallel loops to minimize cache coherency traffic on shared-memory multiprocessors. While several previous papers have looked at hyperplane partitioning of iteration spaces to reduce communication traffic, the problem of deriving the *optimal* tiling parameters for minimal communication in loops with general affine index expressions and multiple arrays has remained open. Our paper solves this open problem by presenting a method for deriving an optimal hyperparallelepiped tiling of iteration spaces for minimal communication in multiprocessors with caches. We show that the same theoretical framework can also be used to determine optimal tiling parameters for data and loop partitioning in distributed memory multiprocessors without caches. Like previous papers, our framework uses matrices to represent iteration and data space mappings and the notion of uniformly intersecting references to capture temporal locality in array references. We introduce the notion of data footprints to estimate the communication traffic between processors and use lattice theory to compute precisely the size of data footprints. We have implemented a subset of this framework in a compiler for the Alewife machine.

1 Introduction

Cache-based multiprocessors are attractive because they seem to allow the programmer to ignore the issues of data partitioning and placement. Because caches dynamically copy data close to where it is needed, repeat references to the same piece of data do not require communication over the network, and hence reduce the need for careful data layout. However, the performance of cache-coherent systems is heavily predicated on the degree of temporal locality in the access patterns of the processor. Loop partitioning for cache-coherent multiprocessors is an effort to increase the percentage of references that hit in the cache.

The degree of reuse of data, or conversely, the volume of communication, depends both on the algorithm and on the partitioning of work among the processors. (In fact, partitioning of the computation is often considered to be a facet of an algorithm.) For example, it is well known that a matrix multiply computation

distributed to the processors by square blocks has a much higher degree of reuse than the matrix multiply distributed by rows or columns.

Loop partitioning can be done by the programmer, by the run time system, or by the compiler. Relegating the partitioning task to the programmer defeats the central purpose of building cache-coherent shared-memory systems. While partitioning can be done at run time (for example, see [1, 2]), it is hard for the run time system to optimize for cache locality because much of the information required to compute communication patterns is either unavailable at run time or expensive to obtain. Thus compile-time partitioning of parallel loops is important.

This paper focuses on the following problem in the context of cache-coherent multiprocessors. Given a program consisting of parallel do loops (of the form shown in Figure 1 in Section 2.1), how do we derive the optimal tile shapes of the iteration-space partitions to minimize the communication traffic through the network.

1.1 Contributions and Related Work

This paper develops a unified theoretical framework for loop partitioning in cache-coherent multiprocessors, or for loop and data partitioning in multicomputers with local memory.[1] The central contribution of this paper is a method for deriving an optimal hyperparallelepiped tiling of iteration spaces to minimize communication. The tiling specifies both the shape and size of iteration space tiles. Our framework allows the partitioning of doall loops accessing multiple arrays, where the index expressions in array accesses can be any affine function of the indices.

Our analysis uses the notion of uniformly intersecting references to categorize the references within a loop into classes that will yield cache locality. This notion helps specify precisely the set of references that have substantially overlapping data sets. Overlap produces temporal locality in cache accesses. A similar concept of uniformly generated references has been used in earlier work in the context of *reuse* and iteration space tiling [3, 4].

The notion of data footprints is introduced to capture the combined set of data accesses made by references within each uni-

---

[1]This paper, however, focuses on loop partitioning, but indicates the modifications necessary for data partitioning.

formly intersecting class. (The term *footprint* was originally coined by Stone and Thiebaut [5].) Then, an algorithm to compute the total size of the data footprint for a given loop partition is presented. While general optimization methods can be applied to minimize the size of the data footprint and derive the corresponding loop partitions, we demonstrate several important special cases where the optimization problem is very simple. The size of data footprints can also be used to guide program transformations to achieve better cache performance in uniprocessors as well.

Although there have been several papers on hyperplane partitioning of iteration spaces, the problem of deriving the optimal hyperparallelepiped tile parameters for general affine index expressions has remained open.

For example, Irigoin and Triolet [6] introduce the notion of loop partitioning with multiple hyperplanes which results in hyperparallelepiped tiles. The purpose of tiling in their case is to provide parallelism across tiles, and vector processing and data locality within a tile. They propose a set of basic constraints that should be met by any partitioning and derive the conditions under which the hyperplane partitioning satisfies these constraints, but they do not address the issue of automatically generating the tile parameters. Our paper describes an algorithm for automatically computing the partition based on the notion of cumulative footprints, derived from the mapping from iteration space to data space.

Abraham and Hudak [7] considered loop partitioning in multiprocessors with caches. However, they dealt only with index expressions of the form index variable plus a constant. They assumed that the array dimension was equal to the loop nesting and focused on rectangular and hexagonal tiles. Furthermore, the code body was restricted to an update of $A[i, j]$. Our framework, however, is able to handle much more general index expressions, and produce parallelogram partitions if desired. We also show that when Abraham and Hudak's methods can be applied to a given loop nest, our theoretical framework reproduces their results.

Ramanujam and Sadayappan [8] deal with data partitioning in multicomputers with local memory and use a matrix formulation; their results do not apply to multiprocessors with caches. Their theory produces communication-free hyperplane partitions for loops with affine index expressions when such partitions exist. However, when communication-free partitions do not exist, they can deal only with index expression of the form variable plus a constant offset. They further require the loop dimension to be equal to the loop nesting. In contrast, our framework is able to discover optimal partitions in cases where communication free partitions are not possible, and we do not restrict the loop nesting to be equal to array dimension. In addition, we show that our framework correctly produces partitions identical to theirs when communication free partitions do exist.

Gupta and Banerjee [9] also address the problem of automatic data partitioning, but they do not address general hyperparallelepiped data distribution, they do not consider caches, and they deal with simple index expressions of the form $c_1 * i + c_2$ and not a general affine function of the loop indices.

Our work complements the work of Wolfe and Lam [3] and Schreiber and Dongarra [10]. Wolfe and Lam derive loop trans-

formations (and tile the iteration space) to improve data locality in multiprocessors with caches. They use matrices to model transformations and use the notion of equivalence classes within the set of uniformly generated references to identify valid loop transformations to improve the degree of temporal and spatial locality within a given loop nest. Schreiber and Dongarra briefly address the problem of deriving optimal hyperparallelepiped iteration space tiles to minimize communication traffic (they refer to it as I/O requirements). However their work differs from this paper in the following ways: (1) Their machine model does not have a processor cache. (2) The data space corresponding to an array reference and the iteration space are isomorphic. These restrictions make the problem of computing the communication traffic much simpler. Also, one of the main issues addressed by Schreiber and Dongarra is the *atomicity requirement* of the tiles which is related to dependence vectors. This paper is not concerned with that requirement as it is assumed that the iterations can be executed in parallel.

Ferrante, Sarkar, and Thrash [11] address the problem of estimating the number of cache misses for a nest of loops. This problem is similar to our problem of finding the size of the cumulative footprint, but differs in these ways: (1) We consider a tile in the iteration space and not the entire iteration space; our tiles can be hyperparallelepipeds in general. (2) We partition the references into uniformly intersecting sets, which makes the problem computationally more tractable, since it allows us to deal with only the tile at the origin. (3) Our treatment of coupled subscripts is much simpler, since we look at maximal independent columns, as shown in Section 3.4.1.

1.2 Overview of the Paper

The rest of this paper is structured as follows. Section 2 states our system model and our program-level assumptions. Section 3 first presents a few examples to illustrate the basic ideas behind loop partitioning; it then develops the theoretical framework for partitioning and presents several additional examples. A subset of the framework including both loop and data partitioning has been implemented in the compiler system for the Alewife multiprocessor. The implementation produces rectangular partitions only. The implementation of our compiler system and a sampling of results is presented in Section 4, and Section 5 concludes the paper.

2 Problem Domain and Assumptions

We address the problem of partitioning loops in cache-coherent shared-memory multiprocessors. Partitioning involves deciding which loop iterations, referred to as a *loop tile*, will run collectively in a thread of computation such that the resulting the volume of communication generated in the system is minimized. This section describes the types of programs currently handled by our framework and the structure of the system assumed by our analysis.

```
Doall (i1=l1:u1, i2=l2:u2, ..., il=ll:ul)
   loop body
EndDoall
```

Figure 1: Structure of a single loop nest

2.1 Program Assumptions

Figure 1 shows the structure of the most general single loop nest that we consider in this paper. The statements in the *loop body* have array references of the form $A[\vec{g}(i_1, i_2, \ldots, i_l)]$, where the index function is $\vec{g} : \mathcal{Z}^l \rightarrow \mathcal{Z}^d$, l is the loop nesting and d is the dimension of the array A. We assume that all array references within the loop body are unconditional.

We address the problem of loop and data partitioning for index expressions that are affine functions of loop indices. In other words, the index function can be expressed as,

$$\vec{g}(\vec{i}) = \vec{i}\mathbf{G} + \vec{a} \qquad (1)$$

where $\mathbf{G}$ is a $l \times d$ matrix with integer entries and $\vec{a}$ is an integer constant vector of length d, termed the *offset vector*. Note that $\vec{i}$, $\vec{g}(\vec{i})$, and $\vec{a}$ are row vectors. We often refer to an array reference by the pair $(\mathbf{G}, \vec{a})$. (An example of this function is presented in Section 3.1). Similar notation has been used in several papers in the past, for example, see [3, 4]. All our vectors and matrices have integer entries unless stated otherwise. We assume that the loop bounds are such that the iteration space is rectangular. However, we note that our methods can still be used to derive reasonable partitions when this condition is not met. Loop indices are assumed to take all integer values between their lower and upper bounds, i.e, the strides are one.

Given P processors, the problem of loop partitioning is to divide the iteration space into P tiles such that the total communication traffic on the network is minimized with the additional constraint that the tiles are of equal size, except at the boundaries of the iteration space. The constraint of equal size partitions is imposed to achieve load balancing. We restrict our discussions to parallelepiped tiles with emphasis on rectangular tiles. Rectangular tiles are important because it is easy to produce efficient code when the tile boundaries are simple expressions.

Like [7, 8], we do not include the effects of synchronization in our framework. Synchronization is handled separately to ensure correct behavior. For example, in Figure 1, one might introduce a barrier synchronization after the loop nest if so desired. We also note that fine-grain data-level synchronization can be used within a parallel do loop to enforce data dependencies and its cost approximately modeled as slightly more expensive communication than usual. See [12] for more details.

2.2 System Model

We assume that the system comprises a set of processors, each with a coherent cache. Cache misses are satisfied by global memory accessed over an interconnection network or a bus. The memory can be implemented as a single monolithic module (as is commonly done in bus-based multiprocessors), or in a distributed fashion. The memory modules might also be implemented on the processing nodes themselves (data partitioning for locality makes sense only for this case). In all cases, our analysis assumes that the cost of a main memory access is much higher than a cache access, and that the cost of the main memory access is the same no matter where in main memory the data is located.

The goal of loop partitioning is to minimize the total number of main memory accesses. For simplicity, we assume that the caches are large enough to hold all the data required by a loop partition, and that there are no conflicts in the caches. When caches are small, the optimal loop partition does not change, rather, the size of each loop tile executed at any given time on the processor must be adjusted so that the data fits in the cache (if we assume that the cache is effectively flushed between executions of each loop tile). We assume that cache lines are of unit length. The effect of larger cache lines can be included easily as suggested in [7].

3 Loop Partitions and Footprints

This section develops a framework for compile time analysis of loops for performing optimal loop partitioning for a cache-coherent multiprocessor. After presenting two illustrative examples, we present the definition of a loop partition. The notion of the loop partition is similar to the clustering basis of Irigoin and Triolet [6]. We then introduce the notion of a *footprint* of a loop partition with respect to a data reference in the loop. Footprints specify the data elements referenced within a loop partition. We then present the concept of uniformly intersecting references and a method of computing the cumulative footprint for a set of uniformly intersecting references. We develop a formalism for computing the volume of communication on the interconnection network of a multiprocessor for a given loop partition, and show how loop tiles can be chosen to minimize this traffic. We briefly indicate how the cumulative footprint can be used to derive optimal data partitions for multicomputers with local memory.

3.1 Examples

This section presents examples to illustrate some of our definitions and to motivate the benefits of optimizing the shapes of loop tiles. As mentioned previously, we deal with index expressions that are affine functions of loop indices. In other words, the index function can be expressed as in Equation 1. Consider the following example to illustrate the above expression of index functions.

Example 1 *The reference* $A[i_3 + 2, 5, i_2 - 1, 4]$ *in a triply nested loop can be expressed by*

$$(i_1, i_2, i_3) \begin{bmatrix} 0 & 0 & 0 & 0 \\ 0 & 0 & 1 & 0 \\ 1 & 0 & 0 & 0 \end{bmatrix} + (2, 5, -1, 4)$$

In this example, the second and fourth column of $\mathbf{G}$ are zero indicating that the second and fourth subscript of the reference

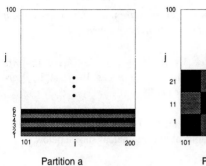
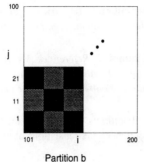

Figure 2: Two simple rectangular loop partitions in the iteration space.

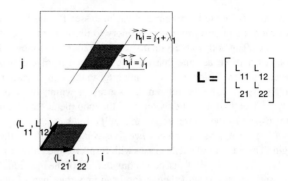

Figure 3: Iteration space partitioning is completely specified by the tile at the origin.

is independent of the loop indexes. In such cases, we show in Section 3.4.1 that we can ignore those columns and treat the referenced array as an array of lower dimension. In future, without loss of generality, we assume that the $\mathbf{G}$ matrix contains no zero columns.

Now, let us introduce the concept of a loop partition by examining the following simple example.

Example 2

Doall (i=101:200, j=1:100)
  A[i,j] = B[i+j,i-j-1]+B[i+j+4,i-j+3]
EndDoall

Let us assume that we have 100 processors and we want to distribute the work among them. There are 10,000 points in the iteration space and so one can allocate 100 of these to each of the processors to distribute the load uniformly. Figure 2 shows two simple ways of partitioning the iteration space into 100 equal tiles.

Let us compare the two partitions in the context of a system with caches and uniform access memory by computing the number of cache misses. For each tile in partition **a**, the number of cache misses can be shown to be 104 whereas the number of cache misses in each tile of partition **b** can be shown to be 140. Although not obvious from the source code, partition **a** is a better choice if our goal is to minimize the cost of memory accesses. In fact, partition **a** has minimum memory traffic. In addition to producing coherence-traffic-free partitions when they exist, our analysis will discover hyperparallelepiped partitions that minimize traffic when such partitions are non-existent as well.

3.2 Loop Tiles in the Iteration Space

Loop partitioning results in a tiling of the iteration space. Loop partitioning is sometimes termed iteration space partitioning. A specific hyperparallelepiped loop tile is defined by a set of bounding hyperplanes. Similar formulations have also been used earlier [6].

Definition 1 *Given a l dimensional loop nest $\vec{i}$, each tile of a hyperparallelepiped loop partition is defined by the hyperplanes given by the rows of the $l \times l$ matrix $\mathbf{H}$ and the column vectors $\vec{\gamma}$ and $\vec{\lambda}$ as follows. The parallel hyperplanes are $\vec{h}_j \vec{i} = \gamma_j$ and $\vec{h}_j \vec{i} = \gamma_j + \lambda_j$, for $1 \leq j \leq l$. An iteration belongs to this tile if it is on or inside the hyperparallelepiped.*

We consider only hyperparallelepiped partitions in this paper; rectangular partitions are special cases of these. Furthermore, we focus on loop partitioning where the tiles are homogeneous except at the boundaries of the iteration space. Under these conditions of homogeneous tiling, the partitioning is completely defined by specifying the tile at the origin, namely $(\mathbf{H}, \vec{0}, \vec{\lambda})$, as indicated in Figure 3; the rest of this paper will deal with producing the shape of this tile. Under homogeneous tiling, the concept of the tile at the origin is similar to the notion of the clustering basis in [6]. For notational convenience, we denote the tile at the origin as $\mathbf{L}$. The rows of matrix $\mathbf{L}$ correspond to the vectors defining the tile at the origin as shown in Figure 3. See [12] for a more precise definition.

3.3 Footprints in the Data Space

For a system with caches and uniform access memory, the problem of loop partitioning is to find an optimal matrix $\mathbf{L}$ that minimizes the number of cache misses. The first step is to derive an expression for the number of cache misses for a given tile $\mathbf{L}$. Because the number of cache misses is related to the number of unique data elements accessed, we introduce the notion of a *footprint* that defines the data elements accessed by a tile. The footprints are regions of the *data space* accessed by a loop tile.

Definition 2 *The **footprint** of a tile $(\mathbf{H}, \vec{\gamma}, \vec{\lambda})$ of a loop partition with respect to a reference $A[\vec{g}(\vec{i})]$ is the set of all data elements $A[\vec{g}(\vec{i})]$ of A, for $\vec{i}$ an element of the tile.*

The footprint gives us all the data elements accessed through a particular reference from within a tile of a loop partition. Because we consider homogeneous loop tiles, the number of data elements accessed is the same for each loop tile.

We will compute the number of cache misses for the system with caches and uniform access memory to illustrate the use of footprints. The body of the loop may contain references to several variables and we assume that aliasing has been resolved; two references with distinct names do not refer to the same location. Let $A_1, A_2, \ldots, A_K$ be references to array A within the loop body, and let $f(A_i)$ be the *footprint* of the loop tile at the origin with respect to the reference A_i and let $f(A_1, A_2, \ldots, A_K) = \bigcup_{i=1,\ldots,K} f(A_i)$ be the *cumulative footprint* of the tile at the origin. The number of cache misses with respect to the array A is $|f(A_1, A_2, \ldots, A_K)|$. Thus, computing the size of the individual footprints and the size of their union is an important part of the loop partitioning problem.

To facilitate computing the size of the union of the footprints we divide the references into multiple disjoint sets. If two footprints are disjoint or mostly disjoint then the corresponding references are placed in different sets, and size of the union of their footprints is simply the sum of the sizes of the two footprints.

However, references whose footprints overlap substantially are placed in the same set. The notion of *uniformly intersecting references* is introduced to specify precisely the idea of "substantial overlap". Overlap produces temporal locality in cache accesses, and computing the size of the union of their footprints is more complicated. The notion of uniformly intersecting references is derived from definitions of intersecting references and uniformly generated references.

Definition 3 *Two references $A[\vec{g}_1(\vec{i})]$ and $A[\vec{g}_2(\vec{i})]$ are said to be intersecting if there are two integer vectors $\vec{i}_1, \vec{i}_2$ such that $\vec{g}_1(\vec{i}_1) = \vec{g}_2(\vec{i}_2)$. For example, $A[i + c1, j + c2]$ and $A[j + c3, i + c4]$ are intersecting, whereas $A[2i]$ and $A[2i + 1]$ are non-intersecting.*

Definition 4 *Two references $A[\vec{g}_1(\vec{i})]$ and $A[\vec{g}_2(\vec{i})]$ are said to be uniformly generated if*

$$g_1(\vec{i}) = \vec{i}\mathbf{G} + \vec{a}_1 \text{ and } g_2(\vec{i}) = \vec{i}\mathbf{G} + \vec{a}_2$$

where G is a linear transformation and $\vec{a}_1$ and $\vec{a}_2$ are integer constants.

The intersection of footprints of two references that are not uniformly generated is often very small. For non-uniformly generated references, although the footprints corresponding to some of the iteration-space tiles might overlap partially, the footprints of others will not overlap. Since we are interested in the worst-case communication volume for any set of footprints, we will assume that the communication generated by two non-uniformly intersecting references is essentially the sum of the individual footprints.

However, the condition that two references are uniformly generated is not sufficient for two references to be intersecting. As a simple example, $A[2i]$ and $A[2i + 1]$ are uniformly generated, but the footprints of the two references do not intersect. For the purpose of locality optimization through loop partitioning, our definition of reuse of array references will combine the concept of uniformly generated arrays and the notion of intersecting array references. This notion is similar to the equivalence classes within uniformly generated references defined in [3].

Definition 5 *Two array references are* uniformly intersecting *if they are both intersecting and uniformly generated.*

Example 3 *References $A[i, j], A[i + 1, j - 3], A[i, j + 4]$ are uniformly intersecting, while the references $A[i, j], A[2i, j]$ are not. For more examples, see [12].*

Footprints in the data space for a set of uniformly intersecting references are translations of one another, as shown below.

Proposition 1 *Given a loop tile at the origin $\mathbf{L}$ and references $r = (G, \vec{a}_r)$ and $s = (G, \vec{a}_s)$ belonging to a uniformly generated set defined by $\mathbf{G}$, let $f(r)$ denote the footprint of $\mathbf{L}$ with respect to r, and let $f(s)$ denote the footprint of $\mathbf{L}$ with respect to s. Then $f(s)$ is simply a translation of $f(r)$, where each point of $f(s)$ is a translation of a corresponding point of $f(r)$ by an amount given by the vector $(\vec{a}_s - \vec{a}_r)$. In other words,*

$$f(s) = f(r) \vec{+} (\vec{a}_s - \vec{a}_r).$$

This follows directly from the definition of uniformly generated references. Recall that an element i of the loop tile is mapped by the reference $(G, \vec{a}_r)$ to data element $\vec{d}_r = \vec{i}G + \vec{a}_r$, and by the reference $(G, \vec{a}_s)$ to data element $\vec{d}_s = \vec{i}G + \vec{a}_s$. The translation vector, $(\vec{d}_s - \vec{d}_r)$, is clearly independent of i.

Proposition 2 *The number of iterations in the tile $(\mathbf{H}, \vec{\gamma}, \vec{\lambda})$, is approximately the sum of the volume of the tile and one half the number of iteration points on the boundaries of the tile. The volume of the tile is $|\det \mathbf{L}(\mathbf{H})|$, which is the same as the volume $|\det \mathbf{L}|$ of the tile $\mathbf{L}$ at the origin.*

Although, the number of iteration points in two different tiles may not be identical for parallelogram partitions, they are equal from a practical viewpoint. Further, for rectangular partitions the tiles are identical except for translation. Also, from a practical viewpoint the size of the footprint is the same for all the loop tiles except for the ones at the boundaries. So we can focus on the loop tile $\mathbf{L}$ at the origin.

The volume of cache traffic imposed on the network is related to the size of the cumulative footprint. We describe how to compute the size of the cumulative footprint in the following two sections as outlined below.

- First, we discuss how the size of *the footprint for a single reference* within a loop tile can be computed. In general, the size of the footprint with respect to a given reference is not the same as the number of points in the iteration space tile.

- Second, we describe how the size of *the cumulative footprint for a set of uniformly intersecting references* can be computed. The sizes of the cumulative footprints for each of these sets are then summed to produce the size of the cumulative footprint for the loop tile.

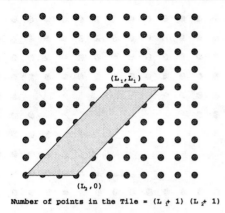

Number of points in the Tile = $(L_1 + 1)$ $(L_2 + 1)$

Figure 4: Tile $\mathbf{L}$ at the origin of the iteration space.

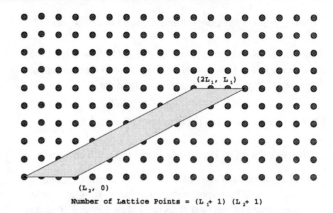

Number of Lattice Points = $(L_1 + 1)$ $(L_2 + 1)$

Figure 5: Footprint of $\mathbf{L}$ wrt $B[i + j, j]$ in the data space.

3.4 Size of a Footprint for a Single Reference

This section shows how to compute the size of the footprint (with respect to a given reference and a given loop tile $\mathbf{L}$) efficiently under certain conditions on $\mathbf{G}$. Although we believe that these conditions cover a majority of practical cases, we summarize in Section 3.7 how to extend the techniques presented in this section with weaker conditions on $\mathbf{G}$. We begin with a simple example to illustrate our approach.

Example 4

Doall (i=0:99, j=0:99)
  A[i,j] = B[i+j,j]+B[i+j+1,j+2]
EndDoall

Assume the loop tile at the origin $\mathbf{L}$ is $\begin{bmatrix} L_1 & L_1 \\ L_2 & 0 \end{bmatrix}$.

Figure 4 shows this tile at the origin of the iteration space. The reference matrix $\mathbf{G}$ is $\begin{bmatrix} 1 & 0 \\ 1 & 1 \end{bmatrix}$.

The footprint of the tile at the origin with respect to the reference $B[i + j, j]$ is shown in Figure 5. The matrix

$$\mathbf{f}(B[i + j, j]) = \mathbf{LG} = \begin{bmatrix} 2L_1 & L_1 \\ L_2 & 0 \end{bmatrix}$$

describes the footprint. The integer points on or inside the parallelogram specified by $\mathbf{LG}$ is the footprint of the tile. So the size of the footprint is $|\det(\mathbf{LG})|$ plus one half the number of integer points on the boundary of the parallelogram. This computation reduces to $L_1 L_2 + L_1 + L_2$. In the rest of our discussion, for brevity, we will drop explicit mention of the integer points on the boundary of the parallelograms[2] and use

$$Size\ of\ the\ footprint\ defined\ by\ \mathbf{LG} = |\det(\mathbf{LG})| \quad (2)$$

The product $\mathbf{LG}$ appears often and we denote it by the matrix $\mathbf{D}$.

---

[2] As we show in Section 4, the number of points on the boundary can be calculated precisely, but their impact is small for most practical partitions.

The above example leads to the following questions. In general, is the footprint exactly the integer points on or inside $\mathbf{D} = \mathbf{LG}$? If not, how do we compute the footprint? The first question can be expanded into the following two questions.

- Is there a point in the footprint that lies outside the hyperparallelepiped $\mathbf{D}$? It follows easily from linear algebra that it is not the case.

- Is every integer point inside of $\mathbf{D}$ an element of the footprint? It is easy to show this is not true and a simple example corresponds to the reference $A[2i]$.

We first study the simple case when the hyperparallelepiped $\mathbf{D}$ completely defines the footprint. Precise definition of the set $S(\mathbf{D})$ of points defined by the matrix $\mathbf{D}$ is as follows.

Definition 6 *Given a matrix $\mathbf{D}$ whose rows are the vectors $\vec{d_i}$, $S(\mathbf{D})$ is defined as the set*

$$\{\vec{x} = a_1 \vec{d_1} + a_2 \vec{d_2} + \ldots + a_l \vec{d_l} | 0 \le a_i \le 1\}.$$

$S(\mathbf{D})$ defines all the points on or inside the hyperparallelepiped defined by $\mathbf{D}$.

So for the case where $\mathbf{D}$ completely defines the footprint, the footprint is exactly the integer points in $S(\mathbf{D})$. One of the cases where $\mathbf{D}$ completely defines the footprint is when $\mathbf{G}$ is unimodular as shown below.

Lemma 1 *The mapping of the iteration space to the data space as defined by $\mathbf{G}$ is one to one if and only if the rows of $\mathbf{G}$ are independent.*

Proof: *$\vec{i_1}\mathbf{G} = \vec{i_2}\mathbf{G}$ implies $\vec{i_1} = \vec{i_2}$ if and only if the rows of $\mathbf{G}$ are linearly independent.*

Lemma 2 *The mapping of the iteration space to the data space as defined by $\mathbf{G}$ is onto if and only if the columns of $\mathbf{G}$ are independent and the g.c.d. of the subdeterminants of order equal to the number of columns is 1.*

Proof: *Follows from the Hermite normal form theorem [13].*

Theorem 1 *The footprint of the tile defined by* **L** *with respect to the reference* **G** *is identical to the integer points on or inside the hyperparallelepiped* $\mathbf{D} = \mathbf{LG}$ *if* **G** *is unimodular.*

Proof: *Immediate from the above two lemmas.*

We make the following two observations about Theorem 1.

- **G** is unimodular is a sufficient condition; but not necessary. An example corresponds to the reference $A[i + j]$. Further discussions on this theorem is beyond the scope of this paper and will be dealt with in a separate paper.

- One may wonder why **G** being onto is not sufficient for **D** to coincide with the footprint. Even when every integer point in **D** has an inverse, it is possible that the inverse is outside of **L**. The one to one property of **G** guarantees that no point from outside of **L** can be mapped to inside of **D**. The reason for this is the one to one property is true even when **G** is treated as a function on reals.

We can also apply Theorem 1 to compute the size of a footprint when the columns of **G** are not independent, as shown below.

3.4.1 Footprint Size when Columns of G are Dependent

We derive a $\mathbf{G}'$ from **G** by choosing a maximal set of independent columns from **G**, such that $\mathbf{G}'$ is unimodular. If a unimodular $\mathbf{G}'$ exists, then it completely specifies the footprint, because all the points within the hyperparallelepiped defined by $\mathbf{D}' = \mathbf{LG}'$ will be accessed. We can now apply Theorem 1 to $\mathbf{D}'$ to compute the size of the footprint.

Example 5 *Consider the reference* $A[i, 2i, i + j]$ *in a doubly nested loop. The columns of the* **G** *matrix*

$$\begin{bmatrix} 1 & 2 & 1 \\ 0 & 0 & 1 \end{bmatrix}$$

are not independent. We choose $\mathbf{G}'$ *to be* $\begin{bmatrix} 1 & 1 \\ 0 & 1 \end{bmatrix}$. *Now* $\mathbf{D}'$ *completely specifies the footprint.*

It is possible that none of the maximal independent columns satisfy the conditions in Theorem 1. Such cases are dealt with in Section 3.7.

3.5 Size of the Cumulative Footprint

The size of the cumulative footprint F for a loop tile is computed by summing the sizes of the cumulative footprints for each of the sets of uniformly intersecting references. First, let us present a method for computing the size of the cumulative footprint for a set of uniformly intersecting references.

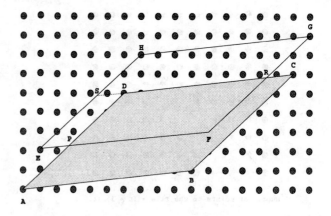

Figure 6: Data footprint wrt $B[i + j, j]$ and $B[i + j + 1, j + 2]$

The complexity of the computation of the size of the cumulative footprint for a set of uniformly intersecting references depends on the reference matrix **G** that defines the set and the loop partition **L**. We first focus on the types of references for which the conditions stated in Theorem 1 are true, that is, when **G** is unimodular.

Let us start by illustrating the computation of the cumulative footprint for Example 4. The references to array B form a uniformly intersecting set and are defined by the following **G** matrix.

$$\mathbf{G} = \begin{bmatrix} 1 & 0 \\ 1 & 1 \end{bmatrix}$$

Suppose that the loop partition **L** is given by $\begin{bmatrix} L_{11} & L_{12} \\ L_{21} & L_{22} \end{bmatrix}$. Then **D** is given by $\begin{bmatrix} L_{11} + L_{12} & L_{12} \\ L_{21} + L_{22} & L_{22} \end{bmatrix}$.

$ABCD$ and $EFGH$ shown in Figure 6 are the footprints of the tile **L** with respect to the two references ($B[i + j, j]$ and $B[i + j + 1, j + 2]$ respectively) to array B. In the figure, $\vec{AB} = (L_{11} + L_{12}, L_{12})$, $\vec{AD} = (L_{21} + L_{22}, L_{22})$, and $\vec{AE} = (1, 2)$.

The size of the cumulative footprint is the size of footprint $ABCD$ plus the number of data elements in $EPDS$ plus the number of data elements in $SRGH$. We can approximate the number of data elements by the area $ABCD + SRGH + EPDS$. Ignoring, the areas of the two triangles APE and HSD, we can approximate the total area by

$$\begin{vmatrix} L_{11} + L_{12} & L_{12} \\ L_{21} + L_{22} & L_{22} \end{vmatrix} + \begin{vmatrix} L_{11} + L_{12} & L_{12} \\ 1 & 2 \end{vmatrix} + \begin{vmatrix} 1 & 2 \\ L_{21} + L_{22} & L_{22} \end{vmatrix}$$

This approximation is reasonable if we assume that the constant terms in a uniformly intersecting set of references are small compared to the tile size. The first term in the above equation represents the area of the footprint of a single reference, i.e., $|\det(\mathbf{D})|$. The second and third terms are the determinants of the **D** matrix in which one row is replaced by the offset vector $\vec{a} = (1, 2)$.

We can generalize this idea of computing the cumulative footprint from multiple overlapping footprints when there are many references in a set of uniformly intersecting references as follows. Let the index vector of reference r from a set of uniformly intersecting references be $\vec{g}(\vec{i}) = \vec{i}\mathbf{G} + \vec{a}_r$, where $\vec{i}$ corresponds to the iteration space (as defined in Equation 1). The index vector has length d, where as before d is the dimension of the array.

Let the *spread* of the offset vectors of a set of uniformly intersecting references with respect to a basis for the data space be captured by a vector $\hat{a}$ as defined here. That is, let $\hat{a} = \text{spread}(\vec{a}_1, \ldots, \vec{a}_R, \mathbf{B})$, where R is the number of references in the uniformly intersecting set, and the rows of $\mathbf{B}$ is a basis for the data space. Recall that $\mathbf{D}$ is a basis for the data space when $\mathbf{G}$ is unimodular. The purpose of $\hat{a}$ is to define a single offset vector that captures the communication in a cache-coherent system resulting from all the offset vectors. The kth component of $\hat{a}$ reflects the communication per unit area across the hyperplane defined by the rows of $\mathbf{D}$ excluding the kth row.

In the definition below, $\vec{b}_r$ is the representation of offset vector $\vec{a}_r$ using $\mathbf{D}$ as the basis.

Definition 7 *Let* $\vec{b}_r = \vec{a}_r \mathbf{D}^{-1}, \forall r \in 1, \ldots, R$ *and let* $\hat{b}$ *be a vector of length* d *such that the* k^{th} *component of* $\hat{b}$ *is*

$$\hat{b}_k = \max_r(b_{r,k}) - \min_r(b_{r,k}), \forall k \in 1, \ldots, d.$$

Then $\hat{a} = \text{spread}(\vec{a}_1, \ldots, \vec{a}_R, \mathbf{D}) = \hat{b}\mathbf{D}$.

Looking at the special case where $\mathbf{D}$ is rectangular helps in understanding the definition.

Proposition 3 *If* $\mathbf{D}$ *is rectangular then*

$$\hat{a}_k = \max_r(a_{r,k}) - \min_r(a_{r,k}), \forall k \in 1, \ldots, d.$$

For caches, we use the $max - min$ formulation (or the spread) to calculate the amount of communication traffic because the data space points corresponding to the footprints of references whose offset vectors have values somewhere between the max and the min lie within the cumulative footprint calculated using the spread.[3]

Theorem 2 *Given a hyperparallelepiped tile* $\mathbf{L}$, *and a unimodular reference matrix* $\mathbf{G}$, *the size of the cumulative footprint with respect to a set of uniformly intersecting references specified by the reference matrix* $\mathbf{G}$ *and a set of offset vectors* $\vec{a}_1, \ldots, \vec{a}_R$, *is approximately*

$$|\det \mathbf{D}| + \sum_{i=1}^{d} |\det \mathbf{D}_{i \to \hat{a}}|$$

where $\hat{a} = \text{spread}(\vec{a}_1, \ldots, \vec{a}_R, \mathbf{D})$ *and* $\mathbf{D}_{i \to \hat{a}}$ *is the matrix obtained by replacing the* ith *row of* $\mathbf{D}$ *by* $\hat{a}$.

---

[3]For data partitioning in distributed-memory multiprocessors, however, the formulation must be modified slightly. Because data partitioning assumes that data from other memory modules is not dynamically copied locally (as in systems with caches), we replace $\hat{a}$ by the *cumulative spread* a^+ of a set of uniformly intersecting references, whose k^{th} element is given by, $a_k^+ = \sum_r | [a_{r,k} - med_r(a_{r,k})] |, \forall k \in 1, \ldots, d$. Here, $med_r(a_{r,k})$ is the median of the offsets in the kth dimension. A more detailed description appears in [12].

As before, we can deal with the case when the columns of $\mathbf{G}$ are not independent by choosing a maximal set of independent columns. We have sharper estimates on communication volume when the tile is rectangular; we can also weaken the conditions on $\mathbf{G}$ when the tile is rectangular as shown in Section 3.7.

Finally, as stated earlier, the total communication generated by non-uniformly intersecting sets of references is the sum of the communication generated by the individual cumulative footprints. Example 6 in Section 3.6 discusses an instance of such a computation.

3.6 Minimizing Cumulative Footprint Size

We now focus on the problem of finding the loop partition that minimizes the size of the cumulative footprint, and hence the total number of cache misses. Let us illustrate this problem through the following example in which there are three sets of uniformly intersecting references: one for A, one for B, and one for C. The basic idea here is that the three sets contribute additively to traffic.

Example 6

Doall (i=1:N, j=1:N)
```
A[i,j]=B[i-2,j] + B[i,j-1] + C[i+j-1,j] +
       C[i+j+1,j+3]
```
EndDoall

Because A has only one reference, its footprint size is independent of the loop partition, given a fixed total size of the loop tile, and therefore need not figure in the optimization process.

For the purpose of making it easy to follow this example, let us assume that the tile $\mathbf{L}$ is rectangular and is given by $\begin{bmatrix} L_1 & 0 \\ 0 & L_2 \end{bmatrix}$. Because $\mathbf{G}$ for the references to array B is the identity matrix, the $\mathbf{D} = \mathbf{L}\mathbf{G}$ matrix corresponding to references to B is same as $\mathbf{L}$, and the $\hat{a}$ vector is $(2, 1)$. Thus, the size of the corresponding cumulative footprint according to Theorem 2 is

$$\begin{vmatrix} L_1 & 0 \\ 0 & L_2 \end{vmatrix} + \begin{vmatrix} 2 & 1 \\ 0 & L_2 \end{vmatrix} + \begin{vmatrix} L_1 & 0 \\ 2 & 1 \end{vmatrix}.$$

Similarly, $\mathbf{D}$ for array C is $\begin{bmatrix} L_1 & 0 \\ L_2 & L_2 \end{bmatrix}$, and using Definition 7 $\hat{a}$ can be shown to be $(4, 3)$, and the size of the cumulative footprint with respect to C is

$$\begin{vmatrix} L_1 & 0 \\ L_2 & L_2 \end{vmatrix} + \begin{vmatrix} 4 & 3 \\ L_2 & L_2 \end{vmatrix} + \begin{vmatrix} L_1 & 0 \\ 4 & 3 \end{vmatrix}.$$

The problem of minimizing the size of the footprint reduces to finding the elements of $\mathbf{L}$ that minimizes the sum of the two expressions above subject to the constraint the area of the loop tile $|\det \mathbf{L}|$ is a constant to ensure a balanced load. For example, if the loop bounds are I, J, then the constraint is $|\det \mathbf{L}| = IJ/P$, where P is the number of processors.

The total size of the cumulative footprint simplifies to $2L_1L_2 + 4L_1 + 3L_2$. The optimal values for L_1 and L_2 can be shown to

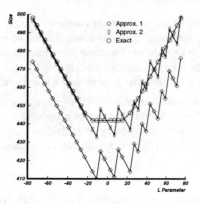

Figure 7: Actual and computed footprint sizes.

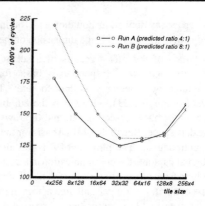

Figure 8: Running times in cycles for different aspect ratios.

satisfy the equation $4L_1 = 3L_2$, using the method of Lagrange multipliers.

Thus far, our analysis has been concerned with minimizing the cumulative footprint size, which at first glance, appears to minimize only the number of first-time cache misses. However, the same optimization process minimizes the number of coherence-related invalidations and misses as well [12].

3.7 Rectangular Tiles and a General G

Rectangular loop tiles are useful to consider because: (1) they lead to very accurate estimates of communication volume, (2) they allow us to handle much more general forms of $\mathbf{G}$, and (3) they allow easy code generation. We use lattice theory to compute sizes of data footprints when $\mathbf{G}$ is not unimodular for most practical cases of $\mathbf{G}$. Details are presented in [12].

4 Implementation and Results

In this section we present cumulative footprint size measurements from an algorithm simulator and execution time measurements from an actual compiler implementation on a multiprocessor simulator.

We have written a simulator of partitioning algorithms that measures the exact cumulative footprint size for any given hyper-parallelepiped partition. The simulator also presents analytically computed footprint sizes using the following two approximations for the cumulative footprint size.

- The first approximation uses the formula presented in this paper in Theorem 2. This approximation ignores the points on the surface of the cumulative footprint.

- The second approximation adds the surface points to the cumulative footprint computed by Theorem 2. For the two dimensional data space considered in these experiments, the surface corresponds to the edges of a parallelogram and the number of integer points along an edge vector (x, y) is $1 + \gcd(x, y)$.

We compare the results for the two approximations with the exact measurements from an algorithm simulator. Our experiments use $\mathbf{G} = \begin{bmatrix} 1 & 0 \\ 0 & 1 \end{bmatrix}$, an $\mathbf{L}$ matrix of the form $\begin{bmatrix} L_{11} & L_{12} \\ L_{21} & L_{22} \end{bmatrix}$, and offset vectors $\vec{a_1} = (0,0)$, $\vec{a_2} = (1,1)$, $\vec{a_3} = (0,2)$. In all cases the size of iteration space tile is kept constant (to within a small integer truncation error). Although [12] analyzes several partition types, we summarize the results for one case in this paper.

This case assumes that $L_{11} = 0$, $L_{21} = L_{12} = 19$, and L_{22} varies from -75 to 75, and results in parallelogram partitions. Figure 7 plots the exact and the computed footprint sizes for the two approximations. L_{22} is shown as the X-axis parameter. It is easy to see from the graphs that the accurate formulation yields a very close estimate of the footprint size, and that both approximations yield the optimal tiling parameters.

We have also implemented some of the ideas from our framework in a compiler for the Alewife machine [14] to understand the extent to which good loop partitioning impacts end application performance, and the extent to which our theory predicts the optimal loop partition. The languages accepted at present are Mul-T, a parallel Lisp language, and Semi-C, a parallel version of C. The Alewife machine implements a shared global address space with distributed physical memory and coherent caches. The nodes are configured in a 2-dimensional mesh network.

Although we have implemented loop and data partitioning with rectangular partitions, the results in this paper will focus on loop partitioning. To isolate the effect of reducing the number of cache misses through loop partitioning, we modify the Alewife simulator to provide roughly uniform-cost cache misses. Otherwise, isolating the effect of cache misses is difficult because changing the loop partition alters both the number of non-local memory references and the number of cache misses.

We ran several versions of a parallel loop nest on the Alewife machine simulator. The program uses the same $\mathbf{G}$ matrix as before, but uses two arrays with offset vectors $\vec{a_1} = (0,0)$, $\vec{a_2} = (1,0)$, $\vec{b_1} = (0,2)$ and $\vec{b_2} = (0,-2)$. The program is run on 64 processors with each array being 256 elements on a side. The compiler, using the algorithms in this paper, chooses rectangular loop partitions with an aspect ratio of 4:1, i.e., a tile size of

64x16. We also generate code using wrong partitions with tile sizes ranging from 4x256 to 256x4. The arrays are stored in row major order and to simplify things we use a non-unit stride in the X dimension to give one element per cache line. This set of executions is labeled run A.

We also ran the program using a different set of offset vectors that give an expected aspect ratio of 8:1 (run B). This results in a desired tile size between 64x16 and 128x8 with the compiler choosing 64x16. Thus, we expect the minimum running time to be at roughly 16x4 for both cases.

Figure 8 shows the running times for the different tile sizes. These times are for the second iteration of the loop nest and so represent coherence traffic rather than first-time cache misses. There is a fair amount of noise in these figures because the communication latency of remote and local cache misses is not exactly uniform and there can be other spurious cache effects as well, but the basic shapes of the curves correspond to our expectations. The main reason that the actual differences are not very large is that the code generated for index calculations for array references is not very good right now. To get an idea what impact that has, we ran the same program, but with a no-op replacing the actual loads and stores for the array references. Running times in this case differed by a factor of three from best to worst.

5 Conclusions

This paper presented a theoretical framework to derive the shapes of the iteration-space partitions of the do loops to minimize the communication traffic in multiprocessors with caches. The framework allows the partitioning of doall loops into optimal hyperparallelepiped tiles where the index expressions in array accesses can be any affine function of the indices. The same framework can also yield optimal loop and data partitions for multicomputers with local memory.

Our analysis uses the notion of uniformly intersecting references to categorize the references within a loop into classes that yield cache locality. The notion of data footprints is introduced to capture the combined set of data accesses made by references within each uniformly intersecting class. Then, an algorithm to compute the total size of the data footprint for a given loop partition is presented. Once an expression for the total size of the data footprint is obtained, standard optimization techniques are applied to minimize the size of the footprint and derive optimal loop partitions.

A subset of the framework, which handles rectangular partitioning of loops and data, has been implemented in the compiler system for the Alewife multiprocessor.

6 Acknowledgments

This research is supported by Motorola Cambridge Research Center and by NSF grant # MIP-9012773. Partial support has also been provided by DARPA contract # N00014-87-K-0825, in part by a NSF Presidential Young Investigator Award. We are grateful to Rajeev Barua for pointing out an error in an earlier formulation of the footprint size, and for implementing partitioning using affine index expressions. Gino Maa helped define and implement the compiler system and its intermediate form. We acknowledge the contributions of the Alewife group for supporting the Alewife simulator and runtime system used in obtaining the results.

References

[1] Constantine D. Polychronopoulos and David J. Kuck. Guided Self-Scheduling: A Practical Scheduling Scheme for Parallel Supercomputers. *IEEE Transactions on Computers*, C-36(12), December 1987.

[2] E. Mohr, D. Kranz, and R. Halstead. Lazy Task Creation: A Technique for Increasing the Granularity of Parallel Programs. *IEEE Transactions on Parallel and Distributed Systems*, 2(3):264–280, July 1991.

[3] M. Wolf and M. Lam. A data locality optimizing algorithm. In *Proceedings of the Conference on Programming Language Design and Implementation*, pages 30–44, 1991.

[4] D. Gannon, W. Jalby, and K. Gallivan. Strategies for cache and local memory management by global program transformation. *Journal of Parallel and Distributed Computing*, 5:587–616, 1988.

[5] Harold S. Stone and Dominique Thiebaut. Footprints in the Cache. In *Proceedings of ACM SIGMETRICS 1986*, pages 4–8, May 1986.

[6] F. Irigoin and R. Triolet. Supernode Partitioning. In *15th Symposium on Principles of Programming Languages*, pages 319–329, January 1988.

[7] S. G. Abraham and D. E. Hudak. Compile-time partitioning of iterative parallel loops to reduce cache coherency traffic. *IEEE Transactions on Parallel and Distributed Systems*, 2(3):318–328, July 1991.

[8] J. Ramanujam and P. Sadayappan. Compile-Time Techniques for Data Distribution in Distributed Memory Machines. *IEEE Transactions on Parallel and Distributed Systems*, 2(4):472–482, October 1991.

[9] M. Gupta and P. Banerjee. Demonstration of Automatic Data Partitioning Techniques for Parallelizing Compilers on Multicomputers. *IEEE Transactions on Parallel and Distributed Systems*, 3(2):179–193, March 1992.

[10] Robert Schreiber and Jack Dongarra. Automatic Blocking of Nested Loops. Technical report, May 1990. RIACS, NASA Ames Research Center, and Oak Ridge National Laboratory.

[11] J. Ferrante, V. Sarkar, and W. Thrash. *On Estimating and Enhancing Cache Effectiveness*, pages 328–341. Springer-Verlag, August 1991. Lecture Notes in Computer Science: Languages and Compilers for Parallel Computing.

[12] Anant Agarwal, David Kranz, and Venkat Natarajan. Automatic Partitioning of Parallel Loops for Cache-Coherent Multiprocessors. MIT/LCS TM-481, December 1992.

[13] A. Schrijver. *Theory of Linear and Integer Programming*. John Wiley & Sons, 1990.

[14] A. Agarwal et al. The MIT Alewife Machine: A Large-Scale Distributed-Memory Multiprocessor. In *Proceedings of Workshop on Scalable Shared Memory Multiprocessors*. Kluwer Academic Publishers, 1991. Available as MIT/LCS Memo TM-454, 1991.

Techniques to Enhance Cache Performance Across Parallel Program Sections

J.K. Peir [1]
Computer & Communication Lab.
Industr. Tech. Res. Inst., Taiwan ROC

K. So, J.H. Tang [2]
IBM Advanced Workstation Systems
Austin, Texas 78758

Abstract

Private caches are critical components in high performance multiprocessor systems. However, it has been found that, when executing a parallel program, individual processors are very difficult to attain high cache hit ratio from one program section to another; therefore sophisticated software coherence schemes are not cost effective. In this study, trace-driven simulation has been used to evaluate various less sophisticated compiler and software techniques which can enhance this inter-section locality in parallel executions.

We found that the locality can be substantially improved through the following ways of altering the scheduling of iterations in parallel DO loops among the executing processors: i) assignment of iterations in chunks, ii) reversed execution of parallel loops, and iii) interchange inner and outer loops. These can be done manually by a programmer or automatically by a parallelizing compiler. Moreover, we also propose a software coherence scheme which can attain the maximum inter-section locality for read-only shared data.

1. Introduction

In a shared memory multiprocessor (MP) system, the memory controls have to guarantee a processor (CPU) to receive the latest version of a piece of data for every memory access. This essentially requires a global search of all remote caches for every CPU's cache miss, and there are data transfer among the caches when there are data sharing among the CPU's. A full hardware support of these actions has been known as a *hardware coherence scheme*.

A highly parallel system, e.g., [Pfi+86], instead may offer its software a few cache instructions to control the cache content and leave it to the software to take care of the coherence of data sharing. The approach has been known as *software coherence schemes* [Vei86] [Che&Vei88] [Che&Vei89] [Min&Bae89] [Lou&Sun92]. Individual CPU under these schemes tries to optimize the hit ratio during the execution of parallel programs without any global information of how shared data are being read or modified by any remote CPU.

A parallel program can be viewed as a sequence of parallel and serial sections such that each parallel section contains a multiple pieces of work assignable to the participating CPU's through accesses to shared variables. Synchronization points (or barriers) are inserted at the end of each section to maintain proper data dependence. An efficient software coherence scheme not only maintains the cache coherence through a few cache control mechanisms, it also tries to minimize cache invalidation so that valid cache lines can be re-used across parallel sections. The easiest software coherence scheme is to invalidate the entire cache whenever the CPU fetches a piece of data potentially modified in a remote cache. For example, an invalidation can take place at the beginning of a section. Apparently, this indiscriminate invalidation causes unnecessary cache misses.

In [Pei+91], the cache hits at a processor in an SPMD (Single-Program-Multiple-Data) computation are partitioned into two categories: the inter-section hits are the first-time hits to the cache lines which have been left in the processor's cache at the end of a previous section; the rest of the cache hits are identified as intra-section hits. If a cache can hold n lines, the *inter-section reuse hit ratio*, or called inter-section locality, at a section is defined as the ratio of the number of inter-section hits to the maximum of { n and the number of distinct lines in the cache before the end of the previous section}. All the software coherence schemes rely on how well they are taking care of inter-section locality of parallel programs. Unfortunately, the working set size in a section is usually too large for a cache of reasonable size to keep and the overall hit ratio is dominated by the intra-section locality. Therefore the inter-section re-use ratio is in general quite low and the overall increase in hit ratios under a sophisticated software scheme may not be significantly different from that of a simple one.

In this report, we present two less sophisticated approaches to enhance the inter-section locality of parallel programs. The approaches and their results are summarized below. The first approach use simple programming techniques to modify the execution order of parallel DO loops; they are:

---

1 Participated the work since the author was in IBM TJ Watson Research Center
2 On assignment from IBM TJ Watson Research Center

1. Iteration assignment by chunk- Instead of assigning one iteration at a time to a processor, we assign a chunk of consecutive iterations to a processor to execute. Since consecutive iterations tend to use data in close-by area, this arrangement normally reduces the working set size of a CPU in a section. For example, applying chunking on parallel application Weather, this approach can increase substantially the inter-section hit ratios from 10 to 60% and 50% to 80% for 16-way MP and 64-way MP, respectively.

2. Reversed execution of parallel DO loops - If parallel loops A and B are executed in sequence and both of them are operating on one or a few matrices in increasing indices, alter loop B such that its iterations are executed in decreasing indices. The idea is based on the observation that, because DO loops in a parallel program are mostly executed in one direction, say from lower indices to the higher, the elements in a matrix under such a computation very likely follows one specific direction. Therefore a reversed execution of the second loop can potentially maximize the chance to re-use the last part of the data in the cache left by the previous loop. For various cache sizes, our study indicates that the technique can improve the inter-section locality of applications Simple and Weather by 50 to 100%.

3. Loop interchange - A parallel program typically contains one sequential outer loop with a parallel inner loop. This will leave many small parallel sections of small working sets. In the case when the outer and inner loops are interchangeable [Wol89], it will leave a single section of larger working set size. A parallel program with small sections in general has a higher inter-section locality. The experiment of loop interchange with our application Tomcat shows that the inter-section locality for the large and small sections are about 10 and 50% respectively.

The above programming techniques are evaluated under an optimistic software scheme such that we are able to find out the ultimate improvement in locality under the techniques. The optimistic scheme simply does not invalidate any cache lines across sections. Although the programming techniques do enhance inter-section locality quite substantially, the improvement can only be considered as an upper bound for other realistic software schemes for which a certain amount of cache invalidation may be needed in order to maintain the cache coherence. To this regard, we introduce a software coherence scheme, called *read-only vector (ROV) scheme*, as our second approach to enhance inter-section locality.

ROV scheme can attain the maximum useful inter-section locality for read-only shared data through the use of a *read-only vector* for each access to a shared variable.

The read-only vector indicates how many sections the accessed cache line can stay in cache before it must be invalidated to maintain coherence. We found that the scheme can attain an inter-section hit ratio 30% to 50% with reasonably large caches. In contrast to the global access pattern defined in [Ben+90], our scheme can dynamically keep track of the read-only status of a variable in each section.

The programming techniques and the proposed software coherence scheme are described respectively in details in section 2 and 3. Section 4 contains the performance results and their implications. Our conclusion and future directions are given in Section 5. Descriptions of our tracing technique for SPMD programs, the inter-section locality, and parallel applications Simple, Weather, and Tomcat can be found in [Pei+91].

2. Programming Techniques to Enhance Inter-Section Locality

In [Pei+91] the inter-section locality of an application is found to be largely determined by the following three factors: i) the working set of each program section, ii) the referencing order of cache lines, e.g. the degree of overlapping of lines used in successive sections, and iii) the cache size. Our programming techniques of reducing the working set and rearranging the order of execution of parallel loops are based on the above observation.

A parallel loop is selected from Weather code and shown in Figure 1 as an example. This loop will be used to illustrate all three programming techniques.

```
     DO 10 I = 1, IM
       DO 910 J = 1, JNP
       PT(I,J) = P(I,J)
       DO 65 N = 1, 4
         NJ = (N-1) * JNP + J
         DO 65 L = 1, NLAY
         QT(I,L,NJ) = Q(I,L,NJ) * DXYP(J) * PT(I,J)
  65     CONTINUE
 910   CONTINUE
  10 CONTINUE

(a) Parallelizing DO 10 loop with chunk=1 (default)
    PARDO 10  I = 1, IM
      or
    PARDO 10  I = 1, IM  CHUNK=1

(b) Parallelizing DO 10 loop with chunk=n, n> 1
    PARDO 10  I = 1, IM  CHUNK=n

(c) Reverse execution of DO 10 loop
    PARDO 10  I = IM, 1, -1
```

Figure 1.   An example of a loop from Weather code

2.1. Iteration assignment by chunks

When the data set used in one or a few iterations in an application is much larger than the capacity of a CPU's cache, domain decomposition, or called blocking, is a well known parallelizing technique to preserve program locality within each CPU. This technique has been explored on cache-based multiprocessor systems, e.g., [ESSL90]. In order to enhance program locality for a particular cache structure, certain restructuring of the original algorithm may be necessary.

In this paper, we report the usefulness of one simple aspect of domain decomposition to increase the inter-section locality. Instead of changing the original algorithm, we simply group loop iterations into chunks. The idea is that consecutive iterations usually use close-by data elements in an array; therefore assigning iterations in chunks to an executing CPU reduces the CPU's working set size and hence increases the inter-section locality.

2.2. Reversed Execution of Parallel Loops

The second major factor in determining inter-section locality is the effect of the order of memory references among the shared arrays or within a shared array. In most of the Doall loops in the three workloads in our study, several shared arrays are normally accessed at the same time. So when the cache is small, the arrays are contenting the tight cache space and eventually only a portion of each of these arrays remains in the cache at the end of a section. On the other hand, the accessing order within each array tends to follow a same direction. For example, as a consequence that loop iterations are normally indexed from low to high order, all references to a shared array tend to start from the lower addresses at the beginning of the loop to the higher addresses of the array at the end of the loop. If a CPU's cache is small and the same array is used in consecutive sections, the portion of the referenced array (high addresses) left in the cache at the end of a section not necessarily get reused at the beginning of the next section; they may be replaced before the second section started and reloaded to the cache during the execution of the second section.

The basic idea of reversed execution of parallel loops is to allow the shared arrays to be accessed in revere across adjacent sections to capture the inter-section locality even when the cache can not hold all the arrays. Without a sophisticated restructuring compiler [Eig+91], we test the value of this programming technique by manually reversing the indices of some program loops in the applications studied. This change is limited only to the Doall loops so that the data dependences between iterations can not be violated. An example with reversed execution of parallel loops is shown in Figure 1c.

In Fortran, each array in memory is stored according to the column-major scheme. For example in a three-dimensional array A, $A(2, 1, 1)$ is located next to $A(1, 1, 1)$ in memory, while $A(1, 1, 2)$ can be far away from $A(1, 1, 1)$ depending on the magnitude of the first two dimensions. In order to achieve an ideal access ordering of shared arrays, a freedom of rearranging the nested loops (i.e. loop interchanging in [Wol90]) as well

as reordering the execution of the loop indices are desirable. In the cases such flexibilities are not provided, the preference of reverse indexing was given to the right-most dimension in this study. This is because the adjacent indices of the right-most dimension defines two 'small' adjacent regions within the shared array.

2.3. Loop Interchange

In computations with mostly Doall loops the granularity of parallelism is mainly determined by the sizes of work assigned to a CPU at a time. Generally speaking, the larger they are, the less overhead during computation. However, in certain applications, a large number of small sections may be inevitable due to various reasons. The most typical one is a small parallel DO loop nested inside a large sequential loop. When loop interchange is not allowed, such parallel program construct creates a large number of small program sections. There are two important aspects related to the inter-section locality with small program sections:

1. The inter-section locality becomes more important with respect to the overall hit ratio.

2. It is easier to capture inter-section locality because of the smaller working set of each section.

We will use application Tomcat to study these effects.

3. A Software Coherence to Enhance Inter-section Locality

Software coherence schemes allow incoherent cache lines exist in multiple caches as long as each CPU does not access any stale data in the lines. At any synchronization point, these cache lines are invalidated to avoid stale data access in subsequent sections. However, it is hard to precisely invalidate these modified lines because there is no global information kept at a CPU regarding the past and future usage of shared cache lines in other CPU's.

In this section, we describe a software coherence scheme with proper hardware support. With a fast selective invalidation of modified cache lines at the synchronization points, this scheme can have a higher inter-section locality over the *global read-only scheme* [Che&Vei88], in which the read-only attribute is associated with a variable throughout the execution of the program.

A few terminologies are needed for our software scheme. A simple or array variable X is said to be *read-only* in a section if X is not modified throughout the section. A *read-only vector* of X in a section, denoted by $x_0, x_1, ..., x_{n-1}$, consists of a sequence of bits which indicate if X is read-only or modified through n contiguous sections. A read-only bit is '1' if X is read-only or not used in the corresponding section; otherwise it is '0'. The first bit x_0 indicates the status of X in the current section; x_1 is for the 1st section following the current section; and so on for the rest of the bits. However, once X is modified, the status of X in all the subsequent sections are considered as modified. This read-only vector is used to guide the selective invalidation at the synchronization points. The number of sections that a read-only vector can cover is the length of the read-only vector.

Two new instructions are introduced, *Shared Read (S-Read)* and *Shared Write (S-Write)*, for reading and writing to shared variables. Both of them have an attached read-only vector. The read-only vector for S-Write can be omitted since it has all '0's. This is because the variable is modified in the current section. In case for private variable accesses, the read-only vectors are all '1's.

Figure 2 shows an example containing five program sections, where X is a shared variable. The S-Read, S-Write of X and their associated 4-bit read-only vectors are illustrated. The cache lines associated with the read of X(I,J,K) in the first section will not be invalidated until after section 4 as indicated by the read-only vector '1110'. Note that the read-only vector for the read of X(I,J,K) in section 3 is '1000' instead of '1010' for the referenced cache lines could be modified by other CPU's at section 4. Therefore, the lines should be invalidated thereafter. For the same reason, if the attribute of a variable is not read-only, its read-only vector is set to '0000' disregard of its usage in subsequence program sections. Moreover, the absence of X in sector 2 is treated as a read-only access.

In the cache directory, a read-only vector is added to each line entry to maintain the validity information of the cache line up to certain number of subsequent sections. This information is used at synchronization points for selective invalidation. As usual, there is a valid bit in associate with each line to indicate if the line is valid for access. Assume that the length of bit vector is fixed at 4, the software coherence scheme works as follows:

1. The compiler generates a bit-vector for each S-read or S-write statement in the program. For example, the bit-vector '1110' in the first section of Figure 2 indicates that the shared variable is read-only in the first 3 sections and will be read-write in the 4th section.

2. Whenever an S-Read or an S-Write causes a cache miss, the read-only vector in the instruction replaces the read-only vector in the corresponding entry in the cache directory.

3. At the end of each section, a selective invalidation is performed. Such invalidation is simply a *shift* operation. The content of the read-only vector is

shifted left one position. The first bit of the vector is shifted into the valid bit and a '0' is moved into the last bit of the vector.

For example, at the end of section 1 in Figure 2, the read-only vectors of the cache lines for X(I,J,K) is '1110'. After invalidation, the lines stay in cache by shifting the first bit '1' into the valid bit. The read-only vector becomes '1100' at the beginning of section 2.

There are two major advantages with the selective invalidation method using compiler generated read-only vectors. First, the read-only status of a shared variable is defined at each section instead of the entire program. This will dramatically increase the re-use rate of a read-only data. For example, it is very common in a scientific computation that a large array is initialized only once at the beginning of the program and then used without any modification throughout the rest of the program [Ben+90]. The second advantage is its ability to capture inter-section locality across multiple sections for the n-bit vector allows a cache line to stay in cache up to n subsequent sections. This is especially effective when the program consists of many small sections.

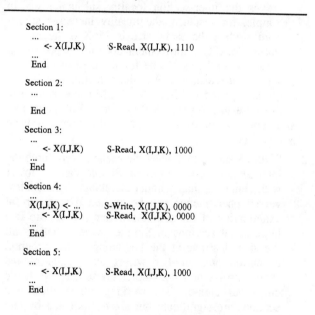

Section 1:
...
  <- X(I,J,K)   S-Read, X(I,J,K), 1110
End

Section 2:
...
End

Section 3:
...
  <- X(I,J,K)   S-Read, X(I,J,K), 1000
End

Section 4:
...
X(I,J,K) <- ...   S-Write, X(I,J,K), 0000
  <- X(I,J,K)   S-Read, X(I,J,K), 0000
...
End

Section 5:
...
  <- X(I,J,K)   S-Read, X(I,J,K), 1000
End

Figure 2.   Example of the read-only vectors for S-Read, S-Write

4. Performance Evaluation

The results of the inter-section locality of the three applications with the proposed enhancement techniques are presented in this section. The optimistic scheme, which does not invalidate any cache lines, is used for the comparison. We will present the results of the three programming techniques under the optimistic scheme.

These will serve as an upper bound for any realistic software scheme. The measurement of the advantage of the read-only vector scheme over the global read-only scheme is presented at the end.

In the performance study, the cache size varies from 16KB to 2MB and the line size is fixed at 128 bytes. LRU replacement algorithm is used in all the caches. The UP, 16-way MP, and 64-way MP cache systems are simulated based upon the UP and MP traces generated.

4.1. Assignment by Chunks

In this study, PSIMUL produces two types of traces per application: traces with chunk=1 and traces with chunk=n, where n is the number of iterations in a parallel loop divided by the number of executing CPU's. In the sequel, we will simply call them traces with no *chunking* and with chunking, respectively. Each of the traces is used to drive our cache model of a specific software scheme for comparison. The advantage of chunking over that of no chunking for the optimistic scheme is shown in Figure 3 and Figure 4 for Simple and Weather applications. There are a number of interesting observations:

1. For both programs, the scheme with chunking improves the inter-section locality substantially. In Simple, for instance, the locality increases significantly when the cache size is 256K or larger. A similar locality improvement does not happen until a 1M (Mega-bytes) cache for the case of chunk=1. The major reason is due to the reduction of working set as depicted in Figure 5. A 2M cache is used in this case in a 64-way MP system. With chunking, a reduction of over 50% in working set has been observed.

2. Figure 3 shows that when the cache is very large the inter-section re-use ratio of Simple can be lower with chunking than without chunking. This controversial phenomenon can be explained with careful examination of the source program. There are two large serial sections in Simple. One initializes all the shared arrays at the beginning of the program execution and another writes the results of the shared arrays and does other house-keeping at the end of execution. The working sets of these two sections are significant but are not affected by any chunk size. In fact, the inter-section references in these sections cause many misses disregard of any cache size or chunk size.

It is also observed that using chunking reduces the working set of each section to a factor of about 1/4 to 1/3 except for the two serial sections. For example in a 64-way MP system with chunking, the working set of all but the two serial sections can fit into a 256K cache. But without chunking, a 1M cache is needed. When the cache size is larger than 1M, there is very little benefit for both partitioning

methods. As a result, the inter-section misses created by the two serial sections becomes more predominant for chunking than no chunking. In fact, the overall reuse misses for chunk=n in a 2M-cache, 64-way MP system are 30% less than those of chunk=1. But, the total reuse references, the nominator in the re-use hit ratio, dropped by more than 45%.

3. A similar argument explains why Simple experiences a drop of locality with very large caches when chunking is used. Again, comparing two cache sizes: 1M and 2M in a 64-way MP system using chunking, the total inter-section references and misses produced by all the sections are very similar except for the two serial ones. For the serial sections, both the inter-section references and misses for the smaller cache are about half of those of the larger cache. Consequently, the total misses from the two serial sections are less dominating when smaller caches are used. The above discrepancy does not exist in Weather. With a 2M cache and chunking, the inter-section hit ratio can reach 61.1% and 81.5%, respectively, for the 16-way and 64-way.

4. For a 64-way MP system with chunking, the locality of Simple increases significantly with a 256K cache. A similar locality increase for Weather does not happen until the cache is 512K or larger. This is due to the fact that the working set of Weather is generally larger than that of Simple.

4.2. Reversed Execution of Parallel Loops

Figures 6 and 7 show that reversed execution of parallel loops can improve the inter-section locality for applications Simple and Weather in UP systems. For this study, there are totally 10 and 5 parallel loops which have been manually reversed in Simple and Weather, respectively.

The inter-section locality is quite low at the original UP system but the improvement is significant due to our reversed execution of parallel loops. It is expected that this improvement can be higher with more aggressive rearrangement of the execution sequence. Another interesting fact from Figures 5 and 6 is that the inter-section locality is not significantly sensitive to the cache size after reversing the loop execution. This can be explained by the fact that ideally the cache content left from the previous section will be accessed first independent of the cache size.

4.3. Loop Interchange

The inter-section locality of Tomcat in Figure 8 is very different from that of Simple or Weather. Its reuse hit ratios are considerably higher (40-50%) and the inter-section locality seems not very sensitive to the cache size and the number of processors used. We observed from the source program that the kernel of the program is a nested loop of a relatively small Doall loop inside a sequential loop. The working set of each small Doall loop can fit into a cache as small as 32 Kbytes. Therefore, larger cache does not help improving the inter-section locality.

Loop interchange of Tomcat can move the Doall loop to the outer loop without violating the data dependence. Consequently, a larger number of small sections, created by the original sequential outer loop, are combined into one big section. In the original program construct, the inter-section locality is much easier to capture because of the small working set of each section. As shown in Figure 8, after loop interchange the inter-section locality reduced significantly from 40% to 10%.

4.4. Evaluation of the Read-Only Vector Scheme

The evaluation so far has been focused on the optimistic scheme. To demonstrate the real gains in inter-section locality, two realistic software coherence schemes: the read-only vector and the global read-only schemes are evaluated. When the length of the read-only vector for the simulation is fixed 8, Figure 9 shows that the reuse hit ratio can be as high as 20% for application Simple. On the other hand, since all the shared arrays are initialized at the beginning of the program execution (i.e. no read-only variable in the program), no inter-section locality can be captured by the global read-only scheme,

The inter-section locality for Weather is somewhat higher than that of Simple. With a 2M cache and chunking, the reuse hit ratios for both 16-way and 64-way MPs are slightly over 25%. The global read-only scheme captures less inter-section locality: they are only 9.9% and 11.5%, respectively, as shown in Figure 10. The length of the read-only vector plays an important role in capturing the inter-section locality. Figure 10 also demonstrates the results for a 16-way MP with chunking using various lengths of the read-only vector. For example, with a 1M cache, the reuse hit ratios for the read-only vector of length 1, 2, 4, and 8 are respectively 4.4%, 8.3%, 11.7%, and 18.8%. The importance of the read-only vector length drops when the cache is getting too small because the chance to keep the useful data in the cache across multiple sections is relatively small. The figure also indicates that the global read-only scheme shows a comparable performance with our ROV scheme of bit-vector length of 2 when the cache is very large.

The useful locality can be captured by the read-only vector scheme for Tomcat is just above 10%. This is due to the presence of large amount of small sections such that the re-use hit ratios are not sensitive to the cache size or the number of processors. The length of the read-only vector is also insignificant because all the data left in

cache for one iterative small Doall loop will be used in the next iterative loop.

5. Conclusion

Three programming techniques to enhance inter-section locality and a new software coherence mechanism to capture such a locality are introduced in this study. The effectiveness of these schemes is measured by the inter-section cache reuse hit ratio through trace-driven simulation. The results can be used to estimate the potential performance gain of software coherence schemes and to provide a better understanding of the behavior of the shared data in parallel programs.

Domain decomposition is a promising technique to increase inter-section locality in parallel computation. The experiment of assigning loop iterations by chunks for Simple and Weather applications has shown significant improvement to the inter-section locality. This is mainly due to the reduction of working set of each parallel program section using the chunk assignment method. Obviously this technique can also improve the intra-section locality as well.

Without the help of a parallelizing compiler, it is difficult to manually rearrange the referencing order of shared variables in each parallel program section to enhance inter-section locality. In this study, we only rearranged the execution sequence of parallel loops for the UP configuration such that the data dependence of the original program would not be violated. This technique can improve the inter-section locality when the cache size is not large enough with respect to the size of the working set. Further study is needed for more aggressive program restructuring to experiment the full benefit of this technique.

Granularity of parallelism, i.e, the average size of tasks assigned to a CPU, is another important factor influencing the inter-section locality. Our demonstration with application Tomcat shows the impact of selection of large and small program sections on the inter-section locality. Loop interchange technique was used to vary the section size. The results have shown that the inter-section locality is much higher with small sections. However, with smaller program sections, the synchronization overhead becomes larger and the execution of parallel programs may be slower. Loop interchange not only alters the inter-section locality, it also changes the intra-section locality as well. Further study is required to fully understand the most effective arrangement.

In addition to the three programming techniques, we also introduced a new software coherence scheme, called read-only vector, to capture useful inter-section locality for read-only variables. It is shown to be superior over the global scheme.

6. References

[Ben+90] J.K. Bennett, J.B. Carter, and W. Zwaenepoel, "Adaptive Software Cache Management for Distributed Shared Memory Architectures," Proceedings, 17th Ann. Int'l Symp. on Computer Architecture, 1990, pp 125-134.

[Che&Vei88] H. Cheong and A. V. Veidenbaum, "A cache coherence scheme with fast selective invalidation," Proc. of the 15th Annu. Int. Sym. on Computer Architecture, p. 299-307, June 1988.

[Che&Vei89] H. Cheong and A.V. Veidenbaum, "A Version Control to Cache Coherence," Proceedings of 1989 Int'l Conf. on Supercomputing, p 322-330, June 1989.

[Eig+91] R. Eigenmann, J. Hoeflinger, G. Jaxon, Z. Li, and D. Padua, "Restructuring Fortran Programs for Cedar," Proc. 1991 Int'l Conf. on Parallel Processing, pp I57-66.

[ESSL90] IBM *Engineering and Scientific Subroutine Library, Guide and Reference*, order number SC23-184, 1990

[Lou&Sun92] A. Louri and H. Sung, "A Compiler Directed Cache Coherence Scheme with Fast and Parallel Explicit Invalidation," Proceedings of 1992 Int'l Conf. on Parallel Processing, pI2-I9, Aug. 1992.

[Min&Bae89] S. L. Min and J.-L Baer, "A Timestamp-based Cache Coherence Scheme," Proceedings of 1989 Int'l Conf. on Parallel Processing, p23-32, Aug. 1989.

[Pei+91] J.-K. Peir, K. So, and J.-H. Tang, "Inter-section Locality of Shared Data in Parallel Programs," Proc. 1991 Int'l Conf. on Parallel Processing, pp I278-286.

[Pfi+85] G.F. Pfister, W.C. Brantley, D.A. George, S.L. Harvey, W.J. Kleinfelder, K.P. McAuliffe, E.A. Melton, V.A. Norton, and J. Weiss, "The IBM Research Parallel Processor Prototype (RP3): Introduction and Architecture, " Proceedings, 1985 Int'l Conf. on Parallel Processing, pp 764-771, Aug. 1985.

[So+88] K. So, F. Darema, D. George, A. Norton, and G. Pfister, "PSIMUL- A System for Parallel Simulation of Parallel Programs," in *Performance Evaluation of Supercomputers*, published by North-Holland, June 1988.

[Vei86] A. V. Veidenbaum, "A Compiler-assisted Cache Coherence Solution for Multiprocessors," Proc. 1986 Int'l Conf. on Parallel Processing, pp 1029-1036.

[Wol89] M. Wolfe, "Optimalizing Supercompilers for Supercomputers," MIT Press, 1989.

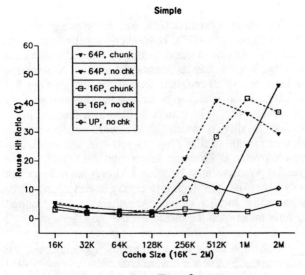

Figure 3

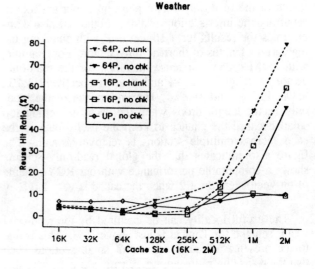

Figure 4

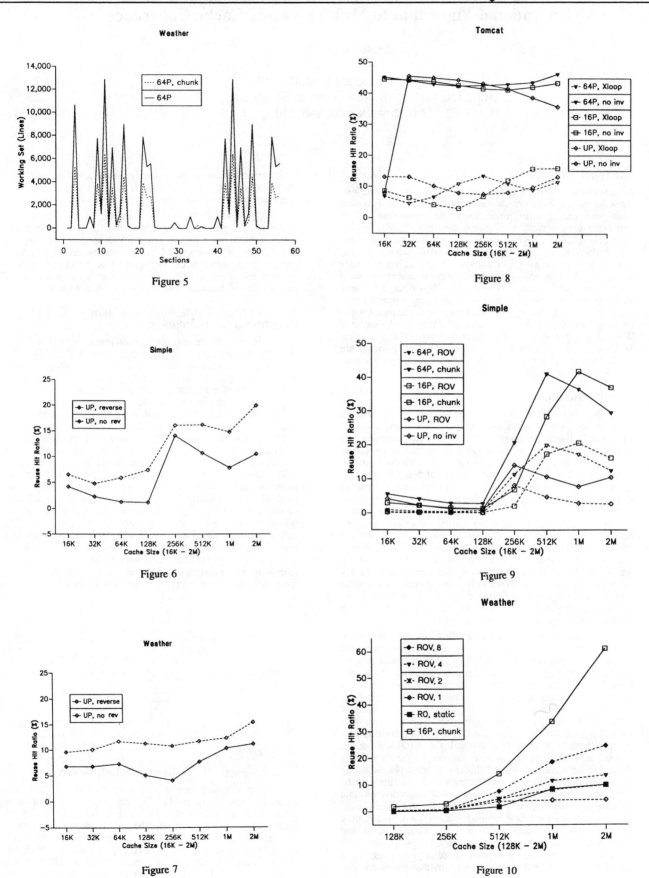

Figure 5

Figure 6

Figure 7

Figure 8

Figure 9

Figure 10

A Generational Algorithm to Multiprocessor Cache Coherence

Tzi-cker Chiueh

Computer Science Department
State University of New York at Stony Brook
chiueh@sbcs.sunysb.edu

Abstract

In view of the growing gap between processor speeds and network latency, it becomes increasingly expensive to maintain multi-processor cache consistency via run-time inter-processor communication. Software-controlled cache coherence schemes have the advantage of simplified hardware and the reduction of inter-processor communication traffic. Among previously proposed software-based schemes, those based on the concept of version/timestamp show the most aggressive performance potential. Unfortunately these methods have several implementation and performance problems that prevent them from being practical implementation choices. In this paper, we discuss these problems and describe a generational cache coherence algorithm that eliminates all of these problems. Moreover, the new algorithm can exploit inter-level temporal locality of parallel programs with significantly less hardware support.

1. Introduction

Multi-processor cache consistency maintenance schemes are generally classified into hardware and software categories. Hardware-based cache coherence schemes use run-time inter-processor notifications to update or invalidate stale cached data. For bus-based shared-memory multiprocessors such as Sequent's Symmetry, these notification messages usually account for a large portion of the bus traffic. In some instances, the percentage of consistency-related to total traffic could be as high as 80% [EGGE89]. In larger-scale multiprocessor systems where buses are easily saturated by inter-processor communication traffic, using run-time notification to maintain cache consistency becomes rather inefficient due to the lack of a broadcast medium. On the other hand, in hardware-based cache-consistency schemes, consistency maintenance actions are taken only when inconsistency actually arises at run time, e.g., a shared data write operation. Consequently, the hardware approach typically won't induce over-invalidations or unnecessary updates.

Among the software-controlled cache coherence schemes proposed in the literature, the version-controlled [CHEO89] or time-stamped [MIN89] cache coherence scheme shows the most aggressive performance potential. Since the basic idea of these two schemes is identical, we will only discuss the version-controlled scheme in the following. Cheong's version-controlled cache coherence algorithm is based on the *single assignment* principle. This principle dictates that a variable be assigned only once throughout its lifetime. Whenever a variable is updated, the variable's current version is abandoned and a new version of the variable is created to hold the new value. Under this principle, detecting stale cached data becomes straightforward: A variable's cache copy becomes stale only when there is a write to that variable in other processors. But by the single assignment principle the variable will assume a new name after the write. Consequently the next access to that variable will use its new name and cause a cache miss on the original cached copy due to name mismatch. Despite its conceptual simplicity, this algorithm has certain implementation and performance problems that need to be solved before it can be put to practical use. The major contribution of this paper is a new software-controlled cache coherence algorithm that is both simpler and free of these performance problems.

The rest of this paper is organized as follows. To motivate the proposed algorithm, we discuss the version-controlled algorithm and its problems in Section Two. In Section Three, a generational cache coherence algorithm is presented, together with the required architectural support. Section Fiour concludes this paper with a summary of the major ideas of this work and pointers for future work.

2. Version-Controlled Cache Coherence Schemes and Its Problems

Before describing the generational algorithm, let's describe our assumed parallel computation model [CHEO89] [MIN89]. A program is either explicitly (i.e., by programmers) or implicitly (i.e., by the compiler) partitioned into tasks. A task is a unit of computation that gets scheduled and assigned to processors for execution at run time. There are data dependency relationships (*true, output,* and *anti* dependency) among tasks. Tasks that are not related by data dependencies are said to be *independent* of each other and can be executed in parallel without synchronization. The data dependency relationships among a program's tasks can be expressed by a directed graph called the task graph [CHEO89]. A task graph's nodes represent tasks while its directed edges represent dependencies. If there is an edge E_{ij} from Node i to Node j, then Task j depends on Task i. Figure 1 shows an example task graph, where, for example, Task Five depends on Task Four. The compiler uncovers independent tasks by *levelizing* the task graph through topological sorting. Nodes at the same level of a levelized task graph are guaranteed to be independent and therefore can be dispatched simultaneously. In Figure 1, Task Two and Task Three are independent of each other.

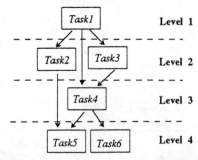

Figure 1   A Levelized Task Graph

If a compiler inserts barrier instructions at the boundaries of consecutive levels of a task graph, only *Doall* loop parallelism is exploited. To exploit the parallelism of *Doacross* loops, a system must be able to dispatch in parallel the tasks at different levels. However, in this case, synchronization instructions must be inserted appropriately to synchronize the execution behavior of concurrent tasks. Although the underlying idea of version-control and time-stamp algorithms are the same, Min's algorithm can exploit *Doacross* loop parallelism by adding extra mechanisms, whereas Cheong's simply can't. In contrast, the proposed generational algorithm can handle both *Doall* and *Doacross* loops in a single framework.

2.1. The Algorithm

Data variables are classified into private, shared read-only, and shared read-write three categories. Only shared read-write variables could cause cache consistency problems. From now on a variable is assumed to be shared read-write unless stated otherwise. A variable can be a scalar, a structure, or an array. Associated with each variable in a program are two numbers: a *current version number* (CVN) and a *birth version number* (BVN). The CVN represents the most up-to-date version number of a variable. The effective address used to access a variable is formed by concatenating the variable's CVN with its original memory address. Each node of a shared-memory multiprocessor maintains a local memory called the *variable ID table*, which keeps track of the CVN's of *all* the variables of a program. Every memory access in the programs has to specify both a variable ID and its address as shown in Figure 2, where the ID is used to index the variable ID table to locate the variable's CVN and to form an effective cache access address.

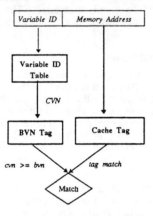

Figure 2   Formation of Effective Cache Access Address

Accordingly, each cached copy of a variable has a BVN in the cache's tag memory that designates the version number immediately after the cache copy is created. A variable's cache copy can be created in two ways: written by a store instruction or brought from main memory upon cache misses. The BVN of a variable's cache copy is set to the variable's CVN if the cache copy is brought from main memory, and is set to CVN+1 if the copy is updated by a write instruction. In a parallel task execution model where barrier instructions are inserted at the level boundaries, a variable can be written by at most one task at each level, possibly more than once. Only the last write to a variable needs an adjustment of the variable's BVN in the cache tag. Before reaching a level barrier, *every* task must increment the CVN's of all the variables that have been modified at that level. The set of variables that are modified at any level are assumed to be known at compile time, and these CVN increment instructions are inserted to a program statically.

A cache access is a hit when the cache tag matches the memory address and BVN ≥ CVN. Cheong's algorithm can ensure that accesses to stale cache copies lead to cache misses because stale copies , by construction, have smaller BVN's than the CVN's used to access shared variables. In addition, a variable's cache copy updated at one level by a processor could be accessed by the same processor without cache miss at subsequent levels if no writes to that variable occur in the interim on other processors. Therefore the algorithm can exploit inter-level temporal locality.

Because each node maintains the CVN's on its local variable ID table, every node is supposed to update the CVN's at the level barrier. As a result, Cheong's algorithm may not be able to guarantee cache consistency when some nodes did not participate in processing tasks at the i-th level and are assigned a task at the j-th level, where j ≥ i. This could take place when the i-th level has fewer independent tasks than the j-th level. In general, this so-called *level skipping* problem can occur when the levels of consecutive tasks assigned on a node are i and i+n (n > 1), respectively. [CHEO89] solves this problem by keeping track of the levels skipped by a node and updating the node's variable ID table *before* assigning a task to that node. Let's use Var_i to denote the set of variables modified at the i-th level. Then the set of variables whose CVN's need to be updated are the set of variables modified from Level i+1 to Level i+n-1, denoted by Mod(i+1, i+n-1):

$$\mathbf{Mod\,(i+1,\ i+n-1)} = \bigcup_{j=i+1}^{j=i+n-1} \mathbf{Var_j}$$

where $\cup$ is the multi-set union operator, i.e., duplicates are NOT eliminated. If a variable occurs in Mod(i, i+n) X times, its CVN must be incremented X times. The computation of Mod(i, i+n) and the associated CVN updates must occur at run time since processors are dynamically assigned to tasks. In the above discussion, we are assuming an acyclic task graph, the update procedure for cyclic task graphs are even more complex and can be found in [CHEO89].

2.2. Problems

In addition to the level-skipping problem, Cheong's algorithm has several other problems that prevent it from being a practical implementation choice. This prompts us to search for a better approach. The first problem is the run-time overhead associated with updates of current version numbers. In Cheong's algorithm, a task at one level must update the CVN's of all shared variables written at that level, even if the task itself only modifies a subset of the written variables. In addition, *level skipping* could result in more CVN updates *before* a task is assigned to a node. Consequently, although the Cheong's algorithm avoids run-time inter-processor consistency-maintenance communication, a program's tasks have to perform extra work. When this overhead is relatively small, this may seem to be a good tradeoff in the light of the increasing speed gap between processors and interconnection networks.

> $A[0]=1;$
>
> $For\ i = i\ to\ A\_Dim$
>
> $A[i] = A[i-1] * 2;$
>
> ⋮

Figure 3   A Doacross Loop

The second problem associated with Cheong's algorithm is that it only works for Doall loops but not for Doacross loops. Consider the Doacross loop in Figure 3, Variable A is written in Iteration I and read in Iteration I+1, and therefore Iteration I+1 is flow-dependent on Iteration I. Suppose Iteration I and I+1 are assigned to Node N and Node M respectively. To exploit Doacross loop parallelism, iterations at different levels, in this case, Iteration I and I+1, are allowed to be dispatched simultaneously. Further assume that there is a copy of A in Node M before this Doacross loop is initiated. If the BVN of A in Node N is l after the write operation in Iteration I, A's CVN in Node M is still $l-1$ when the read access in Iteration I+1 is made. This is because CVN updates are only performed at the end of the tasks that are dispatched simultaneously, and by this time A's new CVN has not yet propagated from Node N to Node M. As a result, when Node M accesses A, the stale copy of A seems to be up to date even though it isn't, thus leading to a false hit. Note that synchronization primitives only guarantee the execution order of dependent tasks but don't prevent false hits. Unless special notification mechanisms are provided, Cheong's algorithm can not handle Doacross loops.

The third problem of Cheong's algorithm is related to the finiteness of the version numbers of shared data variables. When a shared variable's version number is overflowed, it becomes impossible to use its version number to detect staleness of the variable's cache copy. One possibility is to reset the overflowed variable's version number to zero. However, this requires a system-wide coordination that resets the variable's CVN's and BVN's in all nodes and in main memory. Exactly how overflow cases are handled are not discussed in [CHEO89] and [MIN89]. There is a fundamental tradeoff between the number of bits allocated to version numbers and the frequency of version number reset operations. The shorter version numbers are, the more frequently version numbers overflow. On a similar vein, Cheong's algorithm also assumes that the variable ID table in each node is always big enough to hold the version numbers of all the shared variables in a given program. Since the number of shared data variables varies from program to program, it is not clear how big the variable ID table should be, and what will happen when variable ID tables are not big enough.

On the performance side, Cheong's algorithm is necessarily conservative. Consider the example in Figure 4, where the entire array A is modified at Level I, one element of A is modified at Level J, and then at Level K, only those elements whose value is greater than 1000 are modified. At compile time, there is no information as to which element of A is modified at level J, and the values of A's elements. Consequently Cheong's algorithm has to make pessimistic assumptions by incrementing the CVN's of A's cached elements in all nodes at the end of Level I. Similar things occur at the end of Level K even though it may well be the case that only a small portion of A are actually modified. Since CVN increments render stale those cache copies that have smaller BVN's, unnecessary CVN updates essentially correspond to over-invalidation in hardware-based cache coherence schemes.

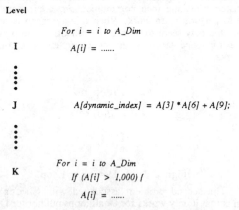

Level

I
```
For i = i to A_Dim

    A[i] = ......
```

J
```
A[dynamic_index] = A[3] *A[6] + A[9];
```

K
```
For i = i to A_Dim
    If (A[i] > 1,000) {

        A[i] = ......
```

Figure 4   An Example Program

There are two factors that contribute to cache over-invalidation. First, because an aggregate data structure such as an array is treated as an indivisible unit and is allocated a single variable ID, the version numbers of all array elements are implicitly incremented even though only the version numbers of a small portion of the array need to be adjusted. One possible solution is to allocate an ID to each array element, thus allowing a finer-granularity control over the version numbers of individual array elements. However, this approach could potentially aggravate the finite variable ID table problem

described above. Second, due to a lack of run-time information, state-of-the-art data flow analysis techniques can't identify the array elements that are actually modified at a particular level. As a result, over-invalidation is inevitable even if each array element has its own variable ID. This problem can only be solved by more aggressive compile-time data dependency analysis techniques.

Last but not least, Cheong's algorithm requires each memory access in a program to be specified in terms of a memory address AND a variable ID (as shown in Figure 2) This calls for a modification to the instruction set architecture of the host processor.

3. Generational Cache Coherence Algorithm

Through a closer examination, we found that almost all the problems associated with Cheong's algorithm are rooted in the fact that each shared data variable is allowed to have a separate version numbering dynamics. That is, at any point in the program execution, each shared variable is allowed to have a different current version number. As a result, each node has to maintain a variable ID table to keep track of each variable's CVN's, and pays the overhead of updating the CVN's. Suppose, on the other hand, all shared variables conform to a single version numbering scheme, there is no need to maintain individual variable's current version numbers, and the CVN update would be greatly simplified. The proposed generational cache coherence scheme is exactly based on this observation.

3.1. An Example

In Cheong's algorithm, each shared variable has a CVN in the variable ID table and a BVN in the corresponding cache line when it is cached. Because a variable's BVN is incremented each time the variable is written, the version numbers of different variables progress at different paces, depending on the update frequency of each individual variable. The basic idea of our generational cache coherence algorithm is to impose a single version numbering scheme upon a program's shared variables.

Let's call each level of the task execution graph a *generation*, each of which has an associated *generation number*. When a task is dispatched to a node, the task is also assigned a *current generation number* (CGN), which is the number of the generation to which this task belongs. Figure 5(a) shows the set of tasks and the set of variables modified in each generation. Note that each variable is either a scalar or an aggregate data structure. For example, *a, b, c,* and *d* are modified by T1 and T2 in the first generation. In our algorithm, the generation number is a replacement of the CVN's in Cheong's algorithm. In addition, each variable's cache line has a *valid generation number* (VGN), which designates the next generation in which the variable will be modified. Figure 5(b) shows the VGN's of *a, b, c, d,* and *e* at the end of each generation.

Initially the VGN's of all shared variables are zero; therefore the accesses to *a, b , c,* and *d* in the first generation cause cache misses, which bring a copy of each of these variables into the nodes. At the end of the first generation, because *a, b, c,* and *d* are modified in this generation, their VGN's need to be adjusted. However, unlike Cheong's algorithm, the VGN's are not simply incremented. Instead a variable's VGN after it is updated is assigned to the number of the next generation in which that variable *may* be written. Intuitively the VGN of a variable denotes the last generation at which the current value of that variable is guaranteed to be up to date. In this example, because *a* is next written in the third generation, *a*'s VGN becomes 3 at the end of the first generation. Similarly the VGN's of *b, c,* and *d* become 2, 4, and 3, respectively. Because *e* is not modified in the first generation, its VGN remains unchanged. Following the same reasoning, one can deduce that the VGN's of *a, b, c, d,* and *e* are <3,3,4,3,4> and <5,5,4,4,4> at the ends of the second and third

Generation	Tasks	Modified Variables
1	T1, T2	a, b, c, d
2	T3	b, e
3	T4, T5, T6	a, b, d
4	T7, T8	c, d, e
5	T9	a, b
6	END	

Figure 5(a) An Example Task Graph and Its Modified Variable Sets

Variable / Generation	a	b	c	d	e
0	0	0	0	0	0
1	3	2	4	3	0
2	3	3	4	3	4
3	5	5	4	4	4
4	5	5	6	6	6
5	6	6	6	6	6

Figure 5(b) Assignment of Valid Generation Numbers for Modified Variables in Figure 5(a)

generations, respectively, as shown in Figure 5(b).

When a program accesses a shared data variable, the system concatenates the current generation number with the memory address to form the cache access address. A cache access is a hit if the memory address matches a tag and VGN ≥ CGN. Essentially the current generation number replaces the current version numbers in Cheong's algorithm and is shared by all data variables in the programs. If, for instance, the write to *a* in the third generation is preceded by a read to *a*, and the node assigned to do this is the same one that writes to *a* in the first generation, then the read access is a cache hit because *a*'s VGN is equal to the current generation number. However, if these two writes are performed by different nodes, the read access will be a cache miss.

3.2. The Algorithm

More formally, the generational cache coherence algorithm works as follows. Given a task execution graph as explained in Section One, the compiler levelizes the graph and assigns a generation number to each level. Tasks at a level use the level's generation number as their current generation number (CGN). When a task is dispatched to a node, the task's generation number is loaded into the node's current generation number register. For each variable modified at a particular level, the last write to the variable at that level entails an update of the variable's valid generation number (VGN) in the cache. After a variable is updated, its VGN becomes the generation number of the level at which the next write to that variable may take place. A read access to the cache memory is a hit if the memory address matches the tag *and* the cache line's VGN is greater than or equal to the node's CGN. A write

access to the cache memory is a hit if the memory address matches the tag.

In the case of a cache miss, the requested cache line is either brought from the main memory or from another CPU. If it is from main memory, then the cache line's VGN is assigned the node's CGN; if from another CPU, then both the cache line and its VGN are copied. When a cache line is replaced, only the data is written back to main memory, but not its VGN. Although this decision could degrade the effectiveness of data caching, there are two reasons for not maintaining VGN's in main memory. First, maintaining VGN's in main memory could significantly increase the hardware requirements of main memory directories because every cache line now needs an extra field for its VGN. Second, by keeping main memory directories as stateless as possible, it is easier to handle VGN overflow as will be explained in the next section. Since a task's generation number is determined at compile time, there is no need to update CGN's. Moreover, since all variables share a CGN at any point in program execution, the hardware overhead for maintaining each individual variable's current generation number is avoided. Only a current generation number register is needed for each node.

Note that the generational algorithm *doesn't* require more data-flow knowledge than Cheong's algorithm. They both require that the set of potential modified variables at each level be known at compile time. For Cheong's algorithm, this knowledge is needed to insert CVN update instructions into the tasks. For our algorithm, it is needed to update the VGN's of the modified variables.

3.3. Discussion

Because our algorithm doesn't need a variable ID table in each node, the problem of finite ID tables and the overhead of updating CGN's are eliminated automatically. However, for a given word size, presumably it is easier for generation numbers than for version numbers in Cheong's algorithm to overflow. This is because the dynamic range of a generation number is dependent on the update frequencies of a program's *all* variables while that of a version number depends on the update frequency of *one* variable. As a result, the word size of generation numbers needs to be larger than version numbers, which in turn implies larger cache tags. On the other hand, because only a single generation number is used for all variables of the tasks at a particular level, it becomes feasible to manage the generation numbers at compile time, which will be discussed in the next section.

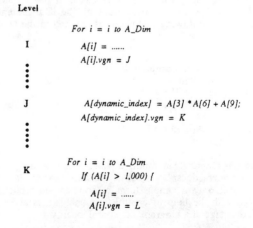

Figure 6   An Example of VGN Update

Most of the performance problems associated with Cheong's algorithm are eliminated by our generational algorithm. We discuss each of them in detail next. First, since all variables share a current generation number, this algorithm doesn't need to pay the overhead of updating individual variable's CGN. Second, this algorithm can exploit the Doacross loop parallelism without extra mechanisms. Consider a Doacross loop as shown in Figure 3, with each iteration flow-dependent on its preceding one except the first one. Let's further assume that the first iteration is assigned the generation number 1, and the second iteration the generation number 2, and so on. At run time, the dispatcher can dispatch as many iterations of this loop as the number of nodes in the system. With proper synchronization, the first read access from Iteration I+1 to the shared variable written in Iteration I will cause a cache miss. The reason is that the shared variable, A[I], if previously cached in the node assigned to run Iteration I+1, should have a VGN equal to I, the generation number of the I-th iteration. But the current generation number register holds I+1, therefore a cache miss occurs. By assigning CGN's to the tasks at compile time, the generational algorithm can exploit parallelism beyond Doall loops without run-time inter-processor communication.

The problem of over-invalidation in Cheong's algorithm can also be solved by the generational algorithm *without* extra hardware. Consider Figure 6, which is similar to Figure 4 except that it is augmented with instructions to update the VGN's of the shared variables. Again A is modified at Level I, J, and K, and $K > J > I$. At level J, only one element of A is modified but its index is not known at compile time. At the time of the last updates to the array elements of A at Level I, the VGN's of these array elements, in theory, should be J, as shown in Figure 6. However, it is possible to change the VGN's to K and insert the following instruction to the tasks at level I after the update of A[i]'s.

```
If (i == dynamic_index)
    A[i].vgn = J
else
    A[i].vgn = K
```

Note that the compiler doesn't need to know which A's element it is. By moving the VGN's from J to K, more temporal locality in the programs could be exploited. For example, if there are read accesses to A's other elements than *dynamic_index* at Level J+1, they may well hit on the cache because their VGN's are K rather than J.

Similarly, in the case of Level K, where the value of A's element determines whether it should be modified , in this case, the element's value must be greater than 1000. Then the VGN's of A's array elements could still be K at the end of Level I if the following instruction is inserted to the tasks at level I.

```
If (i == dynamic_index)
    A[i].vgn = J
else If (A[i] > 1000)
    A[i].vgn = K
    else
    A[i].vgn = L
```

where L is the next generation in which A is modified. Again the compiler doesn't need to acquire extra data-flow knowledge; only a simple form of source-code replication, perhaps with the help of control-flow analysis, is needed to take advantage of a program's temporal locality.

Just as the version-controlled scheme needs to increment BVN's, the generational scheme needs to update VGN's. However, there is a difference. Unless there is a special hardware support, incrementing BVN's requires a read-modify-write instruction sequence, whereas updating VGN's only requires a single write. In addition, since all shared variables use a generational numbering system, elements of an aggregate data structure, e.g., an array, can have different VGN's simultaneously. In other words, unlike the version-controlled scheme, the granularity of cache coherence is decoupled from high-level data structures in the program. This allows a more aggressive exploitation of cache locality.

4. Conclusion

In this paper, we have describe the performance and implementation problems of the by far most aggressive software-controlled cache coherence scheme, version-controlled [CHEO89] and time-stamped [MIN89]. A closer examination of the problems reveals that they are all rooted in the fact that every shared data variable is assigned a separate version number. Based on this observation, we propose a generational algorithm that imposes a uniform version numbering scheme upon the shared data variables of a parallel program. Because of this unified version numbering, most of the hardware cost and performance overhead associated with Cheong's algorithm are eliminated. Moreover, the generational algorithm can exploit the Doacross loop parallelism and can maintain consistency at the granularity of individual array elements without extra hardware costs, thus achieving a better performance over Cheong's algorithm.

Due to space constraints, we didn't report on a new generation number management algorithm and its architectural support [CHIU93]. This algorithm successfully handles the finiteness problem of generation numbers by combining a special generation encoding scheme and a fast reset hardware. Both [CHEO89] and [MIN89] failed to address this issue. Consequently they have to assume a large version number field in each cache line, or to pay a serious performance penalty due to a system-wide invalidation when version numbers overflow.

Because the proposed scheme involves compile-time transformation, a trace-driven simulation study won't provide a meaningful performance evaluation. Currently we are developing a cycle-by-cycle MIMD machine simulator based on the R2000/3000 architecture, which incorporates the hardware features described above, and a parallel compiler that can exploit program parallelism and perform affinity scheduling, as well as manage cache consistency. With these tools in place, we can start to evaluate the performance of the generational algorithm and compare with other hardware/software cache consistency techniques.

REFERENCE

[CHIU93]   T. Chiueh, "A Generational Approach to Software-Controlled Cache Coherence," Technical Report, Computer Science Department, SUNY at Stony Brook, 1993.

[CHEO89] H. Cheong, *Compiler-Directed Cache Coherence Strategies for Large-Scale Shared Memory Multiprocessor Systems*, UILU-ENG-89-8018, CSRD Report No. 953, Ph.D. Thesis, Center for Supercomputing Research and Development, University of Illinois, Urbana-Champaign, December 1989.

[EGGE89]   S. Eggers, *Simulation Analysis of Data Sharing in Shared-Memory Multiprocessors*, Report No. UCB/CSD 89/501, Ph.D. Thesis, Computer Science Division (EECS), University of California, Berkeley, April 1989.

[MIN 89]   S. Min, *Memory Hierarchy Management Schemes in Large-Scale Shared-Memory Multiprocessors*, TR 89-08-07, Ph.D. Thesis, Department of Computer Science and Engineering, University of Washington, Seattle, August 1989.

[MIN 89]   S. Min, J. Baer, H. Kim, "An Efficient Caching Support for Critical Sections in Large-Scale Shared-Memory Multiprocessors," IBM RC 15311, 1/4/90, IBM Research Division, T.J. Watson.

SEMI-UNIFIED CACHES

Nathalie Drach, André Seznec

IRISA/INRIA

Campus de Beaulieu

35042 Rennes Cedex, FRANCE

e-mail: drach@irisa.fr, seznec@irisa.fr

Abstract – *The purpose of the semi-unified on-chip cache organization, is to use the data cache (resp. instruction cache) as an on-chip second-level cache for instructions (resp. data). Thus the associativity degree of both on-chip caches is artificially increased, and the cache spaces respectively devoted to instructions and data are dynamically adjusted. The off-chip miss ratio of a semi-unified cache built with two direct-mapped caches of size S is equal to the miss ratio of a unified two-way set associative cache of size 2S; yet, the hit time of this semi-unified cache is equal to the hit time of a direct-mapped cache. Trace driven simulations show that using a direct-mapped semi-unified cache organization leads to higher overall system performance than using usual split instruction/data cache organization.*

1 Introduction

Since the gap between main memory access time and processor cycle time is continuously increasing, processor performance dramatically depends on the behavior of caches, and particularly on the behavior of small on-chip caches. Most of the recently introduced microprocessors have relatively small on-chip caches [1, 5, 7, 9]. High clock frequency and/or parallel instruction issuing are used; yet main memory access time remained approximately constant for ten years (around 250 ns) and relative miss penalties become very high even when second-level caches are used. As cache misses induce long pipeline stall, reducing the miss ratio has become an important challenge for microprocessors designers [6].

In section 2, we present a new organization for on-chip caches: the *semi-unified* cache organization. This cache organization uses the data cache (resp. instruction cache) as an on-chip second-level cache for instructions (resp. data). The semi-unified cache organization was introduced in order to maintain a short cache hit time while introducing some on-chip cache

associativity. Trace driven simulations presented in section 3 show that the performance of a microprocessor system may dramatically be improved when a semi-unified cache is used in place of a split instruction/data cache.

2 Semi-unified caches

A semi-unified cache organization (figure 1) consists of two distinct instruction/data caches where the instruction cache (resp. data cache) is used as the secondary cache for data (resp. instructions). This cache organization has been defined for on-chip caches. The cache C1 (resp. C2) is the main cache for instructions (resp. data) and the secondary cache for data (resp. instructions). For sake of simplicity, we shall now consider that the semi-unified cache lies on the microprocessor chip, while higher level caches are external.

2.1 Sequencing a request

Let us now describe the sequencing of an instruction request to address A. The request is presented to cache C1: on a miss in cache C1, the request is presented to cache C2. This first-level cache miss is called an *on-chip miss*. On a hit in cache C2, the requested line is brought back to cache C1: a line must be rejected from cache C1, the replacement of this line is discussed in the next section. In general, the lines are swapped between cache C1 and cache C2. But on a miss in cache C2, the request is presented to an external cache or the main memory. We shall call this second-level cache miss, an *off-chip miss*.

2.2 Replacement strategy

We now focus our attention on a semi-unified cache built with two direct-mapped caches of equal sizes, the *direct-mapped semi-unified cache* organization.

Off-chip misses: E.g. let us consider an off-chip instruction miss. The requested instruction line L must be stored in physical cache line $L(A)$ (where A is the instruction request address) of C1, then the content $L1$ of line $L(A)$ must be rejected from cache C1. But this rejected line $L1$ may be stored in cache C2 in physical cache line $L(A)$. Finally the rejected line from the whole on-chip direct-mapped semi-unified cache must be chosen among a set of two lines, $L1$, the content of the targeted physical line $L(A)$ in cache C1 and $L2$ the content of physical line $L(A)$ in cache C2: if $L2$ is rejected, then $L1$ is moved from cache C1 to cache C2. The strategy for chosing the rejected line may be random or LRU as on a classical two-way set-associative caches.

On-chip misses: Let us consider an on-chip instruction miss which hits on the secondary on-chip cache. The requested line L lies in line $L(A)$ of cache C2; this line must be brought back to line $L(A)$ in cache C1. The content L1 of line $L(A)$ in cache C1 must be removed from C1, but since line $L(A)$ in cache C2 is now empty, it may be stored in this location (L1 and L2 are swapped between cache C1 and cache C2): no external traffic is induced, global content of the on-chip cache has not changed.

Cache coherency: For an agent external to the microprocessor, the on-chip semi-unified caches must be seen as an on-chip unified cache.

2.3 Assets and drawbacks

The direct-mapped semi-unified cache organization built with two direct-mapped caches of equal sizes S have most of the advantages of both two-way set-associative cache organization and direct-mapped cache organization; it has also the assets of both unified cache and split cache organizations. The cache hit time [4] on a direct-mapped semi-unified cache is the same as when using split direct-mapped instruction/data caches of equal sizes S. But the off-chip miss ratio of the direct-mapped semi-unified cache is the same as that of a unified two-way set-associative cache of size 2S. And parallel access to instructions and data is provided; yet the on-chip cache spaces respectively devoted to instructions and data are dynamically adjusted as on a unified two-way set-associative cache.

Implementing such a semi-unified cache organization induces some additional hardware cost over an usual split cache organization: essentially, a data path is needed for swapping the lines between cache C1 and cache C2. Notice that an on-chip miss induces a miss penalty even when the request hits on the secondary on-chip cache.

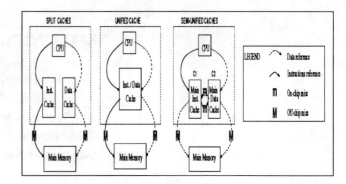

Figure 1: Cache organizations

In the next section, we use trace driven simulations to compare performance of microprocessors using a direct-mapped semi-unified cache organization or using classical cache organizations.

3 Performance evaluation

We consider a split cache organization and a semi-unified cache organization built with two direct-mapped caches of equal sizes S and a unified cache organization built with one two-way set-associative cache of size 2S.

3.1 Simulation model

For sake of simplicity in our simulation, we suppose that on split or semi-unified cache organization, a new instruction is issued on each cycle and that on a unified cache organization, a new request is issued on each cycle. And we assume a write back strategy with a write buffer of two cache lines and a LRU replacement for direct-mapped semi-unified and unified two-way set-associative caches. A cycle-by-cycle simulation was performed.

Servicing off-chip misses: on an off-chip miss, *latency* cycles are needed for getting back the first 64 bits word from the memory or the off-chip extra-level cache; then 4 cycles (resp. 2) more are needed for obtaining each extra 64 bits word in the line from the memory (resp. the off-chip extra-level cache). We also assume that the first word coming back from memory or off-chip extra-level cache is the missing word and that the execution may then be restarted as soon as this word is available.

Write back: when no off-chip extra-level cache is used, we assume that the write back of a line busies the memory for the same time as the service of a miss; this approximatively corresponds to the behavior of dynamic RAM.

When an off-chip extra-level cache is used, we assume that the writes may be pipelined with other accesses:

the write back of a line busies the off-chip cache for $2 * LineSize/64$ cycles. And when off-chip extra-level cache is used, second-level misses are not considered.

Miss penalty: The miss penalty for an on-chip miss which hits on the second-level on-chip may be quite low: from 2 to 4 cycles depending on the pipeline for accessing the cache. Pointed out the off-chip miss penalty is higher, particularly when no external cache is implemented. All simulation results reported in this paper were obtained assuming an on-chip miss penalty of 3 cycles, while several off-chip miss penalties were simulated.

Performance metric: As pointed out in the previous section, off-chip misses on a direct-mapped semi-unified cache are exactly the same as on a unified two-way set-associative cache with corresponding sizes. In order to compare the performance of systems using semi-unified caches with classical systems, we need a more accurate metric than off-chip miss ratio. As a metric, we use the average number of clock cycles per instruction (CPI) [3].

Traces: We have simulated a single process and traces include user mode only. The first set was composed with the three traces from the Hennessy-Patterson software obtained from the DLX simulator [3] (gcc, TeX and Spice). Seven other traces were generated using the SparcSim simulator facility [8]; this set was composed with: RESEAU (the simulator of a particular interconnection network), POISSON (a Poisson solver), STRASSEN (a matrix-matrix multiply using the Strassen algorithm), LINPACK (part of the LIN-PACK benchmark), CACHE (the cache simulator itself), CPTC (a Pascal-to-C translator) and MM (a sparse matrix vector multiply).

3.2 Miss ratio

Figure 2 illustrates on-chip miss ratio per instruction and off-chip miss ratio per instruction for split and semi-unified caches for each benchmark. The number of off-chip misses per instruction is lower for semi-unified caches than for split caches. The total first level miss ratio on a semi-unified cache organization is generally slightly higher than the miss ratio on a split direct-mapped cache organization. But as the on-chip miss penalty is lower than the off-chip miss penalty, this may result in lower CPI as it will be shown below. Notice also that only off-chip misses result in effective off-chip accesses. Using a semi-unified cache organization in place of a classical split cache organization significantly reduces the memory traffic; this may be very interesting in order to build low-end single-bus shared memory multiprocessor system.

3.3 Simulation results

By the end of 1992, on-chip cache sizes vary between 12 and 36 Kbytes [1, 5, 7, 9]. We have studied CPI performance as a function of cache organization (split, unified, semi-unified caches), benchmark, on-chip cache size (8, 16, 32 Kbytes), line size (16, 32, 64 bytes) and off-chip penalty (4, 5, 7 CPU cycles with external second-level caches and 14, 24, 34, 44, 54 CPU cycles with memory only); complete results are given in [2]. Figure 3 shows the CPI for different off-chip penalties: 5 CPU cycles with off-chip secondary cache and 14, 24, 54 CPU cycles without off-chip secondary cache (32 bytes line and two 8Kbytes caches or one 16 Kbytes cache). CPIs for direct-mapped semi-unified cache organization are generally better than for direct-mapped split and unified two-way set-associative cache organizations. This performance advantage is quite insignificant when a very fast external second-level cache is used, but becomes very important when the off-chip latency increases. As the gap between CPU cycle and off-chip miss penalty will continue to increase, these results clearly indicate that implementing an on-chip direct-mapped semi-unified cache organization in a microprocessor is worthwhile for microprocessors dedicated to highend systems as well as for microprocessors dedicated to medium end and low end systems.

4 Conclusion

As the integration density increases, first level caches may be integrated on the microprocessor chip. Yet, on-chip cache sizes remain limited, therefore the available space must be cautiously used. Microprocessor clock cycle, and therefore theoretical peak performance, is mostly determined by the cache hit time. Lower cache hit time is possible if direct-mapped caches are used. Yet, a higher degree of associativity would reduce the miss ratio and finally the overall pipeline stall delays for servicing the misses.

A semi-unified cache consists of a split instruction/data cache organization where the data cache (resp. instruction cache) is used as a second-level cache for instructions (resp. data). As it may hit on the on-chip data cache (resp. instruction cache), a first-level instruction (resp. data) miss does not necessarily lead to an *off-chip* transaction. The *off-chip* miss ratio on a semi-unified cache built with two direct-mapped caches of size S, is equal to the miss ratio on a unified two-way set associative cache of size 2S. Yet, the hit time of this semi-unified cache is equal to the hit time of a direct-mapped cache of size S. Moreover, both

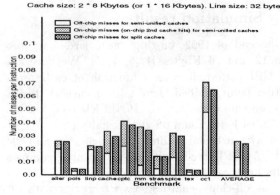

Figure 2: On-chip and off-chip misses

instructions and data may be accessed in parallel as for a split instruction/data cache organization. Using a semi-unified cache organization in place of a usual split instruction/data cache organization slightly increases the first-level cache miss ratio and significantly decreases the off-chip miss ratio. Yet, as the penalty of an on-chip miss is lower than the penalty of an off-chip miss, the overall system performance is increased as shown by simulation results.

Since the gap between off-chip miss penalty and on-chip internal delays will continue to grow, using a direct-mapped semi-unified cache organization is very attractive because it allows fast clock cycle and optimizes the off-chip miss ratio.

References

[1] *DECChip 21064-AA RISC Microprocessor, Preliminary Data Sheet* Digital Equipment Corporation, 1992

[2] N. Drach, A. Seznec, "Semi-unified caches", IRISA report, Jan.1993

[3] J.L. Hennessy, D.A. Patterson, "Computer Architecture a Quantitative Approach", Morgan Kaufmann Publishers, Inc. 1990

[4] M.D.Hill, A.J. Smith, "Evaluating Associativity in CPU Caches", IEEE Trans. on Comp., Dec. 1989

[5] *i860: 64-bit microprocessor hardware reference manual* Intel, 1990

[6] N.P. Jouppi, "Improving Direct-Mapped Cache Performance by the Addition of a Small Fully-Associative Cache and Prefetch Buffers", Proceedings of the 17th Inter. Symp. on Comp. Arch., June 1990

[7] G. Kane, J. Heinrich, "MIPS RISC Architecture", Prentice-Hall, 1992

[8] *SparcSim Manual* SUN Inc, Dec. 1989

[9] *TMS390Z55 Cache Controller, Data Sheet* Texas Instrument, 1992

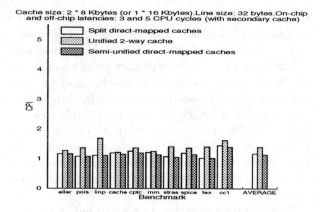

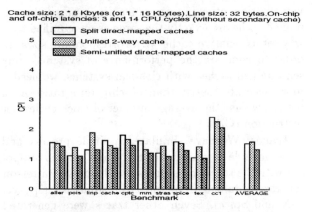

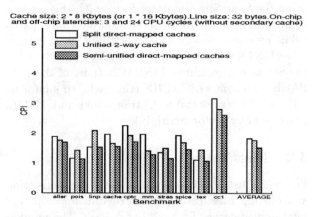

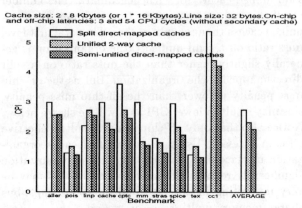

Figure 3: Performance

SESSION 2A

PROCESSOR AND COMMUNICATION ARCHITECTURE

Dependence Analysis and Architecture Design for Bit-Level Algorithms

Weijia Shang
Center for Advanced Computer Studies
University of Southwestern Louisiana
Lafayette, LA 70504
sw@cacs.usl.edu

Benjamin W. Wah
Coordinated Science Laboratory
University of Illinois, Urbana-Champaign
Urbana, IL 61801
wah@manip.crhc.uiuc.edu

Abstract:. *In designing application-specific bit-level architectures and in programming existing bit-level processor arrays, it is necessary to expand a word-level algorithm into its bit-level form before dependence analysis can be performed. In this paper, we consider dependence structures of bit-level algorithms as functions of three components — dependence structures of word-level algorithms, dependence structures of the arithmetic algorithms implementing word-wise operations, and algorithm expansions. Based on these components, we can derive dependence structures of bit-level algorithms without using time consuming general dependence analysis methods. To illustrate our approach, we derive two dependence structures for bit-level matrix multiplication and apply a method developed earlier [5,6,10] to design two bit-level architectures. One of these architectures is $O(p)$ times faster than the best word-level architecture, where p is the word length. The speedup we found here is true in general because a bit in a bit-level architecture goes to the next processor for processing as soon as it is available.*

1. INTRODUCTION

Bit-level architectures can exploit parallelism at the bit-level that is often ignored in word-level processor arrays. This is true because a bit in such a structure does not have to wait for other bits to finish before propagating to the next processor for processing. In this paper we study the design of application-specific bit-level architectures and the programming of existing bit-level processor arrays.

A general method involves three steps. A word-level algorithm of the application can first be expanded into a bit-level algorithm [8]; this is followed by an analysis of the dependence relations of the bit-level algorithm; finally, based on the bit-level dependence structure, the algorithm is mapped to a bit-level processor array.

This paper addresses the dependence analysis of an expanded bit-level algorithm and proposes a method for deriving the bit-level dependence structure without using time consuming general dependence analysis procedures. We discuss the

mapping of bit-level algorithms and the design of special-purpose bit-level architectures. We further show that bit-level architectures can be $O(p)$ times faster than the corresponding word-level structures, where p is the word length. The solution presented in this paper is an important step towards systematically programming or designing bit-level processor arrays.

Many methods have been proposed for deriving dependence structures of algorithms with nested loops [9]. These methods generally involve finding all integer solutions of a set of linear Diophantine equations, followed by a verification to see if the integer solutions are inside the *index set* or *iteration space* of the algorithm. In an exact analysis, the time complexity of these methods is exponential with respect to the number of nested loops (or the *algorithm dimension*).

The general dependence analysis methods discussed above can be applied to find bit-level dependence structures. However, we can substantially reduce the complexity in finding bit-level dependence structures if we exploit properties of bit-level algorithm expansions in our dependence analysis. For instance, we can view a bit-level dependence structure as a function of the corresponding word-level dependence structure, the dependence structures of the arithmetic algorithms implementing the word-wise operations (such as multiplication, division and addition), and the algorithm expansions. This allows us to derive a bit-level dependence structure without representing the algorithm at the bit-level and using general dependence-detection procedures.

The idea presented above is discussed in detail in Section 3. Basically, our approach is to expand a word-level algorithm based on its word-level dependence structure and the dependence structures of the underlying arithmetic algorithms. (Two of these algorithms are presented in Section 3.) Since many word-level algorithms involve a limited number of word-level arithmetic algorithms, the dependence structures of these algorithms need to be derived only once. For example, word-level algorithms, such as matrix multiplications, LU decompositions and convolutions, involve only a limited number of arithmetic algorithms for multiplication, addition and division (such as add-shift multiplication and carry-save multiplication).

Once the bit-level dependence structure is known, the next step is to either design a bit-level architecture based on the dependence structure, or map the dependence structure to a bit-level processor array. This can be carried out by an extension of the design method we have developed earlier [4,5,6,10]. We illustrate our approach by showing the design of two special-purpose bit-level architectures for matrix multiplication. One of

---
Research of the first author was supported in part by Louisiana Education Quality Support Fund under contract number LEQSF(1991-93)-RD-A-42, and in part by the National Science Foundation under Grants MIP 91-10940. Research of the second author was supported by the Joint Services Electronics Program Contract N00014-90-J-1270.

the architectures presented is time optimal; that is, it has the minimum total execution time and is $O(p)$ times faster than the best word-level architecture, where p is the word length. This improvement is expected to be true in general for bit-level architectures, as a bit can propagate to the next processor for further processing as soon as it is available.

This paper is organized into five sections. After defining basic terminology in Section 2, we show in Section 3 our bit-level algorithm dependence analysis. Section 4 shows how to design bit-level architectures based on the dependence structures obtained in Section 3. Two bit-level architectures are presented for matrix multiplication. Section 5 concludes the paper.

2. TERMINOLOGY AND DEFINITIONS

Throughout this paper, *sets*, *matrices* and *row vectors* are denoted by capital letters; *column vectors* are represented by lower-case symbols with an overbar; and *scalars* are shown as lower-case letters. The *transpose* of a vector $\bar{v}$ is denoted as $\bar{v}^T$. Vector $\bar{0}$ denotes a row or column vector whose entries are all zeroes. The dimension of vector $\bar{0}$ and whether it denotes a row or column vector are implied by the context in which it is used. The rank of matrix A is denoted *rank(A)*. The set of integers is denoted Z. The notation $|C|$ and $|\alpha|$ represents the cardinality of set C and the absolute value of scalar α, respectively. Let $\bar{v}$ and $\bar{u}$ be two vectors. Then $\bar{v} \geq \bar{u}$ means that every component of $\bar{v}$ is greater than or equal to the corresponding component of $\bar{u}$.

Algorithms considered in this paper are represented by a special kind of Fortran-like nested Do loops having the following form.

$$DO\ (\ j_1 = l_1,\ u_1;\ j_2 = l_2,\ u_2;\ ...,\ j_n = l_n,\ u_n\ )$$
$$S_1(\bar{j})$$
$$S_2(\bar{j})$$
$$...$$
$$\tag{2.1}$$
$$S_q(\bar{j})$$
$$END$$

Column vector $\bar{j} = [j_1, j_2, ..., j_n]^T$ is the index vector (also called the *index point*). $S_1(\bar{j})$, $S_2(\bar{j})$, ..., $S_q(\bar{j})$ are q assignment statements in iteration $\bar{j}$ having the form $x_k(g(\bar{j})) = f(x_1(h_1(\bar{j})), ..., x_t(h_t(\bar{j})))$, where $1 \leq k \leq t$, and $g()$, $h_i()$, $i = 1, ..., t$, are linear functions of $\bar{j}$. The lower and upper bounds of the ith nested loop, $1 \leq i \leq n$, are denoted l_i and u_i, respectively. Algorithms with n nested Do loops are called *n-dimensional algorithms*.

In general nested Do loops, cross-iteration dependences may exist. If iteration $\bar{j}$ depends on iteration $\bar{j}'$, then this dependence can be described by a pair $(\bar{j}, \bar{d})$, where $\bar{d} = \bar{j} - \bar{j}'$ is the vector difference of the index vectors of these two iterations. Vector $\bar{d}$ is called a *dependence vector* and is said to be *valid* at index point $\bar{j}$. Assuming that iteration $\bar{j}$ depends on iteration $\bar{j}'$, there are three types of dependences [1]. The first type is called *flow dependence* (or read-after-write dependence) where an input variable of the computation in $\bar{j}$ is an output variable of the

computation in $\bar{j}'$. The second is called *anti-dependence* (or write-after-read dependence) where an output variable of the computation in iteration $\bar{j}$ is an input variable of the computation in iteration $\bar{j}'$. The third is called *output dependence* (or write-after-write dependence) where an output variable of the computation in iteration $\bar{j}$ is an output of the computation in iteration $\bar{j}'$.

In this paper, we assume that every variable in the program is written only once during the entire execution of the algorithm; therefore, there is no output dependence. To illustrate this idea, consider the following word-level matrix multiplication algorithm.

Example 2.1 (matrix multiplication). Consider the matrix multiplication $Z = X \cdot Y$ as follows.

$$DO\ (j_1 = 1,\ u;\ j_2 = 1,\ u;\ j_3 = 1,\ u)$$
$$z(j_1, j_2) = z(j_1, j_2) + x(j_1, j_3)\, y(j_3, j_2)$$
$$END$$

Variable $z(j_1, j_2)$, $j_1, j_2 = 1, ..., u$, is written more than once during the execution of the algorithm. This program can be transformed to the following equivalent one where every variable is written only once.

$$DO\ (j_1 = 1,\ u;\ j_2 = 1,\ u;\ j_3 = 1,\ u)$$
$$z(j_1, j_2, j_3) = z(j_1, j_2, j_3 - 1) + x(j_1, j_3)\, y(j_3, j_2) \tag{2.2}$$
$$END$$

where $z(j_1, j_2, 0) = 0$, $j_1, j_2 = 1, ..., u$, and the $z_{i,j}$ entry of the product matrix is $z(i, j, u)$, $i, j = 1, ..., u$. □

In program (2.2), datum $x(j_1, j_3)$ is needed as an input by the n computations at index points $[j_1, 1, j_3]^T$, $[j_1, 2, j_3]^T$, ..., $[j_1, u, j_3]^T$. In other words, we need to broadcast datum $x(j_1, j_3)$ to these n index points if these n computations were to be executed simultaneously. Usually, broadcasting is not preferred in VLSI implementations because it incurs additional area on a chip and longer clock cycles. If we do not allow broadcasts, then data can be pipelined to those computations that need the data using the method developed by Fortes and Moldovan [2]. After eliminating broadcasts, program (2.2) can be transformed into the following form [2].

$$DO\ (j_1 = 1,\ u;\ j_2 = 1,\ u;\ j_3 = 1,\ u)$$
$$x(\bar{j}) = x(\bar{j} - [0, 1, 0]^T)$$
$$y(\bar{j}) = y(\bar{j} - [1, 0, 0]^T) \tag{2.3}$$
$$z(\bar{j}) = z(\bar{j} - [0, 0, 1]^T) + x(\bar{j})\, y(\bar{j})$$
$$END$$

where $x(\bar{j}) = x(j_1, j_2, j_3)$, $y(\bar{j}) = y(j_1, j_2, j_3)$ and $z(\bar{j}) = z(j_1, j_2, j_3)$. Intuitively, data $x(j_1, j_3)$ are pipelined along the j_2 axis through index points $[j_1, 1, j_3]^T$, $[j_1, 2, j_3]^T$, ..., $[j_1, u, j_3]^T$. Similarly, $y(j_3, j_2)$ are pipelined along the j_1 axis. Initially, $x(j_1, 0, j_3) = x_{j_1, j_3}$ and $y(0, j_2, j_3) = y_{j_3, j_2}$. The dependence structure of the matrix multiplication algorithm in (2.3) can also be obtained by using Banerjee's technique [1].

Consider a dependence vector $\bar{d}$. If for any two arbitrary index vectors $\bar{j}_1, \bar{j}_2 \in J$ such that $\bar{j}_2 - \bar{j}_1 = \bar{d}$ and that dependence vector $\bar{d}$ is valid at $\bar{j}_2$, then this dependence vector is *uniform*. If all dependence vectors are uniform, then the algorithm is a *uniform dependence algorithm*. For the purpose of this paper, an algorithm can be characterized by a triplet (J, D, E) where J is the index set, D is the dependence matrix containing all distinct dependence vectors as its columns, and E contains all different computations in all iterations. For uniform dependence algorithms, because all dependence vectors are valid at every index point, it is not needed to specify the points they are valid at. As an example, the matrix multiplication algorithm in (2.3) is a uniform dependence algorithm because dependence vectors $\bar{d}_1$, $\bar{d}_2$ and $\bar{d}_3$ are uniform and can be characterized by the triplet $A = (J, D, E)$ where

$$J = \left\{ \begin{bmatrix} j_1 \\ j_2 \\ j_3 \end{bmatrix} : 1 \leq j_1, j_2, j_3 \leq u, j_1, j_2, j_3 \in Z \right\} \quad D = \begin{bmatrix} \overset{y}{1} & \overset{x}{0} & \overset{z}{0} \\ 0 & 1 & 0 \\ 0 & 0 & 1 \end{bmatrix} \quad (2.4)$$

and

$$E = \{ x(\bar{j}) = x(\bar{j} - \bar{d}_2), y(\bar{j}) = y(\bar{j} - \bar{d}_1),$$
$$z(\bar{j}) = z(\bar{j} - \bar{d}_3) + x(\bar{j}) y(\bar{j}) \} .$$

The symbol on the top of each column in D indicates the variable that causes the dependence. This algorithm in (2.3) is *computationally uniform* because all iterations have the same computation.

3. BIT-LEVEL DEPENDENCE ANALYSIS

This section presents our method for finding dependence structures of bit-level algorithms. Section 3.1 discusses dependence structures of arithmetic algorithms for the multiplication and addition of two integers. Section 3.2 shows two algorithm expansions for implementing word-wise operations by bit-wise operations. Finally, we show how to obtain the bit-level dependence structure directly from word-level dependence structures, dependence structures of arithmetic algorithms, and the corresponding algorithm expansion.

3.1. Dependence Structures of Arithmetic Algorithms

Consider the *add-shift* [3] arithmetic algorithm that multiplies two nonnegative integers $s = a \times b$, where $s = s_{2p-1} s_{2p-2} \dots s_1$, $a = a_p a_{p-1} \dots a_1$, and $b = b_p b_{p-1} \dots b_1$. As is illustrated in Fig. 1a, the multiplication of two integers $s = a \times b$ can be implemented by adding p integers $(a_p \wedge b_i)(a_{p-1} \wedge b_i) \dots (a_1 \wedge b_i)$, $i = 1, \dots, p$, with the ith integer shifted $i - 1$ positions to the left.

For instance, in computing point $a_2 b_2$ in Fig. 1a, one partial sum bit s and one carry bit c are to be produced using input variables a_2, b_2, the carry bit from the east (computing point $a_1 b_2$), and the partial sum bit from the north (computing point $a_3 b_1$). The output variables are the carry bit to be sent to the west (computing point $a_3 b_2$) and the partial sum bit to be sent to the south (computing point $a_1 b_3$).

If the parallelogram of the data distribution in Fig. 1a is reshaped to the square shown in Fig. 1b, then the operations performed are similar except that data are input and output in different directions. The two algorithms in Fig. 1a and 1b are

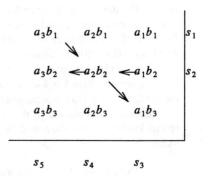

(a) $s = a \times b$ implemented by the add-shift algorithm.

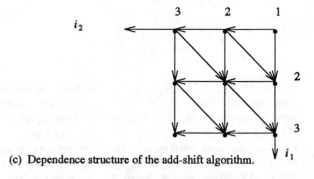

(b) The square obtained by reshaping the parallelogram in (a).

(c) Dependence structure of the add-shift algorithm.

Figure 1. The add-shift arithmetic algorithm.

equivalent. The general add-shift arithmetic algorithm as illustrated in Fig. 1b can be described by the following code.

$$DO \ (i_1 = 1, p; \ i_2 = 1, p)$$
$$c(\bar{i}) = g(a(i_2) \wedge b(i_1), c(i_1, i_2 - 1), s(i_1 - 1, i_2 + 1))$$
$$s(\bar{i}) = f(a(i_2) \wedge b(i_1), c(i_1, i_2 - 1), s(i_1 - 1, i_2 + 1)) \quad (3.1)$$
$$END$$

where $\bar{i} = [i_1, i_2]^T$, $c(\bar{i}) = c(i_1, i_2)$ is the carry bit, $s(\bar{i}) = s(i_1, i_2)$ is the partial sum bit, and Boolean functions g and f are defined as follows.

$$g(x_1, x_2, x_3) = (x_1 \wedge x_2) \vee (x_2 \wedge x_3) \vee (x_3 \wedge x_1)$$
$$f(x_1, x_2, x_3) = x_1 \oplus x_2 \oplus x_3 . \tag{3.2}$$

Then initial values are $s(0, i_2) = 0$, $i_2 = 2$, ..., $p+1$, $s(i_1, p+1) = 0$, $i_1 = 1, ..., p-1$, and $c(i_1, 0) = 0$, $i_1 = 1, ..., p$. The final results are $s_i = s(i, 1)$ for $1 \leq i \leq p$, and $s_i = s(p, i-p+1)$ for $p < i \leq 2p-1$.

Every variable in program (3.1) is written only once. Broadcasts exist because variable $a(i_2)$ is needed by index points $[1, i_2]^T, ..., [p, i_2]^T$, and $b(i_1)$ is needed by index points $[i_1, 1]^T, ..., [i_1, p]^T$. By applying Fortes and Moldovan's technique [2] for eliminating broadcasts, program (3.1) can be transformed into one without any broadcast as follows.

$$DO \ (i_1=1, p; \ i_2=1, p)$$
$$a(\bar{i}) = a(\bar{i} - \bar{\delta}_1)$$
$$b(\bar{i}) = b(\bar{i} - \bar{\delta}_2)$$
$$c(\bar{i}) = g(a(\bar{i}) \wedge b(\bar{i}), c(\bar{i} - \bar{\delta}_2), s(\bar{i} - \bar{\delta}_3)) \tag{3.3}$$
$$s(\bar{i}) = f(a(\bar{i}) \wedge b(\bar{i}), c(\bar{i} - \bar{\delta}_2), s(\bar{i} - \bar{\delta}_3))$$
$$END$$

where $\delta_1 = [1,0]^T$, $\delta_2 = [0,1]^T$, and $\delta_3 = [1,-1]^T$.

In iteration $\bar{i}$, $a(\bar{i} - \bar{\delta}_1)$ produced in iteration $\bar{i}' = \bar{i} - \bar{\delta}_1$ is input to produce $a(\bar{i})$; hence iteration $\bar{i}$ depends on iteration $\bar{i}'$. This flow dependence is uniform and is described by the pair $(\bar{i}, \bar{\delta}_1)$, $\bar{i} \in J$. Similarly, for the statement generating $b(\bar{i})$, there is a uniform dependence described by the pair $(\bar{i}, \bar{\delta}_2)$, $\bar{i} \in J$. For statements generating $c(\bar{i})$ and $s(\bar{i})$ in iteration $\bar{i}$, variables $c(\bar{i} - \bar{\delta}_2)$ and $s(\bar{i} - \bar{\delta}_3)$ are needed as inputs, which are generated in iterations $\bar{i}' = \bar{i} - \bar{\delta}_2$ and $\bar{i}'' = \bar{i} - \bar{\delta}_3$, respectively. Hence there are two uniform dependences $(\bar{i}, \bar{\delta}_2)$ and $(\bar{i}, \bar{\delta}_3)$, $\bar{i} \in J$. Since all dependence vectors are uniform, this algorithm is a uniform dependence algorithm. In short, there are three distinct dependence vectors, and the add-shift algorithm in (3.3) is described by the triplet $A_{add-shift} = (J_{as}, D_{as}, E_{as})$ where

$$J_{as} = \{\bar{i}: 1 \leq i_1, \text{ and } i_2 \leq p, i_1, i_2 \in Z\}, \tag{3.4}$$

$$D_{as} = [\bar{\delta}_1, \bar{\delta}_2, \bar{\delta}_3] = \begin{matrix} a & b,c & s \\ \begin{bmatrix} 1 & 0 & 1 \\ 0 & 1 & -1 \end{bmatrix} \end{matrix}$$

Note that δ_2 in D_{as} is involved in the computations of both $b(\bar{i})$ and $c(\bar{i})$.

Fig. 1c shows the index set and the dependence structure of the arithmetic algorithm in (3.3) when $p=3$. As an example, index point $[2, 2]^T$ represents the computation where a_2, b_2, carry $c(2,1)$ and partial sum $s(1, 3)$ are the four input bits. The computation is to sum three bits $a_2 \wedge b_2$, $c(2,1)$ and $s(1,3)$ to produce carry bit $c(2,2)$ to be sent to $[2, 3]^T$ and partial sum bit $s(2,2)$ to be sent to $[3,1]^T$. Dependence vector $[1,0]^T$ is due to pipelining a_2 from index point $[1, 2]^T$ and dependence vector $[0, 1]^T$ is due to pipelining b_2 from index point $[2, 1]^T$.

Due to space limitation, the dependence structure of an algorithm for adding two integers is not included here [7].

3.2. Algorithm Expansions and Bit-level Dependences

In this subsection we discuss algorithm expansions, namely, how to implement word-wise operations by bitwise operations. We study two expansions with the algorithm model in (2.1) restricted to the following form.

$$DO \ (j_1=l_1, u_1; \ j_2=l_2, u_2; \ ...; \ j_n=l_n, u_n)$$
$$x(\bar{j}) = x(\bar{j} - \bar{h}_1)$$
$$y(\bar{j}) = y(\bar{j} - \bar{h}_2) \tag{3.5}$$
$$z(\bar{j}) = z(\bar{j} - \bar{h}_3) + x(\bar{j}) \cdot y(\bar{j})$$
$$END$$

The triplet describing this word-level program is (J_w, D_w, E_w) where

$$D_w = \begin{matrix} x & y & z \\ [\bar{h}_1, \bar{h}_2, \bar{h}_3] \end{matrix} \quad J_w = \{\bar{j}: l_i \leq j_i \leq u_i, j_i \in Z, i=1,...,n\} \tag{3.6}$$

This model can describe applications such as matrix multiplication, convolution, matrix-vector multiplication, discrete cosine transform, and discrete Fourier transform. More general models are under investigation currently.

In each iteration of (3.5), two numbers $x(\bar{j})$ and $y(\bar{j})$ are multiplied, and the result added to the number $z(\bar{j} - \bar{h}_3)$ computed at index point $\bar{j} - \bar{h}_3$. Consider that each index point in J_w is replaced by a 2-dimensional index set J_{as} in Fig. 1c for multiplying $x(\bar{j})$ and $y(\bar{j})$. Then each index point $[\bar{j}^T, i_1, i_2]^T$ has $n+2$ indexes.

Fig. 2 illustrates two ways to add $z(\bar{j} - \bar{h}_3)$ to $x(\bar{j}) \cdot y(\bar{j})$. Let $x(\bar{j}) = x_p x_{p-1} ... x_1$, $y(\bar{j}) = y_p y_{p-1} ... y_1$ and $z(\bar{j} - \bar{h}_3) = z_{2p-1} z_{2p-2} ... z_1$. As is shown in Section 3.1, $z_{2p-1}, z_{2p-2}, ..., z_1$ are produced at boundary points $[(\bar{j} - \bar{h}_3)^T, i_1, i_2]^T$ where $i_1 = p$ or $i_2 = 1$. As is illustrated in Fig. 2a (Expansion II), the $2p-1$ bits of $z(\bar{j} - \bar{h}_3)$ are added to the product $x(\bar{j}) \cdot y(\bar{j})$ at boundary index points in iteration $\bar{j}$ where $i_1 = p$ or $i_2 = 1$; i.e., at points $[\bar{j}^T, p, i_2]^T$, $i_2 = 1, ..., p$ and $[\bar{j}^T, i_1, 1]^T$, $i_1 = 1, ..., p-1$. As an example, at index point $[\bar{j}^T, p, 2]^T$, the bits that need to be summed are the partial sum bit from $[\bar{j}^T, p-1, 3]^T$, z_{p+1}, $x_2 \wedge y_p$, and the carry bit from index point $[\bar{j}^T, p, 1]^T$.

In reality, we do not have to add $z_{2p-1}, ..., z_1$ of $z(\bar{j} - \bar{h}_3)$ in order to produce $x(\bar{j}) \cdot y(\bar{j})$ as is shown in Fig. 2a. Instead, as is illustrated in Fig. 2b (Expansion I), we can add the partial sum bits $z(\bar{j} - \bar{h}_3, i_1, i_2)$, $i_1, i_2 = 1, ..., p$, of $z(\bar{j} - \bar{h}_3)$ generated at $[(\bar{j} - \bar{h}_3)^T, i_1, i_2]^T$ to $x(\bar{j}) \cdot y(\bar{j})$ at $[\bar{j}^T, i_1, i_2]^T$. In other words, partial sum bit $z(\bar{j} - \bar{h}_3, i_1, i_2)$ is not sent to $[(\bar{j} - \bar{h}_3)^T, i_1 + 1, i_2 - 1]^T$, but is sent to $[\bar{j}^T, i_1, i_2]^T$ instead.

As an example, let $p=3$ as is shown in Fig. 2b. At index point $[\bar{j}^T, 2, 2]^T$, three bits, the carry bit from index point $[\bar{j}^T, 2, 1]^T$, $z(\bar{j} - \bar{h}_3, 2, 2)$ from iteration $\bar{j} - \bar{h}_3$, and $x_2 \wedge y_2$ are summed to produce a new partial sum bit $z(\bar{j}, 2, 2)$ to be sent to index point $[(\bar{j} + \bar{h}_3)^T, 2, 2]^T$, and a new carry bit $c(\bar{j}, 2, 2)$ to be sent to index point $[\bar{j}^T, 2, 3]^T$.

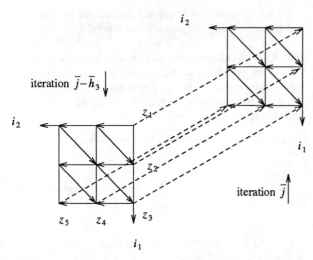

(a) Expansion II: add $z_5 \ldots, z_1$ at southern and eastern boundaries.

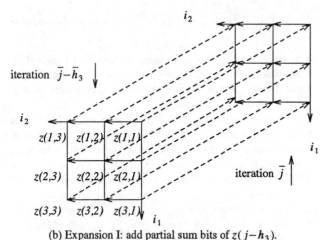

(b) Expansion I: add partial sum bits of $z(j-h_3)$.

Figure 2: Two ways to add $z(j-h_3)$ to product $x(j) \cdot y(j)$.

In short, each of the two ways described above for adding $z(\bar{j}-\bar{h}_3)$ to $x(\bar{j}) \cdot y(\bar{j})$ corresponds to an algorithm expansion for implementing word-wise operations by bit-wise operations. To get more insight on the dependence structures of expanded bit-level algorithms, consider the following simple 1-dimensional algorithm.

$$DO \ (\ j=l, \ u)$$
$$x(j) = x(j - h_1)$$
$$y(j) = y(j - h_2) \qquad (3.7)$$
$$z(j) = z(j - h_3) + x(j) \cdot y(j)$$
$$END$$

Since this is 1-dimensional, j, h_1, h_2, and h_3 are scalars instead of vectors. The index set and the dependence structure of this 1-dimensional word-level algorithm in shown in Fig. 3a. Without loss of generality, we assume that $h_1 = h_2 = h_3$.

Fig. 3b shows the dependence structure where the p^2 partial

sum bits of $z(j-h_3)$ are added to the corresponding partial sum bits of $x(j) \cdot y(j)$ (corresponding to Expansion II in Fig. 2b). Fig. 3c shows the dependence structure of the bit-level algorithm where the $2p-1$ final sum bits of $z(j-h_3)$ are added to $x(j) \cdot y(j)$ at boundary index points where $i_1 = p$ or $i_2 = 1$ (corresponding to Expansion I in Fig. 2a).

In Fig. 3b, each index point in Fig. 3a is replaced by the 2-dimensional index set for the multiplication of two integers in Fig. 1c. Hence, each index point in Fig. 3b has three indexes j, i_1, and i_2. The p bits of $x(j)$ are pipelined along the j axis at index points where $i_1 = 1$. In other words, the dependence caused by pipelining x_k, the kth bit of $x(j)$, is valid only at index points where $i_1 = 1$ and $i_2 = k$ and is, therefore, not uniform. All bits of $x(j)$ are also pipelined along axis i_1 in order to compute $x(j) \cdot y(j)$. The dependence caused by pipelining along direction j is described by the pair ($[j, 1, i_2]^T$, $\bar{d}_1 = [h_1, 0, 0]^T$), where $l \le j \le u$, and $1 \le i_2 \le p$. On the other hand, the dependence caused by pipelining along direction i_1 can be described by the pair ($[j, i_1, i_2]^T$, $\bar{d}_4$) where $i_1 \ne 1$ and $\bar{d}_4 = [0, 1, 0]^T$ (i.e., dependence vector $\bar{\delta}_1$ in (3.4) is prefixed by a zero corresponding to the j axis). Dependence vector $\bar{d}_4$ is not uniform and is valid only at index points where $i_1 \ne 1$.

Similarly, the p bits of $y(j)$ are pipelined along axis j at index points where $i_2 = 1$, and the dependence caused by pipelining y_k, the kth bit of $y(j)$, is valid only at boundary index points where $i_1 = k$ and $i_2 = 1$. This dependence is, therefore, not uniform. All bits of $y(j)$ are also pipelined along axis i_2 in order to compute $x(j) \cdot y(j)$. The dependence caused by pipelining along direction j is described by the pair ($[j, i_1, 1]^T$, $\bar{d}_2 = [h_2, 0, 0]^T$), where $l \le j \le u$, and $1 \le i_1 \le p$. The dependence caused by pipelining along direction i_2 can be described the pair ($[j, i_1, i_2]^T$, $\bar{d}_5$), where $i_2 \ne 1$ and $\bar{d}_5 = [0, 0, 1]^T$ (i.e., $\bar{\delta}_2$ in (3.4) is prefixed by a zero corresponding to axis j). Dependence vector $\bar{d}_5$ is not uniform and is valid only at points where $i_2 \ne 1$.

Variable $z(j-h_3)$ causes a flow dependence because it is to be added to $x(j) \cdot y(j)$ after it is generated. In Fig. 3b, because the p^2 partial sum bits $z(j-h_3, i_1, i_2)$, $i_1, i_2 = 1, \ldots p$, of $z(j-h_3)$ produced at index points $[j-h_3, i_1, i_2]^T$, $i_1, i_2 = 1, \ldots, p$, are sent to index points $[j, i_1, i_2]^T$, $i_1, i_2 = 1, \ldots, p$, respectively, the flow dependence is described by the pair ($\bar{q}$, $\bar{d}_3$) where $\bar{q} = [j, i_1, i_2]^T$, $i_1, i_2 = 1, \ldots, p, j = l, \ldots, u$, and $\bar{d}_3 = [h_3, 0, 0]^T$. This dependence is uniform. At index points where $j=u$, all partial sum bits have to be added. Partial sum bit $z(u, i_1, i_2)$ generated at index point $[u, i_1, i_2]^T$ will be sent to index points $[u, i_1+1, i_2-1]^T$ instead of index point $[j+h_3, i_1, i_2]^T$ as is in the cases when $j \ne u$. This flow dependence can be described by the pair ($[u, i_1, i_2]^T$, $\bar{d}_6 = [0, 1, -1]^T$) (where $\bar{d}_6$ is obtained from $\bar{\delta}_3$ in (3.4) by prefixing it by a zero corresponding to j axis). This dependence is not uniform and is valid only when $j = u$.

Carry bit c flows in parallel along with the i_2 axis. The corresponding dependence vector $\bar{d}_5 = [0, 0, 1]^T$ is not uniform and is not valid at index points where $i_2 = 1$ because at these points, carry c is zero and is not added. Therefore, the pair describing the dependence caused by c is ($[j, i_1, i_2]^T$, $\bar{d}_5$), $i_2 \ne 1$, which is the same as the one describing pipelining $y(j)$

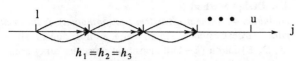

(a)   Index set and dependence structure of program (3.7).

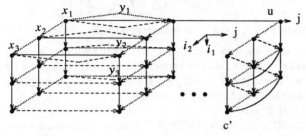

(b)   Dependence structure using expansion I.

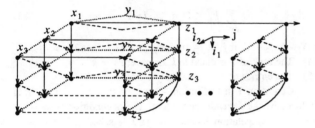

(c)   Dependence structure using expansion II.

Figure 3. In (b) and (c) , x is pipelined along solid lines; y is pipelined along dotted lines and z flows along dashed lines.

along the i_2 axis.

In Fig. 3b, three bits are summed at some index points. In these cases, one carry bit and one partial sum bit have to be generated. At other index points, more than three bits have to be summed; hence, we need to generate at least two carry bits and one partial sum bit. For example, at index point $[u, 2, 2]^T$, we need to sum at least four bits: $x_2 \wedge y_2$, two partial sum bits $z(u-1,2,2)$ and $z(u,1,3)$, and carry bit $c(u,2,1)$. Three output bits should be generated: the partial sum bit $z(u,2,2)$ to be sent to index point $[u, 3, 1]^T$, and two carry bits $c(u,2,2)$ to be sent to index point $[u, 2, 3]^T$ and $c'(u,2,2)$ to be sent to index point $[u, 2, 4]^T$ (if $p = 3$, this carry goes out of the index set and is not useful; however, if $p > 3$, this carry is useful). If four of these input bits are one, carry c' will be one. If two and not more than three are ones, then carry c will be one. The dependence vector corresponding to the second carry c' is $\bar{d}_7 = [0, 0, 2]^T$, which is valid at index points where $j = u$ and $i_2 \neq 1,2$, or $j = u$ and $i_1 \neq 1$.

In Fig. 3c (corresponding to Expansion I in Fig. 2a), all the dependence vectors can be obtained in a similar way. The two bit-level dependence structures in Fig. 3b and 3c have the same index set. Let J, D_I, and D_{II} be the index set, the dependence matrix for Fig. 3b, and the dependence matrix for Fig. 3c, respectively. Then

$$J = \{[j, i_1, i_2]^T : l \leq j \leq u, \ 1 \leq i_1, i_2 \leq p, \ j, i_1, i_2 \in Z\}$$

$$D_I = \left[\bar{d}_1, \bar{d}_2, \bar{d}_3, \bar{d}_4, \bar{d}_5, \bar{d}_6, \bar{d}_7 \right] \qquad (3.8)$$

$$= \begin{bmatrix} h_1 & h_2 & h_3 & 0 & 0 & 0 & 0 \\ 0 & 0 & 0 & 1 & 0 & 1 & 0 \\ 0 & 0 & 0 & 0 & 1 & -1 & 2 \end{bmatrix}$$

$$i_1=1 \ \ i_2=1 \ \ \bar{q} \ \ i_1 \neq 1 \ \ i_2 \neq 1 \ \ j=u \ \ \bar{q}_1$$

$$D_{II} = \left[\bar{d}_1, \bar{d}_2, \bar{d}_3, \bar{d}_4, \bar{d}_5, \bar{d}_6, \bar{d}_7 \right] \qquad (3.9)$$

$$= \begin{bmatrix} h_1 & h_2 & h_3 & 0 & 0 & 0 & 0 \\ 0 & 0 & 0 & 1 & 0 & 1 & 0 \\ 0 & 0 & 0 & 0 & 1 & -1 & 2 \end{bmatrix}$$

$$i_1=1 \ \ i_2=1 \ \ \bar{q}_2 \ \ i_1 \neq 1 \ \ i_2 \neq 1 \ \ \bar{q} \ \ i_1 \neq p$$

where $\bar{q}_1 = [u, i_1, i_2]^T$, $i_1 \neq 1$ or $i_2 \neq 1,2$, $\bar{q}_2 = [j, i_1, i_2]^T$, $i_1 = p$, or $i_2 = 1$, and $\bar{q}$ is an arbitrary index point in J. The information at the bottom indicates those index points where the corresponding dependence vector is valid at. For example, $\bar{d}_1$ is valid only at points where $i_1 = 1$ as is indicated by "$i_1 = 1$" at the bottom of $\bar{d}_1$. Vector $\bar{d}_3$ is uniform in Expansion I and $\bar{d}_6$ is uniform in Expansion II.

We now extend the above analysis on 1-dimensional algorithms in (3.5) to n-dimensional algorithms in (3.5). Let

$$\bar{q} = \begin{bmatrix} \bar{j} \\ \bar{i} \end{bmatrix} \quad \text{where } \bar{j} = [j_1, ..., j_n]^T \text{ and } \bar{i} = [i_1, i_2]^T , \quad (3.10)$$

and $\bar{d}_1 = [\bar{h}_1^T, 0, 0]^T$, $\bar{d}_2 = [\bar{h}_2^T, 0, 0]^T$, $\bar{d}_3 = [\bar{h}_3^T, 0, 0]^T$, $\bar{d}_4 = [\bar{0}, \bar{\delta}_1^{\ T}]^T$, $\bar{d}_5 = [\bar{0}, \bar{\delta}_2^{\ T}]^T$, $\bar{d}_6 = [\bar{0}, \bar{\delta}_3^{\ T}]^T$, and $\bar{d}_7 = [\bar{0}, 0, 2]^T$ where $\bar{h}_1$, $\bar{h}_2$ and $\bar{h}_3$ are as defined in (3.5) and $\bar{\delta}_1$, $\bar{\delta}_2$ and $\bar{\delta}_3$ are as defined in (3.4). Also let D_{as} and D_w be as defined in (3.4) and (3.6), respectively. Let J and D_I and D_{II} be the index set and dependence matrices of Expansions I and II described in Fig. 3b and 3c, respectively. Then the following theorem describes how D_I, D_{II} and J are related to the word-level dependence structure, the arithmetic algorithm, and algorithm expansions.

Theorem 3.1:

$$J = \left\{ \bar{q} = \begin{bmatrix} \bar{j} \\ \bar{i} \end{bmatrix} : \bar{j} \in J_w, \ \bar{i} \in J_{as} \right\}, \qquad (3.11a)$$

$$D_I = \begin{bmatrix} D_w & 0 & \bar{0} \\ 0 & D_{as} & \bar{\delta}_4 \end{bmatrix} = \left[\bar{d}_1 \ \bar{d}_2 \ \bar{d}_3 \ \bar{d}_4 \ \bar{d}_5 \ \bar{d}_6 \ \bar{d}_7 \right] \qquad (3.11b)$$

$$= \begin{bmatrix} \bar{h}_1 & \bar{h}_2 & \bar{h}_3 & \bar{0} & \bar{0} & \bar{0} & \bar{0} \\ \bar{0} & \bar{0} & \bar{0} & \delta_1 & \delta_2 & \delta_3 & \begin{matrix} 0 \\ 2 \end{matrix} \end{bmatrix}$$

$$i_1=1 \ \ i_2=1 \ \ \bar{q} \ \ i_1 \neq 1 \ \ i_2 \neq 1 \ \ j_n=u_n \ \ \bar{q}_1$$

$$D_{II} = \begin{bmatrix} D_w \mathbf{0} & \overline{0} \\ \mathbf{0} & D_{as}\overline{\delta}_4 \end{bmatrix} = \begin{bmatrix} \overline{d}_1 & \overline{d}_2 & \overline{d}_3 & \overline{d}_4 & \overline{d}_5 & \overline{d}_6 & \overline{d}_7 \end{bmatrix} \quad (3.11c)$$

$$= \begin{array}{c} \begin{array}{ccccccc} x & y & z & x & y,c & z & c' \end{array} \\ \begin{bmatrix} \overline{h}_1 & \overline{h}_2 & \overline{h}_3 & 0 & \overline{0} & \overline{0} & \overline{0} \\ \overline{0} & \overline{0} & \overline{0} & \delta_1 & \overline{\delta}_2 & \overline{\delta}_3 & 0 \\ & & & & & & 2 \end{bmatrix} \\ \begin{array}{ccccccc} i_1{=}1 & i_2{=}1 & \overline{q}_2 & i_1{\neq}1 & i_2{\neq}1 & \overline{q} & i_1{=}p \end{array} \end{array}$$

where $\overline{q}_1$ is defined such that $(i_1{\neq}1$ or $i_2{\neq}1,2)$ and $j_n = u_n$, $\overline{q}_2$ is defined such that $i_1 = p$ or $i_2 = 1$, $\overline{q}$ is an arbitrary point in J, $\overline{\delta}_4 = [0, 2]^T$, and $\mathbf{0}$ is a matrix with proper dimension and has all zero entries.

Proof. The proof is omitted due to space limitation [7]. □

Example 3.1. Consider the matrix multiplication in (2.3). Its dependence matrix and index set at the word-level is shown in (2.4). According to Theorem 3.1, the dependence matrix and index set of the corresponding bit-level matrix multiplication algorithm derived by Expansion II (Fig. 3c) are as follows.

$$D = \begin{array}{c} \begin{array}{ccccccc} y & x & z & x & y,c & z & c' \end{array} \\ \begin{bmatrix} 1 & 0 & 0 & 0 & 0 & 0 & 0 \\ 0 & 1 & 0 & 0 & 0 & 0 & 0 \\ 0 & 0 & 1 & 0 & 0 & 0 & 0 \\ 0 & 0 & 0 & 1 & 0 & 1 & 0 \\ 0 & 0 & 0 & 0 & 1 & -1 & 2 \end{bmatrix} \quad (3.12) \\ \begin{array}{ccccccc} & & & & & & i_1{=}p \\ i_1{=}1 & i_2{=}1 & \text{or} & i_1{\neq}1 & i_2{\neq}1 & \overline{q} & i_1{=}p \\ & & & i_2{=}1 & & & \end{array} \end{array}$$

$$J = \{[\,j_1,j_2,j_3,i_1,i_2\,]^T \in Z^5 : 1 \leq j_1,j_2,j_3 \leq u, 1 \leq i_1,i_2 \leq p\} \quad (3.13)$$

Expansion II is slower than Expansion I because, as is indicated in Fig 3c, the computation at $\overline{j}$ has to wait for the final results at $\overline{j} - \overline{h}_3$. In Expansion I, partial sum bits in $\overline{j} - \overline{h}_3$ are sent to $\overline{j}$ and takes less time. Further, Expansion I is more computationally uniform because at all points, except when $j = u$ in Fig 3b, at most three bits are to be summed; in contrast, in Expansion II, four or five bits have to be summed on the hyperplane $i_1 = p$ (the southern boundary points in the 2-dimensional index set of the add-shift operation). This may cause unbalanced load distribution. More discussions on algorithm expansion can be found in the reference [7].

4. DESIGN OF BIT-LEVEL ARCHITECTURES

Based on the dependence structures presented in Section 3, we discuss in this section the design of bit-level architectures. After summarizing in Section 4.1 a design method we have developed earlier [5,6], we show in Section 4.2 the application of the design method to design two bit-level architectures for matrix multiplication.

4.1. Design Method

Definition 4.1 (Linear algorithm transformation): A *linear algorithm transformation* maps an *n*-dimensional algorithm (J, D, E) into a $(k-1)$-dimensional processor array according to the mapping:

$$\tau: J \to Z^k, \ \tau(\overline{j}) = T\overline{j}, \forall \overline{j} \in J$$

where $T = \begin{bmatrix} S \\ \Pi \end{bmatrix} \in Z^{k \times n}$ is the *mapping matrix*, $S \in Z^{(k-1) \times n}$ is the *space mapping matrix*, and $\Pi \in Z^{1 \times n}$ is the *time mapping vector* or *linear schedule vector*. The computation indexed by $\overline{j} \in J$ is executed at time $\Pi\overline{j}$ and at processor $S\overline{j}$. The mapping τ must satisfy the following conditions:

(1) $\Pi D > \overline{0}$.

(2) $SD = PK$ where $P \in Z^{(k-1) \times r}$ is the matrix of interconnection primitives of the target machine, and $K \in Z^{r \times m}$ is such that

$$\sum_{j=1}^{r} k_{ji} \leq \Pi \overline{d}_i, \quad i = 1, ..., m. \quad (4.1)$$

(3) $\forall \overline{j}_1, \overline{j}_2 \in J$, if $\overline{j}_1 \neq \overline{j}_2$, then $\tau(\overline{j}_1) \neq \tau(\overline{j}_2)$, or $T\overline{j}_1 \neq T\overline{j}_2$.

(4) The rank of T is equal to k, or $rank(T) = k$.

(5) The entries of T are relatively prime.

Condition 1 in Definition 4.1 preserves the partial ordering induced by the dependence vectors. If this condition is satisfied, then the computation indexed by $\overline{j} \in J$ is scheduled to execute only after the computations indexed by $\overline{j} - \overline{d}_i \in J$, $i = 1, ..., m$. In this case $\Pi D > \overline{0}$ and the dependence relation is, therefore, satisfied.

The matrix of interconnection primitives P describes the connection links of processors in the processor array. For an array with each processor connected to its four nearest eastern, southern, western and northern neighbors, it has four interconnection primitives $[0, 1]^T$, $[0, -1]^T$, $[1, 0]^T$ and $[-1, 0]^T$ and matrix $P = \begin{bmatrix} 0 & 0 & 1 & -1 \\ 1 & -1 & 0 & 0 \end{bmatrix}$. Condition 2 in Definition 4.1 guarantees that the space mapping can be implemented in a systolic architecture with interconnection primitive matrix P. The summation on the left hand side of inequality (4.1) is the number of times that the interconnection primitives have been used to pass a datum according to dependence vector $\overline{d}_i$ from its source to its destination. The item on the right is the time units between the source usage and the destination usage of that datum. Assuming that it takes one time unit for a datum to travel one interconnection primitive, then the inequality must be satisfied in order to have the datum arrive before it is used.

Condition 3 defines the condition for avoiding computational conflicts. If the condition were not true, that is, $\tau(\overline{j}_1) = \tau(\overline{j}_2)$, then the computations indexed by $\overline{j}_1$ and $\overline{j}_2$ are mapped to the same processor at the same time, and a conflict occurs. Condition 4 guarantees that the algorithm is to be mapped into a $(k-1)$-dimensional array but not a q-dimensional array, where $q < k-1$. When $rank(T) = q+1 < k$, there are exactly $q+1$ linearly independent rows in T, and all other rows of T are linear combinations of these $q+1$ linearly independent rows. Let T' be the matrix consisting of these $q+1$ linearly independent rows. T

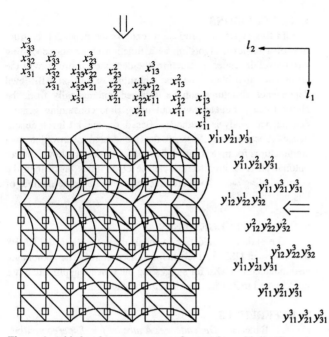

Figure 4. A bit-level processor array for matrix multiplication with $p = n = 3$ corresponding to T in (4.2).

---

can now be transformed by a linear transformation to T', which means that the algorithm is actually mapped into a q-dimensional processor array. Condition 5 is used to guarantee that at any time during the execution, at least one processor is busy. For detailed description of the mapping model and optimization method, please see the references [5,6].

4.2. Bit-Level Architectures for the Matrix Multiplication

In this subsection, based on the dependence structure obtained by Expansion II, two bit-level architectures for matrix multiplication are presented. The dependence matrix and index set of the bit-level matrix multiplication by Expansion II are shown in (3.12) and (3.13), respectively

Theorem 4.5: The following mapping matrix

$$T = \begin{bmatrix} S \\ \Pi \end{bmatrix} = \begin{bmatrix} p & 0 & 0 & 1 & 0 \\ 0 & p & 0 & 0 & 1 \\ 1 & 1 & 1 & 2 & 1 \end{bmatrix} \quad (4.2)$$

is both feasible and time optimal.

Proof: The proof is omitted due to space limitation [7]. □

As is discussed in the reference [7], one feasible matrix of interconnection primitives and the corresponding K matrix are

$$P = \begin{bmatrix} p & 0 & 0 & 1 & 0 & 1 \\ 0 & p & 0 & 0 & 1 & -1 \end{bmatrix}, K = \begin{bmatrix} 1 & 0 & 0 & 0 & 0 & 0 \\ 0 & 1 & 0 & 0 & 0 & 0 \\ 0 & 0 & 1 & 0 & 0 & 0 \\ 0 & 0 & 0 & 1 & 0 & 0 \\ 0 & 0 & 0 & 0 & 1 & 0 & 2 \\ 0 & 0 & 0 & 0 & 1 & 0 \end{bmatrix} \quad (4.3)$$

The architecture described by mapping matrix T in (4.2) is shown in Fig. 4, where P and K are used and $p = u = 3$. The timing and connections are specified completely by

$$TD = \begin{bmatrix} y & x & z & x & y, c & z & c' \\ p & 0 & 0 & 1 & 0 & 1 & 0 \\ 0 & p & 0 & 0 & 1 & -1 & 2 \\ 1 & 1 & 1 & 2 & 1 & 1 & 2 \end{bmatrix} \quad (4.4)$$

In Fig. 4, item $x_{ij}{}^k$ ($y_{ij}{}^k$) represents the kth bit of x_{ij} (y_{ij}), Data x_{ij} flow from top to bottom and y_{ij} flow from right to left. Data z_{ij} are stationary and the final results are stored at the eastern and southern boundary points of each small block. There is a buffer on the interconnection primitive $[1, 0]^T$ because $S\bar{d}_4 - [1, 0]^T$ and $\sum_{t=1}^{6} k_{t4} - 1 < \Pi \bar{d}_4 = 2$. Note that interconnection primitives $[p, 0]^T$ and $[0, p]^T$ require long wiring with distance p.

The total execution time required by mapping matrix T in (4.2) is [6]

$$t = \max\{ \Pi(\bar{q}_1 - \bar{q}_2): \bar{q}_1, \bar{q}_2 \in J\} + 1 \quad (4.5)$$
$$= [1, 1, 1, 2, 1] ([u, u, u, p, p]^T - [1,1,1,1,1]^T) + 1$$
$$= 3(u-1) + 3(p-1) + 1 .$$

The total number of processors required is

$$s = |\{ \bar{l}: S\bar{q} = \bar{l}, \bar{q} \in J\}| = u^2 p^2 .$$

Consider another mapping matrix

$$T' = \begin{bmatrix} S \\ \Pi' \end{bmatrix} = \begin{bmatrix} p & 0 & 0 & 1 & 0 \\ 0 & p & 0 & 0 & 1 \\ p & p & 1 & 2 & 1 \end{bmatrix} . \quad (4.6)$$

Let

$$P' = \begin{bmatrix} 1 & 0 & 1 & 0 \\ 0 & 1 & -1 & 0 \end{bmatrix} \text{ and } K' = \begin{bmatrix} p & 0 & 1 & 0 & 0 & 0 \\ 0 & p & 0 & 1 & 0 & 0 \\ 0 & 0 & 0 & 0 & 1 & 0 \\ 0 & 0 & 0 & 0 & 0 & 1 \end{bmatrix} . \quad (4.7)$$

Then, $SD = P'K'$ and $\Pi'\bar{d}_i \geq \sum_{t=1}^{4} k'_{ti}$, $i = 1, ..., 7$. It can be shown similarly that T' is feasible; the corresponding architecture is shown in Fig. 5. The total execution time in this case is

$$t' = \max\{ \Pi'(\bar{q}_1 - \bar{q}_2): \bar{q}_1, \bar{q}_2 \in J\} + 1 \quad (4.8)$$
$$= [p, p, 1, 2, 1]([u,u,u,p,p]^T - [1,1,1,1,1]^T) + 1$$

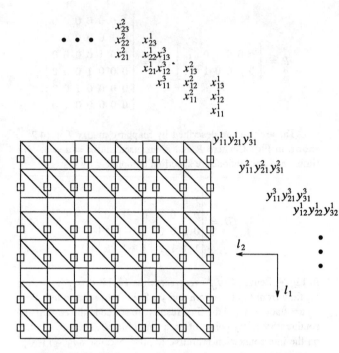

$$x_{23}^2$$
$$\cdots \quad x_{22}^2 \quad x_{23}^1$$
$$x_{21}^2 \quad x_{22}^1 x_{13}^3$$
$$x_{21}^1 x_{12}^3 \quad x_{13}^2$$
$$x_{11}^3 \quad x_{12}^2 \quad x_{13}^1$$
$$x_{11}^2 \quad x_{12}^1$$
$$x_{11}^1$$

$$y_{11}^1 y_{21}^1 y_{31}^1$$

$$y_{11}^2 y_{21}^2 y_{31}^2$$

$$y_{11}^3 y_{21}^3 y_{31}^3$$
$$y_{12}^1 y_{22}^1 y_{32}^1$$

l_2

l_1

Figure 5. A bit-level architecture for matrix multiplication described by (4.6).

$$= (2p-1)(u-1) + 3(p-1) + 1$$

and the total number of processors is $(u \cdot p)^2$.

The main disadvantage of the design in Fig. 5 is that its total execution time is longer than that in Fig. 4. In this case, data x_{ij} and y_{ij} are pipelined at one speed instead of two different speeds as is shown in Fig. 4. However, long wires are not needed in Fig. 5, whereas in Fig. 4, long wires are used to pipeline x_{ij} and y_{ij} at different speeds.

We can compare the time optimal bit-level architecture in Fig. 4 with the best word-level architecture for matrix multiplication described in the literature [4]. The total execution time of the best word-level architecture for matrix multiplication with index set $J = \{ [j_1, j_2, j_3]^T : 1 \le j_i \le u, \ j_i \in Z, \ i = 1,2,3 \}$ is $(3(u-1)+1) \cdot t_b$, where t_b is the time for multiplying two integers and adding two integers. Suppose the add-shift arithmetic algorithm is used to multiply two integers, then the multiplication time is $O(p^2)$, and the total time for word-level matrix multiplication is $O(p^2) \cdot (3(u-1)+1)$. If this number is compared with the total execution time in (4.5), then the speedup of our bit-level architecture over the word-level architecture described above is $O(p^2)$ if $u > p$ is assumed. Intuitively, a bit-level architecture is faster because, unlike in word-level architectures, a bit does not have to wait for all other bits to finish before going to next computation. In practice, faster arithmetic algorithms such as carry-save multiplication with complexity $t_b = O(p)$ can be used to multiply two integers. In this case the speedup of our bit-level architecture is $O(p)$.

5. CONCLUSIONS

In this paper, we propose to express the dependence structure of a bit-level algorithm as a function of its corresponding word-level dependence structure, the dependence structure of the arithmetic algorithm, and its algorithm expansion. Bit-level dependence structures can be obtained automatically from the above three parameters without using time consuming general dependence analysis methods. Based on the bit-level dependence structure, we illustrate the the design of two bit-level architectures for matrix multiplication. Our design shows that an optimal bit-level architecture can be $O(p)$ times faster than the corresponding word-level architecture, where p is the number of bits.

6. ACKNOWLEDGEMENTS

The authors would like to thank the students in the CMPS 639 course at USL for their helpful discussions. The authors are also indebted to L. Xu for suggesting the space mapping matrix S in (4.2), and to Z. Chen for drawing all the figures.

7. REFERENCES

[1] U. Banerjee. *Dependence Analysis for Supercomputing*, Kluwer Academic, 1988.

[2] J. A. B Fortes and D. I. Moldovan, "Data Broadcasting in Linearly Scheduled Array Processors," *Proc. 11th Annual Symp. on Computer Architecture*, 1984, pp. 133-149.

[3] K. Hwang, *Computer Arithmetic: Principles, Architecture, and Design*, John Wiley, New York, 1979.

[4] G.-J. Li and B. W. Wah, "The Design of Optimal Systolic Arrays," *IEEE Trans. Computers*, Vol. C-34, Jan. 1985, pp. 66-77.

[5] Z. Yang, W. Shang and J. A. B. Fortes, "Conflict-Free Scheduling of Nested Loop Algorithms on Lower Dimensional Processor Arrays," *Proc. 6th IEEE Int'l Parallel Processing Symposium*, March 1992, Beverly Hills, CA, pp. 156-164.

[6] W. Shang and J. A. B. Fortes, "On Mapping of Uniform Dependence Algorithms into Lower Dimensional Processor Arrays," *IEEE Trans. on Parallel and Distributed Systems*, Vol. 3, No. 3, May 1992, pp. 350-363.

[7] W. Shang and B. W. Wah, *Dependence Analysis and Architecture Design for Bit-Level Algorithms*, Technical Report 93-3-1, The Center for Advanced Computer Studies, Univ. of SW Louisiana, Lafayette, LA 70504, April, 1993.

[8] V. E. Taylor and J. A. B. Fortes, "Using RAB to Map Algorithms into Bit-level Systolic Arrays," *Proc. of Int'l Conf. on Supercomputing*, May 1987.

[9] Z. Xing and W. Shang, *Polynomial and Exact Data Dependence Analysis*, Technical Report 92-3-5, The Center for Advanced Computer Studies, Univ. of SW Louisiana, Lafayette, LA 70504, Dec., 1992.

[10] K. Ganapathy and B. W. Wah, "Synthesizing Optimal Lower Dimensional Processor Arrays," *Proc. Int'l Conf. on Parallel Processing*, Aug. 1992, pp. 96-103.

ATOMIC:

A Low-Cost, Very-High-Speed, Local Communication Architecture

Danny Cohen, Gregory Finn,
Robert Felderman, Annette DeSchon
USC/Information Sciences Institute
4676 Admiralty Way
Marina del Rey, CA 90292
atomic@isi.edu

Abstract-- *ATOMIC[1] is an inexpensive O(gigabit) speed LAN built by USC/ISI. It is based upon Mosaic technology developed for fine-grain, message-passing, massively parallel computation. Each Mosaic processor is capable of routing variable length packets, while providing added value through simultaneous computing and buffering. ATOMIC adds a general routing capability to the native Mosaic wormhole routing through store-and-forward. ATOMIC scales linearly, with a small interface cost. Each ATOMIC channel has a data carrying capacity of 500Mb/s. A prototype ATOMIC LAN has been constructed along with host interfaces and software that provides full TCP/IP compatibility. Using ATOMIC, 1,500 byte packets have been exchanged between hosts at an aggregate transfer rate of more than 1Gb/s. Other tests have demonstrated throughput of 5.25 million packets per second over a single Mosaic channel. This paper describes the architecture and performance of ATOMIC.*

1.0 Overview

ATOMIC is an O(gigabit) speed, low cost, LAN built by USC/ISI. It relies on the repetitive application of Caltech's Mosaic chip [17] (see Section 3.0), that serves as a fast and smart switching element. It is capable of routing variable length packets while providing added value through simultaneous computing and buffering. ATOMIC scales easily by adding point to point channels and has a low interface cost.

At present, a prototype of ATOMIC is operational including an IP interface for the BSD UNIX[2] environment. The developmental VME Host Interface (HI) boards have four memoryless Mosaic chips creating four network interfaces. Each interface is equipped with 25MHz SRAM memory chips that are accessible to the host Sun-3 workstation via the VME bus. These HI boards were able to obtain the following performance: a single interface acting as a source can send 381 megabits per second (Mb/s) if 1,500 byte packets are used. Longer packets allow us to approach 400Mb/s, the limit of the 25Mhz

---

1. ARPA supports ATOMIC through Ft. Huachuca contract No. DATB63-91-0001 with USC/ISI. The views and conclusions contained in this document are those of the authors and should not be interpreted as representing the official policies, either expressed or implied, of the Advanced Research Projects Agency or the U.S. Government.

2. UNIX is a registered trademark of ATT.

memory. A packet rate as high as 5.25 million packets per second (Mpkt/s) has been achieved over a single channel. The bit error rate of ATOMIC channels is still unknown. Transfers of test patterns both intra-board and inter-host of over 1,000 Terabits (1×10^{15}) resulted in neither bit errors nor lost packets.

ATOMIC is a switch-based local area network and is constructed using HIs and switches ("concentrators"), each of which may be connected to HIs or to other switches. Each switch is a perfect crossbar and every port is bidirectional with two independent channels, in and out. HIs have two such ports. If traffic analysis so suggests, multiple ports may be used between switches. Multiple hosts may be connected to one another without employing switches by creating any strongly-connected topology including linear, ring, dual-ring, mesh etc.

1.1 ATOMIC Attributes

ATOMIC possesses attributes not commonly seen in current LANs.

- Hosts do not have absolute addresses. Packets are source routed, relative to senders' positions. At least one host process in an ATOMIC LAN is an Address Consultant (AC). It "learns" the LAN's topology and can provide a source route to all the hosts on that LAN by mapping IP addresses to source routes. The network topology can be determined dynamically by the AC.

- ATOMIC consists of multiple interconnected clusters of hosts. This is unlike Ethernet and FDDI in theory, but similar to them in practice as Ethernets are often interconnected via gateways and routers.

- In general, there may be many alternate routes between a source and a destination. The rich routing topology may be exploited by an AC to provide bandwidth guarantees or to minimize switch congestion for high bandwidth flows, such as for video.

- Aggregate performance of the entire network is limited only by its configuration, since ATOMIC traffic flows do not interfere with each other unless they share links. This is unlike bus and ring LANs, such as Ethernet and FDDI, where all the traffic flows compete with each other for the same total bandwidth (of 10 and 100Mb/s respectively).

- Each Mosaic processor is a powerful general purpose computer, thus allowing the network itself to perform complex functions such as encryption or protocol conversion.
- Topological flexibility and programmability make ATOMIC suitable for any application. ATOMIC supports IP, but it could support any protocol by modifying the software. The system is infinitely variable.

ATOMIC is an early example of a Very-High-Speed LAN (VHSLAN) that is characterized by total transfer rates in excess of a gigabit per second. ATOMIC has a low interface cost and is capable of hosting a broad range of digital video, multimedia and large-scale, high data-rate distributed applications.

2.0 Introduction

Workstations and associated applications can now overload traditional local area networks. Ethernet and token ring have proven to be extremely capable networking technologies for interconnecting relatively slow machines using file transfers, electronic mail and remote procedure calls. 100 MIPs workstations are just over the horizon and a single such processor could overload the capacity of existing LANs. These faster workstations, coupled with high bandwidth applications such as multimedia editors, desktop video conferencing and virtual reality, will be hampered if interconnected using current LANs. FDDI only presents a near term solution with its 100Mb/s shared bandwidth limitation.

ATOMIC is an effort to create an extensible LAN technology to address the local area networking needs of the future. Rather than settle for an incremental improvement in current designs, ATOMIC leverages recent advances in parallel computing technology to achieve several orders of magnitude performance improvement over current local area networks.

We believe ATOMIC is unique in its architecture, performance capability and potential for low-cost deployment. The remainder of this paper discusses the design and performance of the prototype ATOMIC network that has been operational at USC/Information Sciences Institute since October 1991. We postpone a discussion of related work until Section 9.0.

3.0 Architecture Overview

The goal of the ATOMIC project is to take technology originally developed for intra-computer communication on parallel machines and extend it so that it also may operate as a local area network. We chose the Mosaic technology from Caltech as our base technology due to its high performance, low-cost and simplicity. Certain architectural choices were fixed once we focused on Mosaic technology e.g. wormhole source routing[17][18] (see Figure 1). Each Mosaic chip contains a general purpose processor, RAM, ROM, a DMA channel interface and self-timed routing hardware that supports a two-dimensional source-routing topology. Eight simplex channels are supported at a nominal rate of 500Mb/s each. A full-duplex (FDX) point-to-point host link con-

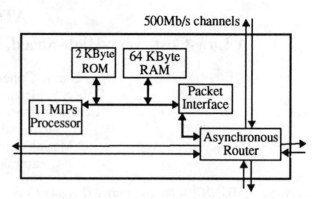

FIGURE 1. Mosaic Processor Chip

structed from a pair of these channels provides 1 Gigabit per second (Gb/s) of data transfer capacity. There is also a memoryless version of the Mosaic chip that uses external memory by supporting only four channels. The channel pins are scavenged to provide address and data bus output from the chip.

Mosaic chips route variable-length messages via a limited form of two-dimensional hop-by-hop source routing. Each message contains a source route prefix: a delta-X (ΔX)-byte followed by a delta-Y (ΔY)-byte. Each byte can take on a value from -127 to +127 hops, with the sign controlling the West/East direction for the ΔX-byte or the South/North direction for the ΔY-byte. As a packet passes through Mosaic nodes, its leading prefix is decremented until it reaches zero. All X-direction routing occurs before Y-direction routing. This restriction avoids deadlock [6]. The Mosaic chip is the building block for both the HI boards and the switches. Other chip types are needed only for the long distance transmission necessary for construction of a practical LAN.

3.1 ATOMIC

The Mosaic chips provide an excellent base from which to create a LAN, but alone, they do not provide a sufficiently general network. Mosaic was designed for Caltech's 128-by-128 mesh super computer. In a rectangular mesh, the limited X-then-Y routing is sufficient for each node to reach any other. In an ATOMIC LAN, the interconnection geometry will not be that regular. It will consist of smaller meshes attached to one another and to hosts. Often, a single X to Y transition will be insufficient. A more general routing mechanism is necessary and is described in Section 5.2.

Host interfaces and cables also must be developed to create a LAN. Also, an address resolution mechanism for converting between IP addresses (host names) and source routes needs to be in place. Therefore, ATOMIC requires several enhancements to Mosaic to construct a viable LAN, including interfaces, extended routing, switches, cables and address resolution. We briefly describe all of these in the following sections, then provide details later.

3.1.1 Interfaces.
The first requirement for building a functioning LAN is the addition of host interface hardware and software. Caltech has designed and built

the Program Development/Host Interface Board (HI) to prototype Mosaic software. This board is capable of acting as an interface for the ATOMIC LAN. Section 4.1 describes the board in detail. To seamlessly integrate the hardware with the operating system, a BSD UNIX interface and a device driver were written for use in Sun-3 workstations (see Section 5.1). ATOMIC currently supports the IP protocol and therefore all the communication protocols above it, such as UDP, TCP, ICMP, TELNET, FTP, SMTP, etc.

3.1.2 Extended Routing. We have created a network-layer protocol (ATOMIC) that runs in the Mosaic processors to provide more general routing by using a store-and-forward capability (see Section 5.2). ATOMIC source routes consist of multiple pairs of X-Y Mosaic routes such as $(X_1,Y_1)(X_2,Y_2)$. Mosaic's native wormhole routing is used for each pair, and ATOMIC forwarding between the pairs.

In addition to allowing more general network topologies, ATOMIC routing can exploit multiple routes between hosts.

3.1.3 Switches. If deemed necessary to provide extra bandwidth or multiple paths between hosts, 8-by-8 meshes of Mosaic nodes acting like switches will be interconnected in a multi-star configuration. Each 8-by-8 mesh is a single board approximately 8 inches square. A mesh is capable of acting as a 16-port crossbar interconnect or as a 32-port switching fabric with blocking. These meshes will be inexpensive (approximate manufacturing cost $5000), because they are being mass-produced for Caltech's Mosaic supercomputer. Section 4.2 describes the mesh and LAN configuration in more detail.

3.1.4 Cables. Long cables are necessary for connecting hosts to switches. Gigabit-speed fiber optic interfaces practical for use in low-cost LANs have not yet been developed. Section 8.0 discusses the issues involved in creating cables for ATOMIC.

3.1.5 Address Resolution. There are no absolute addresses in the ATOMIC network. All packets are source routed from a particular sender to a particular receiver. In order for a sender to know the route along which to send a message to a specific destination (known to it only by its IP address), we have created an "Address Consultant" (AC) user process to map the network and to provide routes between hosts. As the AC maps the network and discovers hosts, it provides each host with a host→AC route. When host A needs to send a packet to host B, it first sends a request to the AC. The AC returns an A→B source route to A. Then host A is able to send packets directly to host B without the intervention of the AC. These routes are cached locally at each host. The AC also handles the coordination of multicasting and broadcasting. Since ATOMIC is a point-to-point network, we cannot rely on the hardware to implement broadcasting. The AC sets up multicast trees using idle Mosaic processors in the network to distribute messages to all recipients. Section 6.0 provides more details on the

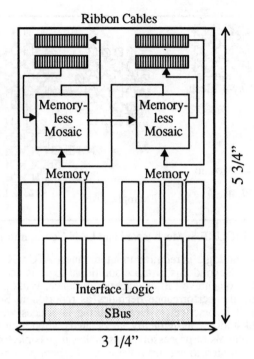

FIGURE 2. SBus Host Interface Board

operation of the AC.

4.0 Network Hardware

4.1 Prototype Host Interface Board
Caltech created two versions of the HI board to prototype Mosaic software. One attaches to a VME bus, the other to an SBus. The ATOMIC project utilizes HI boards for its initial host interfaces to demonstrate the capabilities of the LAN, the underlying point-to-point Mosaic technology, and new architectural and communication software innovations.

An ATOMIC host interface is small, composed of four 32KByte memory chips, the memoryless Mosaic chip, plus clock and bus interface logic chips. The memory chips impose a performance ceiling whenever a chip is either the source or the destination for a packet as the memory in the existing interfaces is slower than the Mosaic channels. The VME bus HI uses 25Mhz (16 bit) memory thus yielding 400Mb/s while the SBus interface uses 20Mhz memory to provide 320Mb/s.

Each HI board contains several (VME = 4, SBus = 2) complete ATOMIC host interfaces and 4 external channels, 2 in and 2 out. Each host requires at least one interface, though two are preferable, since one may be dedicated for network input and one for network output, doubling potential throughput and providing parallel processing and management of network traffic. The Sbus HI board is depicted in Figure 2.

The HI board demonstrates another attribute of this networking technology. Interfaces may be directly connected to one another, as are the Mosaics on an HI board. Unless Mosaic-to-Mosaic distance exceeds two feet, no

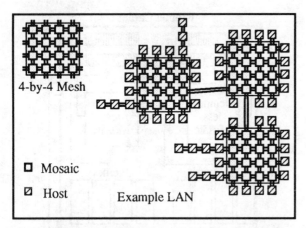

FIGURE 3. Mesh Board and LAN Configuration

external logic is required to interconnect ATOMIC interfaces. Beyond this distance additional logic is needed in order to achieve full channel performance. Without such logic the performance degrades as reported in Section 8.0. It is practical to have multiple interfaces within a workstation, individually associated with high-speed peripherals or processors. This makes it possible for data to be routed to and from these interfaces at gigabit rates independently of buses or backplanes [9].

4.2 Mesh Routers

Arrays of Mosaic chips can be interconnected with one another in a two dimensional mesh. In an ATOMIC LAN, meshes of Mosaic chips function as crossbar switches. A typical mesh board will contain 64 Mosaic chips organized in an 8-by-8 matrix (for illustration 4-by-4 mesh boards are shown in Figure 3). 8-by-8 meshes are being mass produced at low cost using Tape Automated Bonding on Multi-Chip Module packaging technology for Caltech's Mosaic computer of 128-by-128 nodes. An n-by-n mesh has $4n$ full-duplex Mosaic channel pairs available at its edges. Each channel pair may be used to connect one or a chain of hosts to the mesh. They may also be used to connect to other switches. At present, we are using a 3-by-3 version while the 8-by-8 meshes are being debugged by Caltech. An 8-by-8 mesh can support 16 hosts in non-blocking "crossbar" mode or 32 hosts with blocking.

5.0 Network Software

5.1 UNIX/Sun OS Modifications

The current prototype of ATOMIC is implemented for Sun-3 and Sun-4 workstations running a BSD UNIX (SunOS) operating system. The ATOMIC LAN driver is implemented in the BSD kernel and interfaces to the standard BSD socket mechanism. This provides IP level access to the ATOMIC network, allowing higher level protocols, such as UDP, TCP and ICMP, and higher level network services, such as telnet, file transfer and electronic mail, to be run transparently. The ATOMIC LAN driver looks functionally similar to the BSD Ethernet drivers.

5.2 ATOMIC Network Layer

The ATOMIC network layer extends the native Mosaic wormhole routing and allows more complicated source routes by providing a store-and-forward capability. ATOMIC routes are created by repeated application of Mosaic routes. In Figure 4 suppose host A needed a route to host B. Mosaic routing with only a single X to Y transition would be insufficient to connect A and B. By catenating multiple Mosaic routes together we can create a path (several paths) from A to B. One example path is the highlighted one. We denote this path as (+6,+1)(+4,+3)(+1,0)(0,0).The path from host B to host A travelling along the reverse links would be denoted as: (-1,-3)(-4,-1)(-6,0)(0,0).

ATOMIC network layer software runs in each of the Mosaic nodes of the network. A packet generated at A and destined for B would be prepended with the ATOMIC address listed above. The packet would be injected into the network using Mosaic routing and would only appear at the Mosaic processor labeled T1 after travelling (+6,+1). The ATOMIC network layer software would recognize that the packet needed forwarding because the leading bytes would be (+4,+3). After travelling its next Mosaic route it would appear at the Mosaic processor labeled T2. This processor would

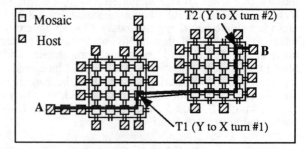

FIGURE 4. Example Network.

forward the packet (+1,0) where it would arrive at Mosaic processor B. This processor would deliver the packet to the host upon detection of the (0,0) at the head of the packet. Note that at each Y to X turn the entire packet is stored in the Mosaic processor before being forwarded along its next path. Native Mosaic routing uses wormhole routing to avoid store-and forward delay. ATOMIC uses native Mosaic routing when it can, but resorts to store-and-forward for complex routes.

Though space precludes a full discussion, we note that it is possible to connect hosts and switches such that no ATOMIC forwarding is needed.

6.0 Address Consultant

The Address Consultant (AC) is responsible for providing routes between hosts. In traditional LANs, such as Ethernet, each host is associated with a unique (fixed) address, and host IP addresses (or names) are translated into unique hardware addresses. In ATOMIC all packets must be source routed. In order for a host to convert a host name or IP address into a source route, it needs to "know" the topology of the network. It would be inefficient for every Mosaic node or even every host to maintain the topology of the entire LAN. We have chosen instead to designate one (or more) hosts to perform this topology discovery as a user process known as the Address Consultant or AC.

The AC performs two basic functions. The first is to map (and remap) the network so as to maintain a consistent picture of the LAN. Its second function is to use this map to provide routes between hosts upon request. An added benefit of the AC is its ability to balance flows throughout the network. When a source host requests a path to a sink host, the source may also provide some information to the AC about the type of connection to be opened. If the connection were a video stream, the AC would know to return a source route that avoided other connections and to avoid creating new, conflicting paths.

To increase fault tolerance, any host may become an AC if it cannot find one in the network. In fact, every host may be an AC simultaneously without affecting the correct operation of the network. In a large network, it may make sense for multiple ACs to be running in different parts of the network so that requests from hosts need not travel large distances to get to an AC.

6.1 AC Mapping

The AC maps the network by discovering its own neighbors and then recursively discovering neighbors of neighbors. A node is "discovered" if an ATOMIC message returns from it. For example, in a simple network with bidirectional links, the AC would check for a neighbor on its right by sending a message to (+1,0)(-1,0) (right, then left). If a Mosaic processor exists to the AC's right, then the message will return. The AC then probes left, up and down. If any nodes are discovered, the procedure is recursively repeated from each of the discovered nodes, thus implementing a breadth-first search over the network. The mapping operation requires that all the Mosaic nodes in the network be prepared to forward ATOMIC packets, but the nodes do not need to perform any further processing. The AC, in effect, sends packets to itself by looping through potentially existing Mosaic nodes. If a packet returns, the AC knows that Mosaic nodes exist along the path. If a packet doesn't return, the AC discovers the absence of a node. On hypothetical LAN topologies, it appears that eight messages per Mosaic node are sufficient to completely map the network. The exact number is topology dependent, and we have not yet derived a tight bound on the number of messages required.

6.1.1 Network Configurations.
ATOMIC allows hosts and switches to be connected in arbitrary topologies. For purposes of AC mapping, we define two parameters to classify various networks; *symmetric* and *consistent*. Symmetric implies that each channel has a matching channel in the reverse direction. Consistent describes a network where channels do not change direction of travel: East output is connected to West input and North output is connected to South input, etc. (see Figure 5).

In a symmetric and consistent network, the AC's network mapping is straightforward. Probe (loop) messages can be constructed by reversing the outgoing path to create a

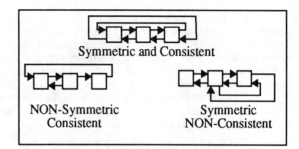

FIGURE 5. Network Configurations.

return path. For example, (+2,0)(-2,0) sends a message right two hops, then left two hops. In a symmetric and consistent network, with nodes at (+1,0) and (+2,0) from the AC, the message will return. In a non-symmetric or non-consistent network, no such guarantee is possible and a more complicated algorithm, using "modified flooding", is required. Non-symmetric nets potentially reduce the number of channels. Non-consistent connections are useful to avoid store-and-forward delay. A connection going out on a Y channel and in on an X channel creates the "Y-to-X" turn that native Mosaic routing lacks. Non-symmetric network connections are useful to improve the performance of the LAN, but they add other mapping complications.

We have implemented an AC for mapping any topology including the non-symmetric and non-consistent networks. The algorithm for non-symmetric nets needs the Mosaic processors to flood mapping messages in a controlled manner through the network and therefore requires more complex code in the network Mosaic nodes. An example network is shown below including six hosts (2 Sbus, 4 VME) and the 3-by-3 switch with one node acting as a broadcast server.

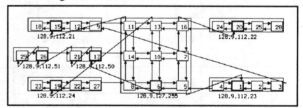

FIGURE 6. A Sample Network.

The number of messages needed to map an N node network is dependent on topology, but we do have some worst case estimates. To discover each link requires that a message traverse each link. That will require $4*N$ messages. Another message is used to collect remaining topology information from each node. At most four more messages are used to determine the direction of each link. This results in $O(k*N)$ messages with k approximately equal to nine. These messages are often being sent by various Mosaic processors in parallel, so the time complexity is more difficult to determine, but it should be smaller than the message complexity.

7.0 Performance

The Sun processors, the host bus, and UNIX socket processing are too slow to determine ATOMIC's performance. We have measured a single TCP connection between two Sun-4 machines at 17Mb/s using 1500 byte packets over ATOMIC. This is 20 times slower than the network can support. Therefore, traffic generation and monitoring code were installed in the memory that is shared by the Sun and the Mosaic processors, thus avoiding the Sun performance limitations. Results reported below are ATOMIC network measurements in the absence of UNIX software overhead.

When 1,500 byte packets (typical FTP packet size) are transferred across a Mosaic channel, a single interface limits the flow to 381Mb/s. Longer packets approach 400Mb/s which is the limit of the HI's 25MHz memory. Similarly, the flow of 54 byte packets (such as ATM) is limited to 171Mb/s. The limit for short packets (4 bytes) is a rate in excess of 657 thousand packets per second (Kpkt/s). Note, these are interface (not Mosaic channel) limitations.

When two flows compete for the same Mosaic channel, the measured performance is 405Mb/s for 500 and 1,500 byte packets, 343Mb/s for 54 byte packets, and 1.3Mpkt/s for the short packets. The latter figure is still interface limited and is not even close to the channel capability. With eight flows competing over the same Mosaic channel on a 4.5 foot cable, 5.25Mpkt/s for 4 byte packets was measured. The data are summarized in the tables below.

TABLE 1. Single flow measured performance.

Byte/pkt	Kpkt/s	Mb/s
4	657	21
54	396	171
1,500	31	381

TABLE 2. Multiple flows competing for a single channel.

Byte/pkt	Kpkt/s	Mb/s	Flows
4	5,250	168	8
54	793	343	2
1,500	33	405	2

8.0 Active Cables

Mosaic technology was designed to communicate over short distances. However, a practical LAN requires that it be possible for hosts to be separated from their switches by distances of hundreds of meters. Mosaic channels operate asynchronously using a byte-by-byte flow control of request-acknowledgment signals. Therefore, as cables get longer, the propagation time of these signals may become the dominant bottleneck in packet transfer. Caltech has created "Slack" chips that employ FIFOs to eliminate this bottleneck. These chips buffer the incoming bytes and "fake" the acknowledgment signal. The Mosaic channels are then able to transfer data at the normal rate of 500Mb/s[18]. Using Slack chips and drivers,

we are developing bi-directional twisted pair cables that we expect to run up to 100 feet. That should be sufficient for most office-type installations. For a practical LAN, we would need cables of a few hundred meters. This strongly suggest the use of serial fiber-optic cable technology and we are beginning research in this area.

9.0 Previous Work

Having presented the details of the ATOMIC LAN it is now possible to compare it to previous work in this area.

9.1 ATOMIC vs. Switch-Based LANs

On the surface, an ATOMIC 8-by-8 mesh router may resemble a switch. It bears some similarity to a space division crossbar switch with a self-routing property [19]. But unlike traditional switch elements, each Mosaic-C node contains its own CPU and program store. Each node can buffer and perform a function on the messages that it receives. A Mosaic-C node router is symmetric in two dimensions and independent from the processor. It may transmit or receive in any of four external directions simultaneously and may do that while node computation continues without interruption.

A mesh can behave much like a crossbar with no internal buffering. This produces head-of-line (HOL) blocking, where the horizontal mesh input channels vie for access to vertical output channels. With no use made of mesh internal buffering, if k packets contend for the same output channels, k-1 input channels will remain blocked and so k-1 output channels remain idle. There may exist packets destined for one of the idle output channels that are queued behind blocked packets. Though unable to remove the problem completely, mesh flexibility can be used in conjunction with the AC to alter the network routing paths in response to congestion.

In the table below, the expected performance of a 64-processor ATOMIC mesh using Mosaic-C and the upcoming Mosaic-T[17] is compared to both the Autonet switch[16] and Nectar HUB[1]. For each switch we list

TABLE 3. A Comparison of Various Switches.

Network switch (# ports)	Pkt/s (M)	latency (ns)	channel BW (Mb/s)	switch BW (Gb/s)
Autonet (12x12)	2	2000	100	1.2
Nectar (16x16)	14	700	100	1.6
ATOMIC (6x6)	31	125	500	3.0
ATOMIC (16x16)	80	375	500	8.0
ATOMIC (16x16) Mosaic-T	320	100	1600	25.6

the maximum achievable packets per second through the switch, the latency across the switch, the channel bandwidth and the aggregate switch bandwidth. The advantage of distributed routing used in an 64-processor mesh is pointed out by comparing mesh performance to that of an Autonet switch. A mesh contains 64 Mosaic chips, each with a router, each independent, while an Autonet switch contains one router. A 64-processor dual-connected mesh is externally similar to a 16-by-16 Nectar HUB crossbar.

Some of the details of the Autonet project [16] at DEC Systems Research Center are similar to ATOMIC. The Autonet LAN is also a switch-based network, and it uses point-to-point links of 100Mb/s over coaxial cable. One difference between the two projects lies in packet addressing. Autonet uses a scheme described as somewhere between source routing and routing by unique identifier. ATOMIC uses strict source routing inherited from the Mosaic chips [3][17][18]. Another difference is that Autonet's switches are responsible for reconfiguration and route selection in a distributed manner, while ATOMIC opted for a centralized address consultant process to perform this function. This difference is mainly due to the design of the Mosaic chips and the source routing used in the network. Our "switches" are actually distributed processing networks. Packets may be source routed through a mesh without ever appearing at a Mosaic processor in the mesh. Native Mosaic routing is performed entirely by the routers on each chip in the mesh without interrupting the CPU. All the "switching decisions" can be made external to the mesh by the host generating a source routed packet. We have plans to move the AC out of user space, although we have not settled on the eventual configuration.

The fact that our meshes and host interfaces have general purpose processors associated with each router makes ATOMIC far more general and flexible than Autonet or any other similar network.

9.1.1 ATM-Based LANs. Asynchronous Transfer Mode (ATM) has been suggested as an implementation technology for local as well as wide-area networks [2][4][12]. ATM messages are of fixed length. Each is 53 bytes long with five bytes reserved for header, leaving a 48 byte payload. Several teams are creating prototype ATM host interfaces and networks[2][7][20]. The Autonet follow-on, AN2, will have ATM switches [14].

ATOMIC sends variable length packets and it uses the distributed computational and routing capability of a mesh. This sets ATOMIC well apart from ATM-based LANs. The fragmentation and reassembly required when ATM carries higher layer traffic are not required in ATOMIC. That is one reason why ATOMIC host interfaces are small, inexpensive and fast. However, nodes could be programmed to implement the AAL (ATM Adaptation Layer) for IP and associate source routes with circuit identifiers, if desired. ATOMIC is more general than all of the ATM work, which is specifically designed to support only ATM. ATOMIC supports IP, but with a software change could support ATM or any network protocol.

One additional point should be stressed. The low cost of nodes coupled with their programmability makes it practical to utilize them in workstation architecture [9]. The network is then extended into the workstation. This allows internal devices to have their own independent Gb/s interconnect in addition to direct network access. It has also been suggested that ATM could be used for this, although switching would be external [4][10].

9.1.2 HIPPI and Fibre Channel LANs. Cray has created a HIPPI LAN at its facility in Minnesota[13] and LLNL is using Fibre Channel as the basis for its circuit-switched local area network[15]. These two networks and ATOMIC will look similar in implementation since they will all use point-to-point links and switches or concentrators. Cray's network is operational and connects 11 Cray machines using two switches. The LLNL work is still in the prototype stage but they expect to create a switch with 2048 ports with link speeds of 800Mb/s.

There are two main differences between HIPPI [11] or Fibre Channel [8] and ATOMIC, the first is cost and the type of machines that will be on the network and the second is connection setup time. The HIPPI and FC networks are designed with large, expensive machines in mind. ATOMIC is designed to provide low-cost, high-speed connectivity to all classes of machines. In fact, we advocate using many network interfaces in a single machine interconnected with ribbon cables[9]! HIPPI and Fibre Channel LANs are connection-oriented and optimized for large transfers between super-computer-class machines. Packet transmission requires an explicit connect operation. ATOMIC is a packet-switched technology and is better suited to LAN environments with less expensive workstations and smaller transfers between hosts. The HIPPI LAN designers state the cost issue explicitly [13]: *"We have demonstrated that an effective 800 megabit LAN can be built from HIPPI-SC switches, at a cost that is reasonable for large systems"*.

The time to set up a connection in each network is shown below, and we can see how ATOMIC outperforms the other switches/LANs in packet switching performance.

TABLE 4. Network Comparisons.

Network	Conn. Setup (ns)	Pkts/sec	Switch Size
Fibre Channel (LLNL)[15]	200,000	5,000	32 2,048 expected
HIPPI (Cray) [13]	10,000	100,000	32
ATOMIC	190	5,250,000	16

10.0 Summary

ATOMIC is an operating very-high-speed local area network. It has already demonstrated over one gigabit per second of aggregate throughput between two hosts using

1,500 byte packets, and more than 5.25Mpkt/s over a single channel with small packets (4 bytes).

ATOMIC was built by using hardware designed for message-based multi-processing and adding software layers to manage inter-host communication in the IP style. ATOMIC is completely flexible and extensible both in topology and protocols. Any interconnection topology is possible and an existing topology can be changed at will. Switches are not necessary, though they provide more network bandwidth and connectivity if needed. All the components are programmable, so other protocols could easily be supported. We are now investigating the implementation of ATM over ATOMIC and expect that we'll be able to provide both IP and ATM switching in the same network.

ATOMIC's low cost and high speed makes it a very attractive LAN technology.

11.0 Acknowledgments

Chuck Seitz of Caltech developed the Mosaic technology. Both he and Wen-King Su (of Caltech) taught us how to use Mosaic and provided us with much needed hardware and software utilities. ARPA supports ATOMIC through Ft. Huachuca contract No. DATB63-91-0001 with USC/ISI. The views and conclusions contained in this document are those of the authors and should not be interpreted as representing the official policies, either expressed or implied, of the Advanced Research Projects Agency or the U.S. Government.

12.0 References

[1] Arnould, E., Bitz, F., Cooper, E., Samsom, R., Steenkiste, P. "The Design of Nectar: A Network Backplane for Heterogenous Multicomputers", in *Proceedings, ASPLOS-III*, pp. 205-216, April 1989.

[2] Biagioni, E., Cooper, E., Sansom, R. "Designing a Practical ATM LAN", *IEEE Network*, Vol 7, No. 2, March 1993, pp. 32-39.

[3] Cheriton, D. R. "Sirpent^TM: A High-Performance Internetworking Approach", *in Proceedings of Sigcomm-89*, pp. 158-169.

[4] Clark, D. D., Tennenhouse, D. L. Research Program on Distributed Video Systems, Personal communication.

[5] Cohen, D., Finn, G., Felderman, R., DeSchon, A., "ATOMIC: A High-Speed, Low-Cost Local Area Network", Information Sciences Institute Technical Report ISI/RR-92-291, Sept., 1992.

[6] Dally, W. J., Seitz, C. L. "Deadlock-Free Message Routing in Multiprocessor Interconnection Networks", *IEEE Transactions on Computers*, Vol. C-36, No. 5, May 1987.

[7] Davie, B. S. "An ATM Network Interface for High-Speed Experimentation", *IEEE Workshop on the Architecture and Implementation of High-Performance Communication Subsystems* HPCS '92.

[8] Fibre Channel: Physical and Signalling Interface (FC-PH) Rev. 2.2, Working draft, Proposed American National Standard for Information Systems, January 24, 1992.

[9] Finn, Gregory G. "An Integration of Network Communication with Workstation Architecture", *ACM Computer Communication Review*, Oct 1991.

[10] Hayter, M., McAuley, D., "The Desk Area Network", ACM *Transactions on Operating Systems*, October 1991, pp. 14-21.

[11] Hughes, J.P. "HIPPI", in *Proceedings of the 17th Conference on Local Communication Networks*, pp. 346-354, Sept. 1992.

[12] Leslie, I., McAuley, D. "Fairisle: An ATM Network for the Local Area", *Proceedings of SIGCOMM-91*, pp. 327-336, August 1991.

[13] Renwick, John K. "Building a Practical HIPPI LAN", in *Proceedings of the 17th Conference on Local Communication Networks*, pp. 335-360, Sept. 1992.

[14] Rodeheffer, Thomas, "A Self-Configuring Local ATM Network", Presentation at the 4th Annual Workshop on Very High Speed Networks, Baltimore, Maryland, March 1993.

[15] Rupert, Paul. "Status of LLNL's Prototype Gigabit/Sec LAN", Presentation at the 4th Annual Workshop on Very High Speed Networks, Baltimore, Maryland, March 1993.

[16] Schroeder, Michael D., et. al. "Autonet: A High-speed, Self-Configuring Local Area Network Using Point-to-Point Links", *IEEE Journal on Selected Areas in Communications*, Vol. 9, No. 8, October 1991.

[17] Seitz, C.L., Boden, N., Seizovic, J., Su, W. "The Design of the Caltech Mosaic C. Multicomputer". To appear in the *Proceedings of the Washington Symposium On Integrated Systems*, Seattle, Wash., 1993.

[18] Seitz, C.L., Su, W. "A Family of Routing and Communication Chips Based on the Mosaic", To appear in the *Proceedings of the Washington Symposium On Integrated Systems*, Seattle, Wash., 1993.

[19] Tobagi, F. A. "Fast Packet Switch Architectures for Broadband Integrated Services Digital Networks", *Proceedings of the IEEE*, Vol. 78, No. 1, January 1990, pp. 133-166.

[20] Traw, S.B.C, Smith, J.M. "A High-Performance Host Interface for ATM Networks", *Proceedings of SIGCOMM-91*, pp. 317-325, August 1991.

RECONFIGURABLE BRANCH PROCESSING STRATEGY IN SUPER-SCALAR MICROPROCESSORS

Terence M. Potter, Hsiao-Chen Chung, and Chuan-lin Wu
Department of Electrical and Computer Engineering
The University of Texas at Austin, Austin, TX 78712

Abstract -- *In this paper, we develop a model for measuring branch performance on super-scalar processors. This model takes the form of CPI equations for different branch processing strategies. This model is the basis for a proposed design wherein the branch processing strategy in a processor can be reconfigured for optimal performance over a varying application load. There are nine parameters which are required to describe performance of a given branch processing method -four are dependent primarily on the processor architecture, two depend on the application program, and three depend primarily on the branch instance. The highest performance branch prediction algorithm can be determined by a combination of application profiling and runtime statistic gathering.*

INTRODUCTION

In super-scalar processors, efficient branch processing is extremely important. The idea behind super-scalar processing is to execute as many instructions per cycle as possible (i.e. design for very few cycles per instruction - CPI). However, it has been shown that the average size of basic blocks of code (code between branch instructions) are typically only 3-4 instructions long [4]. This implies that even a modest super-scalar processor can expect to see a branch every cycle, and aggressive designs may see multiple branches per cycle.

In addition, branch processing performance has been shown to depend on a number of application specific characteristics, thus it is difficult to design a non-adaptive strategy which performs well over a wide range of applications. This leads us to believe that in order to obtain good performance from aggressive super-scalar designs, an adaptive (reconfigurable) branch processing strategy must be included.

PERFORMANCE MODEL

Processor Model

We use a generalization of the processor model presented in [1]. This base model was a pipelined uni-scalar processor (i.e. single instruction dispatch). Our model can dispatch multiple instructions per cycle (D instructions). The position of a taken branch within the D instructions which are dispatched on a cycle will determine the number of those instructions which

actually complete - thus the CPI depends on this parameter. We will simply give a CPI range (the lowest CPI is when the branch is the first of the instructions dispatched, while the highest CPI is when the branch is the last of the instructions dispatched). The significant pipeline stages for branch processing are as follows: the fetch stage (f), the prediction stage (n"), the target address resolution stage (n'), and the branch direction resolution stage (n). These stages are relative to the fetch initiation stage (stage zero). When we say n=x, we mean that x cycles after a fetch for a branch has been initiated, that branch can be resolved. We assume that the execution units are all fully pipelined and we do not model data dependencies.

Program Model

We model programs as a series of basic blocks, each following a single branch and terminated by a single branch (basic blocks can have a length of zero). Branch instructions are each associated with two basic blocks: the target block and the sequential block, corresponding to taken and not-taken paths respectively.

The average basic block length for a program is determined by dividing the total number of instructions (measured dynamically) by the number of branch instructions and subtracting one.

BRANCH PROCESSING STRATEGIES

We will now briefly discuss some different branch processing strategies. Some of these are super-scalar generalizations of those presented in [7]; we will compare the CPI equations when this is the case.

Branch Prediction

Branch prediction is a method of reducing the latency associated with fetching target instructions of a conditional branch. In general, both the target address and the direction of a branch must be predicted, and there are different accuracies associated each of these predictions. Both predictions occur at stage n", but the target address is resolved at stage n' while the direction is resolved at stage n. Different prediction algorithms may result in different accuracies for a given application, and a given prediction algorithm may result in different accuracies for different applications. We believe that there does not exist a single prediction algorithm which yields high performance for all applications. Also, different prediction algorithms may have different stages for n" and n'.

In general, n' and n may be different for different types of branches and even for different instances of the same type of branch. Examples of some algorithms for predicting branch target addresses and branch directions are given in [3][4][5][7].

Basic Branch Prediction. The first branch processing strategy we will study is basic branch prediction. In this strategy, branches are predicted at stage n" and the predicted path is the only path which is executed. If a branch is predicted taken, then the sequential path is purged immediately and the target path is fetched. If the prediction of the target address is incorrect, then at stage n', the target would be corrected (re-fetched). Finally, if at stage n, the branch direction is resolved to have been predicted incorrectly, then the direction is corrected, and the appropriate instructions are fetched.

Branch Prediction With Delayed Purge. This strategy is similar to the *basic branch prediction* strategy above, except that when a branch is predicted taken, the sequential path is *not* purged immediately. Instead, the sequential instructions are purged when the target instructions actually come into the pipeline. This means that if the prediction is resolved as incorrect before target instructions get to the pipeline, then the sequential instructions will still be there and can be kept.

Multiple Path Execution

Another method for reducing the penalty for conditional branches is to execute both paths of a branch then discarding the results of the unused (incorrect) path.

Branch Target Buffers

A third method for reducing the CPI of branches is to use a branch target buffer. A branch target buffer stores either instructions which are the targets of branches or the target addresses of branches [4]. If instructions are stored, we call it a branch target instruction buffer or a branch target cache. When branch target addresses are stored we call it a branch target address buffer.

When a branch is taken, the fetcher accesses the branch target address buffer to get the target address; thus, this is a form of branch target prediction. Once the target address has been predicted or is known, the branch target cache is accessed. If the target instructions are contained in the branch target cache, then they are fetched into the pipeline (depending on the other branch processing methods used). Typically a branch target cache can be small and fast relative to a general instruction cache or mixed cache.

CPI EQUATIONS

In this section, we will develop the CPI equations for the two branch prediction strategies described above. These CPI equations will be used in the next section to develop the framework for an adaptive branch processing strategy. We develop the CPI equations for branch prediction as an example, There are many other methods for reducing branch CPI which can be modelled with their own CPI equations; although many other strategies can be modelled by affecting parameters of the prediction CPI equations. We will briefly discuss some other methods at the end of this section.

Branch Prediction

There are three major components of the CPI of a system which uses a predicted branch strategy: non-branch component, taken branch component, and non-taken branch component. Non-branch instructions can complete at a rate of D instructions per cycle - equation (2) gives the non-branch CPI component.

$$CPI_{nb} = \bar{r}\left(\frac{1}{D}\right) \qquad (1)$$

For taken branches, there are three cases: the predicted direction and target address are both correct, the predicted direction is correct but the predicted target address is incorrect, and finally, the predicted direction is incorrect. These components depend on the position of the branch within the D instructions dispatched on the given cycle. This dependence is shown in the variable P - see equation (3). In equations (4),(5), and (6), A_{ta} is the target address prediction accuracy, and A_d is the direction prediction accuracy.

$$P = 1 - \frac{1}{\rho}, \qquad 1 \leq \rho \leq D \qquad (2)$$

$$CPI_{bt1} = A_{ta}A_d(n" - P) \qquad (3)$$

$$CPI_{bt2} = \overline{A_{ta}}A_d(n' - P) \qquad (4)$$

$$CPI_{bt3} = \overline{A_{ta}}\,\overline{A_d}(n - P) \qquad (5)$$

The total CPI for taken branches is the sum of the three components multiplied by the probability that an instruction is a taken branch. We define T as the probability that a branch is taken.

$$CPI_{tb} = rT(CPI_{tb1} + CPI_{tb2} + CPI_{tb3}) \qquad (6)$$

The CPI for non-taken branches contains two components: the case where the direction is predicted correctly, and the case where the direction is predicted incorrectly. The second of these cases is the only one in which there is a difference between *basic branch prediction* and *branch prediction with delayed purge*. This is because in the delayed purge case, the sequential instructions are still contained in stages n through (n-n") when the branch is resolved. Thus the only holes are between stage (n-n") and stage 1 (the stage following the fetch request stage). The equations for the different components of non-taken branch CPI are given in equations (8), (9), and (10).

$$CPI_{bn1} = A_d(\frac{1}{D}) \qquad (7)$$

$$CPI_{bn2a} = \overline{A_d}(n - P) \quad \text{(basic prediction)} \qquad (8)$$

$$CPI_{bn2b} = \overline{A_d}(n - n" + \frac{1}{D}) \quad \text{(delayed purge)} \qquad (9)$$

The CPI for non-taken branches using the branch prediction strategy is given in equations (11), and (12).

$$CPI_{bn} = r\overline{T}(CPI_{bn1} + CPI_{bn2a}) \quad \text{(basic)} \qquad (10)$$

$$CPI_{bn} = r\overline{T}(CPI_{bn1} + CPI_{bn2b}) \quad \text{(delayed purge)} \qquad (11)$$

Now we can write the total CPI for the branch prediction strategies:

$$CPI_{pred} = CPI_{nb} + CPI_{bt} + CPI_{bna} \text{ (basic)} \qquad (12)$$

$$CPI_{pred} = CPI_{nb} + CPI_{bt} + CPI_{bnb} \text{ (del. purge)} \qquad (13)$$

We can see from the CPI calculations above that the true meaning of the n", n', and n stage numbers is the number of cycles they are from the fetch initiation cycle. We can use this to calculate the value of n" (for example) including cache hit rates and various memory latency parameters. Another parameter which is worth discussing is D. In the CPI equations above, we assumed that there were no data dependencies - an unrealistic assumption for most applications. There were also no restrictions on the types of instructions which are dispatched - an unrealistic assumption for most processors. The D parameter can be modified to account for these effects by using a statistically calculated effective D. We will now give an example application of the CPI equations presented above.

The IBM/Motorola 601[a] processor is a super-scalar microprocessor which implements a static branch prediction scheme. The 601 is capable of dispatching up to 3 instructions per cycle, one to each of three execution units - integer unit, branch unit, and floating point unit. The integer unit handles address calculation for load and store instructions. For the 601 pipeline, n"=n'=2, and n=4 (on average).

We will use EQNTOTT (one of the '89 SPECmark applications) for our example. EQNTOTT compiled for the POWER[b] architecture contains about 34% branch instructions, 66% integer instructions, and

---

(a) IBM, Motorola, and 601 are trademarks.
(b) POWER is a trademark of the IBM corporation.

no floating point instructions. Additionally, about 15% of the load instructions have first order dependencies. Thus for D we will use 1.3. This is calculated by first accounting for the dispatch limitations - half the time the 601 can dispatch a branch and an integer instruction, the other half of the time it can dispatch only an integer instruction. This number is then downgraded by 15% for first order dependencies (a first order dependency will cause a one cycle stall). For P, we will assume that branches are the first instruction half the time and the second instruction the other half of the time (P=.33). For the 601, $A_{ta}=1$, and for EQNTOTT running on the 601, $A_d=.9$. Using the above numbers with equation (13) gives a CPI=.95. This CPI can be further downgraded for system overhead (5%), and finite cache effects (1%)[6] to yield a final CPI of about 1. This is slightly optimistic compared to the actual CPI of the 601 processor on this benchmark. The error is due to the fact that integer ALU dependencies, TLB effects, and multi-cycle instructions were not considered.

CPI equations for some other branch processing techniques can be modelled by changing the parameters of the prediction CPI equations. For instance, branch target address buffers are used to predict the target address of a branch; Thus they affect the A_{ta} parameter. Since a branch target address cache may be accessed earlier than when the branch target can be calculated, it may also affect n". Branch target caches reduce fetch latency. This will reduce n", n' and n (equally) because fetch latency contributes to all of them.

There are other branch processing techniques which alter the form of the CPI equation (possibly adding parameters). Executing both paths of a conditional branch is a method of ensuring that the correct path is executed; although the incorrect path must be canceled (aborted) in order to maintain correctness. The performance of this strategy depends on how many paths can be handled at once, and the fetch policy for the multiple paths. Clearly processing multiple paths has a different CPI equation from the branch prediction case.

ADAPTIVE BRANCH PROCESSING STRATEGY

In an adaptive branch processing strategy, multiple branch processing strategies are employed in order to obtain high performance in many different environments. Continuing our example from the CPI equations above, we have identified nine parameters:

- r: probability that an instruction is a branch,
- T: probability that a branch is taken,
- D: the *effective* number of execution units,
- P: the position variable (relating to the

position of the branch in the D instructions),
- A_{ta}: target address prediction accuracy,
- A_d: direction prediction accuracy,
- n": number of cycles after fetch initiation in which a branch can be predicted,
- n': number of cycles after fetch initiation in which the target address of a branch is known, and
- n: number of cycles after fetch initiation in which the direction of a branch is known.

Note that r and T are application dependent, D, n", n', and n depend on a combination of application and processor architecture. Finally, P, A_{ta}, and A_d depend largely on the branch instance.

A simple approach to developing an adaptive branch processing strategy is to implement several non-adaptive strategies and keep a history for each one. This history can be used in conjunction with their respective CPI equations to determine, on a branch-by-branch basis, which of the strategies has the best performance. The required precision for the CPI calculations is fairly low - probably 8 bit precision would be enough to meaningfully distinguish between algorithms. Assuming limited precision requirements, the circuitry required for the CPI calculations will be reasonable for VLSI microprocessors of the near future.

Another form of adaptive branch strategy could be to use the framework of the CPI calculations presented here to apply limited resources in a branch processing unit more effectively. For instance, it may not be worth putting certain branches into a branch target address cache (e.g. return branches), or some branches may not need to use the full resources of a dynamic prediction history table (e.g. loop branches). By using resources more efficiently, the branch unit can get the benefit of larger caches, or more effective (and more expensive) prediction algorithms without paying the silicon penalty; obviously the added silicon to handle the adaptability must be weighed against the cost of these resources.

CONCLUSION

As super-scalar designs become more aggressive, branch performance becomes a gating factor in performance - the CPI contribution from branches becomes very important. Furthermore, cycle times of processors are decreasing much more rapidly than the latencies of most memory hierarchies. This means that main memory is becoming further and further away from the processor core. Better branch prediction is required

to increase the performance of instruction prefetching without overloading the limited bandwidth of most memory systems.

We feel that adaptive branch processing strategies warrant further study. The possibility of developing a branch strategy which performs very well across many applications is an enticing one, and we feel that the framework presented herein may be useful in such a strategy.

There are many problems related to adaptive branch processing which require further research, including: the precision requirements of on-chip runtime CPI calculations; the optimal non-adaptive strategies to include in the adaptive strategy (for the simple case discussed above); and the cost-performance in silicon, cycle time, and verification time associated with an adaptive strategy, relative to the cost-performance of an equivalent non-adaptive design.

REFERENCES

[1] H. Cragon, Branch Strategy Taxonomy and Performance Models, IEEE Computer Society Press, (1992).

[2] M. Johnson, Superscalar Microprocessor Design, Prentice-Hall, Eaglewood Cliffs, NJ, (1991).

[3] D. R. Kaeli, and P. G. Emma, "Branch History Table Prediction of Moving Target Branches Due to Subroutine Calls," 18th Annual International Symposium on Computer Architecture, (1991), pp. 34-41.

[4] J. K. F. Lee, and A. J. Smith, "Branch Prediction Strategies and Branch Target Buffer Design," IEEE Computer, Vol. 17, No. 1, January, 1984, pp. 6-22.

[5] S-T. Pan, K. So, and J. T. Rahmeh, "Improving Dynamic Branch Prediction Using Branch Correlation," The 5th Conference on Architectural Support for Programming Languages and Operating Systems, pp. 76-84.

[6] D. N. Pnevmatikatos, and Mark D. Hill, "Cache Performance of the Integer SPEC Benchmarks on a RISC," technical report, Computer Sciences Department, University of Wisconsin - Madison, 1210 W. Dayton Street, Madison, WI 53706.

[7] T-Y. Yeh and Y. N. Patt, "Two-Level Adaptive Branch Prediction," The 24th ACM/IEEE International Symposium and Workshop on Microarchitecture, (Nov. 1991), pp. 51-61.

EXPLOITING SPATIAL AND TEMPORAL PARALLELISM IN THE MULTITHREADED NODE ARCHITECTURE IMPLEMENTED ON SUPERSCALAR RISC PROCESSORS*

D. J. Hwang
Dept. of Information Eng.
SungKyunKwan University
djhwang@simsan.skku.ac.kr

S. H. Cho, Y. D. Kim, S. Y. Han
Dept. of Computer Science and Statistics
Seoul National University
shcho@pandora.snu.ac.kr

Abstract

In most multithreaded node architectures motivated by the dataflow computational model, spatial parallelism could not be exploited at the thread level due to the resource deficit incurred by their internal organization. So we proposed a node architecture exploiting both spatial and temporal parallelism of a program. A multi-port non-blocking data cache is incorporated into our design to cope with the excessive data bandwidth required in parallel execution of multiple threads. The proposed node architecture may contribute to greatly reducing communication latency through the interconnection network. Simulation results show that parallel loops can be executed on this architecture more efficiently than on other competitive ones.

1. INTRODUCTION

Based on the locality among instructions, von Neumann architectures[1] have improved the performance of a single-thread processing by temporally overlapping instruction executions in their pipelines. Recent superscalar processors overcome the limitations of conventional RISC processors and further reduce the average of CPI (Clock cycles Per Instruction) less than 1 by widening instruction streams of their pipeline. These attempts are likely to have limits to performance improvement because of extracting parallelism relied on a single instruction stream. On the contrary, dataflow architectures can spontaneously exploit asynchronous parallelism in a program execution according to data availability but have a weakness of not exploiting locality of computations.

Multithreaded architectures[2,4] synthesizing both the advantages of von Neumann and dataflow architectures have two distinct features for parallel processing: tolerating latencies, efficient synchronization of remote data. In *T[4] recently being developed by MIT in collaboration with Motolora, threads assigned to different nodes can only be executed simultaneously following the dataflow firing

rule, while threads activated in the same node are to be executed in sequence. The spatial parallelism among threads assigned to the same node should be sacrificed since all enabled threads are to be mapped onto a single thread processor in *T.

Multiple thread processors of the node architecture proposed in this paper make it possible to further improve the performance of extant multithreaded architectures. Whereby, the parallelism among threads is exploited by the multiple superscalar RISC processors and the parallelism within a thread by the pipeline and multiple functional units of the superscalar processor.

In section 2, we classify the types of parallelism in a program into two categories: spatial parallelism, temporal parallelism. In section 3, we describe the node architecture featuring the multi-port non-blocking data cache incorporated into our design. In section 4, we present the analysis of the simulation results and in section 5, the conclusion.

2. TYPES OF PARALLELISM

A program contains two types of parallelisms: spatial parallelism and temporal parallelism. Figure 1 shows that instructions i3, i5, i10, and i12 in thread 2 should be executed according to data dependency. Temporal locality among those instructions can be exploited by temporally overlapping their executions in a pipeline, which is called *temporal parallelism*. On the other hand, instructions i5, i6, and i7 have no data dependency among them, so they can be simultaneously executed in different functional units only when there is no resource conflicts. Such parallelism is called *spatial parallelism*. We can also find another version of spatial parallelism between thread 1 and thread 2 in this figure.

A program is considered as a set of code blocks consisting of multiple threads; Each thread corresponds to a sequence of related instructions and no interruption is allowed during its execution. Such a hierarchical structure of a program makes it easy to extract the spatial parallelism and the temporal parallelism from a program. With this view, a program is compiled and then partitioned into multithreaded codes by the compiler back-end being developed at SNU which are processed by the node architecture exploiting spatial and temporal parallelisms. Thus

---

*This work is supported by the *Agency for Defence Development* under the contract UD920067BD.

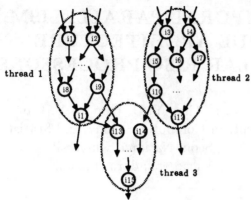

Figure 1: Dataflow graph

all types of parallelism can be achieved in this node architecture.

3. NODE ARCHITECTURE

The proposed node architecture illustrated in Figure 2 consists of six major parts: Multiple Thread Processor Unit(MTPU), Synchronization Processor(SP), Thread Scheduler(TS), Data Memory(DM), Code Memory(CD), and External Interface Unit(EIU).

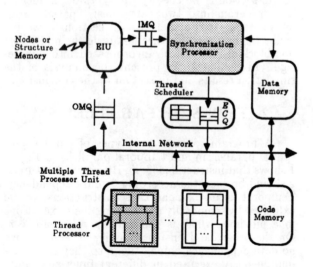

Figure 2: Node architecture

3.1 Synchronization Processor

The SP performs inter-thread synchronization upon receiving messages from the EIU. Synchronization for each thread begins with access to the DM using the frame base address, the synchronization displacement, and the offset in a message. As thread activation mechanism follows the dataflow firing rule, a thread becomes active only when all its input values are available. The continuations of the threads satisfied their synchronization condition are put into the ECQ(Enabled Continuation Queue) in the TS. Otherwise, the threads get back to the dormant state,

waiting for their activation. The instructions related to synchronization commit to overhead operations for parallel processing which are to be processed independently by the SP. While the ECQ is not empty, the performance of the node architecture is not dominated by the synchronization latency of the SP but by the thread processing rate of the MTPU.

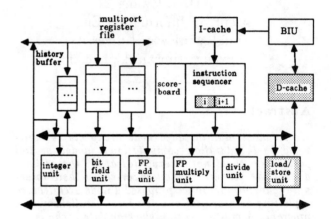

Figure 3: Organization of a superscalar TP

3.2 MTPU

After the TS maps a continuation <fp,ip> (fp: frame pointer, ip: instruction pointer) onto one of the thread processors available in the MTPU, it starts to execute its corresponding thread. The thread generates synchronization messages for other threads or messages having access to structure memory during its execution if necessary. Those messages are delivered to the EIU through the OMQ(Outgoing Message Queue). The organization of a TP illustrated in Figure 3 is quite similar to those of the superscalar RISC processors. Superscalar processor architectures have higher compatibility with existing softwares than the VLIW architectures. Furthermore they are more suitable to the CMOS technology than the superpipelined architectures requiring a fast clock and a deep pipeline[1]. With wide instruction streams, the superscalar processor has several advantages in handling multiple threads of control over the scalar processor.

In spite of the spatial parallelism between thread 1 and thread 2 specified in Figure 1, they are to be processed sequentially on the extant multithreaded architectures. Unlike these architectures, the proposed node architecture can exploit spatial parallelism among threads on multiple thread processors, and spatial and temporal parallelism within a thread in the pipeline and multiple functional units of the superscalar thread processor.

Generally, the multithreaded node architectures need to load registers with the input values of a thread prior to its execution. Large number of load instructions appear in the front part of a thread code, which means that the bandwidth of a data cache becomes much more critical.

3.3 Cache

The cache organization is organized as two-level caches. The first level L1 cache is divided into an instruction cache and a data cache. As a frame is accessed by the multiple threads belonging to it, the L2 cache is organized as a unified cache shared by the multiple TPs. To utilize the parallelism in full scale, read hit and read miss must be handled quickly. So the operation of the TP does not have to be blocked even when a data cache miss occurs. For this purpose, the L1 data cache (unless stated otherwise, all reference to a cache shall imply the data cache from now on) depicted in Figure 4 is composed of P interleaved banks and connected to the L2 cache via a pipelined bus.

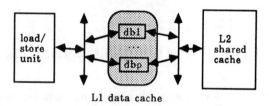

Figure 4: Interleaved cache organization

Each bank has n MSHR's (Miss information/Status Holding Register)[3]. By storing the status information related to a cache miss in that register, a processor can issue the next cache request with no blocking. As the L1 cache has p banks, p hits can be handled simultaneously. We can also reduce the waste of a processor time due to cache misses by means of pipelining the maximum of p*n misses. The bandwidth of the L1 cache featuring the multi-port non-blocking cache with interleaved p banks can be described in equation (1). The first term denotes the bandwidth of the L1 cache, and the second term the bandwidth of a L1-L2 bus.

$$\min\{\frac{H+M}{(H+M)/p}, \frac{H+M}{M*(B+T_r)}\}$$
$$= \min\{p, \frac{1}{m(B+T_r)}\} \qquad (1)$$

where

H : the number of cache hits in the L1 cache.
M : the number of cache misses in the L1 cache.
m : the miss ratio, M/(H+M).
p : the number of ports in the L1 cache.
T_r: the time taken for L2 cache requests to be transferrd on L1-L2 bus caused by a read miss. It is assumed one cycle.
B : The number of words in a cache block(the pipelined L1-L2 bus can transfer one word at each cycle).

In order to provide enough data bandwidth between the load/store unit and cache banks, the TP organized as such seems to be problematic to implement since hardware complexity grows high and the cache hit latency increases. The hit latency delayed may be improved by pipelining the CPU-L1 bus[5].

3.4 Thread Scheduler.

The TS maintains the status of TPs to select a candidate into which the continuation is to be scheduled at the next time. It is desirable that all continuations belonging to the same frame should be scheduled earlier than others belonging to different frames. This can be implemented by searching the corresponding continuation with the same fp as that of the thread just terminated. This scheduling is *a prioritized frame-based scheduling* in a sense. Despite of the searching overheads, the compulsory cache misses can be reduced because the code and the frame of the thread are maintained in the cache after its termination. This effect comes from exploiting locality among threads, which can greatly contribute to reducing the program exection time.

4. PERFORMANCE EVALUATION

The simulation results carried out on Lawrence Livermore Loops(LLL) at the instruction level are described. Considering data dependency and program parallelism, only three of them, that is, LLL3, LLL6, and LLL9 were selected as the benchmarks. We assumed throughout the simulation that the SP and the TP have the same processing capability, and the instruction cycle time used corresponds to that of the Intel i860. Several cache parameters include: a 2-way set-associative organized as 128 slots of 32 bytes each, the access time of the L2 cache and memory in blocking case are 6 cycles and 12 cycles, respectively, the block transfer time of the L2 cache and memory corresponds to 4*(the number of words) cycles and 8*(the number of words) cycles, respectively. In the simulation, instructions are run down in cycle-by-cycle, and the latency through the interconnection network is assumed 10 times longer than the latency of a typical integer operation.

Figure 5 shows the MTPU utilization in accordance with the number of TPs for three loops. While the utilization of the MTPU for LLL6 drastically decreases, those for the LLL3 and LLL9 gradually decreases. The data dependency committed to the degradation of the MTPU utilization in LLL6 while the saturation of the SP did in LLL3 and LLL9.

Figure 6 shows the speedup is proportional to the number of TPs. In cases of LLL3 and LLL9, the speedups increase almost linearly up to the point the utilization of the MTPU starts to drop. But in case of the LLL6, additional TPs do not contribute to the further speedup due to starvation of the parallelism. Thus a resource management scheme is required in order to use resources efficiently especially when programs with lower degree of parallelism are executed. As the degree of parallelism is sensitive to the benchmarks used, it is difficult to decide the optimal number of TPs in a proposed node architecture based on the simulation results. More extensive simulation studies on this problem are needed. Figure 7 shows the execution time of the MTPU with different cache organizations: blocking and non-blocking cache organization. The latter is 1.72-1.95 times faster than the former where the number of TPs is 5 and the loop unfolding factor is 7 for LLL9.

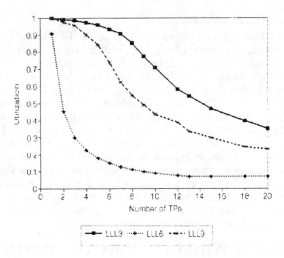

Figure 5: MTPU utilization

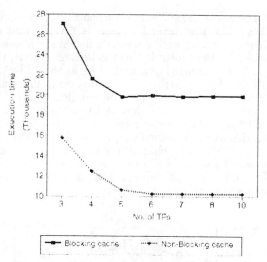

Figure 7: Comparison of cache organizations

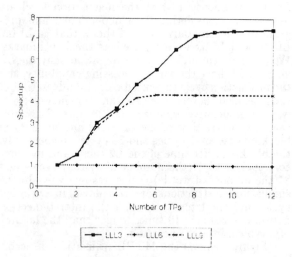

Figure 6: Speedup

5. CONCLUSION

The conventional multithreaded architectures have limitation in exploiting spatial parallelism of inter-threads. The proposed node architecture is indebted to the advances in VLSI technology and its application to the design of superscalar RISC processor. With the features of the node architecture, the loss of parallelisms among threads being experienced in the conventional multithreaded architectures can be addressed. This means that spatial parallelism among threads is exploited by multiple superscalar thread processors incorporated into our node design and the temporal and spatial parallelism of a thread by the pipeline and the multiple functional units of the thread processor. To provide the cache bandwidth compensating for tha data requirements of the node architecture, two-level multi-port non-blocking cache organization is incorporated.

These promising features for parallel processing make the node architecture execute the itera-

tions of a loop more efficiently; different instances of a loop are executed by multiple TPs without inter-node communication, than the extant multithreaded architectures in which synchronizing signals for loop executions should be transferred among nodes through an interconnection network. After the implementation of a multithreaded compiler being developed, more extensive performance evaluation on the node architecture will be taken for the application programs widely used. For the same purpose, further researches on the partitioning strategy of dataflow graphs and on the frame-based scheduling have been under the way.

References

[1] J. Hennessy and D. Patterson, *Computer Architecture - A Quantitative Approach*, Morgan Kaufmann Publishers, Inc., 1990.

[2] D.J. Hwang, et al., "A High Performance Multithreaded Node Architecture: DAVRID," 2nd Computer Science Conference, Hong Kong, pp. 106-112, 1992.

[3] D. Kroft, "Lock-free Instruction Fetch /Prefetch Cache Organization," 8th ISCA, pp. 81-87, 1981.

[4] R.S. Nikhil, G.M. Papadopoulos and Arvind, "*T: Multithreaded Massively Parallel Architecture," 19th ISCA, Australia, pp. 156-167, 1992.

[5] G. Sohi and M. Franklin, "High-Bandwidth Data Memory Systems for Superscalar Processors," 4th ASPLOS, Santa Clara, pp. 53-62, 1991.

SESSION 3A

MEMORY

Fixed and Adaptive Sequential Prefetching in Shared Memory Multiprocessors

Fredrik Dahlgren, Michel Dubois[*], and Per Stenström

Department of Computer Engineering
Lund University
P.O. Box 118, S-221 00 LUND, Sweden

[*]Department of Electrical Engineering-Systems
University of Southern California
Los Angeles, CA90089-2562, U.S.A.

Abstract

To offset the effect of read miss penalties on processor utilization in shared-memory multiprocessors, several software- and hardware-based data prefetching schemes have been proposed. A major advantage of hardware techniques is that they need no support from the programmer or compiler.

Sequential prefetching is a simple hardware-controlled prefetching technique which relies on the automatic prefetch of consecutive blocks following the block that misses in the cache. In its simplest form, the number of prefetched blocks on each miss is fixed throughout the execution. However, since the prefetching efficiency varies during the execution of a program, we propose to adapt the number of prefetched blocks according to a dynamic measure of prefetching effectiveness. Simulations of this adaptive scheme show significant reductions of the read penalty and of the overall execution time.

1 INTRODUCTION

Shared-memory multiprocessors offer significant performance improvements over uniprocessors at an affordable cost while preserving the intuitively appealing programming model provided by a single, linear memory address space. With the advent of ultra-fast uniprocessors and of massively parallel systems, the shared-memory access penalty (i.e., the average time each processor is blocked on an access to shared data) becomes a significant component of program execution time. Private caches in conjunction with hardware-based cache coherence [15] are effective at reducing this latency. In a system with caches, most of the memory access penalty originate from the miss rate of read requests, i.e., the fraction of read requests that miss in the cache. By contrast, the latency of write requests can easily be hidden by appropriate write buffers and weaker memory consistency models [7].

Prefetching is a common technique to reduce the read miss penalty. Good prefetching mechanisms can predict which blocks will miss in the future and bring these blocks in the cache before they are accessed. *Software-controlled* prefetching [12] relies on the user or compiler to insert prefetch instructions in the code. *Hardware-controlled* prefetching [6, 9, 10] usually takes advantage of the regularity of data accesses in scientific computations to detect access strides dynamically and requires complex hardware.

The main theme of this paper is to evaluate two simple hardware-controlled prefetching schemes with a low implementation cost and requiring no compiler support. These two schemes are called *fixed and adaptive sequential prefetching* and are *nonbinding*, meaning that a block prefetched in a cache remains subject to the coherence protocol. In sequential prefetching, the cache controller prefetches on a miss the K blocks following the missing block, where K is the *degree of prefetching*. In fixed sequential prefetching, the degree of prefetching remains constant throughout the execution whereas it varies dynamically in the case of adaptive sequential prefetching.

Section 2 shows the usefulness of the schemes by reviewing the effects of the cache block size on the miss rate and network traffic. We then describe the two hardware schemes in Section 3 and, in Sections 4 and 5, we evaluate their performance with a detailed multiprocessor simulator model and a set of six benchmark programs. We find that fixed sequential prefetching provides significant read miss reduction in some cases in spite of its simplicity. The adaptive technique, although slightly more expensive to implement, yields considerably lower miss rates coupled with low traffic overhead. We finally relate our results to others' in Section 6 and conclude the paper in Section 7.

2 IMPACT OF BLOCK SIZE

In this section, we review the effects of cache block sizes on the miss rate and memory traffic in multiprocessors. This is important in order to understand the effects of sequential prefetching and to compare them to prefetching effects due to larger block sizes. (Increasing the block size is the simplest and best known means of prefetching.)

A number of studies have reported the effects of block size variations on cache miss rates in the context of shared-memory multiprocessors [5, 8]. We need to distinguish between the different types of misses. In the case of

This work was partly supported by the Swedish National Board for Technical Development (NUTEK) under contract number 9001797 and by the National Science Foundation under Grant No. CCR-9115725.

infinite caches, there are two types of misses: *cold misses* (case where the block has never been referenced by the processor) and *coherence misses* (case where the block is missing because it was invalidated). Coherence misses can further be classified as *true sharing misses* and *false sharing misses*. Loosely speaking, false sharing is the sharing of a block without actual sharing of data whereas true sharing is due to the read/write sharing of data items. For finite-size caches another type of miss called *replacement miss* occurs when a block is victimized in the cache and subsequently referenced. Replacement misses are ignored until Section 5.5 where we compare the schemes for finite size caches.

Previous papers (e.g., [5]) have evaluated the effects of the block size on cold misses, true sharing misses, and false sharing misses in the context of infinite caches. In general, as the block size increases, the number of cold misses is significantly reduced for all applications, the number of true sharing misses decreases for most applications, but the number of false sharing misses increases and eventually becomes predominant. Thus the miss rate as a function of the block size follows a U-shaped curve whose minimum depends on the application. The memory traffic is also affected by the block size and consumes memory and interconnection bandwidth. Gupta and Weber [8] have observed that, in the case of infinite caches, the memory traffic as a function of the block size follows a U-shaped curve as well. As we will see in the subsequent sections, sequential prefetching reduces the cold and true sharing miss components in a similar fashion as a larger block size does without increasing the false sharing miss component.

3 SEQUENTIAL PREFETCHING

In this section, we describe to some extent the implementation details of the fixed and adaptive sequential prefetching schemes.

3.1 Processor Node Architecture

The proposed prefetching schemes are applicable to standard processors with blocking loads and, in contrast to software-controlled prefetching, they do not require non-blocking prefetch instructions. As shown in Figure 1, each processing node consists of a processor, a first-level cache (FLC), a second-level cache (SLC), a first- and second-level write buffer (FLWB, SLWB), a local bus, a network interface controller, and a memory module (not shown). The FLC is a direct-mapped write-through cache with no allocation of blocks on write misses, no prefetch and is blocking on read misses. Writes are buffered in the FLWB.

Since the prefetching schemes are non-binding, they behave correctly under sequential consistency. In this study, however, we assume an aggressive cache hierarchy that can hide write latency and relying on release consistency. (In [4] we also consider sequential consistency.) The second-level cache (SLC) is a direct-mapped write-back cache. We only prefetch into the SLC. A second-level write buffer (SLWB) keeps track of outstanding requests (read, prefetch, write, and synchronization requests). In order to exploit the potential of write-penalty reduction allowed by relaxed memory consistency models, the SLWB allows multiple pending requests as long as they are for different blocks. (Normally this means not more than 4 requests other than prefetches.) Moreover, a read miss request may bypass writes in the buffer provided they are for different blocks.

The SLC controller deals with all the complexities of the cache-coherence protocol by snooping on the local bus for consistency actions. If an invalidation to a block residing in the SLC is detected, the copy of the block is invalidated in both the SLC and the FLC. Requests from the local bus have priority over those from the first-level write buffer. These interferences with the processor accesses are tolerable because most accesses hit in the FLC. In Section 4, we further discuss the interface between the processing node and the rest of the multiprocessor memory system.

3.2 Fixed Sequential Prefetching

Sequential prefetching has been extensively studied in the context of uniprocessors [14] but, to our knowledge, has never been considered for general applications on shared-memory multiprocessors. Although many sequential strategies have been proposed for uniprocessors, we have restricted ourselves to prefetching on a miss in the SLC. When a reference misses in the SLC, the miss request is sent to memory, and the cache is searched for the K consecutive blocks directly following the missing block in the address space. The blocks among these K consecutive blocks that are not present in the SLC and have no pending requests in the SLWB are prefetched. Since the processor is stalled, a Prefetch Counter in the SLC controller (see Figure 1) is used to generate the addresses of the K consecutive blocks; for each address missing in the cache, a prefetch request is inserted in the SLWB if no request for the block is already pending. The prefetch requests are issued one at a time, are pipelined in the memory system and thus can be overlapped with the original read request.

The optimum prefetching degree K is dictated by the spatial locality of the application and by the block size and it varies during the execution of most benchmarks, which suggests that K should be adjusted during the execution.

3.3 Adaptive Sequential Prefetching

The adaptive scheme associates two bits per cache line and three counters per cache, defined in Table 1. To adjust the prefetching degree we count the useful prefetches, i.e.

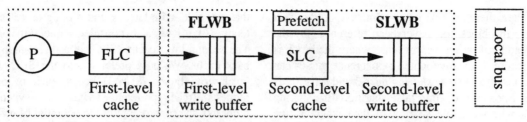

Figure 1: The simulated processor-node with its memory subsystem.

the prefetches such that the prefetched block is referenced during its lifetime in the cache. Initially, all PrefetchBits and ZeroBits are cleared, the LookaheadCounter is set to one, and the PrefetchCounter and the UsefulCounter are cleared. The prefetch mechanism is activated on a read miss as in fixed sequential prefetch but the degree of prefetching is dictated by the LookaheadCounter. The

Table 1: Additional hardware for the adaptive scheme.

PrefetchBit	Used to detect useful prefetches (when prefetching is turned on)
ZeroBit	Used to detect when a prefetch would have been useful (when prefetching is turned off)
LookaheadCounter	Degree of prefetching (K)
PrefetchCounter	Counts the prefetches that are returned
UsefulCounter	Counts the number of useful prefetches

memory system acknowledges the prefetch of each block by sending either the requested block (PreData) or a negative acknowledgment (PreNeg) if the prefetch was not serviced[1]. The PrefetchCounter counts the number of Pre-Data and PreNeg returned. The PrefetchBit of a block is set when it is prefetched into the SLC. When a block is accessed with its PrefetchBit set, the UsefulCounter is incremented and the PrefetchBit is cleared.

Every time the PrefetchCounter reaches its maximum, the value of the UsefulCounter is compared to preset thresholds: if it is higher than the high mark, the Lookahead-Counter is incremented and if it is lower than the low mark, the LookaheadCounter is decremented. In both cases, the UsefulCounter is cleared. In our evaluation, all counters are modulo 16 (4 bits).

When the LookaheadCounter reaches zero, prefetching is turned off. To turn it back on, the following mechanism is then activated. When a block is received on a read miss, the ZeroBit of the block in the SLC is set and the Prefetch-Counter is incremented. On a read miss, after the read request is inserted in the SLWB, a cache lookup is made to the previous block (by address); if it hits and the ZeroBit is

set, the UsefulCounter is incremented. This simple heuristic detects when prefetching would have been useful. Eventually, prefetching is reactivated when the Useful-Counter reaches the low mark.

4 EXPERIMENTAL METHODOLOGY

In this section, we present the simulation environment, the system-level architectural model, and the benchmark programs used in our evaluations. The simulation models are built on top of the CacheMire Test Bench [2], a program-driven multiprocessor simulator. It consists of two parts: (i) a functional and (ii) an architectural simulator. The functional simulator consists of simulated SPARC processors. The architectural simulator models the processing nodes and the memory system.

We consider a CC-NUMA architecture in which internode cache coherence is maintained by a full-map directory-based write-invalidate protocol similar to Censier and Feautrier's [3]: A bit vector is associated with each memory block to point to the processor nodes with copies in their cache. A read miss sends a read request to the home memory module, where the block resides in memory. If the home is the local node and if the memory block is up-to-date, the miss is serviced locally in the node. Otherwise, the miss is serviced either in two or in four network traversals depending on whether the block is up-to-date or dirty (a processor has a modified copy). A write request to a shared or invalid copy sends an ownership request to home. The home node transmits invalidation requests to all nodes with copies, waits for acknowledgments from these nodes, and, finally, sends an invalidation acknowledgment to the requesting node. Acquire and release requests are supported by a queue-based lock mechanism similar to the one implemented in DASH [11].

Throughout all the simulations, we assume a 16-node configuration. We have simulated both finite and infinite FLC and SLC with infinite write buffers under release consistency. The timing model is based on a processor clock rate of 100 MHz. The FLC has the same cycle time as the processor. The access time of the SLC is 30 ns (33 MHz). The SLC is connected to the network interface and the local memory module by a 128-bit wide split transaction bus clocked at 33 MHz. It takes 30 ns to arbitrate for the bus and 30 ns to transfer a request or a block on that bus. The network interface controller is clocked at 33

---

1. For example, if the memory block is in a transient state, the prefetch is not serviced and a PreNeg is returned.

MHz. Furthermore, the memory cycle time is 90 ns. We correctly model contention for all components in the processing node.

Finally, in order to simplify generalizations, we do not simulate a specific interconnection topology. Rather, we assume that the network has infinite bandwidth and a fixed latency of 540 ns to send a message between two arbitrary nodes in the 16 node system. Table 2 shows the processor stall times due to a read request for a processor clock rate of 10 ns and in the case of a conflict-free memory system. Since we model contention for all components in the processing node, all requests normally take longer.

Table 2: Latency numbers (1 pclock = 10 ns).

Latency for Read Requests	Time
Read from FLC	1 pclock
Read from SLC	3 pclocks
Read from Local Memory	27 pclocks
Read from Home (2-hop)	159 pclocks
Read from Remote (4-hop)	315 pclocks

We use six benchmark programs to drive our simulation models. Four of them are taken from the SPLASH suite [13] whereas two (LU and the C-version of Ocean) have been provided to us from Stanford University. We list them in Table 3.

Table 3: Benchmark programs.

Benchmark	Description
MP3D	3-D particle-based wind-tunnel simulator
Water	Water molecular dynamics simulation
Cholesky	Cholesky factorization of a sparse matrix
LU	LU-decomposition of a dense matrix
PTHOR	Distributed time digital circuit simulator
Ocean	Ocean basin simulator

MP3D was run with 10K particles for 10 time steps. LU uses a 200x200 random matrix. Cholesky was run using the bcsstk14 benchmark matrix. Water was run with 288 molecules for 4 time steps, Ocean with the 128x128 grid and tolerance 10^{-7}, and finally PTHOR was run using the RISC-circuit for 1000 time steps. All applications are written in C using the PARMACS macros from Argonne National Laboratory [1], and have been compiled using gcc (version 2.0) with the optimization level -O2. All statistics are gathered in the parallel sections.

5 EXPERIMENTAL RESULTS

This section reports simulation results on fixed and adaptive sequential prefetching. Except from Section 5.1, all simulations are carried out with a block size of 32

bytes. Sections 5.1 through 5.4 assume infinite caches, whereas Section 5.5 concentrates on the effect of cache size. To simplify the wording, we refer to fixed sequential prefetch as *fixed prefetch*, and to adaptive sequential prefetch as *adaptive prefetch* throughout this section.

5.1 Fixed Prefetch

Some preliminary experiments with fixed prefetch showed that it was not useful to consider values of K greater than 1. Although a large K effectively cuts the cold and true sharing miss rates in program phases where the spatial locality is high, these phases are short. On the other hand, a large K means a significant increase of the memory traffic for those phases where the spatial locality is low. Therefore, all results presented in this section are for a degree of prefetching equal to one (K=1). We first compare three designs: (i) no prefetching with a block size B (called NoPrefetch(B)), (ii) fixed prefetch (K=1) with a block size B (called FixedPrefetch(B)), and (iii) no prefetching with a block size 2B (called NoPrefetch(2B)).

As the block size increases, we expect the numbers of cold misses and true sharing misses to decrease while the number of false sharing misses increases. This results in a reduced read-miss penalty (read stall) as long as the false sharing component is small. In Figure 2a, read stall times are shown for the three designs and for two block sizes; B=32 bytes (left diagram) and B=128 bytes (right diagram). For each application, the bars for FixedPrefetch(B) and NoPrefetch(2B) are normalized to NoPrefetch(B). As expected, for B=32, FixedPrefetch(B) and NoPrefetch(2B) exhibit nearly identical read stall times for all applications except for Ocean. For B=128 bytes, however, NoPrefetch(2B) does worse than FixedPrefetch(B) as a result of false sharing. To confirm this, Figure 2b shows the number of coherence misses for the three designs and the two block sizes. We see that for B=32 bytes, the number of coherence misses is about the same for FixedPrefetch(B) and for NoPrefetch(2B) for MP3D, LU, and Water. We speculate that most misses are true sharing misses and FixedPrefetch(B) reduces the number of true sharing misses nearly identically as NoPrefetch(2B). For larger block sizes (B=128 bytes), NoPrefetch(2B) results in a substantially larger number of coherence misses than FixedPrefetch(B) because of false sharing misses. Except for LU, the number of coherence misses in FixedPrefetch(B) is less than in NoPrefetch(B). This verifies that FixedPrefetch(B) reduces the number of true sharing misses about the same as NoPrefetch(2B), but keeps the number of false sharing misses at the same level as NoPrefetch(B). Dubois et. al. [5] showed that the number of true sharing misses in LU goes up slightly with increased block sizes, which explains why fixed prefetch cannot reduce coherence misses for this application.

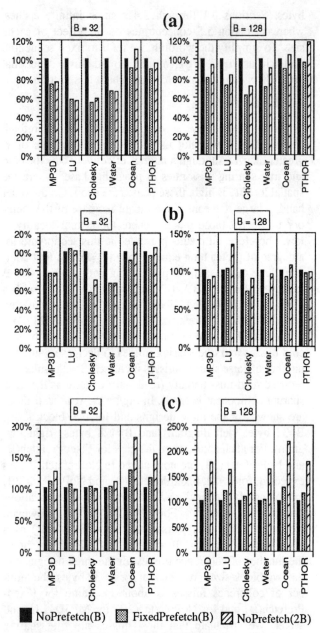

Figure 2: (a) Read stall, (b) Coherence misses, and (c) Traffic consumption.

Figure 2c shows the memory traffic in number of bytes sent through the network during the execution. The traffic for FixedPrefetch(B) and NoPrefetch(2B) are normalized to that of NoPrefetch(B). The left diagram shows the results for B=32 bytes. We see that fixed prefetch never reduces the memory traffic: Whereas the write traffic is about the same, the read traffic (including prefetches) is increased since some prefetches are not useful. However, for MP3D, LU, Cholesky, and Water, the difference is very small, which indicates a high *prefetch efficiency* (fraction of prefetches that are useful). For the case of NoPrefetch(2B), the traffic is reduced for both LU and

Cholesky and is increased for the other programs. For B=128 bytes, these observations hold for Fixed-Prefetch(B). For NoPrefetch(2B), the traffic is much higher for all applications, because of the predominance of false sharing. In summary, FixedPrefetch(B) manages to exploit the spatial locality for reads as does NoPrefetch(2B), while maintaining the false sharing component and write traffic of NoPrefetch(B).

5.2 Adaptive Prefetch

Figure 3 compares the read stall times for systems with 32 byte blocks, for adaptive and fixed prefetch both normalized to the design with no prefetching. The read stall times for LU, Cholesky, and Water are drastically reduced by the adaptive scheme. For MP3D, Ocean, and PTHOR, the read stall time is slightly higher than for fixed prefetch but still lower than for no prefetching.

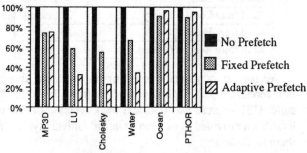

Figure 3: Read stall time.

The maximum degree of prefetching, K, is set to 8 in our simulations. The scheme for changing K (i.e. the value of LookaheadCounter) is the following: After 16 prefetches (PrefetchCounter = 16) the value of Useful-Counter is checked. If UsefulCounter > 9, then K is incremented. If UsefulCounter < 6, then K is decremented (never below 0). If UsefulCounter < 2, then K is shifted right, i.e. set to $K/2$. If $K = 0$, and if UsefulCounter > 5, then K is set to 1. To alleviate the effect of prefetches on memory contention, prefetches arriving to a memory module with a full buffer are acknowledged negative. The effect of this is twofold; (i) the number of messages in the memory buffer is kept low, and (ii) the PreNeg message reduces the fraction of useful prefetches and eventually the degree of prefetching.

Figure 4 compares the number of cold misses and coherence misses. For LU, Cholesky, and Water, the number of cold misses is drastically reduced. The spatial locality for phases with high cold miss rates is high, which makes most prefetches useful and increases the degree of prefetching. The number of coherence misses for Cholesky and Water is also drastically reduced because of the spatial locality of true sharing misses, and of the low fraction of false sharing misses for B = 32 bytes. For Ocean and PTHOR, the number of coherence misses is slightly higher than for fixed prefetch because the adaptive scheme sometimes does not prefetch in the phase of the program with low spatial locality. For PTHOR, we noticed that the number of prefetches issued under the adaptive scheme is cut by 50% as compared to fixed prefetch.

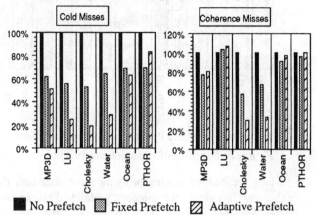

Figure 4: Cold (left) and coherence misses (right).

Figure 5 compares the memory traffic under fixed and adaptive prefetch normalized to no prefetching. The large reduction of the read stall times for LU, Cholesky, and Water comes from a large number of prefetches, and thus an increased traffic. Observe the large reduction of traffic for Ocean and PTHOR for the adaptive scheme as compared to the fixed scheme. This follows directly from a low prefetch efficiency for those programs and thus much less prefetches are issued. The higher number of prefetches in the fixed scheme yields a slight reduction of the read stall time.

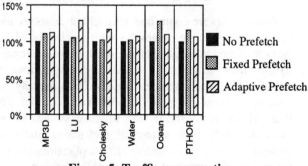

Figure 5: Traffic consumption.

5.3 Dynamic Behavior of Adaptive Prefetching

Figure 6 shows the cold and coherence miss rates during the execution of LU with a block size of 32 bytes. The upper diagram shows the execution with no prefetching, and the lower diagram shows the execution under the adaptive scheme. The left Y-axis shows the miss rate (%) as the total number of misses for shared data for the whole system divided by the total number of shared reads. The right Y-axis in the lower diagram shows the average value of LookaheadCounter for all 16 processors. One sample is taken every 10,000 shared reads; the miss rates are the average rates since the last sample and the value of the LookaheadCounters is taken at the time of the sample.

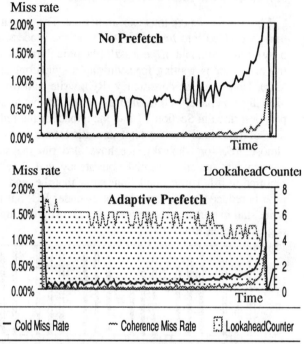

Figure 6: The behavior of adaptive prefetch for LU.

Under no prefetching (the upper diagram), LU has a large cold miss rate at the beginning, which remains the dominant part of the miss rate throughout the whole execution, going up to 3% at the end. The number of coherence misses is negligible most of the time, except for a peak at the end. In the adaptive scheme, the value of LookaheadCounter raises immediately to between 6 and 7, and the prefetches manage to keep the cold miss rate down to below 2% for the first peak (originally 8%). Since the prefetch efficiency is high, the LookaheadCounter stays at a high level. Towards the end, the number of coherence misses increases and the LookaheadCounter goes down (reflecting a smaller fraction of useful prefetches), until prefetching completely stops at the point where the number of coherence misses peaks. The cold misses in the end do not have enough spatial locality to boost up the degree of prefetching.

The choice of thresholds for the adaptive scheme affects the number of prefetches sent to the network. Lower thresholds means that the LookaheadCounter will tend to have a higher value, and that more prefetches are issued. A secondary effect is that the thresholds also affect the prefetch efficiency. Higher thresholds mean that more useful prefetches are required to increase the degree of prefetching and favor higher prefetching efficiency at the possible expense of higher read stalls. This behavior has been verified for all applications. Clearly the threshold values ought to be selected based on the amount of interconnection bandwidth.

5.4 Effects on Total Execution Time

Figure 7 shows the total execution time. The execution time is divided into *busy time* (the time the processor is busy), *read stall,* and *acquire stall* (the time during which the processor is waiting for acquiring a synchronization lock). Under release consistency (RC), write stall time is virtually completely eliminated using the hardware support described in Section 3.1 and the impact of the reduction of read stall on the total execution time becomes almost 50% for Cholesky. We have also run the same experiments under sequential consistency and they are reported in [4]. These simulations show that the read stall time is reduced to the same extent as under RC, but that write stall can increase somewhat due to a slightly higher degree of replication.

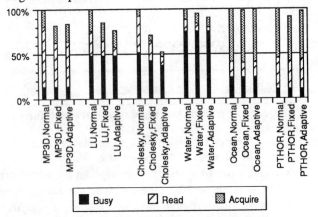

Figure 7: Total execution time under RC.

5.5 Effects of Finite Size SLC

One damaging side-effect of prefetching into small caches is cache pollution, which happens when a prefetched block never accessed during its lifetime in the cache replaces a block that would be referenced. Since prefetching increases the number of blocks loaded into the cache, it also increases the number of replaced blocks, which might lead to an increased miss rate. On the other hand, replacement misses give rise to additional prefetches to those replaced blocks. Figure 8 below reports results from simulations with finite first-level caches of 4Kbyte, and finite second-level caches (SLC) of 4Kbyte and 16Kbyte. It is interesting to see that both fixed and adaptive prefetch show significant reduction in the number of replacement misses for all applications but PTHOR. For all applications but MP3D, we also observed that the replacement misses are the major part of all misses for the 4Kbyte SLC.

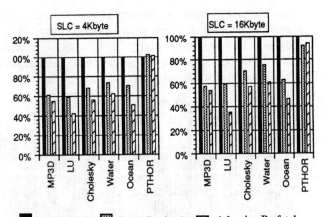

Figure 8: Number of replacement misses.

Since prefetching means bringing more data into the cache, and thus more replacements, prefetched blocks may be replaced before they are accessed. For the adaptive scheme, this means that the prefetch efficiency is reduced, which in turns leads to reduced prefetching. In [4] we also show the effects of cache size variations on the read stall. The main observations are that while the effectiveness of fixed prefetch is not significantly affected by the cache size, the effectiveness of the adaptive prefetch is smaller for a cache size of 4Kbyte than for 16Kbyte. One reason is the reduced prefetch efficiency, and the fact that since replacement misses are dominant, the effectiveness of prefetching is dictated by the spatial locality of replacement misses.

6 COMPARISON WITH RELATED WORK

Other hardware-controlled prefetching schemes have been proposed in the context of multiprocessors. Hagerstens ROT prefetching [9] takes advantage of the regularity of data accesses in scientific computations by dynamically detecting access strides. Unfortunately, this solution requires complex hardware and does not work that well for applications with irregular strides. Fu and Patel [6] study both sequential prefetching and stride prefetching, but only consider vector applications and vector caches. The proposal by Lee et al. [10] relies on data address lookahead in the processor, and its effectiveness is limited by the typical small size of basic blocks.

Several authors have evaluated software-based prefetching [12]. Special prefetch instructions are inserted in the code by the programmer or by the compiler. By focusing on the reduction of the read stall (which does not include the additional overhead due to the execution of inserted prefetch instructions) we can compare Mowry and Gupta's results with ours for a block size of 16 bytes. For MP3D our adaptive scheme reduces the read stall to 61%. They reduced it to 65% with their pf2 implementation and to 44% with their pf4 implementation (heavily optimized by hand). For LU, our adaptive scheme reduced the read stall to as low as 29%, while their most sophisticated prefetching scheme reduced it to 57% only. For PTHOR, which was the application that showed the worst performance, the adaptive scheme only managed to reduce the read stall to 90%. By reorganizing the element record and grouping entries together on the basis of whether they were likely to be modified, they reduced it to 86%. Their most aggressive implementation reduced it to 73%. In summary, while the software approach sometimes was more effective, it required a substantial programmer effort.

7 CONCLUSIONS

The major contribution of this paper is the detailed design and performance analysis of sequential prefetching schemes for multiprocessors: when a read miss occurs, the subsequent K blocks are prefetched if they are not resident in the cache, where K is the degree of prefetching.

We have analyzed the behavior of fixed sequential prefetching with K=1. This simple prefetching technique is shown to cut the number of cold and true sharing misses by about the same amount as in a system with twice the block size, without increasing the number of false sharing misses. Overall this simple prefetching scheme is shown to reduce the read stall by at least 30% for four or five applications out of six (depending on cache size) at a block size of 32 bytes. The read stall is reduced by up to 50%.

We have then proposed an adaptive prefetching scheme which tunes the degree of prefetching based on a dynamic measure of prefetching efficiency. Our simulations show considerable reduction of the read stall for applications where the spatial locality of read misses is high, and show little added traffic for applications where the spatial locality of misses is small. The read stall is reduced by as much as 78% for Cholesky, which yields a reduction of the total execution time of almost 50% under release consistency. These performance improvements are shown to be obtained with a modest increase in hardware complexity.

Acknowledgments

The authors are deeply indebted to Håkan Nilsson for his work on the simulator. We also want to thank the authors of the SPLASH benchmarks [13], Ed Rothberg (for LU), and Steven Woo (for the C-version of Ocean) of Stanford University.

REFERENCES

[1]   Boyle, J. et al. "*Portable Programs for Parallel Processors*". Holt, Rinehart, and Winston Inc. 1987.

[2]   Brorsson, M., Dahlgren, F., Nilsson, H., and Stenström, P. "The CacheMire Test Bench — A Flexible and Effective Approach for Simulation of Multiprocessors", In *Proceedings of the 26th Annual Simulation Symposium*, 1993.

[3]   Censier, L.M., and Feautrier, P. "A New Solution to Coherence Problems in Multicache Systems", In *IEEE Transactions on Computers*, 27(12), pp. 1112-1118, December 1978.

[4]   Dahlgren, F., Dubois, M., and Stenström, P. "*Sequential Prefetching Schemes for Shared Memory Multiprocessors*". Technical Report, Department of Computer Engineering, Lund University, Sweden.

[5]   Dubois, M. et al, "The Detection and Elimination of Useless Misses in Multiprocessors", In *Proc of the 20th Int Symposium on Computer Architecture*, May 1993.

[6]   Fu, J. and Patel, J.H. "Data Prefetching in Multiprocessor Vector Cache Memories". In *Proc of the 18th Int Symposium on Computer Architecture*, pp.54-63, May 1991.

[7]   Gharachorloo, K., Gupta, A., Hennessy, J. "Performance Evaluation of Memory Consistency Models for Shared-Memory Multiprocessors", in *Proc of ASPLOS IV*, pp.245-257, 1991.

[8]   Gupta, A. and Weber, W.-D. "Cache Invalidation Patterns in Shared-Memory Multiprocessors", In *IEEE Transactions on Computers*, Vol. 41, No.7, pp. 794-810, 1992.

[9]   Hagersten, E. "*Toward Scalable Cache Only Memory Architectures*". PhD thesis, Swedish Institute of Computer Science, Oct. 1992 (SICS Dissertation Series 08)

[10]   Lee R., Yew P-C., and Lawrie D. "Data Prefetching in Shared-Memory Multiprocessors", In *Proc 1987 Int Conf on Parallel Processing*, pp.28-31, 1987.

[11]   Lenoski, D.E. et al, "The Stanford DASH Multiprocessor", In *IEEE Computer*, pp.63-79, March 1992.

[12]   Mowry, T. and Gupta, A. "Tolerating Latency through Software-Controlled Prefetching in Scalable Shared-Memory Multiprocessors", In *Journal of Parallel and Distributed Computing*, 1991.

[13]   Singh, J.P., Weber, W.-D., and Gupta, A. "SPLASH: Stanford Parallel Applications for Shared-Memory", In *Computer Architecture News*, 20(1):5-44, March 1992.

[14]   Smith, A.J. "Sequential Program Prefetching in Memory Hierarchies", In *IEEE Computer*, Vol. 11, No. 12, pp.7-21, Dec. 1978.

[15]   Stenström, P. "A Survey of Cache Coherence Scheme for Multiprocessors", In *IEEE Computer*, Vol. 23, No. 6, pp. 12-24, June 1990.

Assigning Sites to Redundant Clusters in a Distributed Storage System*

Antoine N. Mourad   W. Kent Fuchs   Daniel G. Saab

Center for Reliable and High-Performance Computing
Coordinated Science Laboratory
University of Illinois
Urbana, Illinois 61801

Abstract *Redundant Arrays of Distributed Disks (RADD) can be used in a distributed computing system or database system to provide recovery in the presence of disk crashes and temporary and permanent failures of single sites. In this paper, we look at the problem of partitioning the sites of a distributed storage system into redundant arrays in such a way that the communication costs for maintaining the parity information are minimized. We show that the partitioning problem is NP-hard. We then propose and evaluate several heuristic algorithms for finding approximate solutions. Simulation results show that significant reduction in remote parity update costs can be achieved by optimizing the site partitioning scheme.*

1 Introduction

Redundant disk arrays are used for the purpose of providing reliable storage while increasing the I/O bandwidth in high performance systems [1, 2]. Redundant disk arrays can also be used in a distributed setting to increase availability in the presence of temporary site failures, disk failures, or major disasters. Stonebraker and Schloss have proposed the Redundant Arrays of Distributed Disks (RADD) scheme [3] as an alternative to multicopy schemes which are much more costly in terms of storage requirements. Cabrera and Long [4] have proposed the use of redundant distributed disk striping in a high speed local area network to support such I/O intensive applications as scientific visualization, image processing, and recording and play-back of color video. The RADD concept can also be used in multicomputer I/O subsystems such as the one proposed by Reddy and Banerjee [5] for hypercubes. The IDA approach proposed by Rabin [6] provides another way to tolerate failures in distributed storage systems with limited extra storage cost. However, in that approach when a file or table is dispersed over several sites and a portion of it is updated at a given site, the portions on the other sites need to be read in order to recompute the encoding before they are all written back. In the case of RADD, when a block is updated, only one parity block needs to be read and updated.

When RADDs are used, sites are grouped together to form a redundant array containing data and parity and capable of recovering from a single site failure. The size of each array is fixed and is determined by the tradeoff between the availability require-

ments of the system and the cost of the storage overhead. Hence, a large distributed data storage system may have to be divided into several arrays of fixed size. In this paper we look at the problem of partitioning the distributed storage systems into fixed size arrays in such a way as to minimize the cost of remote accesses that have to be performed to update the parity information. This problem is somewhat related to the problem of file allocation and replica placement in a distributed system which has been studied extensively in the literature [7, 8]. However, the two problems are different in nature because, in the RADD case, there is one redundant item for N data items while in the file allocation problem each file is replicated several times. More importantly, in the replica placement problem there is no stringent constraint on the number of sites "sharing" a replica because when the replica becomes unavailable those sites can access the second nearest replica while in the RADD case there is a hard constraint on the number of sites in an array. Note that the assignment of sites to redundant arrays (parity groups) can occur after all decisions on placing the data have been made. Data placement decisions are governed by a different set of criteria and are more influenced by the read access patterns since reads are usually more frequent than updates. Decisions on site assignment to redundant arrays are based on the update rate at each site and the cost of communication between sites and are independent of the read access rate. Changing the assignment of sites to redundant arrays does not change the placement of the data. The purpose of site assignment is to reduce the parity traffic and does not directly affect the data traffic.

In the following section, we describe the RADD organization. In Section 3, we present the model used to formulate the problem mathematically and we prove that the problem is NP-hard. In Section 4, heuristic algorithms for solving the problem are described and results from an experimental evaluation are presented. In Section 5 we develop heuristics with guaranteed bounds on the deviation from the optimal cost. In Section 6 we address the issue of hot spots and non-uniform site capacity and discuss the use of RADD for disaster recovery in OLTP systems.

2 Distributed Redundant Disk Array Organization

The RADD organization is shown in Figure 1. The data at each site is partitioned into blocks. Data blocks from different sites are grouped into a block parity group. The bitwise parity of the data blocks in each parity group is computed and written at a different site. In Figure 1, D_{ij} denotes a data block, P_i denotes a parity

---

*This research was supported in part by the National Aeronautics and Space Administration (NASA) under Contract NAG 1-613 and in part by the Department of the Navy and managed by the Office of the Chief of Naval Research under Grant N00014-91-J-1283.

block	$Site_0$	$Site_1$	$Site_2$	$Site_3$	$Site_4$	$Site_5$
0	P_0	S_1	D_{20}	D_{30}	D_{40}	D_{50}
1	D_{00}	P_1	S_2	D_{31}	D_{41}	D_{51}
2	D_{01}	D_{10}	P_2	S_3	D_{42}	D_{52}
3	D_{02}	D_{11}	D_{21}	P_3	S_4	D_{53}
4	D_{03}	D_{12}	D_{22}	D_{32}	P_4	S_5
5	S_0	D_{13}	D_{23}	D_{33}	D_{43}	P_5

Figure 1: Organization of a distributed redundant disk array ($N = 6$).

block	$Site_0$	$Site_1$	$Site_2$	$Site_3$	$Site_4$	$Site_5$
6	P_0	D_{14}	S_2	D_{34}	D_{44}	D_{54}
7	D_{04}	P_1	D_{24}	S_3	D_{45}	D_{55}
8	D_{05}	D_{15}	P_2	D_{35}	S_4	D_{56}
9	D_{06}	D_{16}	D_{25}	P_3	D_{46}	S_5
10	S_0	D_{17}	D_{26}	D_{36}	P_4	D_{57}
11	D_{07}	S_1	D_{27}	D_{37}	D_{47}	P_5

Figure 2: Alternative placement pattern for parity and spare blocks.

block and S_i denotes a spare block, all at site i. The number under *block* in the first column of the figure denotes the physical block number on disk. Each row in the figure represents a parity group. The position of the parity block is rotated among the sites in order to avoid creating a bottleneck at the site where parity is stored. For every update to one of the data blocks in the parity group, the parity block needs to be updated using the following formula:

$$P_{new} = (D_{old} \oplus D_{new}) \oplus P_{old}.$$

Spare blocks are provided in order to be able to reconstruct data blocks that become inaccessible due to a site failure. The failed data block is reconstructed by XORing all other data blocks and the parity block in its parity group. If K denotes the number of data blocks per parity group then $N = K + 2$ denotes the number of sites in a distributed disk array. The storage overhead for the parity *and* spare blocks required by RADDs is $(200/K)\%$ compared to a 100% overhead for the case of two copy schemes.

3 The Model

We model the distributed computing system by an undirected connected graph $G = (V, E)$ where V is the set of sites and each edge $e \in E$ represents a bidirectional communication link between two sites. For each $e \in E$, w_e denotes the cost of communication over link e. For $e = (u, v)$, w_e could be the actual distance between site u and site v. We assume that if n is the number of sites in V then $n = mN$ for some m. We will assume that the site capacity is uniform. In Section 6.2 we show how to deal with non-uniform site capacity. In the pattern shown in Figure 1, the parity blocks of the $N - 2$ data blocks of site i reside on sites $i + 1 \bmod N$ through $i + N - 2 \bmod N$. Therefore there is no parity update traffic from site i to site $i - 1 \bmod N$. In order to make the problem symmetrical and thus easier to tackle, we assume that for the next set of N blocks the pattern shown in Figure 2 is used. In all, there are $N - 1$ such patterns obtained by changing the distance between the parity block and the spare block on a given row. These $N - 1$ patterns should alternate throughout the range of blocks so that update traffic from a given site is distributed over the remaining $N - 1$ sites. This will also provide more load balancing for the parity update traffic in the array. Let μ_v designate the rate of update accesses to data blocks

at site v. Each update will cause communication between the site where the update took place and the site holding the parity for the given data block. At each site the set of data blocks that have their corresponding parity blocks on the same site is called a data group. To simplify the model, we assume that the $N - 1$ data groups share equally the update rate. This implies that the rate at which site v sends parity update information to each other site in its redundant array is $\lambda_v = \mu_v / (N - 1)$. This assumption is supported by the fact that consecutive data blocks have their parity blocks on different sites which implies that accesses to a heavily used file that is stored on consecutive disk blocks will be spread over different data groups. In Section 6, the above assumption will be removed. The problem of partitioning the sites into arrays of size N in such a way that parity update costs are minimized can be mathematically formulated as follows:

Problem 1 (SP) *Find a partition of V into m disjoint subsets V_1, V_2, ..., V_m of size N such that if $d(u, v)$ denotes the length of the shortest path between u and v then*

$$\sum_{i=1}^{m} \sum_{u \in V_i} \lambda_u \sum_{v \in V_i - \{u\}} d(u, v)$$

is minimum.

Theorem 1 *Problem SP is NP-hard for any fixed $N \geq 3$.*

Proof: We prove that problem SP is NP-hard by showing that there is a polynomial time transformation from the problem of partitioning a graph into cliques of size N to problem SP. The Partition into Cliques of size N (PC) problem can be stated as follows:
Instance: A graph $G = (V, E)$, with $|V| = Nm$ for some positive integer m.
Problem: Is there a partition of V into m disjoint subsets V_1, V_2, ..., V_m such that, the subgraph of G induced by V_i is a clique of size N (complete graph with N nodes)?

PC is NP-complete for any fixed $N \geq 3$ (see Partition into Isomorphic Subgraphs [9]). To transform an instance of PC into an instance of SP, it is sufficient to set $\lambda_v = 1$ for all $v \in V$, and $w_e = 1$ for all $e \in E$. Then graph G can be partitioned into cliques of size N if and only if the cost of the optimal solution to the above instance of problem SP is $n(N - 1)$. □

The cost function $\sum_{i=1}^{m} \sum_{u \in V_i} \lambda_u \sum_{v \in V_i - \{u\}} d(u, v)$ can be rewritten as

$$\sum_{i=1}^{m} \sum_{u,v \in V_i, u \neq v} (\lambda_u + \lambda_v) d(u, v) = \sum_{i=1}^{m} \sum_{u,v \in V_i, u \neq v} D(u, v),$$

where $D(u, v)$ is defined as $D(u, v) = (\lambda_u + \lambda_v) d(u, v)$. In this form the general problem is reduced to a uniform load problem with the distance D replacing d. However, D is not a true distance since it does not necessarily satisfy the triangular inequality.

4 Approximation Algorithms

4.1 Description of the Heuristics

The first heuristic is based on a greedy strategy that consists of satisfying first the sites with the largest update rate. Let Λ be the list of update rates for all sites. When sites are grouped into clusters their update rates are removed from Λ and replaced by a single update rate for the cluster. The cluster update rate is the average update rate of the sites in the cluster.

Algorithm 1:

Step 1. Select the largest value in Λ and let a be the corresponding site (or cluster). Find the site (or cluster) b such that merging a and b results in the smallest increase in the cost function. Merge the two sites (or clusters) if the resulting cluster has less than N sites and the total number of clusters does not exceed m. If the clusters cannot be merged, find the next best choice for b and repeat. Remove the update rates of the merged sites (or clusters) from Λ and replace them with the cluster update rate.

Step 2. Repeat Step 1 until m clusters having N sites each have been formed.

The computational cost of Algorithm 1 is $O(Nn^2)$. But it requires that the all-pair shortest path algorithm be performed first which requires $O(n^3)$ operations.

The second approach consists of two stages: in the first stage m sites are identified to be used as cluster seeds and in the second stage the remaining sites are allocated to the clusters to form m subsets of N sites each.

Algorithm 2:

Step 1. Select the two sites with the largest distance between each other and include them in the set S of cluster seeds.

Step 2. Select the site v with the largest average distance to the sites already in S and add it to S.

Step 3. Repeat Step 2 above until $|S| = m$. Each cluster initially contains one of the m seeds in S.

Step 4. For each of the m clusters, compute the average update rate of the sites in the cluster. In decreasing order of their average update rate, allocate to each cluster the site that is closest to it in terms of the distance metric D.

Step 5. Repeat Step 4 above until all sites have been allocated to the m clusters.

We use the distance metric D in Step 4 because it provides the actual increase in the cost function of a cluster when a node is added to it. The computational cost of the Algorithm 2 is $O(Nn^2)$. It also requires that the all-pair shortest path algorithm be performed first.

The third approach is based on the hierarchical clustering technique [10]. We use the distance matrix whose entries are $d(u, v)$ for all $u, v \in V$. Clusters are formed by merging together sites or smaller clusters that are close to each other. When two sites (or clusters) are grouped together, the distance matrix is modified by eliminating the columns and rows corresponding to the merged sites (or clusters) and replacing them with a single column and a single row reflecting the average distance between the merged sites and other sites (or clusters). The procedure is as follows:

Algorithm 3:

Step 1. Find the smallest entry in the distance matrix and merge the two sites (or clusters) together if the resulting cluster has N sites or less and if the total number of clusters does not exceed m. If any of the latter conditions is not satisfied, select the next smallest entry and repeat. Once two sites (or clusters) have been merged, update the distance matrix and the number of clusters accordingly.

Step 2. Repeat Step 1 above until m clusters having N sites each have been formed.

The complexity of Algorithm 3 is $O(n^3)$.

After an initial partition has been found, the following procedure may be used to improve it.

Procedure Improve:

Step 1. Select the site u with the highest update rate. For each site v outside site u's partition, compute the change in cost $\Delta C(u, v)$ if u and v were swapped. Let v^* be the site corresponding to the minimum change in cost: $\Delta C(u, v^*) = \min_{v \notin V_u} \Delta C(u, v)$. If $\Delta C(u, v^*) < 0$ then swap u and v^*.

Step 2. Repeat Step 1 above for all sites in V in decreasing order of their update rate.

The complexity of the above procedure is $O(n^3)$. The procedure may be repeated several times to improve the total cost. The procedure may also be repeated until a local minimum of the cost function is reached. However, it is not guaranteed that such a local minimum will be reached in finite time. The procedure can also be employed as the basic move in meta-heuristics such as simulated annealing [11] or tabu search [12] that avoid getting trapped in a local minimum.

4.2 Experimental Evaluation

We have conducted experiments to evaluate the approximate solutions obtained using the heuristics and to compare the three proposed approaches for site assignment. In the experiments, we used randomly generated graphs. The distance on each edge in the graph was drawn from a uniform distribution over the interval $[1, K_w]$. The update rates at each site were drawn from a uniform distribution over the interval $[1, K_\lambda]$.

In our experiments we found out that Algorithm 2 performs better when the distance D is also used in the first stage of the algorithm. This can be explained by the fact that using D in the generation of the cluster seeds ensures that edges with large $D(u, v)$ will not be used within a cluster, i.e., sites that have large loads and that are far apart are not placed in the same cluster. The results shown here for Algorithm 2 were obtained using D instead of d.

In the first experiment, we compare the approximate solution provided by the heuristics to the optimal solution. The optimal solution was obtained using exhaustive search. N was taken to be equal to 5 and n equal to 15. Table 1 shows the results for three situations: one where the edge weights vary more widely than the site loads, one where both are picked from the same interval and one where the site loads vary more widely than the edge weights. Each entry represents the average over 100 randomly generated graphs. For all data values reported in this paper, the width of the 90% confidence interval is less than 8% of the average value (typically in the 2–3% range). The costs of the approximate solutions are within 10% of the cost of the optimal solution. In the first column of the table, we have listed the cost of a random solution.

Since, in the first experiment, an exhaustive search was used to find the optimal solution, the number of nodes n could not be very large. In a second experiment, we compared the performance of the three heuristics for larger values of n. Figure 3 shows the results for the second experiment. For clarity of the figure, we plotted the cost of the approximate solution divided by 1000. In the case $N = 10$, Algorithm 3 outperforms Algorithms 1 and 2 for all values of n except when $n = 20$ in which case Algorithm 2 performs better. For the first and second environments Algorithm 1 outperforms Algorithm 2 for large values of n but for the last environment Algorithm 2 outperforms Algorithm 1. For $N = 5$, Algorithm 2 does not do very well except in the last environment in which the range of site loads is much larger than the range of edge weights. Algorithm 3 performs best in the first two environments. The main point that can be deduced from this experiment is that, in spite of the fact that Algorithm 3 does not use any information about site loads, it outperforms the other two algorithms when n and N are relatively large and in the other cases its performance is always very close to that of the best algorithm. This means that, in a large system, it is more important to minimize the sum of the edge weights within each cluster than

Table 1: Comparison between approximate solutions and the optimal solution.

K_w, K_λ	Random	Algorithm 1	Algorithm 2	Algorithm 3	Exhaustive
1000, 10	68400	52439	53462	52649	47475
100, 100	66071	50012	51347	51237	45661
10, 1000	96757	76388	77362	77062	70004

to use the greedy approach that attempts to assign to the sites with large loads their nearest neighbors.

5 Heuristics with Performance Guarantees

The heuristics described in Section 4 provide in general a good approximate solution. However, there is no guarantee that the approximate solution will not diverge significantly from the optimal one in certain cases. In this section, we seek to find a heuristic for which it is possible to establish a bound on the error between the approximate solution and the optimal one. We develop such a heuristic first for the case of a system with balanced load, $\lambda_v = \lambda$, for all $v \in V$, and uniform edge weights, then we look at the more general case of a balanced load system with arbitrary edge weights. Since a problem with arbitrary site loads can always be transformed into a problem with uniform site load as shown in Section 3, then the heuristic for the balanced load case with arbitrary edge weights will also provide performance guarantees for the arbitrary load case.

5.1 Balanced Load and Uniform Edge Weights

The heuristic requires the use of a spanning tree with many leaves. The problem of finding a spanning tree with a maximum number of leaves is NP-hard [9], however, there exist polynomial time algorithms for generating spanning trees with many leaves. Typically these methods guarantee that a certain fraction of the nodes will be leaves. The fraction of leaves is a function of the minimum degree k of the graph. Kleitman and West proved the following result [13]:

Theorem 2 (Kleitman-West) *If k is sufficiently large, then there is an algorithm that constructs a spanning tree with at least $(1 - b \ln k / k)n$ leaves in any graph with minimum degree k, where b is any constant exceeding 2.5.*

It was also conjectured that a spanning tree can be constructed with a larger fraction of leaves. More specifically, Linial conjectured that the number of leaves could be at least $\frac{k-2}{k+1}n + c_k$. This stronger result was proved for $k = 3$ with $c_3 = 2$ and for $k = 4$ with $c_4 = 8/5$ [13].

Algorithm

Step 1. Find a spanning tree with many leaves.
Step 2. Partition the spanning tree into m clusters of N nodes each using procedure **Partition_Tree** described below.

The partition found for the tree will be used for the original graph. In the description of the procedure **Partition_Tree**, we assume that the tree is levelized starting from the root.

Procedure Partition_Tree:

The procedure partitions the tree from the bottom up. As the clusters are built, whenever the size of a cluster reaches N nodes, that cluster is removed from the tree. Starting from the deepest level in the tree, sibling leaves are placed together in a cluster. If all siblings have been used then their parent is included in the cluster. At an internal node v, all subtrees rooted at its siblings

must be processed so that only less than N nodes are left in each subtree. Those subtrees are numbered from 1 to $d(v) - 1$, $d(v)$ being the degree of v. Then the clusters are formed by adding to the nodes of subtree i enough nodes from subtree $i + 1$ to make an N node cluster. If there are not enough nodes in subtree $i + 1$ to form a complete cluster, the nodes of the two subtrees are placed together and the next subtree is used to complete the cluster. If all the subtrees have been used, and there remains an incomplete cluster then the parent node is added to the remaining cluster and the procedure continues at the next level. When adding a portion of the nodes of a given subtree to the preceding subtree(s) to complete a cluster, the nodes at the deepest level in that subtree are used first so that removal of the newly completed cluster will not disconnect the tree.

Theorem 3 *The cost (HEU) of the approximate solution found using a spanning tree with many leaves and the cost (OPT) of the optimal solution satisfy the following relationship:*

$$\frac{HEU}{OPT} \le 2\alpha + (1 - \alpha)\frac{N^2}{N - 1},$$

where α is the fraction of leaves in the spanning tree.

Proof We need to establish an upper bound on the cost of the approximate solution and a lower bound on that of the optimal one. The cost in the graph of the approximate solution is at most the cost of that solution in the tree. We evaluate the cost in the tree by adding up the contributions of each edge in the spanning tree to the overall cost. If an edge connects a leaf node to the tree it will be referred to as a leaf edge otherwise it will be called an internal edge. A leaf edge will be used in only one cluster and it will be used only for communication between the leaf node and the other $(N - 1)$ nodes in the cluster. Therefore the contribution of a leaf edge to the overall cost is $2(N - 1)$. An internal edge will be used in at most two clusters and in each cluster it will be used by i nodes to communicate with the other $N - i$ nodes in the cluster. If α designates the fraction of leaf nodes in the tree, we have:

$$
\begin{aligned}
HEU &\le \alpha n \times 2(N - 1) + (n - 1 - \alpha n) \times 2 \times \\
&\quad \max_{1 \le i \le N-1} 2i(N - i) \\
&\le n(N - 1)(2\alpha + (1 - \alpha)N^2/(N - 1))
\end{aligned}
$$

For the cost of the optimal solution, an obvious lower bound is the cost in a complete graph which is $n(N - 1)$. Hence, $HEU/OPT \le 2\alpha + (1 - \alpha)N^2/(N - 1)$. □

As stated in Theorem 2, for large k, α converges to 1 and the above bound approaches 2. Note that it is reasonable to assume that the minimum degree will be large in practice because the underlying network has to have sufficient connectivity to enable communication under node failures, and hence, has to have a reasonably large minimum degree.

The complexity of the algorithms for generating trees with many leaves [13] is $O(|E|)$. The complexity of the **Partition_Tree** procedure is $O(n)$.

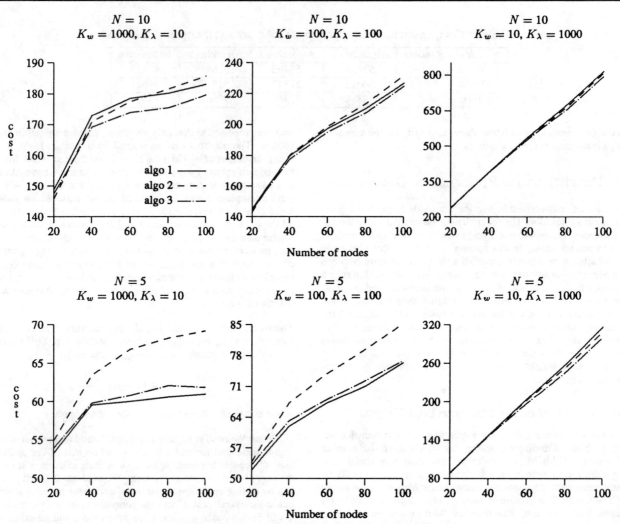

Figure 3: Comparison between the three heuristics.

5.2 Balanced Load and Arbitrary Edge Weights

For arbitrary edge weights the problem of finding a heuristic with guaranteed performance bounds is much harder. In the following we describe a heuristic for which a worst case performance bound can be established. The bound is more significant for systems where link communication costs (edge weights) do not vary widely. The heuristic consists of finding a minimum spanning tree, partitioning the tree into clusters using procedure **Partition_Tree** and using that partition as an approximate solution. The following result will be used to establish a lower bound on the cost of the optimal solution.

Lemma 1 *In a complete graph, the average weight of the edges in a minimum spanning tree is at most the average weight of all edges.*

Proof We use induction on the number of nodes n. The lemma is obviously true for $n = 2$ or $n = 3$. Suppose it is true for graphs with $n - 1$ nodes and consider an n-node graph. Select node v such that the average weight of edges incident on v is at least the average weight of all edges in the graph. Remove v from the graph and find a minimum spanning tree in the remaining $(n-1)$-node graph. Then add to this spanning tree the lightest edge e^*

connecting v to the other nodes to form an n-node spanning tree. Let MST_{n-1} and MST_n be the *total* weights of the $(n-1)$-node and the n-node spanning trees respectively. Let $\mathcal{E}(v)$ be the set of edges incident on v. Using the induction hypothesis, we have:

$$\frac{\text{MST}_{n-1}}{n-2} \leq \frac{\sum_{e \in E - \mathcal{E}(v)} w_e}{(n-1)(n-2)/2}.$$

Therefore

$$\text{MST}_n \leq \text{MST}_{n-1} + w_{e^*} \leq \frac{\sum_{e \in E - \mathcal{E}(v)} w_e}{(n-1)/2} + \frac{\sum_{e \in \mathcal{E}(v)} w_e}{n-1}$$

$$\leq \frac{\sum_{e \in E - \mathcal{E}(v)} w_e}{(n-1)/2} + \frac{\sum_{e \in \mathcal{E}(v)} w_e}{n-1} +$$

$$\underbrace{\frac{\sum_{e \in \mathcal{E}(v)} w_e}{n-1} - \frac{\sum_{e \in E} w_e}{n(n-1)/2}}_{\geq 0}$$

$$= \frac{\sum_{e \in E} w_e}{n/2}.$$

Hence, the average weight of the edges in the minimum spanning tree is $\text{MST}_n/(n-1) \leq \sum_{e \in E} w_e/(n(n-1)/2)$. $\quad\square$

To obtain a lower bound on the cost of the optimal solution, we consider the optimal partition and we build a spanning tree by first finding a minimum spanning tree in each cluster and then replacing each cluster by a single node and connecting each pair of these nodes by the lightest edge linking the initial clusters. An intercluster minimum spanning tree is then found. The intracluster spanning trees along with the intercluster spanning trees form a spanning tree for the entire graph.

Lemma 2 *The list of edge weights of the intercluster minimum spanning tree (ICMST) is included in the list of edge weights of the global minimum spanning tree (GMST).*

Proof Let e be an edge in the ICMST that does not appear in the GMST. Let u and v be its endpoints in the original graph and let w be its weight. The path in the GMST from u to v induces a path in the intercluster graph from the cluster of u to that of v. If the path is a single edge then this edge must have weight w and could replace the edge e in the ICMST. If the induced path has more than one edge then, since the ICMST cannot contain a cycle, some of the edges on the induced path must not appear in the ICMST and at least one of these induced edges that do not appear in the ICMST forms a cycle containing e when added to the ICMST. Let e' be such an edge. e' must have weight at most w otherwise it could be replaced in the GMST by (u, v) to obtain a spanning tree with a smaller cost. In addition, e' cannot have weight less than w because it would then be possible to replace e by e' in the ICMST and obtain a smaller intercluster spanning tree. Hence the weight of e' is w and we could remove e and replace it with e' in the ICMST. This process can be repeated until all edges in the ICMST also appear in the GMST. $\quad\square$

Theorem 4 *The cost (HEU) of the approximate solution found using a minimum spanning tree and the cost (OPT) of the optimal solution satisfy the following relationship:*

$$\frac{\text{HEU}}{\text{OPT}} \leq N \frac{\text{MST}}{\text{MST} - (m-1)\overline{w}},$$

where MST is the total weight of the edges in the minimum spanning tree and $\overline{w}$ is the average weight of the $m-1$ heaviest edges in the minimum spanning tree.

Proof In evaluating an upper bound on the cost of the approximate solution, we follow the same procedure as in the proof of Theorem 3 but we will not distinguish between leaf edges and internal edges. Each edge e in the tree will be used by at most two clusters and the contribution of e to the overall cost is bounded by $2 \times w_e \times \max_{1 \leq i \leq N-1} 2i(N-i)$. Hence, we have $\text{HEU} \leq N^2\text{MST}$.

Let MST_i be the weight of the minimum spanning tree of cluster i for $1 \leq i \leq m$ and MST_c be the weight of the intercluster tree. We have $\sum_{i=1}^{m} \text{MST}_i + \text{MST}_c \geq \text{MST}$. Using Lemma 2, we have $\sum_{i=1}^{m} \text{MST}_i + (m-1)\overline{w} \geq \text{MST}$. Let OPT_i be the contribution to the optimal cost by cluster i. Using Lemma 1 we have $\text{OPT}_i/N \geq \text{MST}_i$ therefore $\text{OPT} \geq N(\text{MST} - (m-1)\overline{w})$. $\quad\square$

Let r be the ratio of the largest edge weight to the smallest edge weight. A looser but simpler bound than the one established in Theorem 4 can be derived using the parameter r:

$$\text{HEU/OPT} \leq N \left(1 + \frac{m-1}{n-m}r\right) \leq N(1 + r/(N-1)).$$

6 Generalization of the Model

6.1 Non-Uniform Load within Site

In our model, we assumed that each site sends parity updates to each other site in its partition at the same rate. This implies a uniform update rate to each of the $N-1$ data groups of a given site that have parity information on each of the $N-1$ other sites. If the update rate information for each data group at each site is available then the model can be refined to account for the difference in the rate of parity update requests issued by a given site and destined to the other sites in the array. The refined model should yield better results in the presence of hot spots. The update rate λ_u of site u is replaced by $N-1$ update rates $\lambda_{u,1}, \ldots, \lambda_{u,N-1}$ corresponding to each of its data groups. In this case, an obvious optimization would be to have the parity of the ith most frequently accessed data group of a given site placed on the ith nearest site in its partition. Note that this can be implemented without having to reshuffle the data on disk by saving the permutation describing the remapping of the $N-1$ data groups for each site and using it to send parity update requests to the proper site. Given the above optimization, the algorithms of Section 4 with some minor modifications can still be used to partition the sites. The site update rate used in Algorithm 1 and 2 is set to the sum of all $N-1$ data group update rates at that site. We have evaluated the three algorithms of Section 4 in the case of the refined model, along with a new greedy strategy that looks at data groups instead of sites and tries to place the parity of the data groups with the largest update rates on the closest sites. Details of the greedy algorithm are provided in the Appendix.

Figure 4 shows the results of the comparison between the four algorithms. The individual data group update rates are chosen randomly from the interval $[1, K_\lambda]$ while the edge weights are chosen from $[1, K_w]$. We found that Algorithms 2 and 3 perform best for $N = 10$ with Algorithm 2 being the winner for lower values of n while Algorithm 3 is better for the high values of n. For $N = 5$ Algorithm 3 performs best in almost all situations. We also found that the parity assignment within a cluster is as important as the problem of partitioning the sites into clusters. The policy that consists of placing the parity of the ith most accessed data group on the ith closest site within the cluster reduces the cost by 15 to 20%.

6.2 Non-Uniform Site Capacity

The case of non-uniform site capacity can be handled in the same fashion as proposed by Stonebraker and Schloss [3]. We assume that the total number of disks is Np for some p[†] and that the number of disks at any given site is at most p. The system could then be partitioned using the following procedure.

Step 1. Select the $N\lfloor|V|/N\rfloor$ sites with the largest number of disks and apply one of the partitioning algorithms described in the previous sections to assign one disk from each of the selected sites to an array.

Step 2. Remove the assigned disks and remove sites with no disks

---

[†]This replaces the assumption that $|V| = mN$.

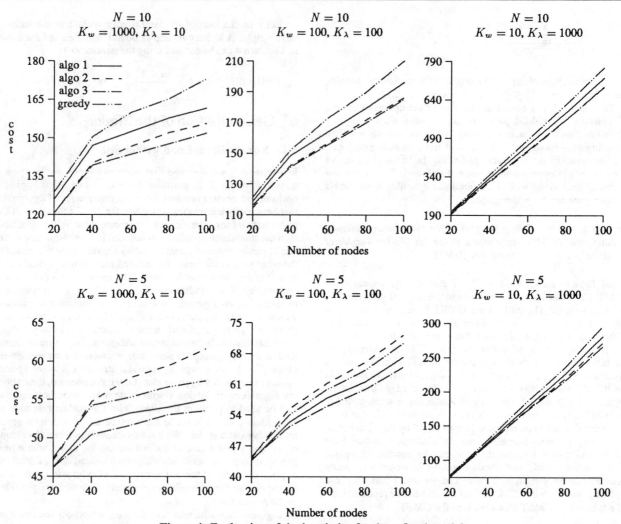

Figure 4: Evaluation of the heuristics for the refined model.

left.

Step 3. Repeat the above steps until all disks have been assigned.

Non-uniform disk capacity can be dealt with by using logical disks of size B blocks such that the site capacities are multiples of B [3].

6.3 Disaster Recovery in OLTP systems

Disaster recovery is an important issue in On-Line Transaction Processing (OLTP) systems [14–16]. However, in such systems, updating the remote parity after each disk update may be too expensive, especially since there are usually stringent requirements on transaction response time in those systems.

Typically, disaster recovery in OLTP systems is implemented by duplicating the data of a given site at a remote backup site and shipping Redo log information to the backup site where the updates are applied to the backup database. There are two approaches used in shipping the log [17]. In the first approach, the log records are shipped asynchronously to the backup site. Therefore transaction response time is not affected by the communication with the backup. However some transactions my be lost in the case of a disaster. This configuration is called *1-safe*. In the second approach, log records are sent to the backup at commit time and the transaction waits for an acknowledgment before it is allowed to commit. No transactions are lost in this case. This configuration is called *2-safe*.

Similar configurations can be implemented using RADD. In a *1-safe* implementation, parity updates (XOR's of old and new data) can be accumulated at the originating site and shipped to the remote parity locations periodically. In a *2-safe* implementation, the parity updates originated by a transaction are grouped according to their destination site and shipped to that site while the transaction waits for an acknowledgment. If the updates performed by the transaction involve only one of the $N - 1$ data groups then only one remote message has to be sent by the committing transaction and the delay will be the same as in the traditional remote backup scheme. The advantage of RADD over the traditional schemes is that it uses much less storage space than full duplication.

Our model can still be used to solve the site assignment problem in both of the above implementations. However, instead of using the update rate at each site, the frequency of the periodic updates should be used in the *1-safe* case and the *update* transaction rate should be used in the *2-safe* case.

Another optimization that might be useful in OLTP environments consists of using the scheme proposed by Bhide and Dias in [18] to reduce the number of random I/O's performed in updating the parity at the remote site. The scheme consists of storing the parity updates in nonvolatile memory or sequentially on a dedicated disk and then periodically propagating them to their permanent locations. The scheme was originally proposed for use with a RAID level 4 organization [1] to reduce the load on the parity disk. When the parity updates are stored sequentially on a dedicated disk, disk sorting is used to apply the parity updates to their permanent location.

7 Summary

We looked at the problem of partitioning the sites of a distributed storage system into redundant disk arrays while minimizing the communication costs for updating the parity information. The problem was shown to be NP-hard in its general form. Several heuristic methods were investigated to obtain approximate solutions to the site partitioning problem. It was found that the heuristic that minimizes the sum of distances between sites within each cluster performs consistently well in all environments especially in large systems with a relatively large array size. In such systems, the above approach outperforms greedy methods that attempt to satisfy first the sites with the largest loads by placing their nearest neighbors in their partition. The solutions produced by this heuristic are also more robust because they provide good performance under different site loads. Guaranteed upper bounds were established on the deviation from the optimal cost for some of the heuristics. It was also found that modifying the parity assignment within each cluster to place the parity of the heavily accessed data groups on the nearest sites within the cluster can significantly decrease the parity update cost. Finally, we discussed implementations of the RADD scheme for disaster recovery in OLTP systems and described various optimizations that can be helpful in those environments.

Appendix

Algorithm Greedy

Let Λ be the list of update rates for all data groups at all sites.
Let p_v be the number of site v's partition. Initially $p_v = -1$ for all $v \in V$.
Let n_i be the number of sites in partition i. Initially, $n_i = 0$. Assume $n_{-1} = 1$ throughout.
Let k be the current number of partitions. Initially $k = 0$.
Let $\mathcal{N}(v) = V - v$, for all $v \in V$.
Let $l = 0$.
Step 1. Select the largest value λ in Λ and let u be the corresponding site. If $n_{p_u} = N$ go to Step 4.
Step 2. Find the site v in $\mathcal{N}(u)$ that is nearest to u and satisfies: p_u or $p_v \neq -1$ and $n_{p_u} + n_{p_v} \leq N$ or if $p_u = p_v = -1$ and $k < m$. If none exist go to Step 4.
Step 3. Remove v from $\mathcal{N}(u)$.
If $p_u = p_v = -1$ set $p_u = p_v = l$, $n_l = 2$, $l = l + 1$, and $k = k + 1$.
If $p_u = -1$ and $p_v \neq -1$ set $p_u = p_v$ and $n_{p_v} = n_{p_v} + 1$.
If $p_u \neq -1$ and $p_v = -1$ set $p_v = p_u$ and $n_{p_u} = n_{p_u} + 1$.
If $p_u \neq -1$ and $p_v \neq -1$, set the partition number for every site in v's current partition to p_u, set $n_{p_u} = n_{p_u} + n_{p_v}$, $n_{p_v} = 0$, and $k = k - 1$.
Step 4. Remove λ from Λ.
Step 5. If $\sum_i n_i < n$, go to Step 1, otherwise stop.

The algorithm is similar to Algorithm 1 in that it tries to satisfy first the nodes with the highest data group update rates. The

complexity of the algorithm is $O(Nn^2)$, but as in the case of Algorithm 1, it requires the all-pair shortest path algorithm.

References

[1] D. Patterson, G. Gibson, and R. Katz, "A case for redundant arrays of inexpensive disks (RAID)," in *Proceedings of the ACM SIGMOD Conference*, pp. 109–116, June 1988.

[2] J. Gray, B. Horst, and M. Walker, "Parity striping of disk arrays: Low-cost reliable storage with acceptable throughput," in *Proceedings of the 16th International Conference on Very Large Data Bases*, pp. 148–161, Aug. 1990.

[3] M. Stonebraker and G. A. Schloss, "Distributed RAID – a new multiple copy algorithm," in *Proceedings of the Sixth IEEE International Conference on Data Engineering*, pp. 430–437, Feb. 1990.

[4] L.-F. Cabrera and D. D. E. Long, "Swift: Using distributed disk striping to provide high I/O data rates," *USENIX Computing Systems*, pp. 405–433, Fall 1991.

[5] A. L. N. Reddy and P. Banerjee, "Design, analysis, and simulation of I/O architectures for hypercube multiprocessors," *IEEE Transactions on Parallel and Distributed Systems*, vol. 1, pp. 140–151, Apr. 1990.

[6] M. O. Rabin, "Efficient dispersal of information for security, load balancing, and fault tolerance," *Journal of the ACM*, vol. 36, pp. 335–348, Apr. 1989.

[7] L. W. Dowdy and D. V. Foster, "Comparative models of the file assignment problem," *ACM Computing Surveys*, vol. 14, pp. 287–313, June 1982.

[8] B. W. Wah, "File placement on distributed computer systems," *Computer*, pp. 23–32, Jan. 1984.

[9] M. R. Garey and D. S. Johnson, *Computers and Intractability – A Guide to the Theory of NP-Completeness*. New York: Freeman, 1979.

[10] M. R. Anderberg, *Cluster Analysis for Applications*. New York: Academic Press, 1973.

[11] S. K. C. D. Gelatt and M. P. Vechi, "Optimization by simulated annealing," *Science*, vol. 220, May 1983.

[12] F. Glover, "Tabu search methods in artificial intelligence and operations research," *ORSA Artificial Intelligence Newsletter*, vol. 1, 1987.

[13] D. J. Kleitman and D. B. West, "Spanning trees with many leaves," *SIAM Journal on Discrete Mathematics*, vol. 4, pp. 99–106, Feb. 1991.

[14] D. L. Burkes and R. K. Treiber, "Design approaches for real-time transaction processing remote site recovery," in *35th IEEE Compcon*, pp. 568–572, 1990.

[15] H. Garcia-Molina and C. A. Polyzois, "Issues in disaster recovery," in *35th IEEE Compcon*, pp. 573–577, 1990.

[16] J. Lyon, "Tandem's remote data facility," in *35th IEEE Compcon*, pp. 562–567, 1990.

[17] J. Gray and A. Reuter, eds., *Transaction Processing: Concepts and Techniques*. San Mateo, California: Morgan Kaufmann, 1993.

[18] A. Bhide and D. Dias, "RAID architectures for OLTP." IBM T.J. Watson Research Center Technical Report, 1992.

BALANCED DISTRIBUTED MEMORY PARALLEL COMPUTERS

F. Cappello, J-L Béchennec, F. Delaplace, C. Germain, J-L Giavitto, V. Néri, D. Etiemble
LRI - UA 410 CNRS, Université Paris Sud,
Bat 490, 91405 Orsay Cedex, France
E-mail : de @ lri.fr

Abstract

Mismatches between on-chip high performance CPU and data access times is the basic reason for the increasing gap between peak and sustained performance in distributed memory parallel computers.

We propose the concept of balanced architectures, based on a network with a dynamic topology and communication patterns determined at compile time. The corresponding processing element is a cacheless CPU, which can achieve a 1 FLOP/clock cycle rate. Network and PE features are presented. An example shows that balanced architectures keep efficiency when scaling.

1. INTRODUCTION

1.1. RISC technology and DMPCs.

The objectives of the "grand challenge" are now well known and it is clear that the distributed memory parallel computers (DMPCs) are candidates for the highly intensive numerical applications. One important feature of DMPCs is the use of standard RISC microprocessors as the building blocks of the processing elements. Although different topologies (fat tree, mesh,...) are used for the interconnection network, any machine of this class dynamically routes the messages between the different PEs. This leads to a significant gap between the processing capability of each PE, and the communicating capability of the interconnection network. Figure 1 gives some insights on this phenomena for three different machines [1]-[3].

The large gap between computation and communication capabilities explains the gap between peak performance, that only relies on the performance of the processing elements, and the sustained performance, that consider the effect of memory hierarchy and interconnection network. The first reason for this gap is that the currently available DMPCs are very new. Next generation of DMPCs to be shipped will probably have interconnection networks with higher bandwidths and smaller latencies. The second one is that the software tools for these machines are still insufficient. Hardware and software improvements should reduce the gap between peak and sustained performance.

However, it is worthy to consider the long term future of the ratio between computation, memorisation and communication. If the gap between computation and communication can only increase (together with the gap

between peak and sustained performance), the effective use of DMPCs could be disappointing compared with the promise. In this case, new architectural approaches would be needed to obtain DMPCs that give scalable sustained instead of scalable peak performance.

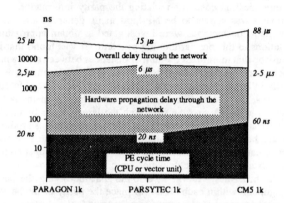

Figure 1 : Time for computation and communication in DMPCs.

1.2. Computation, memorisation and communication : the evolution

The fact that is illustrated in figure 1 comes from the growing rates of computation, memorisation and communication during the last decade. The exponential growth of peak performance, based on technological advances, is mainly concentrated within a single silicon chip, with an internal cache working at the same clock frequency than the CPU core. When the chip must communicate with the external world (for instance with an external secondary cache), everything is slowed down.

When considering the communications between different PEs that use the interconnection network, the situation is still worse. Dynamic routing involves an interpretation level to route messages, which has two major impacts. First, the available network bandwidth is reduced because all messages must embed an address information to the data information. Secondly, the dynamic routing strategy must be distributed over the PEs, leading to a far from optimum conflict resolution. Even if the impact of the interconnection network is application dependant (regularity and locality of data references), it is quite evident that the performance gap between computation and communication can only increase in the future: memory hierarchy and interconnection networks will have more and more impact on the actual performance of DMPCs.

2. NEW BALANCED ARCHITECTURES?

2.1. Floating point operation rates

First use of DMPCs is for highly intensive numerical applications. The FP operation rate is expressed by the formula $FR \ (MFLOPS) = \dfrac{F(MHz)}{CPFO}$ where F is the clock frequency and $CPFO$ is the average number of clock cycles per FP operation. $CPFO$ includes the clock cycles for data accesses, either from local memory or from the network, the duration of FP operations and the clock cycles to execute the integer instructions used for data accesses, sequencing...

The FP operation rate can be rewritten according to (1), where all the terms correspond to the average number of clock cycles to execute an average FP operation. CFO is the clock cycles used in the floating point operators, $NI.CPI$ is the clock cycles to execute the integer instructions. Nma is the number of memory accesses per FP operation and CMA is the average number of cycles per memory access. Nna is the number of network accesses and CNA is the number of clock cycles per network access.

$$FR = \frac{F}{CFO + NI.CPI + Nma.CMA + Nna.CNA} \quad (1)$$

Expression (1) corresponds to the worst possible implementation of a given architecture, as it assumes that there is no overlapping between all the different times.

2.2. Balanced DMPCs.

For floating point computations, the computation time of the FP operators, CFO, is the basic term. The optimal case occurs when data accesses and program sequencing totally overlap with the FP operations to obtain the peak rate. In that case, $NI.CPI + Nma.CMA + Nna.CNA = 0$

$Nna.CNA = 0$ is the most severe constraint. For applications which exhibit spatial locality, the term Nna is small. Generally, the term CNA which corresponds to the latency to get a "remote" data through the network, must be significantly reduced. In section 3, we propose to determine the communication pattern at compile time. This is achieved by using an interconnection network with a dynamic topology, that can be changed at each clock cycle. This scalable network has a low latency and a high bandwidth.

In section 4, we propose an architecture of the PE to realise the condition $NI.CPI + Nma.CMA = 0$. Memory accesses and instruction sequencing totally overlap with FP computation. This is realised with a VLIW format, which specifies up to 3 memory accesses per instruction, the network accesses, and a sequencing field. The computational rate and the data access rate are thus balanced as much as possible.

The network with a static control is intended to minimise $Nna.CNA$ and the PE is designed to get $NI.CPI + Nma.CMA = 0$ to reach one FLOP per clock cycle. A

computer using this approach is called a balanced distributed memory parallel computer (BDMPC).

3. A NETWORK WITH STATIC CONTROL

Using a static routing determined at compile time suppresses all the address overhead that is used for dynamic routing. As data are routed and the conflicts are resolved at compile time, the latency of the network only relies on the hardware resources (switches and wires).

3.1. Compiling communications

Compiling communications is possible only if data references are known at compile time. A first question is: what percentage of data references are known at compile time? If the value is rather high, a communication network with a static control may be considered as equivalent for communications in DMPCs to the RISC philosophy for instructions in sequential processors: efficiently execute the communications known at compile time (the most frequent ones) and execute correctly the communications for data which references are only known at run time. Communications for algorithms using dynamic data references can be implemented by sequencing appropriated communication patterns: in this case, each node acts as a routing unit by performing an appropriate routing program (some optional dedicated hardware may support the routing operations). The communications for dynamic references are more time consuming. But if they are rare, they will not counterbalance the gain obtained on the most frequent static communications.

Data references for numerical applications. Zhiyu et al [4] have already shown that more of 50% of data references within a large set of numerical applications can be known at compile time.

Table 1 [5] gives the percentage of the different types of data references for Linpack. The first column presents the different types of references that we have considered. The second column corresponds to the percentage of each type of reference found in the code. Third and fourth columns correspond to percentages that are found when running the routines for two sizes of data (100 and 1000). Only gather and scatter operations need dynamic references. More than 80% of data references can be known at compile time in Linpack. Moreover, most of the gather and scatter operations that need dynamic references are found in the routines that triangularise matrix by Gauss method. But it is also possible to use a specific "static" algorithm to triangularise matrix without using dynamic references .

The presented results show that we can expect that most of the data references in classical numerical routines can be known at compile time. More benchmarks are needed to confirm these results. However, the problem is also to examine if many dynamic references could be suppressed by rewriting the routines to maximise the static references. Advanced data-flow analysis [6]-[7] together with optimisation techniques can be used to reduce as much as possible the number of dynamic references.

Types of data reference	% in code	% in Linpack 100	% in Linpack 1000
A(i)=f(B(i),...)	16,2	2,1	0,1
A(i)=f(A(i-1,B(i),...)	4,7	30,5	34,6
A(i)=f(A(i),...)	43,9	19,2	16,1
A(i)=f(A(i+1),...)	3,6	10,1	11,5
Exchange	12,3	2,1	0,1
Sum	11	32,5	24,9
Gather	4	0,5	0,1
Scatter	4,3	3	12,6
Total	100	100	100

Table 1 : Data references in Linpack

A communication compiler. Compiling the communication is equivalent to translate the accesses to a data array into communications from PEs to other PEs. Feasibility and performance of such a compiler for a parallel language (as HPF) is the second key point for the static approach. We are currently developing a communication compiler for a subset of HPF [8].

3.2. An interconnection network with high speed serial links.

Compiling communications means using an interconnection network with dynamic topology, which can implement a new pattern for each instruction. The network must have high bandwidth and low latency (5 to 20 machine clock cycles). It should be scalable for hundreds or thousands of PEs.

We use high speed serial links. The required bandwidth to transfer a 64-bit parallel word with a 16.6 MHz clock frequency is equivalent to a high speed serial link at 1.3 Gbit/s with a 4B/5B coding. This is feasible with differential signals and a GaAs technology for emitter and receiver circuits. The building block of the interconnection network is a crossbar switch, which realises all the permutations from inputs to outputs, plus the broadcasts. A new configuration should be realised at each clock cycle (60 ns in the present version). The correct transmission of the 64-bit word is achieved before changing the crossbar permutations. Details on the implementation of the network and the GaAs crossbar switch are presented in [9].

3.3. A network with a dynamic topology

The dynamic topology is implemented with 1/2 Clos Up/Down network. It is built from a standard 3-stage Clos network which is folded because in DMPCs, PEs are connected only on one side of the network. Figure 2 shows the 1/2 Clos Up/Down network for 64 PEs with 4 x

4 switches. Bi-directional channels (switches and wires) allows simultaneous Up and Down transfers. First, data cross the network up to the middle of the second stage through 3 levels of switches and then come down through the first 2 levels. When using 16 x 16 crossbar circuits, the network connects 256 PEs through three levels, 4K PEs through five levels and 64K PEs through seven levels of switches.

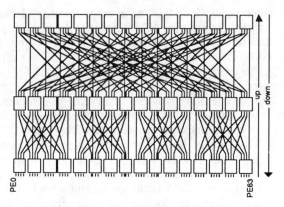

Figure 2: the interconnection network

Each switch has its own private pattern memory which contains all the control patterns. The pattern memory is distributed all over the network. Details on the organisation of the distributed connection pattern memory and of its control are given in [9]-[10].

4. PROCESSING ELEMENT AND STATIC CONTROL

To reach one FLOP per clock, the PE must realise $NI.CPI + Nma.CMA = 0$.

As data accesses must be sequenced at each clock cycle, we cannot use a classical memory hierarchy where data accesses are not deterministic. Moreover, it is totally unrealistic to manage cache coherency at each clock cycle with hundreds or thousands of caches. A processor with a cache is contradictory with a statically controlled communication network.

The total overlap between memory accesses, sequencing and floating point operations is only possible with instructions that specify 3 memory operands and a pipeline execution. We use the VLIW format. At each clock cycle, there are two read and one write operations in a multi-bank memory. The memory constraints are similar to those of vector architectures: data access efficiency is achieved only if bank accesses are conflict-free. Using only multi-banks main memory at reasonable cost means the use of Drams. The 60-ns clock cycle is determined by the execution time of the FP operators, and the Drams are chosen with compatible cycle times.

The core of the PE is the floating point and integer unit (FPIU). Data paths are 64-bit width. The local memory consists of several banks that are simultaneously accessed to sustain the FPIU rate. Figure 3 shows a simplified block diagram of a PE. For a 1 FLOP/cycle rate, a 5-stage VLIW pipeline is used.

As VLIW processors, the PE uses several dedicated functional units that operate simultaneously. A RISC processor is used to manage operations and sequencing, a memory unit computes the data addresses within the memory banks and a network interface controls the communications according to the communication field of the instructions.

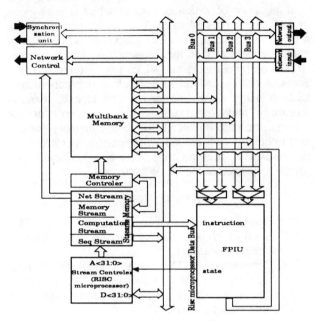

Figure 3: Block diagram of the processing element.

5. MODEL AND PERFORMANCE

5.1. A simple model to compare performance

To compare performance of BDMPCs with the performance of vector supercomputers and those of actual DMPCs, we now define a simple model of the three different architectures, that we use with a benchmark application.

Three classes of architecture. In table 2, we present the basic features of three classes of architecture.

To be coherent with expression (1), the register, memory and network access rates of the different architectures are normalised with the computation rate of floating point operators, considered as 1. Latencies are expressed in processor clock cycles. The registers of vector machines are fed from the memory banks, leading to a memory access rate close to 4/5 of the computational rate. For typical DMPCs, the memory rate is roughly 1/10 of the computational rate (the "load-store" architecture of the RISC processor needs 3 memory accesses for each FP operation, and there are cache misses). The network rate can be estimated to 1/100 of the computational rate. Data for balanced architectures correspond to balanced rates for the three components: computation, memory and network.

		Vector architec.	DMPCs	BDMPCs
Number of processors	N_p	1 - 64	500 - 10000	2500 - 64000
Peak perf. per processor	b_f	200 - 400 MFLOPs	80 -150 MFLOPs	16 MFLOPs
Arithmetic op. latency	l_f	5 - 20	3 - 5	1
Memory latency	l_m	3 -15	10	1
Memory bandwidth	b_m	1	1/10	1
Cache latency	l_c	-	1	-
Cache bandwidth	b_c	-	1/3	-
Network latency	l_n	-	100 - 1000	10
Network bandwidth	b_n	-	1/50 - 1/100	1

Table 2 : Features of vector architectures, DMPCs, and BDMPCs.

A benchmark. It is a simplified version of the parallel implementation of the algorithm of triangularization for a square matrix of size n with N PEs. Gauss method (partial pivot) is used: each PE is associated with a row of the matrix. At each iteration step, one pivot is selected and broadcasted to all PEs; one row is eliminated and a column in the remaining rows is set to zero. At each iteration step, each PE realises two operations on the non-zero terms of its row. The computing intensity of the algorithm is 1, with 2 operations for 2 references. When n $\geq$ N, each PE has a row to compute. The degree of parallelism is N. When the number of rows becomes smaller than N, the degree of parallelism decreases from N to 1. Its average value is 2N/3 [9]

Formulas (2) (3) and (4) intend to approximate the overall FP rate (MFLOPs) of the different architectures for the proposed benchmark, according to the parameters of table 2. (2) assumes only data accesses in the local memory. (3) assumes a data transfer using the interconnection network. (4) consider the cache effect in DMPCs, that gives a third level of locality. For the three architectures, we assume that the accesses are conflict free and we neglect second order effects, as the register load for vector FU.

$$FR_{mem} = \Pi(n,N).E\,(n,N)\,N.p\,[n/(l_m+l_f+n\frac{1}{b_m})] \qquad (2)$$

$$FR_{network} = \Pi(n,N).E(n,N)N.p\,[n/(l_n+l_m+l_f+n\frac{1}{b_n}] \qquad (3)$$

$$FR_{cache} = \Pi(n,N).E(n,N)\,N.p\,[n\,/\,(l_c+l_m+l_f+n.\frac{1}{b_c}] \qquad (4)$$

$\Pi(n,N)$ gives the average parallelism of the application. $E(n,N)$ gives the upper bound of the efficiency of the architecture for the specified application, as it evolves during the process of computation. The third term of each formula gives the efficiency of the PE

Figure 4 plots the values of the FR for the three architectures, according to the size of the problem. The values of the different parameters correspond to the average values in table 2 for vector machines and DMPCs. The BDMPC uses 2500 PEs. For vector architectures, only FR_{mem} is significant. For BDMPC, the balanced rates lead roughly to the same curves for FR_{mem} and $FR_{network}$. For DMPCs, there exist three different curves, corresponding to each component. When running the application, the actual curve is a composition of the three curves, that expresses the impact of the three different levels of locality: it has the "cache" curve as an upper bound, and the "network" curve as a lower bound.

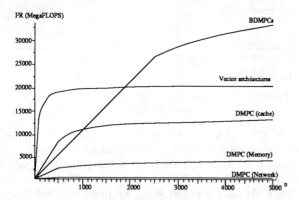

Figure 4 : performance of 3 architectures according to the size of the matrix to triangulate.

5.2. Design of demonstrators

The concept of BDMPCs which is presented in this paper has been experimented within a project called PTAH[1]. First results have already been presented [5], [8]-[11]. The PE instruction set has been defined and validated, and a simulator for a 64-PE machine has been developed. It is currently used to evaluate performance of the architecture on a set of numerical benchmarks. It is also used as the target architecture for a compiler which input language is a subset of HPF.

We are presently developing several hardware and software demonstrators to prove the feasibility and the potential interests of balanced massively parallel architectures, such as a communication compiler and a crossbar switch.

6 CONCLUDING REMARKS

The increasing gap between peak and sustained performance for parallel machines comes first from the gap between the on-chip computational power and the data access constraints (memory hierarchy and

interconnection network). This tendency can be considered one-way, because it relies on fundamental physical and technological features.

We have presented the concept of balanced massively parallel architectures, based on balanced computational, memory access and network access rates. The goal is the scalability of DMPCs, that keeps a reasonable gap between peak and sustained performance. BDMPCs use an interconnection network with dynamic topology, the configurations of which can be switched at each clock cycle. The PEs uses a VLIW-like architecture, suitable for a 1 FLOP/cycle rate. The clock frequency is one order of magnitude slower than the frequency of the most recent RISC microprocessor, but VLIWs with three memory accesses per cycle lead to very close instruction rates without using cache memories. The lack of memory hierarchy makes possible a static control of the communications. Balanced distributed memory parallel architectures, which communication patterns are determined at compile time, can be considered as a parallel extension of the VLIW concept.

REFERENCES

[1] "Paragon XPS Product Overview", Intel Corp., 1991

[2] "Beyond the Supercomputer Parsytec GC", Parsytec Anwendugen GmbH, 1991

[3] R. Ponnusamy, R. Thakur, A. Choudhary, G. Fox, "Scheduling Regular and Irregular Communication Patterns on the CM-5", Proc. Supercomputing'92, p. 394, Minneapolis, Nov 1992.

[4] S. Zhiyu, L. Zhiyuan, Y. Pen-Chung, "An empirical study on array subscripts and data dependencies", Proc. ICPP, 1989.

[5] E. Daugeras, F. Ismond, J-L Béchennec, F. Cappello, "Static Computation of Standard Linear Algebra Subroutines for PTAH", 1993 Euromicro Workshop on Parallel and Distributed Processing".

[6] P. Feautrier, "Dataflow Analysis of Array and Scalar References", International Journal of Parallel Programming, 20(1), February 1991.

[7] A. Lichnewsky, F. Thomasset, "Introducing Symbolic Problem Solving in the Dependence Testing phases of a Vectorizer", 2th Int. Conf. on Supercomputing, St-Malo France, July 1988.

[8] F. Delaplace, F. Cappello, "Data Layouts Impacts on Communications Compilation for Synchronous MSIMD Machine", Proc. Euromicro 92.

[9] V. Néri, J-L Béchennec, F. Cappello, D. Etiemble Hardware features of the static communication network of a parallel architecture, Proc. Euromicro 93.

[10] F. Cappello, J-L Béchennec, F. Delaplace, C. Germain, J-L Giavitto, V. Néri, D. Etiemble, Balanced Distributed Memory Parallel Computers, LRI report.

[11] F. Cappello, J.-L. Béchennec, J.-L. Giavitto, "Introduction to a new parallel architecture for highly numeric processing", Proc. Parallel Architectures and Languages Europe, June 1992, Paris.

[1]This work is partially supported by the french national research program on New Computer Architectures (PRC-ANM) and by DRET under grant #91.168.

A NOVEL APPROACH TO THE DESIGN OF SCALABLE SHARED-MEMORY MULTIPROCESSORS[a]

Honda Shing

Unisys Corporation, MS 10-20
2700 N 1st Street
San Jose, CA 95134-2028

shing@stardust.convergent.com

Lionel M. Ni

Department of Computer Science
Michigan State University
East Lansing, MI 48824-1027

ni@cps.msu.edu

Abstract – *This paper presents the Conflict-Free Memory (CFM) architecture for designing scalable shared-memory multiprocessors. For a single-level CFM architecture, the architecture improves multiprocessor performance by eliminating memory and interconnection network contention and reducing network latency. A hierarchical CFM architecture is introduced in this paper, which presents a new approach in implementing scalable shared-memory multiprocessors. An invalidation-based write-back cache protocol is proposed for the CFM architecture. This cache coherence protocol preserves the low storage overhead of snoopy cache protocols, while it offers the high scalability of directory-based protocols. Furthermore, with the CFM cache protocol, efficient synchronization operations can be implemented.*

1 INTRODUCTION

Scalable shared-memory multiprocessors have been highly demanded in the area of high performance computing. While some software approaches have been proposed/used to provide a single address space, many machines have a direct hardware support, such as BBN TC-2000 [1], DASH [2], IBM RP3 [3], and NYU Ultracomputer [4]. It is hoped that, by using multiple processors, multiplied performance improvement can be obtained; however, ideal (or linear) speedup can hardly be achieved due to a number of performance degradation problems. Two of the most important performance problems observed on shared-memory multiprocessors are shared resource contention and memory access latency.

Accessing shared memory can cause contention, both in memory modules and in the interconnection network, and decrease system performance. In addition to ordinary shared memory accesses, synchronization among concurrent processes can further decrease the performance. Consider the commonly used synchronization mechanism, lock/unlock. Multiple processors may concurrently and repeatedly access the same lock variable and create intensive memory and interconnection network contention. This results in the "hot spot" problem described in [5]. In order to reduce or tolerate the effect of contention, various approaches have been applied.

This paper describes the *Conflict-Free Memory* (CFM) architecture, which is a scalable shared-memory multiprocessor designed to eliminate both memory and interconnection network contention [6]. The CFM architecture is designed based on the block access nature between main memory and caches. For multiprocessors within a certain scale range, the architecture completely eliminates memory and network contention without increasing network latency and overhead. For large scale systems, a hierarchical integration of the CFM architecture improves system performance by reducing memory and network contention and increasing effective memory bandwidth and resource utilization. Moreover, it reduces the setup time and propagation delay caused by message routing in multiprocessors using *Multistage Interconnection Networks* (MIN), such as the BBN Butterfly, the NYU Ultracomputer, and the IBM RP3.

This paper also proposes an invalidation-based write-back cache protocol for the CFM architecture, which preserves the low storage overhead of snoopy cache protocols, while it offers the high scalability of directory-based protocols. With a recursive extension, this CFM cache protocol can be scaled for large scale CFM systems. Furthermore, the protocol explores high degree of pipelining by supporting the weak consistency model in the CFM architecture. With the CFM cache protocol, efficient synchronization operations can be implemented with less states and lower hardware complexity than other cache protocols that support synchronization.

The following section describes the basic concept of the CFM architecture. Section 3 introduces the invalidation-based write-back cache protocol for the CFM architecture. Section 4 presents a hierarchical extension to the CFM architecture as well as the high scalability of the CFM cache protocol. Section 5 concludes the paper.

2 THE CFM ARCHITECTURE

This section gives a brief description to the CFM architecture. Because of the limited size of this paper, readers are referred to [6] for details of the design.

In a shared-memory multiprocessor with a single memory module, the processors share a single address space. Assume the memory module is composed of a number of memory banks. A conventional interleaved memory module can be viewed as a function, M, mapping from its address space A to the range of data elements D. A read operation in the memory module can be depicted by the function $d = M(a \cdot b)$, where d is the data retrieved, and $a \cdot b$ denotes the memory address being accessed. The components a and b represent the address offset in a memory bank and the bank number, respectively. In case of two or more memory accesses to the same memory bank, there

[a]This research was supported in part by the NSF grant ECS-88-14027

are memory conflicts, even if they are not to the same location.

The CFM scheme is based on a deterministic and highly synchronized block accessing model. As in a conventional interleaved memory module, an address in the CFM consists of its offset in a memory bank and the bank number. The bank number, however, is not part of the input to a memory access. Instead, it is defined by the time slot number in which the data is accessed. A time slot is usually the length of a CPU cycle. A constant number (usually the number of memory banks) of time slots compose a time period. The new memory function is now defined as a mapping from its address–time space AT to the range of data elements D. A read operation in this model can be described by the function $d = M(a \cdot t)$, where d is the content at the address offset a in the memory bank defined by the time slot t at which the data is accessed.

In this paper, each set of memory locations with the same offset in all the memory banks of a memory module is defined as a *block*, whose size is determined by cache line size. Memory accesses in the CFM architecture are in blocks, each of which involves retrieving or storing words with the same offset from all memory banks. To simplify the discussion here, let us assume that the memory bank cycle is the same as the CPU cycle. In a real case, the memory cycle is usually longer than the CPU cycle. The CFM design in such a case is described in [6].

Suppose that there are b memory banks, each block access takes b time slots. It is important to note that a block access can start at any time slot. There is **no delay** required before starting a block access. This is because that a block access does not have to start at the first bank, instead, each access starts at the bank defined by the time slot in which the access request is received by the memory module. The CPU and the memory banks are fully synchronized to ensure that each word of a block is transferred between its memory bank and the corresponding section of a CPU line buffer. Again, the implementation of the memory access mechanism can be referenced in [6].

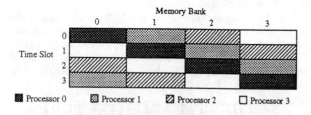

Figure 1. Mutually exclusive subsets in AT–space.

With an appropriate partitioning of the AT–space, four mutually exclusive subsets of the space can be formed and assigned to four different processors in a multiprocessor. Figure 1 demonstrates a partitioning and an assignment of the AT–space. It is shown that, at time slot t, processor p can only access memory bank $((t + p) \bmod 4)$, for $0 \le t, p \le 3$. Each processor occupies an independent subset of the AT–space for a block access, and thus, the memory architecture is guaranteed to be conflict-free.

The above AT–space partitioning and mapping scheme can be implemented with a simple synchronous switch box. Figure 2a shows such a switch box with four input/output ports on each side, connecting four processors and four memory banks. The switch box is similar to an ordinary

crossbar, but much simpler as it requires neither address decoding nor setup delay for routing decisions. Its four routing states, shown in Figures 2*b, c, d* and *e*, are driven by the system clock. At time slot t, input port i is connected to output port $((t + i) \bmod 4)$, for all $0 \le t, i \le 3$. Every four CPU cycles, it completes a time period, which is totally deterministic. This mechanism implements a mutually exclusive partitioning and assignment of the AT-space, which guarantees conflict-free block accesses.

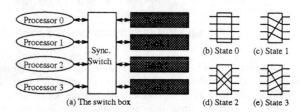

Figure 2. A 4×4 synchronous switch box.

The synchronous switch box shown in Figure 2 serves as the interconnect between processors and memory banks for synchronous block accesses. For a system with a large number of processors, a more sophisticated interconnection scheme is needed. *Multistage Interconnection Network* (MIN) is one of the widely used interconnection schemes in large scale multiprocessors [1, 3]. Like building circuit-switching omega networks with crossbar switches for conventional multiprocessors, synchronous omega networks can be built with synchronous switches for supporting contention-free interconnections, which are basically shift permutation patterns. It has been shown by Lawrie [7] that such mappings can be done with no contention. The entire synchronous omega network can be viewed just as a single synchronous switch with multiplied number of input and output ports. Note that, since all the switches are synchronous, correct connection states for all switches can be set simultaneously for each time slot. There is neither setup time nor propagation delay required for the switches on a path. This is unlike the situations in the BBN Butterfly [1] and the RP3 [3], where setup time and propagation delay are needed for routing and flow control.

Although synchronous omega networks allow more processors and memory modules to be integrated in the CFM architecture, the scalability of the architecture is limited by other factors. With larger numbers of processors and memory banks, the CFM architecture requires a larger block size and introduces a longer block access latency. There are several approaches that can be applied to solve the problems. First, partially synchronous omega networks can be used to construct partially conflict-free memory systems, as presented in [6]. Second, a hierarchical extension to the CFM architecture allows a massively parallel system to be built, which is described in Section 4.

3 THE CFM CACHE PROTOCOL

Cache coherence protocols can basically be divided into two categories, namely, *snoopy cache protocols* and *directory-based cache protocols*. Snoopy cache protocols have the advantages of simplicity, ease of implementation, and lower storage overhead. Their major disadvantage is the strict limitation of scalability. While directory-based cache protocols scale well to larger configurations, they also require more complex hardware and memory overhead. Without broadcast capability, directory-based protocols suffer long communication latency caused by point-to-point invalidation messages and acknowledgements. This section describes how the nature of the CFM architecture allows the CFM cache protocol to accommodate the advantages of both snoopy protocols and directory-based protocols.

3.1 Hardware Configuration

As described in Section 2, the CFM architecture relies on synchronous interconnection networks for data transfer in a pipelined fashion, which, unlike the bus, do not support hardware broadcast. The CFM interconnection, however, has the characteristic that all memory banks are visited in each memory access in a time period. This characteristic allows the cache coherence information of each memory access to be broadcast among memory banks. Cache line state transitions similar to those of snoopy protocols can be determined in each memory bank by the broadcast information. In order for the cache line state information to be usable by processors in maintaining cache coherence, it is desirable for the processors to share cache state information with the memory banks. This is achieved by implementing an additional control connection associating each processor with a memory bank, as shown in Figure 3a.

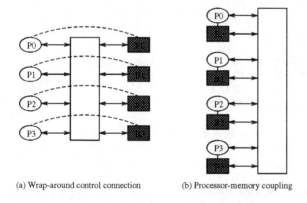

(a) Wrap-around control connection   (b) Processor-memory coupling

Figure 3. Wrap-around control connection and processor-memory coupling.

The CFM architecture can be redrawn as in Figure 3b, which shows each processor associated with a memory bank. This is unlike a message-passing distributed-memory system where each processor has a local memory and remote accesses are accomplished through message-passing. In the CFM architecture, processors access memory banks only through the interconnection network in a synchronous and pipelined fashion. There is no direct data transfer between a processor and its associated memory bank. The control connection between the processor and memory bank in an associated pair simply represents the **sharing of their cache directory**, which contains coherence information. Each shared directory entry contains a *state* field and a *tag*. All the CFM caches are assumed to be direct-mapped throughout this paper, although other approaches can also be used.

3.2 The Protocol and the Primitive Operations

The CFM cache protocol is an invalidation-based protocol with write-back policy. Each cache line can be in one of three states: *invalid*, *valid*, and *dirty*. A valid data block can be shared and can reside in many caches; however, a dirty block cannot exist in more than one cache. This means that the dirty state is exclusive, for there can be at most one dirty copy of a data block in all the caches of a CFM system. Figure 4 shows the state transitions of an invalidation-based write-back protocol.

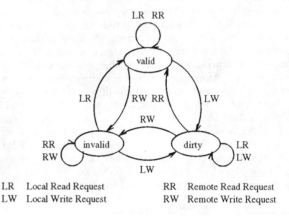

| LR | Local Read Request | RR | Remote Read Request |
| LW | Local Write Request | RW | Remote Write Request |

Figure 4. Invalidation-based write-back protocol.

An invalid cache line becomes valid when a data block is retrieved from memory upon a read request issued by the local processor. As long as the cache line remains valid, subsequent local read requests can be served by the local cache without imposing any memory accesses. A processor issuing a write request must first obtain the exclusive ownership of the target data block. This is accomplished by invalidating all remote copies of the data block and changing the local cache copy of the data block to be in the dirty state. All subsequent write requests issued by the local processor only update the local dirty copy of the data block. The dirty copy of the data block is written back to memory when it is replaced or when an access to the data block is requested by a remote processor. A remote read request causes the dirty copy to become valid after the updated data is written back to memory, while a remote write request causes it to become invalid.

Three primitive operations, *read*, *read-invalidate*, and *write-back*, are used to implement the CFM protocol. Read operations retrieve data from memory to caches. They may cause a data block to be first written back to memory from a remote cache if the block is dirty and owned by the remote cache. Read-invalidate operations are similar to read operations, however, they also obtain exclusive ownership of data blocks by invalidating remote cache copies of the data blocks, if any. Write-back operations flush updated data blocks from caches back to memory.

All the primitive operations follow the block access mechanism described in Section 2. Since all memory banks are visited within each operation, the cache directory of the processor coupled with each memory bank can be checked during the operation to determine actions to be taken. Invalidations imposed by the read-invalidate operation can also be completed in remote caches when each of the memory banks is visited. Unlike snoopy protocols, the state checking and transition do not rely on a broadcast network. The latency of invalidating remote caches is much lower than that of other directory-based protocols, since they are achieved synchronously in a pipelined fashion. No acknowledgement message is required for invalidations. This is unlike other directory-based protocols, such as the DASH protocol [2, 8], where high network overhead is introduced due to point-to-point invalidation messages and required acknowledgements.

Race conditions may occur in the CFM architecture when there are multiple memory operations proceeding concurrently. Without proper access control, they may result in memory inconsistency problems. Since there can be at most one dirty copy of a data block, which is exclusively owned by a cache, there cannot be more than one processor executing write-back on the same data block. No interference between two write-back operations can ever occur. Read-invalidate operations, however, may conflict with each other when competing for the exclusive ownership of a data block. Furthermore, a read-invalidate on a data block may interfere with a write-back operation updating the same block. In this case, the read-invalidate operation must abort and retry later. Detailed discussion and examples of the autonomous access control among the primitive operations can be found in [9].

3.3 Synchronization Supports

The CFM architecture constructs atomic synchronization operations such as "read-modify-write" with the primitive operations. The implementation of atomic synchronization operations using the CFM cache protocol is quite straightforward. As mentioned, before a processor can update a data block, it must first obtain the exclusive ownership of that block. This is achieved by issuing a read-invalidate operation, which also retrieves the data block from memory if it is not yet in the local cache. By modifying the data block and flushing it back to memory with a write-back operation, an atomic read-modify-write is completed. Remotely triggered write-back of this data block is disabled during the modification phase to prevent premature writeback. The read-modify-write operation is atomic, since no other processor can read or update the data block during the period that it is exclusively owned by the local processor. Atomic operations such as swap, test-and-set, and fetch-and-add are special cases of the atomic read-modify-write operation. The atomic operations can be used to implement higher level synchronization mechanisms such as simple lock/unlock and atomic multiple lock/unlock [9].

4 SCALABILITY

The CFM cache protocol is superior to other directory-based protocols in the sense that it is scalable in a consistent and recursive fashion. This section introduces a hierarchical extension to the CFM architecture, which has multiple levels of caches. Accesses to all levels of caches as well as to memory banks are conflict-free. Based on this hierarchical CFM extension, a recursively-defined write-back cache coherence protocol is designed.

4.1 A Hierarchical CFM Architecture

Section 2 describes how the CFM architecture can be implemented by integrating processors and memory banks with a synchronous interconnection network. Section 3 adds to the CFM architecture control connections

through processor-memory coupling for cache coherence control. This architecture can be defined as a conflict-free processor-memory cluster. A larger scale system can be implemented by integrating several conflict-free clusters as well as some global memory banks using a synchronous interconnection network similar to the simple CFM architecture. The memory banks within a cluster can be viewed as a second-level cache local to that cluster. The second-level caches together with the first-level caches and the global memory banks form a three-level memory hierarchy of the two-level hierarchical CFM architecture. The same concept can be applied recursively for implementing larger hierarchical CFM architectures with more processors and cache levels.

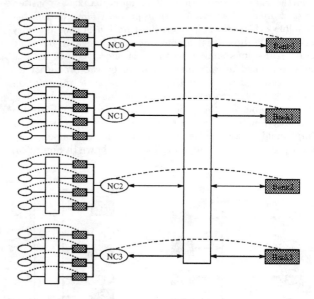

Figure 5. A hierarchical CFM architecture.

Figure 5 shows a two-level CFM architecture with a three-level memory hierarchy. Since the memory banks in a cluster form a second-level cache, for clearer explanation, they are now called *second-level cache banks* or simply *cache banks*. Each block from the cache banks is treated as a second-level cache line, which consists of words with the same address offset in all the cache banks of the cluster. The second-level cache of each cluster has a corresponding cache directory. As in a first-level cache directory, a second-level cache directory entry is composed of a state field and a tag for maintaining cache coherence. Each second-level cache line can also be in one of the three states: invalid, valid, and dirty.

Associated with each conflict-free cluster, there is a network controller. All cache misses in the second-level caches are handled by network controllers. Network controllers operate as pseudo processors accessing the global memory banks through the global synchronous interconnection network, which is similar to that of the simple CFM architecture. A network controller is directly connected to all the cache banks in its associated cluster. It fetches and flushes second-level cache lines from and to the global memory banks by using free time slots of the cache banks or by stealing time slots from the processors in the cluster. The network controller also maintains the second-cache directory for its associated cluster. Like a processor in the simple CFM architecture, each network controller is coupled with a global memory bank by sharing its corresponding second-level cache directory, as represented by the dash lines in Figure 5. The same CFM cache protocol, with little modification, can be applied in this level to maintain cache coherence among the second-level caches and the global memory. By recursively applying the CFM cache protocol, larger scale hierarchical CFM architectures can be constructed with more levels of caches [9].

4.2   Other Issues of the Scalable Cache Protocol

The CFM cache coherence protocol can be applied recursively to hierarchical CFM architectures with more levels of caches. The memory access latency of the worst cache miss situation increases logarithmically with the total number of processors, thus making the scalability of the CFM architecture and cache protocol very attractive. The hierarchical extension approach supports conflict-free accesses in each level of a CFM system. Contention, however, can still occur in a network controller when there are multiple requests from different processors or from its higher-level network controller. By "higher-level", we mean closer to the global memory banks. Each network controller must maintain a queue and serve the requests based on a properly defined priority such that no dead-lock situation can occur. For example, write-back needs to be served first if it is not disabled within a synchronization operation. Also, invalidation requests from the higher-level network controller has higher priority than read-invalidate requests from lower-level network controllers or processors in the associated cluster. This is to ensure that only one exclusive ownership of a data block is granted at any time.

Read operations accessing the same data block can be combined in a network controller. This is especially important in implementing lock/unlock across multiple clusters, since processors in different clusters need to read the same lock variable from memory during lock transfers. Another problem of hierarchical CFM architectures is the interference between a network controller and its associated cluster, since both of them need to check the shared cache directory. One solution to the problem is to duplicate the shared cache directory. The same approach has been used in snoopy cache protocols for reducing processor-cache interference when snooping bus activities. Another solution to the problem is to assign a network controller a free AT-space partition in its associated cluster. In this case, the network controller is in the same position as other processors in the cluster, only it has the special purpose of handling data transfer between the cluster and higher level caches or memory banks and maintaining coherence among different clusters. There are more issues to be investigated concerning the hierarchical extension of the CFM architecture, which are interesting topics for future research.

5   CONCLUSIONS

The CFM architecture improves effective memory bandwidth by eliminating shared memory conflicts as well as interconnection network contention. This paper introduces an invalidation-based write-back cache protocol for the CFM architecture, which preserves the low storage overhead of snoopy cache protocols, while it offers the high scalability of directory-based protocols. Synchronization operations can be efficiently implemented with its three primitive operations: read, read-invalidate, and write. A hierarchical extension to the CFM architecture presented shows that the CFM cache coherence protocol can be recursively defined for large scale CFM systems.

The hierarchical CFM extension is a possible approach to constructing large scale systems. The CFM architecture can also be combined with other approaches in implementing large scale multiprocessors. For example, like the DASH multiprocessor, the CFM architecture can be constructed to connect a number of processor clusters. Within each cluster, processors and memory banks can be connected using a bus. Actually, the CFM architecture can be applied to any levels of a hierarchical multiprocessor and shared-memory structure.

The CFM architecture provides a novel concept to the design of shared-memory multiprocessors. Simulation study is needed for hierarchical CFM architectures in order to evaluate the performance degradation due to possible contention in network controllers.

REFERENCES

[1] BBN Advanced Computers Inc., Cambridge, Massachusetts, *Inside the TC2000 Computer*, 1990.

[2] D. Lenoski, J. Laudon, K. Gharachorloo, A. Gupta, and J. Hennessy, "The directory-based cache coherence protocol for the DASH multiprocessor," in *Proceedings of the 17th Annual International Symposium on Computer Architecture*, pp. 148 – 159, May 1990.

[3] G. F. Pfister *et al.*, "An introduction to the IBM research parallel processor prototype (RP3)," in *Experimental Parallel Computing Architectures* (J. J. Dongarra, ed.), pp. 123 – 140, Elsevier Science Publishers B.V., Amsterdam, 1987.

[4] A. Gottlieb, R. Grishman, C. P. Kruskal, K. P. McAuliffe, L. Rudolph, and M. Snir, "The NYU Ultracomputer — designing an mimd shared-memory parallel computer," in *Proceedings of the 9th Annual International Symposium on Computer Architecture*, pp. 27 – 42, 1982.

[5] G. Pfister and A. Norton, "'Hot spot' contention and combining in multistage interconnect networks," *IEEE Transactions on Computers*, vol. C-34, pp. 943 – 948, Oct. 1985.

[6] H. Shing and L. M. Ni, "A conflict-free memory design for multiprocessors," in *Proceedings of Supercomputing '91 Conference*, pp. 46 – 55, Nov. 1991.

[7] D. H. Lawrie, "Access and alignment of data in an array processor," *IEEE Transactions on Computers*, vol. C-24, pp. 1145 – 1155, Dec. 1975.

[8] J. Lee and U. Ramachandran, "Synchronization with multiprocessor caches," in *Proceedings of the 17th Annual International Symposium on Computer Architecture*, pp. 27 – 37, May 1990.

[9] H. Shing, *A Conflict-Free Memory Design for Multiprocessors*. PhD thesis, Michigan State University, May 1992.

SESSION 4A

GRAPH_THEORETIC INTERCONNECTION STRUCTURES (I)

Incomplete Star Graph : An Economical Fault-tolerant Interconnection Network

C.P. Ravikumar and A. Kuchlous
Electrical Engineering Department
Indian Institute of Technology
New Delhi 110016, INDIA

G. Manimaran
National Informatics Center
Lodhi Road
New Delhi 110003, INDIA

Abstract

A number of existing multiprocessors are based on the hypercube interconnection network. The popularity of the hypercube is due to its small communication diameter, which grows logarithmically with the cube size, its fault-tolerant properties, and its modularity which makes it possible to build a larger cube from smaller subcubes. The *star graph* has been studied as a network topology for fault-tolerant parallel computing. Unfortunately, the size of the network grows too sharply with n to be affordable for values of n larger than 7 or 8. We introduce a novel interconnection network known as the *incomplete star graph*, which overcomes the above problem while retaining the most of the advantages of the star graph. We present the architecture of the incomplete star graph and compare its performance with the full star as well as competing architectures such as the incomplete hypercube and arrangement graphs. We provide routing algorithms for both non-faulty and faulty incomplete-star graphs, and study their performance.

1   Introduction

Interconnection networks play an important role in both SIMD and MIMD parallel computers. The proliferation of massively parallel computers with 4096 processors and more has created a need for networks which have a small communication diameter. The fault-tolerance property of the network is also an important performance metric in massively parallel systems intended for real-time applications. The binary n-cube, or the hypercube, attained popularity in the past decade due to its small communication diameter ($log_2 N$ for a hypercube of N-processors) and its modularity [5]. The hypercube network also offers good fault-tolerance and permits graceful degradation when faults occur. In [8], Akers and Krishnamurthy introduced the *star graph* as an alternative to the hypercube. An n-star graph is a network to interconnect $n!$ processors. The nodes of an n-star are labelled using permutations of $1\ 2\ \cdots\ n$. We shall refer to the label of a node i as $l(i)$. A node i in an n-star is connected to node j if and only if the label of j can be generated by swapping the first symbol of $l(i)$ with any other symbol of $l(i)$. As an example, in a 4-star, the node 1234 is connected to nodes 2134, 3214, and 4231. The properties of the n-star graph are summarized in Table 1. It can be seen from the Table that the n-star is maximally fault-tolerant, i.e., its fault-tolerance index is one less than its degree [2].

The authors of [2] also showed that the *fault-diameter* of the n-star is very close to its communication diameter when the number of faults is tolerable. In particular, when $n > 7$, the n-star outperforms the hypercube with a comparable number of nodes. For instance, a 7-star has 5040 nodes, a communication diameter of 9, a degree 6, an average distance of 5.9, and a fault diameter less than 12. The comparable 12-dimensional hypercube has 4096 processors, communication diameter 12, degree 12, average distance 6, and fault diameter of 13. The main drawback of the star graph is that the number of nodes grows very fast with n, hence making it difficult to expand the network in modest steps. For example, if an n-star must be expanded into an $n + 1$ star, we must add $n! \cdot n$ nodes. In this paper, we are concerned about expanding the n-star by adding a fewer number of processors. We introduce the notion of an *incomplete n-star graph* (ISG) which is a subgraph of the n-star. Figure 1 shows an incomplete star graph of 12 nodes. Naturally, an incomplete star graph does not preserve all the merits of the complete star. In the following sections, we study the performance of the incomplete star graph. In Section 2, we introduce a labelling procedure for incomplete star graphs. We derive properties of the incomplete star, such as the communication diameter, average distance between any two nodes, the number of edges, and the fault tolerance. We present a routing algorithm for the incomplete star in Section 3. We then present a fault-tolerant routing algorithm for *ISG* in Section 4. We have carried out an experimental evaluation of the performance of the incomplete star in order to compare its merits with those of the complete star and the incomplete hypercube; these results are presented in Section 4. We present our conclusions in Section 5.

2 Incomplete Star Graph

An incomplete n-star is a subgraph of the n-star graph. Formally, we define an incomplete n-star as $ISG = (V, E)$, where V is a set of nodes (processors) and E is a set of edges (communication links). The number of nodes in V varies from $(n-1)!+1$ to $n!-1$. The nodes are labelled using permutations on n symbols, similar to a complete star. Note that an ISG is not uniquely defined given the number of nodes m in the ISG. In what follows, we define a canonical incomplete n-star on m nodes. We first introduce some definitions.

Definition 1 r^n_i, $1 \leq i, r \leq n$, *denotes the set of all permutations of* $1\,2\,\cdots\,n$ *with the property that r is fixed in position i.*

For instance, if $n = 4$, then 3^4_4 is the set $\{1243, 2143, 4213, 2413, 4123, 1423\}$. The notation r^n_i is an extension of a similar notation introduced in [7]. The following fact is obvious.

Fact 1 r^n_i *has $(n-1)!$ entries.*

Definition 2 *The generator g_k of a node i is defined as a map from the label of node i to the label of another node j, where $l(j)$ is obtained by swapping the kth element of $l(i)$ with the first element of $l(i)$. E.g., $g_2(12435) = 21435$, and $g_5(12345) = 52341$.*

Definition 3 r^n_i *induces a subgraph in the n-star in a natural way. The subgraph consists of nodes in r^n_i and the edges, if any, among these nodes. For instance, the subgraph induced by 3^4_4 is shown in Figure 2.*

Fact 2 *The subgraph induced by r^n_n is isomorphic to the $n-1$-star.*

Lemma 1 *Let I_n indicate the identity permutation of n symbols $1, 2, 3, \cdots, n$. The node $g_i(g_r(I_n))$ lies on the subgraph induced by r^n_i.*

Proof : Applying g_r on the identity permutation will move the symbol r to the first position. The second generator, g_i, will move the symbol r to the ith position.

■

In this paper, we are especially interested in the subgraphs induced by r^n_n. Based on the Facts 1, 2, and Lemma 1, we now develop a procedure to order the nodes in r^n_n (see Figure 3). In the procedure *PrintSubgraph*, N indicates the starting node in r^n_n; this node is passed as input to the *PrintBFS* procedure of Figure 4, which carries out a breadth first search of the subgraph and prints the nodes in the BFS order. The implementation of the breadth first search procedure is through a queue data structure, and is fairly standard [1].

2.1 Canonical Incomplete Star

In the previous section, we introduced the notion of the canonical incomplete star. We indicate the canonical incomplete n-star graph on m nodes by $CISG(n, m)$. The procedure *BuildCISG* in Figure 5 may be used to print the nodes in $CISG(n, m)$.

Lemma 2 *The incomplete star graph generated by the BuildCISG procedure is a connected graph.*

Proof : Step S1 of the procedure adds the subgraph induced by n^n_n, which is isomorphic to an $n-1$-star (see Fact 2) and is hence a connected graph. The Step S2 adds subgraphs 1^n_n, 2^n_n, and so on, each of which is individually a connected subgraph. We now show that there exist edges between n^n_n and j^n_n, for $j = 1, 2, \cdots, n-1$. The starting node in j^n_n is $g_n(g_j(I))$. The identity permutation I is a node in n^n_n, and hence $g_j(I)$ is also a node in n^n_n. Hence the proof.

■

2.2 Properties of the Incomplete Star

Lemma 3 *There exist at least $m/(n-1)!$ $n-1$-stars in $CISG(n, m)$.*

Proof : Clear from the construction of $CISG(n, m)$.

■

Consider a canonical incomplete star $CISG(n, m)$ where $m = k \cdot (n-1)! + l$. From Lemma 3 above, there are k $n-1$-stars which form subgraphs of $CISG$. In addition, there is an incomplete $n-1$-star on l nodes. We now prove a result regarding this incomplete substar; this result will be useful in obtaining the communication diameter of the $CISG$.

Lemma 4 *In $CISG(n, m)$, it takes no more than $\lfloor \frac{3 \cdot (n-2)}{2} \rfloor$ steps to go from any node of the incomplete substar to the root of the incomplete substar.*

Proof : Recall that the incomplete substar is constructed using breadth-first-search enumeration of node labels (Figure 5). Corresponding to this BFS is a spanning-tree rooted at the starting node of the incomplete substar. The height of the BFS spanning tree cannot be more than diameter of an $n - 1$-star.

■

2.2.1 Communication Diameter

Lemma 5 *$3n - 5$ is an upper bound on the communication diameter d_n of $CISG(n, m)$.*

Proof : Let S be a source node and D be a destination node in $CISG(n, m)$. We can visualize the incomplete star graph in the manner illustrated in Figure 6, where the subgraph $n^n{}_n$ is a "central subgraph" and the subgraphs $1^n{}_n, 2^n{}_n, \cdots$, form satellites around the central subgraph. Since we are developing an upper bound on the communication diameter, we imagine S and D to be nodes on two different satellites and assume that there are no direct links from the satellite of the source to the satellite of the destination. We now show that there exists a path from S to D through the central subgraph. This is clear from Lemma 2, where it was shown that edges exist from $n^n{}_n$ to $j^n{}_n$, for all j. To consider the worst case, we further restrict that S lies on the incomplete sub-star of the $CISG$. From Lemma 4, it is possible to travel from S to a node in the central subgraph in at most $1 + \lfloor \frac{3 \cdot (n-2)}{2} \rfloor$ steps (See Figure 6). We can reach the destination from the central substar in at most $\lfloor \frac{3 \cdot (n-2)}{2} \rfloor$ steps. Hence the proof.

■

2.2.2 Number of edges

Lemma 6 *Let $m = k \cdot (n-1)!$ for $1 \leq k \leq n$. The number of edges in $CISG(n, m)$ is $k \cdot \frac{(n-2) \cdot (n-1)!}{2} + \binom{k}{2} \cdot (n-2)!$.*

Proof : Since m is a multiple of $(n-1)!$, we are focussing our attention on the special class of incomplete star graphs which have k copies of $(n-1)$-stars as building blocks.

Let $T(r)$ be the number of edges in an r-star. When we construct the incomplete star from k copies of the $(n-1)$-star as building blocks, we add $k \cdot T(n-1)$

edges that are internal to the $(n-1)$-stars, and some edges that connect the substars. It is clear that the number of external edges from one substar to another is same for any pair of substars. Let $E(n)$ denote the number of external edges between any two substars. In order to compute $E(n)$, we consider the case when $k = n$, i.e., the case of the complete n-star.

$$T(n) = n \cdot T(n-1) + \binom{n}{2} E(n) \qquad (1)$$

Noting that $T(r) = \frac{(r-1) \cdot r!}{2}$, we have

$$E(n) = (n-2)! \qquad (2)$$

The result now follows.

■

2.2.3 Fault Tolerance

Lemma 7 *The $CISG(n, m)$ is $n - 3$-fault-tolerant if $(n-1)! | m$.*

Proof : Consider first the case where m is a multiple of $(n-1)!$. A node in this network can become isolated in two ways:

1. the external edges fail, disconnecting one sub-star from another, or

2. the internal edges of a substar fail, isolating a node of the substar.

In the former case, we use the result from Lemma 6, namely, that there are $(n-2)!$ external edges connecting one substar to another. Since no two of these edges terminate on the same node in any substar, it will take $(n-2)!$ node failures in one sub-star to isolate it from another substar. Suppose that isolation were to occur due to failure of internal edges. It is easy to see that no node in the $CISG$ has a degree lower than $n - 2$. This is because, each node lies on some $n - 1$-star, whose degree is known to be $n - 2$. Therefore, from [8], the network is k_n-fault-tolerant, where $k_n = (min((n-2)! - 1, n - 3))$. The reader may verify that $k_n = n - 3$.

When m is not an exact multiple of $(n-1)!$, then there may exist a node whose degree is 1, giving a 0-fault-tolerant network. Lemma 7 indicates that the most useful class of incomplete n-star graphs are those whose node count is a multiple of $(n-1)!$. We shall denote these graphs as **Special** Incomplete Star Graphs (SISG). To a good degree, this class of graphs retains the desirable properties of the complete star. Thus, when a network configured as a complete n-star or an incomplete star must be expanded, it is best to do so by adding $(n-1)!$ nodes at a time. This represents a modest increase in node count, compared to the $n \cdot n!$ nodes which would be required to extend the n-star into an $n+1$-star.

3 Routing in an ISG

In this section, we discuss a routing algorithm for the special incomplete star graph. We provide an algorithm that can route from a source node S to the special node which is labelled by the identity permutation. It is easy to see that the same algorithm extends easily for the case where the label of destination D is not the identity permutation (see [8]). The algorithm is a obtained by slightly modifying the routing algorithm for complete stars proposed by Akers and Krishnamurthy [8]. See Figure 7.

Theorem 1 *The algorithm Route finds a path in $SISG(n,m)$ from the source S to the identity permutation I.*

Proof : It is straightforward to see that the algorithm *Route* finally reaches the identity permutation, since the algorithm puts symbols in their right places until the identity permutation is obtained. More importantly, we show that the algorithm never generates an intermediate node T that does not exist in the $SISG$. (A node S exists in SISG if and only if $S[n] \in \{n, 1, 2, \cdots, k-1\}$, where $m = (n-1)! \cdot k$). In order to see this, we show that neither step R1 nor R2 generates a nonexistent node label. First consider the step R1, which moves the symbol '1'. We claim that $j < n$ in step R1. For, if $j = n$, then we get the contradiction that $S[n] \neq n$ and $S[i] = i \forall i < n$. Therefore, $T[n] = S[n]$, ensuring that T exists in the $SISG$ provided S does.

Step $R2$ will necessarily generate a T such that either $T[n] = n$, or $T[n] = S[n]$. In either case, the node T is a valid node in the $SISG$ under consideration.

∎

Theorem 2 *The algorithm Route generates a path of optimal length.*

Proof : Akers and Krishnamurthy gave an optimal routing procedure for the complete star [8]. It is easy to see that our algorithm generates the same path as the routing procedure in [8] for any source S. Since this is the shortest path in the complete n-star, it must be the shortest path in $SISG$ as well, for $SISG(n,m)$ is a subgraph of the n-star.

∎

Lemma 8 *The diameter of $SISG(n,m)$ is $\lfloor \frac{3(n-1)}{2} \rfloor$.*

Proof : For any given source S, a length of the optimal path is identical in both $SISG(n,m)$ and the complete n-star (Theorem 3). The communication diameter of the complete n-star is $\lfloor \frac{3(n-1)}{2} \rfloor$. The result follows.

∎

3.1 Average distance in SISG

Let D_n denote the average distance in a complete n-star. In this section, we derive the average distance of $SISG(n,m)$. Let $k = m/(n-1)!$ denote the number of complete $n-1$-substars in the $SISG$. The average distance $A(n,m)$ of the $SISG$ is given by

$$A(n,m) = \frac{I(n,m) \cdot id(n,m) + E(n,m) \cdot ed(n,m)}{I(n,m) + E(n,m)} \quad (3)$$

where $I(n,m)$ is the number of point-to-point connections within any of the substars ("internal connections"), $E(n,m)$ is the number of point-to-point connections from one substar to another ("external connections"), $id(n,m)$ is the average length of an internal connection, and $ed(n,m)$ is the average length of an external connection. Clearly,

$$I(n,m) = k \cdot \binom{(n-1)!}{2} \quad (4)$$

$$id(n,m) = D_{n-1} \quad (5)$$

$$E(n,m) = \binom{k}{2} \cdot (n-1)!^2 \quad (6)$$

$$I(n,m) + E(n,m) = \binom{k \cdot (n-1)!}{2} \quad (7)$$

Since the complete n-star is a special case of the $SISG$ with $m = n!$, we apply Equation 3 for this special case to derive $ed(n,m)$. The reader may verify that

$$ed(n,m) = \frac{(n!-1) \cdot D_n - ((n-1)!-1) \cdot D_{n-1}}{(n-1) \cdot (n-1)!} \quad (8)$$

The above results can now be used in Equation 3 to get an expression for $A(n,m)$. The reader may verify that $A(n,m)$ is a monotonically increasing function of k.

4 Fault-Tolerant Routing

In this section, we describe how a fault-tolerant routing algorithm can be developed for the incomplete star graph. Our algorithm has the following properties. **(1)** It always finds a path from the source to destination, if a path exists. **(2)** It routes from an arbitrary source S to the Identity permutation. **(3)** It is a distributed routing algorithm; each node has local knowledge about the status (faulty or non-faulty) of each of its neighbors. This information is stored in each node S as a bit vector $lstat$ of length n. $lstat[i]$ is set to 1 if (and only if) the ith neighbor of S, namely, $g_i(S)$, is non-faulty. Note that, an incomplete star can be modelled as a faulty star by suitably storing the link statuses. **(4)** Our algorithm finds a path of optimum length in absence of faults.

The fault-tolerant routing algorithm, which we describe only briefly due to space limitation, is essentially a modification of the routing algorithm of the previous section. Each node S attempts to forward the message to the next node in the path as found by algorithm *Route*. If the next node is faulty, as indicated by $lstat$, the node S tries to route the message through an alternate neighbor which is not faulty. If no such alternate neighbor exists, or if all the neighbors lead to "dead ends", the node S detects that it is a dead end by itself and returns the message to the previous node in the path.

The above fault-tolerant routing algorithm was simulated on the Xenix operating system. We assumed that the non-faulty nodes generate messages according to the Poisson distribution with parameter λ. The parameter λ at any node i is directly proportional to the number of non-faulty neighbors of node i. The destination for a message is chosen uniformly among all non-faulty nodes. The results of simulation are shown in Figure 8. In this figure, we show the average distance of a faulty star graph; we chose the faults to simulate $SISG$ networks. For comparison, we also show the analytically obtained average distance; it can be seen that our routing algorithm is able to achieve this average.

5 Conclusions

In this paper, we have described an interconnection network known as Canonical Incomplete Star Graph. We studied the properties of $CISG(n, m)$ and showed that when $m = k \cdot (n-1)!$ for $1 \leq k \leq n$, the resulting class of graphs have many desirable properties. For this family of graphs, which we call $SISG$, we derived the various performance parameters such as the aver-

Number of nodes	$n!$
Degree	$n-1$
Commn. Diameter	$\lfloor \frac{3 \cdot (n-1)}{2} \rfloor$
Fault Tolerance	$n-2$
Average Distance	$n + \frac{2}{n} + H_n - 4$

Table 1: Properties of the Complete n-Star. H_n is the nth Harmonic number.

Max. Deg.	n-1
Min. Deg.	n-2
Edges	$\frac{k \cdot (n-2) \cdot (n-1)!}{2} + \binom{k}{2} \cdot (n-2)!$
Fault Tol.	$n-3$
Commn. Dia.	$\lfloor \frac{3(n-1)}{2} \rfloor$
Avg. Dist.	$< n + \frac{2}{n-1} + H_{n-1} - 3$

Table 2: Properties of $SISG(n, m)$. $k = \frac{m}{(n-1)!}$.

age distance, communication diameter, and fault tolerance. We developed an efficient routing algorithm for $SISG$. We also showed how this routing algorithm can be extended to work in presence of faults. The properties of $SISG$ are summarized in Table 5 below.

In comparison to the complete star graphs, $SISG$ have a distinct advantage in that the number of nodes in an $SISG$ need not be the factorial of a number. For instance, a 5-dimensional $SISG$ can have 24, 48, 72, 96, or 120 nodes. This allows modest increases in the network size, at the same time maintaining the good properties of the complete star.

Since the star graph was introduced as an alternative to the hypercube, it seems logical to compare the incomplete star with an incomplete hypercube. Incomplete hypercubes were studied by Katseff [6], and Chen and Tzeng [3] who introduced the concept of an Enhanced Incomplete Hypercube to improve the performance of the hypercube. An incomplete hypercube $I^n{}_k$ is formed by a adding a k-dimensional hypercube to an n-dimensional hypercube, $0 < k < n$. The nodes of the two hypercubes are labelled using the first $2^n + 2^k$ binary strings on $n+1$ bits. The authors assumed that each node has $n+1$ ports and enhanced the incomplete hypercube by adding additional edges to improve its communication diameter and fault tolerance. Unlike the incomplete hypercube of [3], the $SISG$ can grow in more than one incremental step. For instance, [3] do not discuss how more nodes could be added to $I^n{}_k$. On the other hand, an $SISG$ of 6 nodes can be gracefully upgraded to 12, 24, 48, 72, 96,

120, 240, $\cdots$ nodes. In Figure 9, we plot the number of edges, the fault tolerance, the communication diameter, and the average distance for the incomplete hypercube and the $SISG$ for varying number of nodes. The dark curves correspond to the incomplete hypercube, and the dotted curves correspond to the $SISG$.

More recently, Day and Tripathi introduced a class of graphs known as *arrangement graphs*, which are a generalization of star graphs [4]. Unlike star graphs, where the nodes are labelled using permutations of n symbols, the nodes of an arrangement graph $A(n, k)$ are labelled using k-arrangements of n symbols. There are $\frac{n!}{(n-k)!}$ nodes in $A(n, k)$, as compared to $n!$ in a complete star. Arrangement graphs were also conceived as a solution to the problem of fast growth in node count in star graphs. However, as k drops from n to 0 in steps of 1, the growth in the node count of $A(n, k)$ is highly nonlinear. For instance, when n is 7, the node count changes as 1, 7, 42, 210, 840, 2520, 5040, 5040 when k goes from 7 to 0. The $SISG$ are superior to arrangement graphs in this aspect.

We recognize that $CISG$ are more generalized than $SISG$, even permitting the addition of a single node to an existing network. This is desirable in a distributed computing environment. However, the general class of $CISG$ do not possess the good properties of star graphs. We are presently working on overcoming this problem through enhancements to $CISG$.

References

[1] A. V. Aho, J. E. Hopcroft, and J. D. Ullman. *The Design and Analysis of Computer Algorithms.* Addison Wesley, Reading, MA, 1974.

[2] S.B. Akers and B. Krishnamurthy. A group-theoretic model for symmetric interconnection networks. *IEEE Transactions on Computers,* 38:555–566, 1989.

[3] H.-L. Chen and N.-F. Tzeng. Enhanced incomplete hypercubes. In *Proceedings of the ICPP,* volume I, pages 270–277, 1989.

[4] K. Day and A. Tripathi. Arrangement graphs – a class of generalized star graphs. Technical Report TR 91-42, Department of Computer Science, University of Minnesota, September 1991.

[5] K. Hwang and P.S. Tseng. An efficient VLSI Multiprocessor for Signal/Image processing. In *Proceedings of International Conference on Computer Design,* pages 172–176, October 1985.

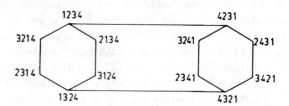

Figure 1: An Incomplete Star on 12 nodes

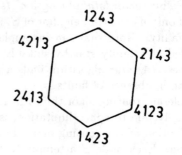

Figure 2: The subgraph of the 4-star induced by $3^4{}_4$.

[6] H. P. Katseff. Incomplete hypercubes. *IEEE Trans. on Computers,* C-37(5):604–608, 1988.

[7] K. Kim and V.K. Prasanna Kumar. An iterative sparse linear system solver on star graphs. In *Proceedings of the ICPP,* volume III, pages 9–16, 1991.

[8] D. Harel S.B. Akers and B. Krishnamurthy. The Star Graph : An attractive alternative to the n-cube. In *Proceedings of the International Conference on Parallel Processing,* pages 393–400, 1987.

```
procedure PrintSubgraph(r, n)
begin
      I := "1 2 3 ··· n";
      N := g_n(g_r(I));
      insert N in the queue q;
      visited[N] := true;
      call PrintBFS(q, n, (n − 1)!);
end
```

Figure 3: Enumerating the nodes in $r^n{}_n$.

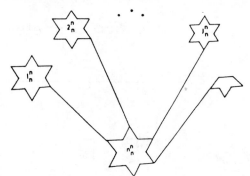

```
procedure  PrintBFS(q, n, count)
begin
        if q is empty return;
        N := delete(q);
        print(N);
        for i := 2 to n − 1
            if  count > 0) then begin
                V := g_i(N);
                if  visited[V] continue;
                visited[V] := true;
                add V in front of queue q;
                count := count − 1;
            endif
            call PrintBFS(q, n, count);
end
```

Figure 4: Breadth First Search of the Induced Subgraph. *count* denotes the number of nodes of the induced subgraph which we desire to print.

```
procedure  BuildCISG(n, m);
begin
S1:     PrintSubgraph(n, n);
(* Include the subgraph n^n_n *)
        d := m div (n − 1)!;
(* How many more n − 1-stars to be included? *)
        l := m mod (n − 1)! − 1;
(* Number of Left over nodes *)
S2:     for  i := 1 to d
            PrintSubgraph(i, n);
(* Include 1^n_n, 2^n_n ⋯ d^n_n *)
        I := "1 2 3 ⋯ n";
(* Include the first l nodes from d + 1^n_n *)
        N := g_n(g_{d+1}(I));
        insert N in the queue q;
        visited[N] := true;
        call PrintBFS(q, n, l);
end;
```

Figure 5: Printing the nodes in the Canonical ISG

Figure 6: Figure for Lemma 5.

```
procedure  Route(S, n, m);
(* Routing in CISG(n, m) *)
Inputs : S is the label of the source node
begin
        if S = '1 2 3 ⋯ n' then return;
        case S[1] of
        1 : (* First symbol is 1 *)
            for j = 2 to n if S[j] ≠ j break;
R1:         T = Swap(S[j], S[1]);
        (* Swap 1 with the first symbol
        not in its right place *)
        else : (* First Symbol is not 1 *)
R2:             T = Swap(S[S[1]], S[1]);
        (* Put the first symbol
        in its right place *)
        endcase
        call Route(T,n,m);
end
```

Figure 7: Routing Algorithm for the Special Incomplete Star

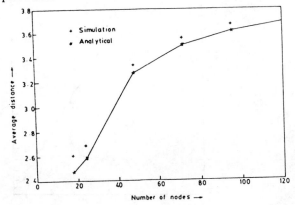

Figure 8: Average Distance in SISG. '+' : Simulation, 'o' : Analytical.

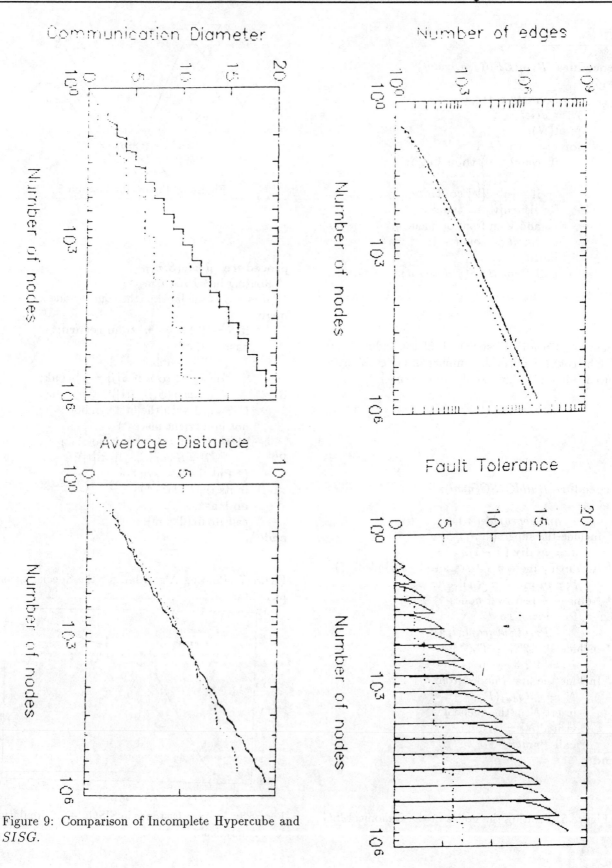

Figure 9: Comparison of Incomplete Hypercube and *SISG*.

THE STAR CONNECTED CYCLES: A FIXED-DEGREE NETWORK FOR PARALLEL PROCESSING

Shahram Latifi
Department of Electrical and Computer Engineering
University of Nevada, Las Vegas
Las Vegas, NV 89154-4026
latifi@jb.ee.unlv.edu

Marcelo Moraes de Azevedo * *and Nader Bagherzadeh*
Department of Electrical and Computer Engineering
University of California, Irvine - Irvine, CA 92717
mazevedo@balboa.eng.uci.edu
nader@balboa.eng.uci.edu

Abstract: *This paper introduces a new interconnection network for massively parallel systems referred to as star connected cycles (SCC) graph. The SCC presents a fixed degree structure that results in several advantages over variable degree graphs like the star graph and the n-cube.*

The description of the SCC graph given in this paper includes issues such as labeling of nodes, degree, diameter, symmetry, fault tolerance and Cayley graph representation. The paper also presents an optimal routing algorithm for the SCC and a comparison with other interconnection networks. Our results indicate that for even n, an n-SCC and a CCC of similar sizes have about the same diameter.

1   Introduction

Over the past years, many interesting graphs such as the n-star and the n-cube have been proposed as interconnection networks for parallel processing applications. Some important properties shown by these graphs are node and edge symmetry, hierarchical structure, maximal fault tolerance and strong resilience [1]. However, the n-star is superior to the n-cube in several areas.

Most graphs studied so far offer a high processor density while keeping the diameter as low as possible. Nevertheless, graphs such as the n-star and the n-cube present a variable degree structure and have low scaleability from the viewpoint of network growth. More specifically, since both the degree of the n-star and the n-cube is $O(n)$, a growing number of communication links is required as n increases. Hence, one disadvantage of variable degree interconnection networks is the large number of I/O communication ports required at each processor in massively parallel systems.

Variable degree interconnection networks also present more complex physical lay-outs and require additional communication ports at each processor to be expanded. In other words, if we want to increase the number of nodes of an existing variable degree parallel system, it might be necessary to substitute all processors in the system, unless unused communication ports are available at each node.

To overcome these difficulties, we propose a new type of interconnection network: the *star connected cycles (SCC)* graph. The SCC offers a fixed-degree structure and can be viewed as an evolution of its counterpart, the cube connected cycles or CCC [2]. The SCC and CCC graphs are formed by connecting cycles or rings of nodes through a particular network communication topology. The underlying topology used to connect the cycles in an n-SCC graph is an n-star, while that of the n-CCC graph is the n-cube. As expected, this results in a fixed-degree interconnection network that is superior to the CCC in several areas.

2   Description of the SCC

The SCC interconnection network is based on the well known star graph [1]. An n-*SCC* graph is obtained by substituting each node of an n-star with a ring of $(n-1)$ nodes. Each ring may be viewed as a *supernode* that can be implemented with a cluster of individual processors or with a single multiprocessor VLSI device. A supernode in the n-SCC graph is connected to $(n-1)$ adjacent supernodes, using lateral links according to the topology of the n-star graph. The nodes inside each ring are identified by a pair of labels (I_j, P_i), where:

- P_i is a permutation obtained using the generators of the n-star graph [3]. We consider that the nodes in an n-star are labeled with a permutation of the digits $\{1, 2, \ldots, n\}$, which allows the labeling of $n!$ different nodes in the n-star. Therefore, each permutation P_i of the n-star labels $(n-1)$ nodes belonging to the corresponding supernode in the n-SCC.

- I_j is a single digit that identifies each particular node inside a ring. The labeling method proposed for SCC consists of assigning to each I_j a label in the range $\{2, 3, \ldots, n\}$, such that I_j corresponds to the label of the lateral link used to connect each node within a ring to other rings in the n-SCC graph. The label of the lateral link is chosen so as to represent the position of the digit in the permutation P_i that is swapped with the first digit of P_i when the lateral link is traversed.

As an example, consider the 4-SCC graph shown in Figure 1. Node 1234 of a 4-star graph is substituted for 3 nodes, labeled respectively as (2,1234), (3,1234) and (4,1234). These nodes are connected to other rings using lateral links 2, 3 and 4 (e.g. node (2,1234) is connected to node (2,2134) via lateral link 2).

<u>Number of nodes:</u> An n-SCC graph can be seen as an n-star graph connecting $n!$ supernodes. Since each supernode contains $(n-1)$ nodes, the total number of nodes in an n-SCC graph is $N = (n-1)n!$.

*This research is supported in part by CNPq – Conselho Nacional de Desenvolvimento Científico e Tecnológico (Brazil), under the grant No. 200392/92-1.

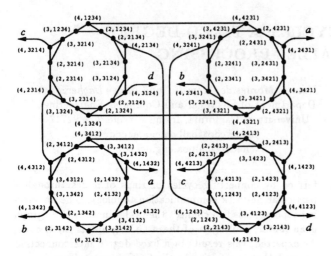

Figure 1: *4*-SCC graph

Degree: The degree of the n-SCC graph is shown in Table 1. Since the degree of every node in the graph is the same, the n-SCC graph is regular. For $n > 3$, every node in a n-SCC graph connects to exactly 3 adjacent nodes using two local links within the same supernode and one lateral link to a node belonging to an adjacent supernode.

A fixed and low degree reduces the communication costs at each node. If each node is implemented as an individual processor, we can use a standard building block with 3 communication links to build any n-SCC graph.

Table 1: Degree and fault tolerance of the n-SCC

Property	$n=2$	$n=3$	$n>3$
Degree (δ)	1	2	3
Fault tolerance (f)	0	1	2

Fault tolerance: A graph is *f-fault-tolerant* if it remains connected when any set of f or fewer nodes are removed from the graph [1]. The *fault tolerance* of a graph corresponds to the largest f for which the graph is f-fault-tolerant. The fault tolerance of a graph with degree δ can be at most equal to $(\delta - 1)$, since if we remove all the neighbors of a node the graph will be disconnected. A graph whose fault tolerance is exactly $(\delta - 1)$ is said to be *maximally fault-tolerant.*

The fault tolerance of the n-SCC graph is shown in Table 1. Clearly, the n-SCC graph is maximally fault-tolerant.

Symmetry: The SCC graph is node symmetric. This is true both from the viewpoint of a single node and from the viewpoint of a supernode. So, given any two nodes a and b there is an automorphism of the graph that maps a to b.

Since not all SCC edges look the same, the SCC graph is not edge symmetric. This implies that the communication load is not uniformly distributed over all communication links. However, if we consider only the lateral links and view the n-SCC as an n-star of supernodes, the edge symmetry properties of the n-star still hold.

Thus, every lateral link in an n-SCC graph is edge symmetric with any other lateral link in the graph. The local links within a ring or supernode are also transitive with any other local link in the graph.

Lemma 1: The n-SCC is not a Cayley graph.

Proof: The generators of a Cayley graph labeled with n digits can either generate S_n or a subgroup of S_n [1], where S_n is the set of all $n!$ possible permutations that can be created with n digits. Each node in an n-SCC graph is actually labeled with $(n+1)$ digits, which means that the labels of an n-SCC graph actually belong to S_{n+1}.

S_{n+1} has $(n+1)!$ different permutations and the n-SCC graph uses only $(n-1)n!$ permutations of S_{n+1}. The relation between these numbers is equal to $(n+1)/(n-1)$.

According to LaGrange's theorem on finite groups [1], the order of any subgroup always divides the order of the group. That's not the case with the ratio above. □

Although the n-SCC graph is not a Cayley graph, its node symmetry property allows that any n-SCC graph can be represented as the quotient of two Cayley graphs [3]. More specifically, we may obtain the quotient graph of an n-SCC graph by identifying subgraphs in the n-SCC and reducing such subgraphs to nodes. The nodes in the resulting quotient graph are connected iff there existed an edge between elements of the corresponding subgraphs. In the case of the n-SCC, each subgraph corresponds to a ring of nodes. Reducing each subgraph to a node results in a Cayley quotient graph: the n-star.

3 Routing in the n-SCC

Routing in the n-SCC is an extension of the routing in the n-star graph and can be seen as two different problems: routing in the lateral links and routing in the local links.

3.1 Routing in the lateral links

Routing in the lateral links uses the same routing techniques already developed for the n-star [1]. Suppose that we want to route from P_s to P_d in an n-star. To find the lateral links connecting P_s to P_d, we can instead find the path from P_{ds} to the identity permutation [3], where $P_{ds} = P_d^{-1} P_s$.

Algorithm 1 (Routing in the n-star):

1. If the first digit in the P_{ds} permutation is 1, move it to any position not occupied by the correct digit.

2. If x (i.e. any digit other than 1) is first, move it to its position.

We may organize the digits of permutation P_{ds} as a set of cycles – i.e. cyclically ordered sets of digits with the property that each digit's desired position is that occupied by the next digit in the set. A permutation $P_{ds} = 26543187$ belonging to an 8-star graph, for instance, consists of the following cycles: (2 6 1), (5 3), (8 7), (4). Note that any digit already in its correct position appears as a *1*-cycle.

Let $C = (i_1 \; i_2 \; \ldots \; i_k)$ be a cycle of length $k \leq n$ in P_{ds}, where $1 \leq i_1 \leq i_2 \leq \ldots \leq i_k \leq n$. The *execution of cycle C* corresponds to a path R in the n-star and can be expressed as a sequence of lateral links as follows [4]:

$$R = (i_2, i_3, \ldots, i_k) \qquad , \; if \; i_1 = 1$$
$$R = (i_1, i_2, \ldots, i_{k-1}, i_k, i_1) \quad , \; if \; i_1 \neq 1$$

Note that in an n-star there are[1] $N_c = c!$ different choices that can be used for an optimal order of execution of cycles of length at least 2 in P_{ds}. If the number of digits in cycle C_i is K_i, $K_i \geq 2$, then there are also N_i different ways to minimally execute cycle C_i, where:

$$N_i = \left\{ \begin{array}{ll} K_i & , \; if \; C_i \; does \; not \; include \; digit \; 1 \\ 1 & , \; if \; C_i \; includes \; the \; digit \; 1 \end{array} \right.$$

If the cycles C_i for which $K_i \geq 2$ are C_1, C_2, ..., C_c then the total number of optimal routing paths in the n-star from P_s to P_d is:

$$T_p = c! \prod_{i=1}^{i=c} N_i$$

3.2 Routing in the local links

Assume that the nodes belonging to the same ring are labeled with $I = 2, \ldots, n$, going counterclockwise. Also, suppose that the lateral link entering the supernode is I_i and the lateral link leaving the supernode is I_j. If D_n is the minimum number of local links between I_i and I_j and $D' = |I_j - I_i|$, then:

$$D_n = min\left(D', n-1-D'\right) \qquad (1)$$

Algorithm 2 (Routing in the local links):

1. Evaluate $L = I_j - I_i$ and $R = |I_j - I_i| \, div \, \lfloor \frac{n+1}{2} \rfloor$, where div is the integer division operator.

2. If $(L > 0$ and $R = 0)$ or $(L < 0$ and $R = 1)$, then take the ring counterclockwise traversing D_n local links; else take the ring clockwise traversing D_n local links.

3.3 Routing algorithm for the n-SCC

We now present an algorithm for routing in the n-SCC graph. Such algorithm is actually a combination of Algorithms 1 and 2 and provides a sequence of lateral and local links as a result.

We recall that Algorithm 1 allows for T_p different optimal paths in an n-star graph. The edges of the n-star graph are the lateral links of the corresponding n-SCC graph. However, not all of the T_p different optimal paths that exist

---

[1] Note that this equation is valid even if the first digit in P_{ds} is not 1. Although Algorithm 1 indicates that the cycle including digit 1 should be executed first, we may actually choose any order to execute the cycles, one at a time. This is possible because the execution of any cycle leaves the position of digits that do not belong to that cycle unchanged.

---

in the n-star result in optimal paths in the n-SCC, since the order of execution of the lateral links affects the number of local links in the routing.

The routing algorithm for the n-SCC graph performs a depth-first search on a weighted tree structure. The algorithm builds the tree by expanding at each step those cycle orderings that seem to result in a minimal number of local links. Backtracking is also performed to enable expansion of previous cycle orderings that seem to be equivalent or better than the most recently expanded orderings.

In the description of the routing algorithm for the n-SCC, we denote the source node by (I_s, P_s) and the destination node by (I_d, P_d). We also define the following:

- S_c is the set of cycles of length at least 2 in P_{ds}.

- S_d is a subset of the digits of P_{ds}, such that:

  - If $(1 \; i_2 \; i_3 \; \ldots \; i_k)$ is a cycle of S_c, then $i_2 \in S_d$ and $1, i_3, \ldots, i_k \notin S_d$.
  - If $(i_1 \; i_2 \; \ldots \; i_k)$ is a cycle of S_c that does not include digit 1, then $i_1, i_2, \ldots, i_k \in S_d$.

The tree structure generated by the routing algorithm has the following characteristics:

- If the number of cycles in S_c is c, then the first level of the tree is 0 and the deepest level is $(c+1)$.

- The label of any vertex in the tree is a pair of digits (f, ℓ), belonging either to S_d or to $\{I_s, I_d\}$. Each vertex located between levels 1 and c in the tree represents one of the cycles of S_c. The label (f, ℓ) is chosen so as to represent the first (f) and last (ℓ) lateral link used during the execution of the cycle.

  If the cycle represented by the vertex is $(1 \; i_2 \; i_3 \; \ldots \; i_k)$, then the vertex is labeled $(f, \ell) = (i_2, i_k)$. If the cycle represented by the vertex does not include digit 1, then the vertex may be labeled as $(f, \ell) = (i_i, i_i)$, where i_i is any of the digits of the cycle.

- The weight of an edge connecting any two vertices (f_i, ℓ_i) and (f_j, ℓ_j) corresponds to the number of local links required to move from ℓ_i to f_j within the same supernode and is given by Equation (1).

- Each vertex (f, ℓ) has an associated data structure consisting of its distance to the root (D_r) and a reduced set of digits $S_d^c = S_d^p - S_i$. The distance D_r is obtained by summing the weights of all edges in the path from the root to the vertex. S_d^p is the set of digits stored in the parent of each vertex and S_i is the set of digits belonging to the cycle that includes digit f.

- The root vertex is $(f, \ell) = (I_s, I_s)$ and has $D_r = 0$ and $S_d^c = S_d$. The vertices located at level $(c+1)$ in the tree are labeled $(f, \ell) = (I_d, I_d)$ and have $S_d^c = \{\}$. The vertices located at level c in the tree have $S_d^c = \{I_d\}$.

- Each vertex also stores an enable/disable bit that informs whether the tree should continue to be expanded from that vertex or not. The root vertex is created with an enabled bit, but all other child vertices are created with a disabled bit. Vertices that have already been expanded also have a disabled bit.

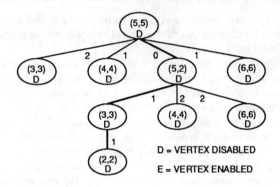

Figure 2: Example of routing tree in the n-SCC

Given the definitions above, the routing algorithm for the n-SCC is as follows :

Algorithm 3 (Routing algorithm for the n-SCC):

1. If $P_s = P_d$, then route inside the ring using Algorithm 2 and exit.

2. If $P_s \neq P_d$, then calculate permutation P_{ds} such that $P_{ds} = P_d^{-1} P_s$. Also identify the cycles of length at least 2 that exist in P_{ds} and create the sets S_c and S_d.

3. Create an enabled root vertex labeled (I_s, I_s) such that $S_d^c = S_d$.

4. Generate child vertices for all enabled vertices, such that the label f for each child corresponds to exactly one of the digits stored in the set S_d^p of each parent vertex. The label ℓ for each child vertex is chosen according to the definitions of the tree structure given earlier in this section.

5. Evaluate D_r and S_d^c for all child vertices. Check if any recently generated child vertex has a distance $D_n = 0$ to its parent. If such vertex exists, enable it. Otherwise, enable all recently generated child vertices that have a distance $D_n = 1$ to their parents. If $D_n > 1$ for all recently generated child vertices, then it is necessary to perform a backtracking search in the tree. Such search enables all vertices that have the smallest *virtual distance* (D_v) to the end of the tree, where $D_v = D_r + c + 1 - h$ and h is the level at which the vertex is located in the tree ($0 \leq h \leq c + 1$).

6. If all enabled vertices are at level $c+1$ of the tree, then an optimal order of execution for the cycles in S_c has already been found. Otherwise, return to Step 5.

7. The optimal order of execution for the cycles in S_c is given by the intermediate vertices existing between the root and any enabled vertex at level $c + 1$ in the tree. This optimal order of execution is actually a sequence of lateral links. Additional routing is required to move between lateral links, but that can be easily done with Algorithm 2.

As an example, consider the routing in an 7-SCC for the case $I_s = 5$, $I_d = 2$ and $P_{ds} = (1\ 5\ 2)(3\ 6\ 4)(7)$. Figure 3 shows the tree built by the routing algorithm. An optimal order of execution for P_{ds} is therefore 5 - 2 - 3 - 6 - 4 - 3.

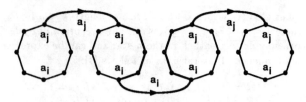

Figure 3: Execution of a cycle $(a_i\ a_j)$ in the n-SCC

4 Diameter

The diameter of the n-SCC graph can be calculated by finding its antipode permutations. An antipode is the farthest node from a given node along the shortest path [4]. We recall that in the n-SCC graph there are $(n-1)$ nodes (I_i, P_1), where I_i is a digit of the set $\{\ 2,\ 3,\ 4,\ \ldots,\ n\ \}$ and P_1 is the identity permutation $123\ldots n$. We must therefore choose one of these $(n-1)$ nodes as the identity node for the n-SCC graph. Let such node be $(I_1, P_1) = (2, 123\ldots n)$. Now, we define an antipode (I_a, P_a) in the n-SCC graph as follows:

- P_a is a permutation chosen such that the node (I_a, P_a) is located $\lfloor 3(n-1)/2 \rfloor$ lateral links away from (I_1, P_1), while keeping the number of local links in the path to the identity node to a maximum.

- I_a is a digit chosen so as to include a maximum number of additional local links in the path to (I_1, P_1).

Lemma 2: The following nodes are antipodes in the n-SCC graph:

For odd n: $(2, (1)(2\ a+1)(3\ a+2)\ldots(a-1\ n-1)(a\ n))$
For even n: $(2, (1\ b+1)(2\ b+2)\ldots(b-1\ n-1)(b\ n))$

where $a = (n+1)/2$ and $b = n/2$.

Reference [5] gives a proof of correctness for these antipodes and lists other possible antipodes.

Theorem 1: The diameter of the n-SCC, $n \geq 2$, is :

$$d_{SCC} = \begin{cases} 2\left(\left\lfloor \frac{n-1}{2} \right\rfloor\right)^2 + \left\lfloor \frac{3(n-1)}{2} \right\rfloor + 2\left\lfloor \frac{n}{2} \right\rfloor - 2 & ,\ n \neq 3 \\ 6 & ,\ n = 3 \end{cases}$$

Proof: Routing from any of the above antipodes requires $d_{lat} = \lfloor 3(n-1)/2 \rfloor$ lateral links. The number of local links can be calculated as follows:

- Permutation P_a has $\lfloor (n-1)/2 \rfloor$ cycles that does not include the digit 1. Execution of each of these cycles requires $2\lfloor (n-1)/2 \rfloor$ local links, as can be seen in Figure 4. Thus, the total number of local links required for execution of all cycles in P_a that do not include digit 1 is $d_{loc}(1) = 2\left(\lfloor (n-1)/2 \rfloor\right)^2$.

- Permutation P_a has $\lfloor n/2 \rfloor$ cycles of length 2 that must be executed in the route to the identity node. The cycles in P_a may be ordered such that only one local link is required to move between the execution of adjacent cycles. This adds $d_{loc}(2) = \lfloor n/2 \rfloor - 1$ local links to the routing from the antipode to the identity.

- Digit I_a in the antipode is such that at most $d_{loc}(3) = \lfloor n/2 \rfloor - 1$ local links may be added to the routing.

- The diameter of the n-SCC graph is the sum of d_{lat}, $d_{loc}(1)$, $d_{loc}(2)$ and $d_{loc}(3)$, so the result follows. $\square$

5   Comparison to other graphs

A comparison between different interconnection network graphs is shown in Table 2. Due to its fixed-degree structure, processors with just 3 communication links can be used to build any n-SCC graph. Clearly, the n-SCC graph has higher scaleability than variable degree graphs like the n-cube and the n-star. Such graphs require a growing number of communication links at each processor as we increase the number of nodes. The result is increased complexity and higher pin count at each processor than required by fixed-degree graphs.

Table 2: Comparison of interconnection networks

Graph	n	Size	Degree	Diameter
n-cube	7	128	7	7
	8	256	8	8
	9	512	9	9
n-star	5	120	4	6
	6	720	5	7
	7	5040	6	9
n-CCC	4	64	3	8
	5	160	3	10
	6	384	3	13
	8	2048	3	18
	9	4608	3	20
n-SCC	4	72	3	8
	5	480	3	16
	6	3600	3	19

One of the trade-offs of fixed-degree graphs is an increased diameter. However, the n-SCC can be built with very high speed buses in the local links. The resulting communication delays may be comparable to the n-star, if we consider that the lateral links often use serial transmission for making their lay-out simpler.

Table 2 also shows another type of fixed-degree interconnection network, namely the cube connected cycles or CCC. An n-CCC graph can be built by substituting each node of an n-cube with a ring of n or more nodes. Table 2 shows typical values for n-CCC graphs containing n nodes in each ring. The number of nodes and diameter of an n-CCC graph formed under such structure are given respectively by $N = n2^n$ and $d_{CCC} = 2n + \lfloor n/2 \rfloor - 2$ [6].

Compared to a CCC graph of similar size, the n-SCC graph presents about the same diameter for the cases where n is even. The diameter of the n-SCC graph shows a sharp discontinuity when n changes from an even to an odd value. Such behavior is due to the presence of the quadratic component in the n-SCC diameter expression.

The diameter of the CCC compares favorably with the n-SCC graph for odd n. However, the underlying topology or quotient Cayley graph used to connect the cycles in the n-SCC (i.e., the n-star) has several advantages over that used in the CCC (i.e., the n-cube) [1]. Among these advantages, we may cite a smaller degree from the viewpoint of the supernodes, as well as a shorter average distance and fault diameter. More specifically, an n-SCC graph requires fewer lateral links and fewer nodes at each ring than a CCC graph with similar number of nodes. Such characteristic reduces the complexity of the supernodes and makes their implementation simpler.

6   Conclusion

The SCC is a fixed-degree graph and has been proposed as an evolution of the previous cube connected cycles or CCC. Aspects such as labeling of nodes, degree, diameter, symmetry, fault tolerance and Cayley graph representation have been presented. We have also developed an optimal routing algorithm for the n-SCC.

We have compared the n-SCC with variable degree graphs such as the star graph and the n-cube. We claim that the n-SCC is an attractive alternative for parallel systems, since it overcomes some disadvantages of variable degree graphs such as the increased requirement of communication ports at each node and the reduced scaleability from the viewpoint of network growth.

We have also compared the n-SCC with another fixed-degree graph, namely the CCC graph. We have shown that the diameter of the n-SCC is close to that of a CCC graph containing a similar number of nodes whenever n is even. However, even for the cases where n is odd, the n-SCC shows some superiority over the CCC, due to the use of the n-star as its quotient graph.

References

[1] S. B. Akers, D. Horel and B. Krishnamurthy, "The Star graph: An Attractive Alternative to the n-cube," *Int'l Conf. on Parallel Processing*, 1987, pp. 393-400.

[2] F. P. Preparata and J. Vuillemin, "The Cube-Connected Cycles: A Versatile Network for Parallel Computation," *Communications of the ACM*, Vol. 24, No. 5, May 1981, pp. 300-309.

[3] S. B. Akers and B. Krishnamurthy, "A Group-Theoretic Model for Symmetric Interconnection Networks," *IEEE Transactions on Computers*, Vol. 38, No. 4, April 1989, pp. 555-566.

[4] S. Latifi, "Parallel Dimension Permutations on Star Graph," *1993 Working Conf. on Architectures and Compilation Techniques for Fine and Medium Grain Parallelism*, January 1993.

[5] S. Latifi, M.M. Azevedo and N. Bagherzadeh, *The Star Connected Cycles: a Fixed Degree Interconnection Network for Massively Parallel Systems*, Dept. of Electrical and Computer Engineering, Univ. of Calif., Irvine, Technical Report ECE 93-02, March 1993.

[6] D.S. Meliksetian and C.Y.R. Chen, "Communication Aspects of the Cube-Connected Cycles," *Int'l. Conf. Parallel Processing*, Vol. 1, 1990, pp. 579-580.

EMPIRICAL EVALUATION OF INCOMPLETE HYPERCUBE SYSTEMS

Nian-Feng Tzeng

Center for Advanced Computer Studies
University of Southwestern Louisiana
Lafayette, LA 70504

Abstract —— *The incomplete hypercube provides far better incremental flexibility than the complete hypercube, whose size is restricted to exactly a power of 2. In this paper, the performance of incomplete hypercube systems is evaluated empirically. The simulation results reveal that mean latency for delivering messages is roughly the same in an incomplete hypercube as in a compatible complete hypercube, unless the message generation rate is extremely high (≥ 0.9). It is also found that mean latency for messages traversing links with heavy traffic can be appreciably larger than the mean latency of overall messages. When the link communication capability is doubled, mean message latency becomes virtually uniform no matter whether or not the message traverses a link with heavy traffic, confirming that the incomplete hypercube can be made congestion-free easily to guarantee high performance under any traffic load.*

1. Introduction

It is desirable that an interconnection scheme allows any sized construction, offering maximum incremental flexibility. A flexible version of the hypercube topology, called the *incomplete hypercube*, makes it possible to construct parallel machines with arbitrary sizes. Simple and deadlock-free algorithms for routing and broadcasting messages in the incomplete hypercube system have been developed by Katseff [1]. The structural properties of a special class of incomplete hypercube systems have been analyzed in [2]. More recently, traffic density over links in an incomplete hypercube with arbitrary nodes under the uniform message distribution has been analyzed and is found to be bounded by 2, regardless of its structural non-homogeneity [5]. However, this analysis does not consider queueing delay or resource contention; it basically investigates the static performance behavior of incomplete hypercube systems.

In this paper, the performance of incomplete hypercube systems is evaluated empirically to get a better insight into this flexible architecture, when the effect of contention at ports/links is taken into account. Peak traffic

density over links is also recorded to verify the correctness of our prior analysis. Mean latency is chosen as the performance measure. Simulation results indicate that an incomplete hypercube, despite its structural non-homogeneity, can deliver messages to their destinations efficiently, creating no congestion points and ensuring good performance always.

2. Preliminaries

2.1. Notations and Background

An n-dimensional complete hypercube, denoted by H_n, comprises 2^n nodes, each with n links connected directly to n nearest neighbors. Nodes in H_n are numbered from 0 to $2^n - 1$, by n-bit binary numbers $(x_{n-1} \cdots x_i \cdots x_0)$ as their *addresses*. A node and a nearest neighbor have exactly one bit differing in their addresses. The incomplete hypercube under consideration is composed of multiple complete cubes. Figure 1 shows an incomplete hypercube comprising three complete cubes H_4, H_3, and H_1. In general, an n-dimensional incomplete hypercube with M nodes, I_n^M, can be defined recursively as follows: I_n^M consists of two components, H_{n-1} and $I_k^{M-2^{n-1}}$ ($k < n$), with a link existing between a node A in H_{n-1} and another node B in $I_k^{M-2^{n-1}}$, if and only if the addresses of A and B differ in bit $n-1$. I_n^M is characterized by a bit vector $V_n^M = <1, x_{n-2}, x_{n-3}, \cdots, x_i, \cdots, x_1, x_0>$ such that x_i equals 1 (or 0) if H_i is present (or absent) in I_n^M. Clearly, the bit vector is the binary representation of M. The incomplete hypercube given in Fig. 1, for example, is characterized by bit vector $V_5^{26} = <1\ 1\ 0\ 1\ 0>$. The *relative address* of two nodes is the bitwise Exclusive-OR of their addresses. A link is said to have *link number* i if it connects two nodes whose addresses differ in the ith bit position, denoted by $\lambda|_i$.

Routing messages from one node to another in the incomplete hypercubes is done as follows [1]: Every issued message carries the relative address of its source node and its destination node as the routing tag. A message is sent over link i only if bit i is the least significant non-zero bit in the tag *and* link i exists. The routing tag is checked leftwards, starting with the least significant bit.

This work was supported in part by the NSF under Grant MIP-9201308 and by the State of Louisiana under Contract LEQSF(1992-94)-RD-A-32.

2.2. Configuration Properties and Analytic Results

Certain nodes in the incomplete hypercube are particular and are referred to as *pivot* nodes. A set of pivot nodes is generally defined in the following. For constituent cubes H_i and H_j, $i > j$, of I_n^M, the *pivot nodes* associated with the two cubes involve the collection of nodes $P_{i,j}$ = {node x | node x in H_i *and* node x has a direct link connected to a node in H_j}. From our construction of incomplete hypercubes, it is clear that the direct link between a node in $P_{i,j}$ and another node in H_j is $\lambda|_i$. Suppose H_i, H_j, and H_k are among the constituent cubes of I_n^M, with $i > j, k$. The two sets of pivot nodes associated respectively with H_i and H_j and with H_i and H_k, $P_{i,j}$ and $P_{i,k}$ satisfy $P_{i,j} \cap P_{i,k} = \varnothing$. Based on this result and the recursive definition of incomplete hypercubes, we arrive at the next observation immediately.

Observation: A node in the constituent cube H_i of I_n^M has the following links present: (i) $\lambda|_x$ for all $x < i$, (ii) $\lambda|_y$ for any $y > i$ such that the yth bit in bit vector V_n^M is "1", and (iii) $\lambda|_i$ only if the node belongs to any pivot set $P_{i,j}$.

Generally speaking, a node in H_i is connected to a node in a "higher" constituent cube H_h, $h > i$, by $\lambda|_h$, whereas it is connected to a node, if any, in a "lower" constituent cube by $\lambda|_i$. For two pivot sets $P_{i,j}$ and $P_{i,k}$ in constituent cube H_i, $i > j > k$, every node in $P_{i,k}$ is connected to a corresponding node in $P_{i,j}$ by using link $\lambda|_j$.

Let $\Phi(< i)$ be the set of all constituent cubes H_j, $j < i$, together with all the intercube links among these cubes inside I_n^M. Notice that $\Phi(< i)$ itself forms an incomplete hypercube. We have an interesting configuration property of the incomplete hypercube as follows: Let Ψ_i be the collection of all pivot nodes associated with H_j, for all $j < i$, plus all the connections among these nodes inside constituent cube H_i, then $\Phi(< i) \equiv \Psi_i$. This property implies that the structure of $\Phi(< i)$ is reflected by Ψ_i inside H_i, and it is essential for the derivation of traffic density. In Fig. 1, for example, $\Phi(< 4)$ has the same structure as Ψ_4 inside H_4, illustrated by bold lines.

Traffic density over links in I_n^M is not fixed and is location-dependent. The highest traffic density is of our major concern, since it tends to dictate the longest time taken to traverse a link, and thus the worst message transmission scenario. The highest traffic density in I_n^M is derived by evaluating traffic density over an arbitrary link and determining its maximum possible value, as presented in [5], where a proof of the next theorem can be found.

Theorem: Traffic density on any link in incomplete hypercube I_n^M under the uniform message distribution is bounded by 2, for all $n > 1$.

3. Empirical Evaluation

3.1. Simulation Model

Simulation studies have been performed to comparatively evaluate the communication behaviors of complete and incomplete hypercubes operating in the packet-switched mode. Every node in a simulated system consists of a processing element (PE) and a hardware router. One of those router links is connected to the PE, and the remaining links are to its immediate neighboring nodes. Messages generated at the PE are sent over the specific link to the router, from which they are delivered through appropriate links to neighboring nodes. Likewise, messages destined for the node, once they arrive at the router, are forwarded over this link to the PE.

A separate buffer is associated with each link. The message at the head of a buffer is transmitted in one cycle to the node connected at the other end of the link and, then, directed to an appropriate buffer chosen following the routing procedure. If multiple messages compete for a link in one cycle, they are all stored in the associated buffer, with the message at the buffer head proceeding over the link in that cycle and the rest being forwarded in sequence during subsequent cycles. A node may send or forward messages simultaneously over all incident links. A PE can receive one message directed to it from the router in a cycle; if multiple messages bound for the same PE arrive in one cycle, a random one is selected to advance to the PE and the rest stay in their original buffers (as only one link exists between a local PE and a router). This is referred to as the *single-accepting strategy* in [4].

In our simulation, messages are generated randomly and independently by all the nodes, and the destinations of messages are evenly distributed across all the nodes (except the originating nodes). A node generates a message with probability r in a cycle, called the message generation rate, and messages are issued by nodes at the beginning of each cycle. The message service discipline is first-come-first-served, and in one cycle, the head message in a buffer advances to the connected neighboring node, whenever possible.

The measure of interest in this experimental study is mean latency. Mean latency (L) is the number of cycles spent by a typical message from its source to its destination, taking the queueing delay into consideration. One additional cycle is included in mean latency to account for the time of transferring messages to the received PEs. When a link experiences heavier traffic, its associated buffers in both sides involve more messages on the average, so that a message traveling through that link tends to encounter higher latency. It is assumed that the buffer associated with a link has an infinite capacity so that no message is dropped and the system is message-preserving, making it possible to gauge message latency contributed by link contention. If traffic becomes excessive, the mean queue length could increase rapidly, giving rise to serious performance degradation. For a given system, mean latency generally grows with the increase of r. When r is low, latency is contributed mainly by the number of hops a typical message makes, because little queueing delay (contention) is involved. As r grows, more contention and longer queueing delay result, yielding higher latency.

3.2. Numerical Results under Uniform Traffic

Mean latency versus message generation rate for various sized systems under uniform message distributions is depicted in Fig. 2. In every incomplete hypercube, L is not increased significantly until r exceeds roughly 0.8, implying that no serious contention exists in any system with low to moderately high traffic. Under extremely heavily loaded situations ($r \geq 0.9$), incomplete hypercubes I_{11}^{1048} and I_{11}^{1114}, like their compatible complete hypercube H_{10}, experience large increases in L. Likewise, L of incomplete hypercube I_{11}^{1818} also grows fast for $r \geq 0.9$. The system can deliver good performance unless the message generation rate is extremely high (≥ 0.9).

The simulated peak traffic density data are also gathered, and they are 1.00, 2.00, 1.96, and 1.32 for H_{10}, I_{11}^{1048}, I_{11}^{1114}, and I_{11}^{1818}, respectively. They are all bounded by 2, as predicted. The degree of contention in a system can be indexed by peak traffic density, and L of a system under the heavily loaded situations is dictated largely by the contention degree. Therefore, L of I_{11}^{1048} should be the largest when r is high. This indeed is the case for $r \geq 0.8$, as illustrated in the figure.

Messages which originate from or terminate at nodes in the "smallest" constituent cube of an incomplete hypercube (e.g., H_3 of I_{11}^{1048}) tend to pass through intercube links with heavy traffic and exhibit a worse case scenario. The mean latency of those messages (called "marked" messages) is recorded, as shown by dashed curves in Fig. 2. It can be seen from the figure that a "marked" message experiences about the same mean latency as any typical message in an incomplete hypercube until r is around 0.7, and thereafter the mean latency of "marked" messages grows considerably faster.

Since traffic density in the incomplete hypercube is bounded by 2, potential traffic congestion would be totally removed if the link capacity is doubled, i.e., during one cycle, up to two messages are allowed to pass through a given link. To confirm this conjecture, simulation runs are conducted to obtain the mean latency amounts of the same set of systems as before, and the results are plotted in Fig. 3. L is never noticeably larger in an incomplete

hypercube than in its complete counterpart for any r. The difference in L between a "marked" message and a typical message in an incomplete hypercube is never pronounced for any message generation rate. Contention virtually disappears if the link capacity is twice as much as the maximum message generation rate. This fact makes the incomplete hypercube a practical and interesting structure.

3.3. Numerical Results under Nonuniform Traffic

There are many forms of nonuniform message distributions, reflecting different types of reference locality. In this study, we focused on a type of nonuniform message distributions called decreasing probability reference [3], which intuitively captures the notion that the probability of sending messages to a node decreases as its distance from the source node increases. This is often expected, as mapping a distributed computation onto a hypercube system desires more frequent message exchanges with physically closer nodes. During a cycle, at most one message may pass through a given link.

Simulation results for the three incomplete hypercube systems given in Fig. 4 are obtained under the reference distribution that the probabilities of message destinations to be 1 hop, 2 hops, 3 hops, and 4 hops away when generated, are 0.34, 0.20, 0.16, and 0.10, respectively, with the remaining probability of 0.20 having messages uniformly destined for all nodes (other than the originating node). Under this reference distribution, the rate for a message to terminate at an immediate neighbor is more than 7 times higher than that at a node 2 hops away, which in turn is about 3.3 times higher than that at a node 3 hops away, which is approximately 2.8 times higher than that at a node 4 hops away. The probability of addressing a node decreases as the distance grows. It is found from Fig. 4 that for every system, L grows slowly as r increases till 0.8, and it then begins to changes swiftly thereafter. An incomplete hypercube is shown to be able to handle messages fluently under this nonuniform traffic, unless generation rate is excessively high (> 0.9). When compared with a corresponding curve in Fig. 2, a curve in Fig. 4 has a lower L for any given $r < 0.9$, as expected, because of reference locality. According to our simulation, the peak traffic density over a link under this decreasing probability reference is less than 1.18.

4. Concluding Remarks

We have evaluated the traffic behaviors of any sized incomplete hypercube under various message distributions empirically. The simulation study indicates that mean latency for transmitting messages is roughly the same in an incomplete hypercube as in a compatible complete hypercube, unless the message generation rate is extremely high (i.e., ≥ 0.9). Our simulation also verifies that link traffic density is upper bounded by 2, independent of the system size, as predicted by our analysis [5].

As the incomplete hypercube is an attractive structure, it is worth pursuing basic communication primitives for such a system. Data distribution and code partition for a given job would be different in an incomplete hypercube, when compared with those in its complete counterpart. The difference results from the desire for taking advantage of an incomplete hypercube in order to execute the job efficiently. Appropriate initial data distribution and code partition as well as optimal task migration (to ensure load balancing and efficient execution) for incomplete hypercubes appear interesting to investigate.

References

[1] H. P. Katseff, "Incomplete Hypercubes," *IEEE Trans. on Computers,* vol. 37, pp. 604-608, May 1988.

[2] N.-F. Tzeng, "Structural Properties of Incomplete Hypercube Computers," *Proc. 10th Int'l Conf. Distributed Computing Systems*, 1990, pp. 262-269.

[3] D. A. Reed and D. C. Grunwald, "The Performance of Multicomputer Interconnection Networks," *IEEE Computer,* vol. 20, pp. 63-73, June 1987.

[4] S. Abraham and K. Padmanabhan, "Performance of the Direct Binary *n*-Cube Network for Multiprocessors," *IEEE Trans. on Computers,* vol. 38, pp. 1000-1011, July 1989.

[5] N.-F. Tzeng and H. Kumar, "Analyzing Link Traffic in Incomplete Hypercubes," Tech. Rep. *TR 92-3*, CACS, Univ. of Southwestern Louisiana, 1992.

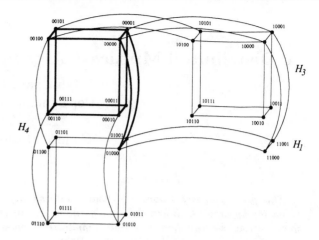

Fig. 1. An incomplete hypercube with 26 nodes, I_5^{26}. (Some connections are omitted intentionally. ψ_3 is shown bold.)

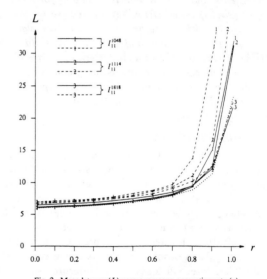

Fig. 2. Mean latency (L) versus message generation rate (r).

(The H_{10} result is shown by the dotted curve. Dashed curves give the results of "marked" messages in incomplete hypercubes.)

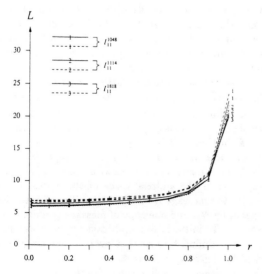

Fig. 3. Mean latency (L) versus message generation rate (r) with the link capacity doubled.

(The H_{10} result is shown by the dotted curve. Dashed curves give the results of "marked" messages in incomplete hypercubes.)

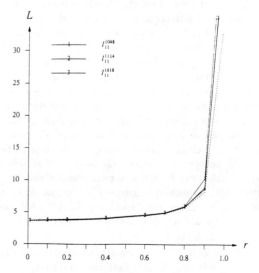

Fig. 4. Mean latency (L) versus message generation rate (r) under decreasing probability reference.

(H_{10} is shown by the dotted curve.)

A Distributed Multicast Algorithm for Hypercube Multicomputers

Jyh-Charn Liu, Hung-Ju Lee

Department of Computer Science

Texas A&M University

College Station, TX 77843–3112

Abstract

This paper proposes a novel distributed multicast algorithm for hypercubes. Based on our algorithm, any node in a multicast-set can directly initiate a multicast to other nodes in the same set. The DImension-PArtition Technique (DIPAT) is the underlying message routing technique. This routing technique is valid only for certain connected graphs, called DIPAT-graphs whose necessary and sufficient conditions are derived. A heuristic algorithm is developed for construction of DIPAT-graphs. In addition, DIPAT is integrated with a simple spanning tree algorithm to form a hierarchical multicasting graph, which retains a higher level of distributed multicasting capability with a similar level of extra traffic as that of the existing single spanning tree based solutions.

1 Introduction

Multicomputer systems are becoming indispensable to solve many computational problems in scientific research and engineering development. To exploit the computing power of a parallel processing system, a computational job is divided into many sub-jobs to be executed in different processors(nodes) concurrently. The processors need to communicate with each other to maintain the consistency of the sub-jobs in order to reach one global computational goal. Therefore, efficiency of the interprocessor communication is critical to performance of the multicomputer systems.

The set of processors that need to receive a multicast message is collectively called a *multicast set*. A multicast set is considered as *open*, if any processor in the system can send a multicast message to members of the multicast set. Otherwise, in a *closed* multicast set, only members in the same multicast set can exchange messages. Although multicast can be implemented by one-to-one (*unicast*) communication mechanisms, this approach has been shown [1] to have much poorer performance than directly supported multicast algorithms. In the conventional single spanning tree based multicasting algorithms, a message to be multicast in an open multicast sets must be first sent to the root, which thus becomes a performance and reliability bottleneck. Moreover, unfairness is a major drawback for the single spanning tree based algorithm to be used in a closed multicast set, because of the different distances of nodes to reach the root.

To overcome the deficiencies of single spanning tree based algorithms, we can distribute the routing information to the nodes in a multicast set so that any node can directly initiate a multicast activity. That is, in an open- or close-multicast set, any node can send an external or internal multicast message to other members, eliminating the performance and reliability bottlenecks. In this way, symmetric, fair multicast in a multicast set becomes possible, making it easy to integrate the communicating activity for coordination, synchronization, or resource access of concurrent processes.

Although multicasting is closely related to broadcast algorithms [2, 3], we will focus on survey of multicast algorithms for limited space. In a multicast set, the nodes need to be interconnected with each other forming a connected graph. It is well known that the complexity of interconnecting nodes in a multicast set with a minimal number of extra nodes is NP-hard [5]. Therefore, most, if not all, of the existing multicasting algorithms are heuristic in their nature. Frank, Wittie and Bernstein [4] investigated the criteria of evaluating multicasting communication and presented a heuristic multicast algorithm on the crossbar computer networks. The performance of multicasting algorithms is measured by the total number of hops from a source node to the destination nodes and the number of extra nodes to interconnect the source to all destinations. Lan, Esfahanian and Ni [6] proposed a graph theoretical model, called Optimal Multicast Tree (OMT) model, for distributed memory multiprocessors. Based on this model, they proposed a heuristic algorithm with minimal message delivery times for hypercubes. Byrd, Saraiya and Delagi [7] presented techniques to resolve the deadlock problem based on different switching mechanisms. Sheu and Su [8] introduced a centralized multicast algorithm for hypercubes based on the notion of lattice ordering.

In this paper, we propose a flexible distributed multicast algorithm to eliminate the performance and reliability bottlenecks in the existing single spanning tree based multicast algorithms. The proposed multicast algorithm is based on the *DImension PArtition Technique (DIPAT)*. In DIPAT, when a message arrives at a node N_x, the *routing space*, which is the address space of the nodes that need to receive the message from N_x, is partitioned into disjoint subspaces by N_x, and one copy of message is routed to each subspace. The initiator of a multicast activity becomes the root of a spanning tree so that every other nodes can be reached for delivery of the message. In this way, each node in the multicast set will receive exactly one copy of the message. DIPAT is applicable only to certain embedded connected graphs in a hypercube. Therefore, we develop a heuristic algorithm to transform an arbitrary multicast set to a connected graph, called a DIPAT-graph, to ensure the mutual reachability between the nodes based on DI-PAT. Finally, a two layered, scalable multicast hierarchy, which can also be commonly found in local (wide) area networks, is proposed based on the DIPAT and a simple spanning tree algorithm. The proposed scheme is found to be quite effective in reducing extra traffic while retaining distributed multicast capability.

The rest of this paper consists of four sections. Section 2 introduces the notion of dimension partition in hypercubes. Section 3 presents a heuristic algorithm to interconnect an arbitrary of nodes to form a DIPAT-graph, and a DIPAT-graph is generalized into a hierarchical one. The paper concludes with section 4.

2 System Model and Distributed Routing

An n-dimension hypercube computer, denoted as Q_n, can be modeled by a graph $G = (V, E)$ consisting of a set of nodes V and edges E. Each node represents a *processor element* (PE), and the edge between two nodes represents the link between the two PEs represented by the two nodes. The number of bits that differ between the addresses of two nodes N_i and N_j is called the *Hamming-distance* H_{ij} of the two nodes. A multicast-set is a set of nodes $G = \{ N_{m_1}, N_{m_2}, \cdots N_{m_k} \}$, in which all nodes $N_{m_i} \in G$ need to receive exactly one copy of a *multicast-message* generated by a node in G. A *DIPAT-graph*, denoted as R_G, for G, is a graph on which any node $N_x \in G$ can multicast a message to all other members in G based on the *DIPAT* routing scheme. Multicast-set i is denoted as G_i, and an n-bit *routing vector* $V_x^{G_i} = (v_{n-1}, \ldots, v_0), v_i \in \{0, 1\}$ is associated with each node N_x in R_{G_i}. v_j is set to 1 if its neighbor connected through L_j^x is in R_{G_i}, and L_j^x is called a *connected-dimension* for G_i. Otherwise, v_j is set to 0, and L_j^x is called a *blocked-dimension* for G_i.

A multicast message consists of three fields, MSG = (G_i, MA, D), where G_i is the identifier of the multicast-set, $MA = (\mu_{n-1}, \mu_{n-2} \cdots \mu_0)$, $\mu_i \in \{0, 1, *\}$, is the *multicast address*(MA) indicating the routing space of the message, and D is the message body. After N_x receives a message, it decides the routing directions of the message as follows. If $\mu_i = 1$ or 0, it implies that dimension i has been "used" for routing the message before it reached N_x, and the message cannot be forwarded to L_i^x. If $\mu_i = *$, it implies that the dimension-i was not used for routing the message, and N_x needs to send a copy of the message to L_i^x if the node connected to L_i^x is on the same multicast-set, as recorded in the routing vector of N_x.

For example, if N_x receives a multicast message with MA = $(\ldots * \ldots * \ldots * \ldots)$, where the *'s are at bits $\mu_{x_{m-1}}, \mu_{x_{m-2}}, \ldots, \mu_{x_0}$, and all other bits are either 0 or 1. N_x needs to forward the message along the m different dimensions marked by *, so that the routing-space of the message is partitioned into m subspaces by N_x. N_x can

	N_0	N_1	N_2	N_8	N_5
$V_x^{G_y}$	1011	0101	0110	1100	1110
	N_6	N_{12}	N_7	N_{13}	N_{14}
$V_x^{G_y}$	1101	0111	0011	1001	1010

Table 1: The routing vectors of nodes in R_{G_y}

partition the routing space in $m!$ different ways depending on the sequence of the message being forwarded in the m different directions.

A *routing tree*, denoted as T_x^G, is the spanning tree generated from $N_x \in G$ to reach other members on G, and N_x is called the root of T_x^G. The DIPAT routing algorithm for a node N_x in a multicast-group G_y is formally presented as follows.

Algorithm I DIPAT-multicast for a node N_x in Q_k

```
If Nx initiates multicast message MSG_Gy = (Gy,MA,D)
  For (i = k − 1 to i = 0) /* from MSB to LSB */
    If (vi = 1)
      set MA as :
        μj = 0, if (vj = 1 and j > i),
        μj = *, if (vj = 1 and j < i),
        μi = 1;
      send the message along Li^x;
If (Nx receives a multicast message)
    MSG_r = (Gy,MA,D) from L_r^x)
  If Nx is a destination node,
    forward MSG_r to its computing processor;
  For (i = k − 1 to i = 0)
    If ((μi = *) AND (i ≠ r) AND (vi = 1))
      set MA as :
        μj = 0, if (vj = 1 and j > i),
        μj unchanged, if (vj = 1 and j < i),
        μi = 1;
      send the message along Li^x;
```

The following example illustrates the principle of our distributed routing algorithm on a DIPAT-graph R_{G_y}, as plotted in Fig.1(a), in which labels on the links denote the dimension numbers of the links. R_{G_y} was constructed for a multicast-group $G_y = \{N_0, N_1, N_2, N_8, N_5, N_6, N_{12}, N_7, N_{13}, N_{14}\}$ in a Q_4. Details on how to derive R_{G_y} from G_y will be clearly explained in next section. The routing-vectors for nodes in G_y are listed in Table 1. Now, if N_{13} initiates a multicast, the message is routed along the connected dimensions of N_{13} in the descending order. Therefore, d_3 is first used for routing the message to N_5, with its $MA = (1 * **)$. Then, a message with $MA = (0 * *1)$ will be sent to N_{12}. After N_{12} receives the message, it first sends a message to N_8, then N_{14} receives exactly one copy of message, and the final routing tree $T_{13}^{G_y}$ is shown in Fig.1(b). Another example on the routing tree generated from N_8 is shown in Fig.1(c). In this example, the message traverses from its source to destinations in an arbitrary order of dimensions, making DIPAT a much more flexible algorithm than the conventional broadcast/multicast algorithms.

3 Construction of DIPAT-Graphs

In this section, we proposed a heuristic algorithm to generate a DIPAT-graph R_G from an arbitrary multicast-

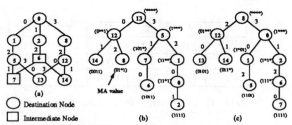

Figure 1: (a)The DIPAT-graph R_{G_y} constructed for a multicast set G_y. (b) and (c) the routing trees $T_{13}^{G_y}$ and $T_8^{G_y}$ generated for the multicast initiated by N_{13} and N_8, respectively.

set G. The basic idea is that, once a DIPAT-graph $R_{G'}$ is formed for a subset of nodes $G' \subset G$, the remaining nodes are added to $R_{G'}$ to augment the DIPAT-graph incrementally. A new DIPAT-graph is formed after each new destination node and the necessary intermediate nodes are added. Note that when a new node N_a is added to $R_{G'}$, we need to consider 1) the mutual reachability between N_a and all the existing nodes in $R_{G'}$, and 2) the potential blocking effect of newly added intermediate nodes to the mutual reachability between the existing nodes in $R_{G'}$. The second condition could occur if the dimensions of the new links are the same as that of the links on $R_{G'}$.

Here, we propose a simple distributed algorithm to guarantee the reachability between nodes on a multicast-set. In this technique, when a new node $N_a \in G - G'$ is added to $R_{G'}$, N_a first finds a shortest path to a node in $R_{G'}$. Then, N_a initiates a *joint-message* to $R_{G'}$. The joint-message has the regular packet format as defined in the previous section, but the message body contains the *routing-path-information* $RI_b^a = (d_{a_1}, d_{a_2}, \ldots, d_{a_m})$, which contains the dimension numbers of the links that the message passes from N_a to a node N_b, and the path that it passed through is denoted as $RP_b^a = (d_{a_1}, d_{a_2}, \ldots, d_{a_m})$.

After N_b received a joint-message, it forwards the joint-message to its neighbors based on the DIPAT, and then N_b forwards the *wait-messages* along its connected-dimensions that were not used for forwarding the joint-message. Each wait message also carries the routing-path-information about the links that it passed through. Every node on $R_{G'}$ will receive at least one joint- or wait-message. If a node N_c receives both joint- and wait-messages, it can directly discard the wait-message. However, if N_c only receives a wait-message but no joint-message, then N_c knows that it cannot be reached by N_a, and thus a new path needs to be found for N_a and N_c. N_c can find such a new path based on routing-path-information $RI_c^a = (d_x, d_y, \ldots, d_z, d_x)$ in the received wait-message, where d_x is the dimension from which the wait message was received. Note that d_x definitely appear twice in RI_c^a, otherwise RI_c^a would be a valid path for N_a to reach N_c based on the DIPAT routing algorithm. A new path between N_a and N_c can always be found as $N_c \overset{d_y}{-} \ldots \overset{d_z}{-} N_a$, since we can form a loop between N_a and N_c, as shown in Fig.2(a). Although N_a can reach

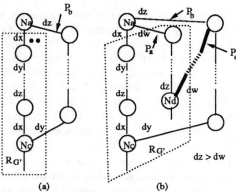

Figure 2: The path P_b is connected from N_a to N_c, as shown in (a). P_a is blocked by P_b, and an additional new path (the thick line) in (b) is added for N_d.

N_c through the newly added path $P_b = N_c \overset{d_y}{-} \ldots \overset{d_z}{-} N_a$, P_b may block some existing paths, as explained below. This example is illustrated through Fig.2(b), where before P_b is added, N_a can reach N_d through the path P_a even though $d_z > d_w$. N_d becomes unreachable from N_a after P_b is constructed to $R_{G'}$, because later when N_a receives a multicast message, it will first use d_z, then d_w for message routing. This means that d_z cannot be used again for message forwarding along P_a. We need to find a new path P_c between N_a and N_d along d_z dimension of N_a, as the thick line depicted in Fig.2(b), so that N_d becomes reachable from N_a again along the new path. This procedure is repeated for every new node added to $R_{G'}$ to ensure that all nodes $N_x \in R_{G'}$ can be reached by each other.

The DIPAT-graph construction algorithm is formally presented in the Algorithm II. Given a multicast set $G = \{N_{m_1}, N_{m_2}, \ldots, N_{m_k}\}$ where the nodes are ordered by their Hamming-weights, and ties are broken based on their values.

Now we show how the DIPAT-graph depicted in Fig.1(a) can be connected from a multicast set $G = \{N_0, N_1, N_2, N_8, N_5, N_{12}, N_{13}, N_{14}\}$ by the DIPAT-graph construction algorithm. It is straightforward to interconnect the first five nodes to form an intermediate DIPAT-graph $R_{G'}$, $G' = \{N_0, N_1, N_2, N_8, N_5\}$. After N_{12} is connected to N_8, which has the shortest distance to N_{12}, N_{12} multicasts a joint-message in $R_{G'}$ as follows. N_0 forwarded the joint-message to its two neighbors N_1 and N_2. N_2 discontinued the message forwarding, since it had only one connected-dimension. Now, N_1 forwards a wait-message to N_5, because d_2 has been used by N_{12} for joint-message routing. After a finite time period, N_5 will know that it cannot be reached by N_{12}, because it only received a wait-message whose routing-path-information is $RI_5^{12}(2, 3, 0, 2)$. That is, N_5 cannot be reached by N_{12}, and thus we need to find a new path for N_5 and N_{12}. The new path between N_5 and N_{12} forms a loop with $RP_5^{12}(2, 3, 0, 2)$. Based on the received routing-path-information, the new path can be randomly selected as either $RP_5^{12}(3, 0)$, or $RP_5^{12}(0, 3)$. If

$RP_5^{12}(0,3)$ is chosen, the new path is as shown in Fig.3(a), and $G' = G' \cup \{N_{12}, N_{13}\}$.

Algorithm II DIPAT-graph Construction

1. $G' = \{N_{m_1}\}$
2. **WHEN** $(G \neq \phi)$ begin
   2.1 Get a node N_{m_i} from G, $G = G - \{N_{m_i}\}$,
   2.2 Find $N_j \in G'$ with min. H_{j,m_i}, and
       Update $V_j^{G'}$, $V_{m_i}^{G'}$, and their intermediate nodes;
   2.3 N_{m_i} multicasts a joint-message to $R_{G'}$;
   2.4 If (N_x receives a joint-message from d_j)
       2.4.1 If $((v_i = 1)$ and $(\mu_i = *))$, append d_i to RI and
              forward the joint-message along d_i;
       2.4.2 If $((v_i = 1)$ and $(\mu_i = 1$ or $0))$, append d_i to RI
              and send the wait message along d_i;
   2.5 If (N_x receives a wait-message)
       If (N_x has received a joint-message)
              discard the wait-message;
       else
              forward the wait-message after RI is updated;
   2.6 After a time-out period, let U be a set of nodes only
       receiving the wait-message
   2.7 **WHEN** $(U \neq \phi)$ begin
       2.7.1 Get $U_i \in U$ with $RI = (d_x, d_y, d_a \ldots, d_z, d_x)$;
       2.7.2 Find a path $U_i \overset{d_y}{-} U_{i_1} \overset{d_a}{-} U_{i_2} \ldots U_{i_m} \overset{d_z}{-} N_{m_i}$;
       2.7.3 Set $v_y = 1$ in $V_i^{G'}$ and $V_{i_1}^{G'}$;
              Set $v_a = 1$ in $V_{i_1}^{G'}$ and $V_{i_2}^{G'}$;
              $\vdots$
              Set $v_z = 1$ in $V_{i_m}^{G'}$ and $V_{m_i}^{G'}$;
       2.7.4 **REPEAT**
              Get $N_d \in G'$ containing d_z in its $RI_d^{m_i}$;
              If $(RI = (d_w, \ldots, d_k, d_z))$ and $(d_z > d_w))$
                 Find a path $N_d \overset{d_z}{-} N_{d_1} \overset{d_k}{-} \ldots N_{d_n} \overset{d_w}{-} N_{m_i}$;
                 Set $v_z = 1$ in $V_d^{G'}$ and $V_{d_1}^{G'}$;
                 $\vdots$
                 Set $v_w = 1$ in $V_{d_n}^{G'}$ and $V_{m_i}^{G'}$;
              UNTIL all nodes in G' reachable;
       2.7.5 $U = U - \{U_i\}$
       end **WHEN**
   2.8 $G' = G' \cup \{N_{m_i}\}$
   end **WHEN**

When N_{13} is added to G', all nodes in G' can receive one copy of the joint message initiated by N_{13}, and thus no intermediate node is needed. Next, we add the node N_{14} to G'. Through a similar communication process, it can be found that N_2 cannot be reached by N_{14}, and N_2 has a wait message with $RI_2^{14}(1,2,3,1)$, as shown in Fig.3(b). Now, N_2 needs to find a new path randomly based on one of the two choices $RP_2^{14}(3,2)$, or $RP_2^{14}(2,3)$, and $RP_2^{14}(3,2)$ is chosen in this case. Finally, we get the DIPAT-graph as shown in Fig.2(a).

From this example, it can be seen that the reachability between nodes on an embedded graph in a hypercube is closely related to the properties of embedded loops, because we often need to create new paths to bypass blocked paths between two nodes. Details on the proof the lemmas can be found in [9].

Lemma 1 *Every node N_j on a DIPAT-graph R_G can be visited at most once when a multicast is initiated by an node $N_i \in R_G$.*

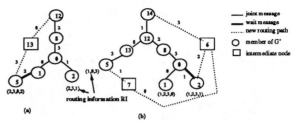

Figure 3: Two snapshots during the construction of a DIPAT-graph when (a) $G' = \{N_0, N_1, N_2, N_8, N_5\}$, and N_{12} is being added, and when (b) N_{14} is being added.

Lemma 2 *If all nodes in $R_{G'}$ become reachable from a new node N_i by the DIPAT-graph construction algorithm, then any node $N_j \in R_{G'}$ can also reach N_i through a unique path.*

Theorem 1 *Any node in a DIPAT-graph R_G can be visited once and only once when a multicast is initiated by an arbitrary node in R_G.*

Proof: *By lemma 1 and 2, result.* ∎

Hierarchical Multicasting

Nodes in a DIPAT-graph of a hypercube can multicast messages to each other in a symmetric manner. A potential limit of (any) multicast algorithms is that significant extra traffic may be generated when the nodes are fairly distant from each other. Since in DIPAT a message always reaches its destinations through the the shortest paths, a fairly large number of intermediate nodes can be generated to ensure reachability between nodes. The hierarchical routing strategies commonly found in the local/wide area networks can be effectively used to reduce extra traffic for a given set of destination nodes. That is, a DIPAT-subgraph is first constructed for a subset of destination nodes located in one particular subcube, and then the remaining nodes are connected to nodes in the DIPAT-subgraph based on spanning trees, resulting a *DIPAS-graph*.

To construct a DIPAT-subgraph for a DIPAS-graph, we first sort nodes in a multicast set based on i of their n address bits. For example, if the first four bits is used as the sorting index for a multicast set in Q_6, then all nodes with addresses $a_0a_1a_2a_3**$ are put in one set, $S_{a_0a_1a_2a_3}$, where $a_i \in \{0,1\}$, and up to 16 different sets will be generated. Then, one node, called a *coordinator*, from each of the sets is interconnected to the other coordinators to form a DIPAT-subgraph. Then, remaining nodes in the multicast set are connected to their coordinators on the DIPAT-subgraph based on a spanning tree algorithm. This simple technique can be easily illustrated by modifying the example given in Fig.1(a) into the ones shown in Fig.4(a) and (b). The sorting indices for Fig.4(a) and (b) are d_3, d_1 and d_3, d_2, d_0, respectively.

To investigate the behavior of different multicast strategies, a series of simulations have been performed. In our simulations, addresses of destination nodes are uniformly

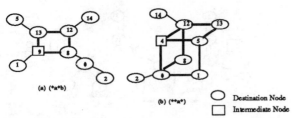

Figure 4: Two DIPAS-graphs for R_{G_y} in Fig.1, if (i) $(d_3\ d_1)$ is the sorting index, and if (ii) $(d_3\ d_2\ d_0)$ is the sorting index.

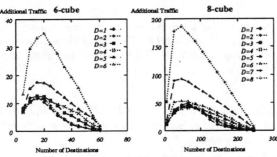

Figure 5: The average additional traffic in different DIPAS-graphs for Q_6 and Q_8.

distributed, representing the worst case performance, and each sample point reported is the average value of at least 500 runs. The amount of extra traffic under this strategy with different DIPAT-subgraph dimensions, which are denoted by D values, is plotted in Fig.5. That is, if $D = i$, it means that i bits are used as the sorting index to generate a DIPAT-subgraph in a Q_i of the host hypercube. As it can be seen that the amount of extra traffic is effectively reduced when the dimension of the DIPAT-graph is reduced by 2 or more, but no further significant reduction can be observed when the height of spanning trees is higher than 3. That is, fairly shallow spanning trees are sufficient to control the amount of extra traffic. The amount of additional traffic in our scheme is very close to that reported in [6]. Most nodes are one or two hops away from the DIPAT-subgraph under all conditions, as shown in Fig.6, implying that the effect of uneven multicast priority for nodes on the spanning trees should be insignificant.

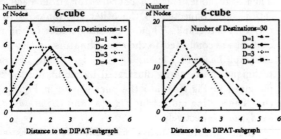

Figure 6: The average distance from nodes on the spanning trees to the DIPAT-subgraphs.

4   Conclusion

In this paper we presented a novel distributed multicast algorithm called DIPAT for hypercubes. Based on DIPAT, any node can directly initiate a multicast to an arbitrary open or closed multicast-set, so that the performance and reliability bottlenecks of the conventional single-spanning tree based multicast algorithms can be effectively eliminated. A heuristic algorithm was proposed to construct the DIPAT-graph whose necessary and sufficient conditions for mutual reachability of nodes for the DIPAT-graph are also derived. To reduce overhead in a fully distributed DIPAT graph, a hierarchical routing strategy is proposed, in which a DIPAT-subgraph is formed within a particular subcube of the hypercube, and the remaining nodes are connected to the DIPAT-subgraph based on spanning trees. It is shown that only very shallow spanning trees are needed to control the amount of extra traffic to be similar to that of existing single spanning tree solutions, with the distributed, symmetric multicasting capability in DIPAT-graphs effectively largely reserved.

References

[1] H. Xu, P.K.Mckinley, and L.M.Ni, "Efficient implementation of barrier synchronization in wormhole-routed hypercube multicomputers," *The 12th Int. Conf. on Distributed Computing Systems*, pp. 118–125, June 1992.

[2] S.L.Johnsson and C.T.Ho, "Optimum broadcasting and personalized communication in hypercubes," *IEEE Trans. Comput.*, vol. C-38, no. 9, pp. 1249–1268, Sep. 1989.

[3] M.-S. Chen and K.G.Shin, "Adaptive fault-tolerant routing in hypercube multicomputers," *IEEE Trans. Comput.*, vol. C-39, no. 12, pp. 1406–1416, 1990.

[4] A.J.Frank, L.D.Wittie, and A.J.Bernstein, "Multicast communication on network computers," *IEEE Software*, vol. 2, no. 3, pp. 49–61, May 1985.

[5] X.Lin and L.M.Ni, "Multicast coomunication in multicomputer networks," *1990 Int. Conf. on Parallel Processing*, vol. 3, pp. 13–17, 1990.

[6] Y.Lan, A.H.Esfahanian, and L.M.Ni, "Multicast in hypercube multiprocessors," *J. of Parallel and Distributed Computing*, pp. 30–41, Jan 1990.

[7] G.T.Byrd, N.P.Saraiya, and B.A.Delagi, "Multicast communication in multiprocessors systems," *1989 Int. Conf. on Parallel Processing*, vol. 1, pp. 196–200, 1989.

[8] J.-P. Sheu and M.-Y. Su, "A multicast algorithm for hypercube multiprocessors," *1992 Int. Conf. on Parallel Processing*, vol. 3, pp. 18–22, 1992.

[9] J.-C. Liu and H.-J. Lee, "A distributed multicast algorithm for hypercube multicomputers," Technical Report 92-027, Dept. of Computer Science, TAMU, Oct. 1992.

A GENERALIZED BITONIC SORTING NETWORK

Kathy J. Liszka and Kenneth E. Batcher
Kent State University
Kent, Ohio 44242-0001
email: kliszka@cs.kent.edu   email: batcher@cs.kent.edu

Abstract -- *The bitonic sorting network will sort $N = 2^m$ keys in $O(log^2 N)$ time with $O(N log^2 N)$ comparators. Developments on the sorter enable the network to sort $N = pq$ keys, a composite number. However, there has been no general method for sorting a bitonic sequence of N keys, N a prime, or N a composite that decomposes into primes larger than 3. The odd-merge method removes this constraint while maintaining the same cost and delay, using a uniform and efficient decomposition.*

1. INTRODUCTION

In 1968, the bitonic sorter was introduced as an $O(log^2 N)$ delay sorting network with $O(N log^2 N)$ comparators[1]. This network, designed for 2^n keys, provides modularity, so the network may easily be partitioned and constructed. In the last two decades, much effort has been devoted to improve the network and map it to various supercomputer architectures. Nakatani et. al. [4], decomposed the bitonic sorter into a k-way sorter, where pk-keys, a composite number, may be sorted. In 1991, Wang, Chen and Hsu [5] mapped a reduced ascending-descending sorter of any size to an incomplete hypercube, based on work done by Batcher [2] in 1990. But this generalization does not work for bitonic sequences in general, only the subset of ascending-descending keys. Until now, there has been no general method for sorting a bitonic sequence of N keys, N a prime, or N a composite that decomposes into primes larger than 3. The odd-merge method presented here removes this constraint while maintaining the same cost and delay, using a uniform and efficient decomposition.

This paper is organized in the following way. Basic definitions are presented in the second section. The odd-merge bitonic sorter is described in section three. Performance is discussed in section four and conclusions are drawn in the last section.

2. TERMINOLOGY

The basic component of the bitonic sorter is a switching element (figure 1). Two input keys, a and b, are compared and output in ascending order with min(a,b) output to L, and max(a,b) output to H. All of the work on the bitonic sorter presented here uses the simple 2x2 comparator. In this paper, the terms switching element, switch, and comparator will be used interchangeably.

The bitonic sorter accepts a bitonic sequence as input to the network and produces a sorted sequence as output. For clarity, an *ascending sequence* (/)is a sequence of keys $k_0, k_1, ..., k_n$ such that $k_0 \leq k_1 \leq ... \leq k_n$. A *descending sequence* (\) is a sequence of keys $k_0, k_1, ..., k_n$ such that $k_0 \geq k_1 \geq ... \geq k_n$. Note that a single key, k, is both an ascending sequence and a descending sequence.

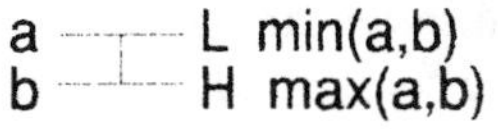

$$a \quad \rule{}{} \quad L \; min(a,b)$$
$$b \quad \rule{}{} \quad H \; max(a,b)$$

Figure 1
A Simple 2x2 Comparator

An ascending-descending sequence ($\wedge$) is an ascending sequence of keys followed by a descending sequence of keys, $k_0, k_1, ..., k_{i-1}, k_i, k_{i+1}, ..., k_n$ such that $k_0 \leq k_1 \leq ... \leq k_{i-1} \geq k_i \geq k_{i+1} \geq ... \geq k_n$. Similarly, for a *descending-ascending sequence* ($\vee$).

A bitonic sequence is the rotation of an ascending and a descending monotonic sequence. For example, (1 2 3 4 5 4 3 2 1) is an ascending monotonic sequence (1 2 3 4 5) and a descending monotonic sequence (4 3 2 1). It is irrelevant if the fifth key is associated with the ascending or the descending sequence. For instance, we could state that the previous sequence is an ascending monotonic sequence (1 2 3 4) and a descending monotonic sequence (5 4 3 2 1). Because this is true, note that either the ascending sequence or the descending sequence may be null.

A significant property of a bitonic sequence is that it remains bitonic if it is split anywhere and the two parts are interchanged. Consider the simplest case, binary bits, where a string of zeros and ones are arranged in an ascending sequence. Then 00011111 is an /-sequence. If it is split between the fifth and sixth keys and the two parts are interchanged, the sequence 11100011 results which is still bitonic. As a check for this bitonic property, one must simply rotate the sequence until a ∧- sequence appears. If one cannot be created by the rotation, then the sequence is not bitonic.

The definition of a *bitonic sorter* is a network which arranges a bitonic sequence into a monotonic sequence. Since any two monotonic sequences may be joined to form a bitonic sequence, as either a strict ∧-sequence or a ∨-sequence, a bitonic sorter may be considered to be a sorting network since it arranges a bitonic sequence into a monotonic sequence. In fact, a simple 2x2 comparator is the smallest bitonic sorter.

The complexity and total number of comparators in a network are factors of cost. When calculating efficiency and speed, one needs to consider delay. The *delay* of a network is defined to be the number of comparators in the longest path through the network. Parallelism is applied when multiple comparators do not conflict in data inputs, and thus count as one step in delay.

Bitonic sorters are modular, meaning that larger networks may be constructed from smaller bitonic sorters. A bitonic sorter for 4 keys is constructed from the 2x2 comparator. A sorter for 8 numbers is constructed from a 4- key bitonic sorter. Figures 2 and 3 show bitonic sorters for N = 4 and 8 keys respectively. Throughout this paper, uppercase N will always refer to the number of keys input to (and output from) the sorting network.

Figure 2
4-Key Bitonic Sorter

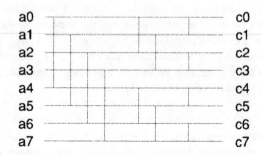

Figure 3
8-Key Bitonic Sorter

The bitonic sorter described thus far, is designed to sort keys where $N = 2^k$. In 1989, Nakatani, Huang, Arden and Tripathi [4] presented work that decomposed the bitonic sorter where N = pq, a composite number. They call it a k-way bitonic sorter, based on a k-way decomposition, instead of a two-way decomposition. Fundamentally, the bitonic sequence may be thought of in a pxq matrix form. A p-key bitonic sorter is applied to each column, then a q-key bitonic sort is applied to the rows.

In general, Batcher's original constructions showed a 2k-key sorter that could be decomposed as k 2-key bitonic sorters or 2 k-key bitonic sorters. Nakatani, et al. showed a pq-key bitonic sort that could be decomposed as p q-key bitonic sorters or q p-key bitonic sorters. The problem with the pq-sorter is one where N decomposes into primes greater than 3. For example, if N=30, then N=2x3x5. There is no bitonic sorter defined for N=5, therefore, there is no bitonic sorter defined for the composite number 30.

3. ODD-MERGE METHOD

The odd-merge method decomposes the keys based on indices as follows where N = 2n+1, an odd number.

1. Divide the bitonic sequence into 2 bitonic sequences based on odd/even indices. By definition, the two subsequences are still bitonic. Simply removing an element does not affect this property.

2. Sort each subsequence. This requires an n-key bitonic sorter and an (n+1)- key bitonic sorter.

3. A simple interleave of the two subsequences will produce a final sorted sequence.

Figure 4 shows an odd-merge bitonic sorter for 7 keys. It is composed of a 4-key bitonic sorter on the keys with even indices, a 3-key bitonic sorter on the keys with odd indices, and a simple interleave of 3 keys with 4 keys.

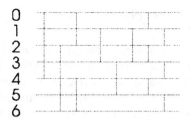

Figure 4
Odd-Merge Bitonic Sorter for N = 7
Bitonic (4) + Bitonic(3) + Interleave
Comparators = 13, Delay = 5

4. PERFORMANCE

The proof of correctness for the odd-merge method must show that the simple interleave will suffice as the completion step for this method. Reference the work done in [3] for details of the proof to show that an N-key bitonic sorter can be constructed from an n-key bitonic sorter, an (n+1)-key bitonic sorter and 2n switches, where N=2n+1.

The odd-merge used recursively with the bitonic merger demonstrates excellent results. It's performance, logically, does not vary significantly from the composite bitonic merger. First, examine the performance of the bitonic merger without the odd-merge method. N is composite, and therefore has a factorization of $N = r_1 \times \ldots \times r_k$, for $k \geq 2$. Select any factorization of $N = pq$ and construct a bitonic merger with q p-key mergers and p q-key mergers. P is a unique set of r_i; q is constructed from the remaining prime factors not included in p.

To construct a p-key merger, p may either be a composite number that must be decomposed further, or it is a terminating value that has a bitonic merger already defined (ex. 2, 4, 8, 16, etc.). If it must be decomposed as a composite, then it may be factored into some unique pair $p_1 q_1$ which are extracted from the set of r_i from the factorization of p. Similarly for a q-key merger, except if q is also a composite number that requires further decomposition, it must be factored into $p_2 q_2$ which are constructed from the remaining factors, r_j of q, not used by p. Therefore, it is irrelevant how N is factored in the composite bitonic merger. The results will be the same. For example, given N = 60, we can choose pq to be (5,12), (3,20), (4,15), etc. If (5,12) is selected, the five q=12 size sorters will decompose further into either (3,4) or (2,6). Each will decompose into 3x2x2. Using the

bitonic merger described in Nakatani [4] means that N will ultimately be reduced to a prime factorization of

$$2^{k1} \times 3^{k2} \times 5^{k3} \times \ldots$$

where each $k_j \geq 0$. To construct a bitonic sorter this way, we need

$$k_1 \times \text{bitonic}(2) + k_2 \times \text{bitonic}(3) + k_3 \times \text{bitonic}(5) + \ldots$$
$$\text{or}$$
$$q \times \text{bitonic}(p) + p \times \text{bitonic}(q)$$

comparators. Therefore, the entire decomposition of N will yield the same results. The performance in terms of delay, may be determined by:

$$\text{bitonic}(p) + \text{bitonic}(q).$$

Remember that the q p-key sorters may be performed in parallel as well as the p q-key sorters. However, there is no bitonic sorter defined for 5, 7, 11, 13, 17, etc. using this method of decomposition alone, so without a generalizing extension, the actual implementation of the sort for N is impossible.

When the odd-merge method is applied to the bitonic merger, for odd N as well as primes, it becomes relevant which factorization is selected for a composite N. The first positive N where this occurs is 105. The proper factors of 105 are 3, 5 and 7. Table 1 shows the results of the different combinations.

factors	cost	delay
3 x 35	468	13
5 x 21	468	13
7 x 15	468	12

Table 1
Results of Alternate Decompositions
of N = 105 Using the Composite Method

In the case where N = 105, the 7 x 15 decomposition yields a smaller delay because the odd-merge requires a delay of 7 for its (7,8) decomposition of 15 while the composite method requires a delay of 8 for its 3x5 decomposition. Optimizing delay, this occurs 22 times for N up to 1000. This means the odd-merge method may not only solve the problem of prime decomposition, it actually improves performance of the composite merger for some N.

The odd-merge method requires a delay for $N=2n+1$ of

$$\max[\text{bitonic}(n), \text{bitonic}(n+1)] + 2.$$

The upper bound for delay is $(2\lceil \log N\rceil)-1$, where $\lceil x\rceil$ is the ceiling function. This may be calculated by looking at the worst case scenario where N decomposes completely by odd numbers. Starting at $N=3$, it requires a delay of 3 to sort. This is our base. To sort $N=5$ or $N=7$ requires bitonic(3)+2 for both, as they decompose into (2,3) and (3,4) respectively. To sort $N=9,11,13,15$ requires either bitonic(5)+2 or bitonic(7)+2, both being (bitonic(3)+2)+2 = bitonic(3)+4. To sort the odds from 17 to 31 will require a delay of 2 plus a bitonic of either 9, 11, 13 or 15, which requires a delay of ((bitonic(3)+2)+2)+2)=bitonic(3)+6. For each range for a power of 2, from $N = 2^m+1$ through 2^{m+1}, we increase delay by 2 for odd numbers. The delay remains constant for values of N within these ranges.

The general formula is $(2\log N)-1$. The odd-merge used as the generalization technique for the bitonic merger is very stable in terms of delay for the upper bound. It is possible to do better than this, case in point being $N=19$. This decomposes into

$$\max[\text{bitonic}(9), \text{bitonic}(10)] + 2$$

which produces a delay of 8. Bitonic(10) requires a delay of 6, and so does bitonic(9) if the composite method is used to decompose it into (3x3) instead of the odd-merge decomposing it into (4,5) with a result of 7. An example of N meeting the upper bound is 31. This requires

$$\max[\text{bitonic}(15),\text{bitonic}(16)]+2$$

which is $2(\lceil \log 31\rceil)-1=9$.

The cost for the composite bitonic merger may be calculated for $N = pq$:

$$p \times \text{bitonic}(q) + q \times \text{bitonic}(p).$$

For the odd-merge, the equation for $N = 2n + 1$ is:

$$\text{bitonic}(n) + \text{bitonic}(n+1) + (N-1).$$

The selected merger is constructed from the minimum result of the above two equations applied recursively.

5. CONCLUSION

For the odd-merge, there is only one choice for decomposition, and therefore, no conflict between optimizing either cost or delay. However, when a table for some range of N is constructed using the composite and odd-merge methods jointly, a decision must be made to minimize either cost or delay. There are cases where the odd-merge may outperform the composite method either in terms of cost or delay for some non-prime N. For example, if $N = 65$, the composite method requires 274 comparators while the odd-merge requires 254 comparators, a savings of 20. For delay, if $N = 15$, the odd-merge will take 7 steps compared to 8 using the composite merger. With exception of $N=2^m$, any N greater than 4 is dependent on efficient decompositions. The results of those decompositions must be selected based on the objectives of saving switching elements or time. The overall performance of the bitonic merger incorporating this generalization remains at $O(N\log^2 N)$ for cost and $O(\log^2 N)$ for delay when applied to sorting inputs of random keys.

REFERENCES

[1]   Batcher, K. E., "Sorting Networks and Their Applications", *1968 Spring Joint Computer Conf.*, *AFIPS Proc.*, Washington, D.C., (1968, Vol. 32), pp. 307-314.

[2]   _____, "On Bitonic Sorting Networks", *Proceedings of the 1990 International Conference on Parallel Processing*, (1990, Vol. 1), pp. 376-379.

[3]   Liszka, K.J., "Generalizing Bitonic and Odd-Even Merging Networks", *Doctoral Thesis*, Department of Mathematics and Computer Science, Kent State University, (1992).

[4]   Nakatani, T., Huang, S., Arden, B. W., and Tripathi, S. K., "K-Way Bitonic Sort", *IEEE Transactions on Computers*, (February 1989,Vol. 32,), pp. 283-288.

[5]   Wang, B. F., Chen, G. H., and Hsu, C. C., "Bitonic Sort with an Arbitrary Number of Keys", *Proceedings of the 1991 International Conference on Parallel Processing*, (1991, Vol. 3), pp. 58-61.

SESSION 5A

HYPERCUBE

A Lazy Scheduling Scheme
for Improving Hypercube Performance*

Prasant Mohapatra, Chansu Yu, Chita R. Das
Dept. of Electrical and Computer Engineering
The Pennsylvania State University
University Park, PA 16802

Jong Kim
Dept. of C. S. E.
POSTECH
P.O. Box 125, Pohang, Korea

Abstract

Processor allocation and job scheduling are complementary techniques to improve the performance of multiprocessors. It has been observed that all the hypercube allocation policies with the FCFS scheduling show little performance difference. A greater impact on the performance can be obtained by efficient job scheduling. This paper presents an effort in that direction by introducing a new scheduling algorithm called lazy scheduling for hypercubes. The motivation of this scheme is to eliminate the limitations of the FCFS scheduling. This is done by maintaining separate queues for different job sizes and delaying the allocation of a job if any other job(s) of the same dimension is(are) running in the system. Simulation studies show that the hypercube performance is dramatically enhanced by using the lazy scheme as compared to the FCFS scheduling. Comparison with a recently proposed scheme called scan indicates that the lazy scheme performs better than scan under a wide range of workloads.

1 Introduction

Processor management in multiprocessors is a crucial issue for improving the system performance. This has become an active area of research for the hypercube computer which has emerged as one of the most popular architectures [1-3]. Hypercube topology is suitable for a wide range of applications and can support multiple users. Judicious selection of processors is essential in a multiuser environment for better utilization of system resources. There are two basic approaches to improve processor management in a multiprocessor system. These are called *job scheduling* and *processor allocation*.

Scheduling decides the job sequence for allocation. Processor allocation is concerned with the partitioning and assignment of the required number of processors for incoming jobs. Structural regularity of the hypercube makes it suitable for partitioning it into independent subcubes. Each user or job is assigned an appropriate subcube by the operating system. It is known that optimal allocation in a dynamic environment is an NP-complete problem [8]. Several heuristic algorithms reported in literature are buddy [4], modified buddy [5], gray code [6], free list [7], MSS [8], tree collapsing [9], and PC-graph [10]. These schemes differ from each other in terms of the subcube recognition ability and/or time complexity.

Comparison of all the hypercube allocation policies shows that the performance improvement due to better subcube recognition ability is not significant [7,11]. This is mainly because of the first-come-first-serve (FCFS) discipline used for job scheduling. In a dynamic environment, FCFS scheduling may not efficiently utilize the system. There are two drawbacks with the FCFS scheme. First, it is more likely that an incoming job with a request for a large cube has to wait until some of the existing jobs finish execution and relinquish the nodes to form a large cube. All arriving jobs are queued during this waiting period. There may be several jobs waiting in the queue with smaller cube requests but they cannot be allocated even though the system can accommodate such jobs. This *blocking* property of the FCFS scheme reduces the system utilization. Second, with the FCFS scheme, the scheduler tries to locate a subcube to allocate a job as soon as it arrives. This *greedy* property creates more fragmentation and makes the allocation of the succeeding jobs difficult. The subcube recognition ability of an allocation policy is thus overshadowed by the limitations of the FCFS scheduling policy.

It is therefore logical to focus attention on efficient scheduling schemes to improve system performance while keeping the allocation complexity minimal. There has been little attention paid towards

*This research was supported in part by the National Science Foundation under grant MIP-9104485.

scheduling of jobs in a distributed system like hypercube. The first effort in this direction was by Krueger et al [11]. They propose a scheme called *scan* which segregates the jobs and maintains a separate queue for each possible cube dimension. The queues are served similar to the c-scan used in disk scheduling. The authors show that significant performance improvement can be achieved by employing the scan policy. However, it turns out that the scheme is suitable for a workload environment where the job service time variability is minimal. For a more general workload, the system fragmentation increases and the performance gain diminishes.

In this paper, we propose a new strategy called *Lazy Scheduling* for scheduling jobs in a hypercube. The main idea is to temporarily delay the allocation of a job if any other job(s) of the same dimension is(are) running in the system. The jobs could wait for existing subcubes rather than acquiring new subcubes and possibly fragmenting the system. The scheduling is not greedy and is thus named *lazy* . The system maintains separate queues for different job sizes. We thus eliminate the blocking problem associated with the FCFS scheme. Waiting time in the queue is controlled by a threshold value in order to avoid discrimination against any job size. The proposed scheme tries to improve throughput by providing more servers to the queues that have more incoming jobs.

A simulation study is conducted to compare mainly three types of processor scheduling schemes for hypercubes. These are FCFS, scan, and lazy. Job allocation is done using the buddy scheme for all the three disciplines. Another simple scheme called static partitioning is also simulated to demonstrate its effectiveness for uniform system load. The performance parameters analyzed here are average queueing delay and system utilization. We show that the lazy scheme provides at least equal performance as that of scan for uniform residence time distribution. For all other workloads, including the most probable residence time distributions like hyperexponential, the lazy scheme out performs scan in all performance measures by 20% to 50%. Thus, the proposed scheduling scheme is suitable and adaptive for a variety of workloads.

The rest of the paper is organized as follows. In Section 2, various hypercube allocation and scheduling policies are summarized. The lazy allocation scheme is described in Section 3. The simulation environment and the workloads are given in Section 4. Section 5 is devoted to the performance evaluation and comparisons of various policies. Conclusions are drawn in Section 6.

2 Allocation and Scheduling Policies

2.1 Allocation Policies

Buddy

The buddy strategy is implemented is based on the buddy scheme for storage allocation [4]. For an n-cube, 2^n allocation bits are used to keep track of the availability of nodes. Let k be the required subcube size for job I_k. The idea is to find the least integer m such that all the bits in the region $[m2^k, (m+1)2^k - 1]$ indicate the availability of nodes. If such a region is found, then the nodes corresponding to that region are allocated to job I_k and the 2^k bits are set to 1. When a subcube is deallocated, all the bits in that region are set to 0 to represent the availability of the nodes. The time complexities of allocation and deallocation are $O(2^n)$ and $O(2^k)$, respectively. The allocation and deallocation complexity can be reduced to $O(n)$ by using an efficient data structure [12].

Modified Buddy

Modified buddy scheme is similar to the buddy strategy in maintaining 2^n allocation bits. Here, the least integer α is determined, $0 \leq \alpha \leq 2^{n-k+1} - 1$, such that α^{n-k+1} is free and it has a pth partner, $1 \leq p \leq (n-k+1)$, α_p^{n-k+1} which is also free. Detail description of the scheme is given in [5]. This scheme has better subcube recognition ability than buddy. The complexities of allocation and deallocation are $O(n2^n)$ and $O(2^k)$, respectively.

Gray Code

The gray code strategy proposed in [6] stores the allocation bits using a binary reflected gray code (BRGC). Here the least integer m is determined such that all the $(i \bmod 2^n)$ bits indicate availability of nodes, where $i \in [m2^{k-1}, (m+2)2^{k-1} - 1]$. Thereafter, the allocation and deallocation are the same as in the buddy scheme. For complete recognition, $\binom{n}{\lfloor n/2 \rfloor}$ gray codes are needed. The complexity of the multiple GC allocation is $O(\binom{n}{\lfloor n/2 \rfloor}2^n)$ and that of deallocation is $O(2^k)$.

Free List

The free list strategy proposed in [7] maintains $(n+1)$ lists of available subcubes, with one list per dimension. A k-cube is allocated to an incoming job by searching the free list of dimension k. If the list is empty, then a higher dimension subcube is decomposed and is allocated. Upon deallocation, the released subcube is merged with all possible adjacent cubes to form the largest possible disjoint subcube(s). The list is updated accordingly. This scheme has the ability to recognize a subcube if it exists in the system. The time complexity of allocation is $O(n)$ and that of deallocation is $O(n2^n)$.

MSS

This strategy is based on the idea of forming a maximal subset of subcubes (MSS), and is described in detail in [8]. The MSS is a set of available disjoint subcubes that has the property of being greater than or equal to other sets of such subcubes. The main idea is to maintain the greatest MSS after every allocation and deallocation of a subcube. The allocation and deallocation complexities are of the order of $O(2^{3^n})$ and $O(n2^n)$, respectively.

Tree Collapsing

Tree collapsing strategy is introduced in [9]. This involves successive collapsing of the binary tree representation of a hypercube structure. The scheme generates its search space dynamically and has complete subcube recognition ability. The complexities of allocation and deallocation are $O(\binom{n}{k}2^{n-k})$ and $O(2^k)$, respectively.

PC-graph

The main idea of this approach is to represent available processors in the system by means of a prime cube (PC) graph [10]. Subcubes are allocated efficiently by manipulating the PC-graph using linear graph algorithms. This scheme has also complete subcube recognition ability. The allocation complexity is $O(\frac{3^{3n}}{n^2})$ and that of deallocation is $O(\frac{3^{2n}}{n^2})$.

2.2 Performance Comparison

It is observed from [7,11] that the performance variations due to different allocation policies are minimal. None of the allocation policies utilizes more than half of the system capacity irrespective of the input load [11]. Fragmentation of the system overides the advantages obtained from better subcube recognition ability. More sophisticated allocation policies, although show little performance improvement, introduce overheads and higher time complexity. It is for this reason we have adopted the buddy allocation scheme which is simple and has low time complexity.

2.3 Scheduling Strategies

A scheduling strategy called *scan* is proposed in [11]. The concept is similar to the disk c-scan policy. The algorithm maintains $(n+1)$ separate queues, one for each possible cube dimension. A new job joins the end of a queue corresponding to its subcube dimension. All jobs of a given dimension are allocated before the scheduler move on to the jobs of next dimension. It is shown that significant performance improvement can be achieved with this scheme compared to the most sophisticated allocation policies.

The problems associated with the scan policy are the following. It tries to reduce fragmentation by al-locating equal-sized jobs. The scheme performs well when the residence time of the jobs has little variation. In this environment, the system at any instant would have jobs of almost the same size and thus fragmentation is reduced. When the residence times of all the jobs are not the same, the allocations and deallocations eventually create a mixture of jobs from all the queues in the system and lead to fragmentation. The performance of the scan scheme is therefore dependent on the workload. The study in [11] is conducted under the assumption of exponential job residence time distribution with a mean of one time unit. Practically, the residence time distribution resembles close to hyperexponential distribution [13]. It will be shown that the scan algorithm does not performs well in this environment. Furthermore, under certain circumstances scan treats jobs unfairly. This happens if there are some queues with large number of jobs and some queues with a few jobs. This is possible with non-uniform arrival pattern. There could be a situation where jobs in a short queue have to wait till all the jobs in the longer queues are processed. In such cases, a job arriving early to a short queue has to wait even longer than the jobs that arrive much later in the longer queues.

3 The Lazy Scheduling Scheme

In this section, we first discuss a simple scheduling algorithm based on static partitioning of the system. Static partitioning scheme is shown to be efficient only for uniform workloads. Next, the concept of *Lazy Scheduling* is discussed. We conclude the section by analyzing the time complexity of the proposed scheme.

3.1 Static Partitioning Scheme

Static partitioning scheme divides the n-cube system into one $(n-1)$-cube, one $(n-2)$-cube,...,one 1-cube, and two 0-cubes. Let S_i denote the partition which accommodates jobs requesting i-subcubes, for $0 \le i \le n-1$. Corresponding to each S_i, there is a queue Q_i. Thus there are n queues in an n-cube system. An incoming job requesting an i-cube joins Q_i. Each queue is scheduled on a FCFS basis. The steps for subcube request and release are given below.

Static Partitioning Algorithm

Static_Request (I_k)

   1. If Q_k is empty then allocate S_k to I_k.

   2. Else enqueue I_k to Q_k.

Static_Release (I_k)

   1. If Q_k is not empty then allocate the header of Q_k to S_k.

3.2 Lazy Scheduling Scheme

Lazy scheduling is based on two key concepts. First, the jobs requiring small subcubes are not blocked behind the large jobs. The scheduler maintains a separate queue for each dimension. This avoids the blockings incurred in the FCFS scheduling. Second, a job tends to wait for an occupied subcube of the same size instead of using a new cube and possibly decomposing a larger subcube. The greedy characteristic of the FCFS policy is thus subdued. Both these issues help in reducing fragmentation. If all the jobs of a dimension wait for a single subcube executing in the system, then eventually the scheduling will resemble the static partitioning scheme. In order to avoid this, we introduce a variable threshold length for each queue. A queue whose length exceeds the threshold value, tries to find another subcube using an allocation policy. This provides more servers for the queues that have more incoming jobs. The threshold value is determined dynamically as explained later.

The lazy scheduling scheme is illustrated in Figure 1. There are $(n + 1)$ queues for an n-cube system represented as Q_k's, for $0 \leq k \leq n$ ($n = 4$ in Figure 1). Each queue has a variable threshold length denoted as N_k, for $0 \leq k \leq n$. N_k is initially set to zero. N_k is incremented with a k-cube allocation and is decremented with a k-cube deallocation. In other words, N_k denotes the number of subcubes of a particular size being occupied at any instant of time. A new job is first enqueued according to its dimension. If the number of jobs waiting, $|Q_k|$, is more than N_k, then the scheduler tries to allocate the job at the head of the queue using an allocation algorithm (we have adopted the buddy scheme although any other scheme could be used). The job is allocated to the system if a suitable subcube is found. Each queue is scheduled on a FCFS basis.

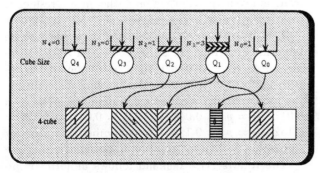

Fig. 1. Queue management in a 4-cube

The determination of the threshold value, N_k, is based on the concept that for every subcube executing in the system, there can be another job waiting

to acquire it. For example, if there are two 1-cubes allocated to the system, N_1 is set to 2. Then, the next two jobs requiring 1-cubes are enqueued. Thus the length of the queue, $|Q_1|$, becomes 2. The jobs waiting in the queue are guaranteed to receive service after the currently executing 1-cube jobs. Any additional request for a 1-cube makes the length, $|Q_1|$, more than the threshold, N_1. Thus, one more 1-cube is searched for allocation. The policy of the scheduler is to create more servers for a queue if the number of requests for the corresponding cube is high. Although the scheduler tries to limit the number of jobs waiting for the existing subcubes, there may be more jobs in the queue at high load.

Some jobs may suffer indefinite postponement under certain distribution of workload. We modify our algorithm to eliminate the possibility of indefinite postponement by using a threshold value for maximum queueing delay that a job can tolerate. Whenever the waiting time of a job reaches this threshold value, it gets priority over all other jobs. No jobs are allocated until the job whose waiting time has reached the threshold value is allocated.

The threshold for the queueing delay could be predefined or computed dynamically. Predefined threshold value is useful in imposing deadline for job completion. A dynamic threshold for the queueing delay is derived based on the following heuristic. Let d_t be the average queueing delay for a job at time t. During this waiting period a job 'sees' the arrival of $(d_t \cdot \lambda)$ jobs to the system, where λ denote the arrival rate. We consider that the maximum delay that a job can tolerate is the processing of these $(d_t \cdot \lambda)$ jobs. This time is equal to $(d_t^2 \cdot \lambda)$ and is used as the threshold value. The average queueing delay and the arrival rate are monitored by the scheduler. The scheduler updates the threshold value periodically or every time a job is allocated to the system.

The formal algorithm for the lazy scheduling is given below. Buddy_Request and Buddy_Release are procedures from the buddy allocation algorithm presented later. The flag *stop_alloc* is used to indicate that the waiting time of a job has reached the threshold value.

Lazy Scheduling Algorithm

Lazy_Request (I_k)

    1. Enqueue I_k to Q_k.

    2. If ($|Q_k| > N_k$) and (*stop_alloc* is FALSE) then call Buddy_Request(header of Q_k).

    3. If succeeds, increment N_k.

Lazy_Release (I_k)

    1. Determine the oldest request. If the waiting

time exceeds the threshold then set *stop_alloc* flag to TRUE and save identity of queue (Q_j).

2. If $(|Q_k| > 0)$ and (*stop_alloc* is FALSE) then allocate the released subsystem to the header of Q_k.

3. If *stop_alloc* is TRUE then call Buddy_Request (header of Q_j). If success then set *stop_alloc* to FALSE.

4. Call Buddy_Release(I_k).

An incoming job is first handled by the scheduler which manages the queues and imposes the algorithmic procedure. Jobs are allocated to the system using the underlying allocation policy.

We adopt the buddy strategy using an efficient data structure as proposed in [12]. This structure maintains a separate list, F_i, for each cube size i. Each list maintains the available subcubes for the corresponding size. Initially all the lists are empty except the list F_n which contains the n-cube. The allocation and deallocation complexities are of $O(n)$. The algorithm is presented below. I_k represents a job that requires a k-cube.

Buddy Strategy

Buddy_Request (I_k)

1. If F_k is not empty, allocate a subcube in F_k to I_k.

2. Otherwise, search $F_{k+1}, F_{k+2}, ..., F_n$, in order until a free subcube is found.

3. If found, decompose it using the buddy rule until a k-cube is obtained and allocate it to I_k. Update the corresponding lists after decomposition.

4. Else enqueue the job to the corresponding queue.

Buddy_Release (I_k)

1. Put the released subcube into the list F_k.

2. If F_k contains the buddy of I_k, merge them and put in the list F_{k+1}.

3. Repeat step 2 until the corresponding buddy is not available.

3.3 Complexity Analysis

For job allocation with the lazy scheduling, when a new job arrives, the queue length is compared with the threshold value and the *stop_alloc* flag is checked. These operations take constant time. Thus, the allocation complexity is the same as the buddy allocation which is of $O(n)$. Determination of the oldest job is of $O(1)$ by keeping track of the waiting time of the earliest generated job. Deallocation with the buddy scheme takes $O(n)$ time. Thus the time complexity of the release process of lazy scheduling is of $O(n)$ for an n-cube.

4 Simulation Environment

A simulation study is conducted to evaluate the performance of the proposed strategy and for comparison with other allocation and scheduling policies. The other schemes simulated are FCFS, scan, and static partitioning. Buddy allocation policy is employed for all the scheduling schemes. The job arrival rate (λ) is based on the system capacity. This is done to avoid saturation by ensuring that the arrival rate to the system does not exceeds the service rate. The observation interval T is 10000 time units which is sufficient to obtain the steady state parameters. The simulation results are obtained by taking average of 1000 iterations.

4.1 Workload

The workload is characterized by the job interarrival time distribution, distribution of the job size (subcube size), and distribution of the job service demands. Job arrival pattern is assumed to follow Poisson distribution with a rate λ. The job size and the total service demand could be either independent or dependent (proportionally related) of each other.

Independent distribution means that a large job (large subcube) has the same distribution of total service demand as that of a small job. The residence time of a job I_k is computed as $x_k/2^k$, where x_k is the required service time of job I_k, and 2^k is the number of processors needed for the k-cube job. x_k is determined from the total service demand distribution. The mean of the total service demand is computed by multiplying the mean job size with the mean residence time.

With dependent distribution, a large job has more total service demand than a small job. In this case, the job size is first obtained using the given distribution. The residence times of the jobs are obtained from a given distribution and the mean residence time, irrespective of the job size.

The distribution of the job size is assumed to be uniform or normal. In a 10-cube system, for example, probability that a request is i-cube (p_i) is $\frac{1}{10}$ with uniform distribution. For normal distribution, the probabilities for a 10-cube system are, $p_0 = p_9 = 0.017$, $p_1 = p_8 = 0.044$, $p_2 = p_7 = 0.093$, $p_3 = p_6 = 0.152$, $p_4 = p_5 = 0.194$. These numbers are used for simulating normal distribution in our study.

The total service demand follows uniform or bimodal hyperexponential distribution. Bimodal hyperexponential distribution is considered as the most probable distribution for processor service time [13]. Mean residence time is assumed to be 5 time units, and the mean job size is assumed (system size/2). For hyperexponential distribution, α is taken as 0.95, and the coefficient of variation (C_x) for the residence time is set to 4.0.

4.2 Performance Parameters

The following parameters are measured during simulation of various n-cubes for T time units.

G : Number of jobs generated during the observation time T.

C : Number of jobs completed during the observation time T.

A : Number of jobs allocated during the observation time T.

R : Total queueing delay of allocated jobs.

S : Sum of total service demand of generated jobs.

The performance parameters obtained from the simulator are utilization (U) and mean queueing delay (M). Mean queueing delay, M, is equal to R/A. The actual job request rate is measured by $S/(2^n \times T)$. The system utilization, U, is computed from $\sum_{i=1}^{A} 2^{|I_i|} t_i / 2^n T$, where t_i is the residence time of job I_i.

5 Results and Discussion

Figure 2 shows the variation of queueing delay with respect to input load for a 10-cube system. The job size is uniformly distributed in Figure 2(a), and Figure 2(b) shows the variations for a normal job size distribution. We compare static, FCFS, scan, and lazy schemes. The delay saturates early for the system employing the FCFS scheduling strategy in both cases. Static partitioning performs very well in case of uniform job size distribution. The system utilization is high and there is almost no fragmentation in static partitioning with uniform distribution. But the performance improvement is not consistent and deteriorates for other distributions. This can be inferred from Figure 2(b). Scan and lazy scheduling show better performance than the FCFS strategy for both uniform and normal distributions. Figures 2(a) and 2(b) show that the average queueing delay with the lazy scheduling is less than that of the scan. Moreover, the lazy scheduling performs close to the static scheme for the uniform job size distribution (Figure 2(a)). This is because by delaying the allocation of a job for which a cube is already busy, the lazy scheme tries to divide the system uniformly for all cube sizes.

It was mentioned in Section 2 that the scan policy is workload dependent. It does not perform well when the job residence time exhibits wide variation. The difference in the delay becomes more prominent when the job residence time is hyperexponentially distributed. This is demonstrated in Figure 3. The job

size is uniformly distributed in Figure 3(a), and is normally distributed in Figure 3(b). The deterioration with the scan scheme is due to the high variability of residence time. Allocation of equal-sized jobs is not maintained in these situations. On the other hand, the lazy scheme performs well under all workloads, particularly with hyperexponential distribution.

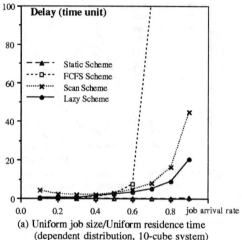

(a) Uniform job size/Uniform residence time
(dependent distribution, 10-cube system)

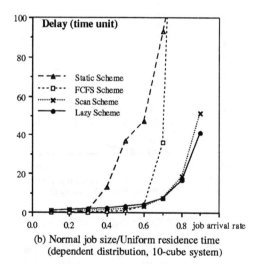

(b) Normal job size/Uniform residence time
(dependent distribution, 10-cube system)

Fig. 2. Average queueing delay for a 10-cube

Figure 4 shows the comparison of system utilization at three different input loads. The job size is uniformly distributed in Figure 4(a) and normally distributed in Figure 4(b). The main observation from this graphs is that the utilization of the static partitioning scheme is very low compared to the other schemes. Because of fixed partitioning, a large part of the system may be empty while there are a number of jobs in the queue for a different partition. This leads to the poor utilization in static scheme. Thus, lower queueing delay does not necessarily means better system utilization. At low loads the system utilization of buddy, scan, and lazy schemes are almost the same. Lazy scheduling can provide better system utilization than others at higher loads.

the FCFS and scan scheduling schemes.

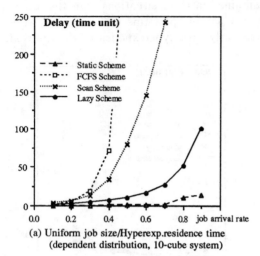

(a) Uniform job size/Hyperexp.residence time
(dependent distribution, 10-cube system)

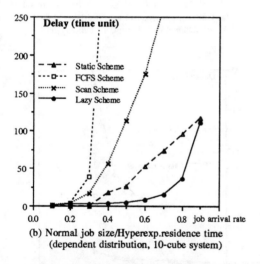

(b) Normal job size/Hyperexp.residence time
(dependent distribution, 10-cube system)

Fig. 3. Average queueing delay for a 10-cube

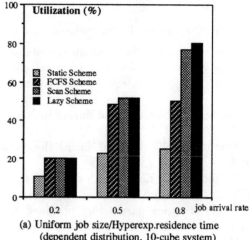

(a) Uniform job size/Hyperexp.residence time
(dependent distribution, 10-cube system)

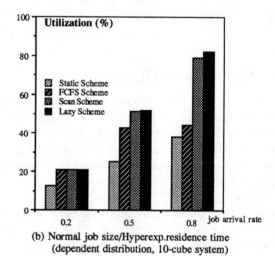

(b) Normal job size/Hyperexp.residence time
(dependent distribution, 10-cube system)

Fig. 4. System utilization vs input load of a 10-cube

The average delay with respect to the observation time is shown in Figure 5. Figure 5(a) is for moderate load ($\lambda = 0.5$), and average queueing delay for heavy load ($\lambda = 0.85$) is shown in Figure 5(b). The graphs indicate that under moderate load, the average queueing delay with the FCFS scheme is very high compared to the other schemes. Average queueing delay with the scan scheduling increases monotonically with time under heavy load, and the delay becomes closer to that of the FCFS scheme. The performance behavior with the static scheme is good because of the uniform job size distribution. Average queueing delay does not increases considerably with time in case of the lazy scheduling. It stays much lower than that of

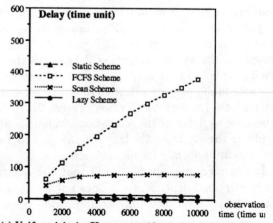

(a) Uniform job size/Hyperexp.residence time (dependent distribution, 10-cube system, job arrival rate of 0.5)

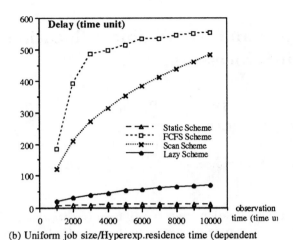

(b) Uniform job size/Hyperexp.residence time (dependent distribution, 10-cube system, job arrival rate of 0.85)

Fig. 5. Delay variation with the observation time

We have done extensive simulations for both dependent and independent workloads. The trends in various performance measures are similar for the two workloads.

6 Concluding Remarks

We have proposed a new scheduling scheme called *lazy scheduling* for assigning jobs in hypercube computers. This scheduling along with the buddy allocation scheme is used to process jobs in a multiuser environment. Prior research has focussed more on efficient allocation policies for hypercubes although they provide only incremental performance gain due to the limitations with the FCFS scheduling. The lazy scheme is proposed as an alternative to the FCFS scheduling to improve the hypercube performance.

It is shown that significant improvement in system utilization and delay can be achieved with the lazy scheduling compared to the FCFS discipline. Our scheme is compared with another technique, called scan for various workload distributions. It is observed that both scan and lazy schemes provide comparable performance under uniform workload distribution. However, for hyperexponential residence time distribution and varied job sizes, the lazy scheme out performs the scan method. In summary, the proposed method is adaptable to all workloads where as scan is workload dependent.

The study argues in favor of exploiting various scheduling schemes as opposed to efficient but complex allocation policies. We are currently investigating the performance tradeoffs due to scheduling and allocation policies in other multiprocessors.

References

[1] J. P. Hayes, T. N. Mudge, *et al*, "Architecture of a Hypercube Supercomputer," Int. Conf. on Parallel Processing, pp. 653-660, Aug. 1986.

[2] L. N. Bhuyan and D. P. Agrawal, "Generalized Hypercube and Hyperbus Structures for a Computer Network," IEEE Trans. on Computers, pp. 323-333, Apr. 1984.

[3] Y. Saad and M. H. Schultz, "Topological Properties of Hypercube," IEEE Trans. on Computers, vol. 37, pp. 867-872, July 1988.

[4] K. C. Knowlton, "A Fast Storage Allocator," Communications of ACM, vol.8, pp. 623-625, Oct. 1965.

[5] A. Al-Dhelaan and B. Bose, "A New Strategy for Processor Allocation in an N-Cube Multiprocessor," Int. Phoenix Conf. on Computers and Communications, pp. 114-118, Mar. 1989.

[6] M. S. Chen and K. G. Shin, "Processor Allocation in an N-Cube Multiprocessor Using Gray Codes," IEEE Trans. on Computers, pp. 1396-1407, Dec. 1987.

[7] J. Kim, C. R. Das, and W. Lin, "A Top-Down Processor Allocation Scheme for Hypercube Computers," IEEE Trans. on Parallel & Distributed Systems, pp. 20-30, Jan. 1991.

[8] S. Dutt and J. P. Hayes, "Subcube Allocation in Hypercube Computers," IEEE Trans. on Computers, pp. 341-352, Mar. 1991.

[9] P. J. Chuang and N. F. Tzeng, "Dynamic Processor Allocation in Hypercube Computers," Int. Symp. on Computer Architecture, pp. 40-49, May, 1990.

[10] H. Wang and Q. Yang, "Prime Cube Graph Approach for Processor Allocation in Hypercube Multiprocessors," Int. Conf. on Parallel Processing, pp. 25-32, Aug. 1991.

[11] P. Krueger, T. H. Lai, and V. A. Radiya, "Processor Allocation vs. Job Scheduling on Hypercube Computers," Int. Conf. on Distributed Computing Systems, pp. 394-401, 1991.

[12] D. E. Knuth, *The Art of Computer Programming, Volume 1, Fundamental Algorithms*, Addison-Wesley, 1973.

[13] K. S. Trivedi, *Probability and Statistics with Reliability, Queuing, and Computer Science Applications*, Prentice-Hall Inc., 1982.

Fast and Efficient Strategies for Cubic and Non-Cubic Allocation in Hypercube Multiprocessors

Debendra Das Sharma [†]   Dhiraj K. Pradhan
Department of Computer Science
Texas A&M University
College Station, TX 77843.

Abstract

A new approach for dynamic processor allocation in hypercube multiprocessors which supports a multi-user environment is proposed. A dynamic binary tree is used for processor allocation along with an array of free lists. Two algorithms are proposed based on this approach that are capable of handling cubic as well as non-cubic allocation efficiently. The time complexities for both allocation and deallocation are shown to be polynomial; orders of magnitude improvement over the existing exponential and even super-exponential algorithms. Unlike the existing strategies, the proposed strategies are best-fit strategies and do not excessively fragment the hypercube. Simulation results indicate that the proposed strategies outperform the existing ones in terms of parameters such as average delay in honoring a request, average allocation time and average deallocation time.

1 Introduction

In an MIMD hypercube which supports multiple users, an incoming task must be allocated the required number of processors for execution. On completion of the task, the processors assigned to the task are released for subsequent allocation, a process known as 'deallocation'. A number of processor allocation schemes have been proposed in the literature [2-9] Most of these, including nCX, the host operating system of the nCUBE series [14], assume the number of processors to be a power of 2. However, many applications, such as embedding of complete & incomplete binary tree and rectangular grid in hypercubes and solving a large number of non-linear equations, do not necessarily require a complete subcube for execution [8, 11, 10, 12]. Implementing a complete sub-

cube allocation strategy will have the drawback of allocating extra processors to the tasks (to form a complete subcube) that will remain idle. This is known as 'internal fragmentation' and may result in higher waiting times for tasks. Strategies for non-cubic allocation proposed in [6, 8] have extremely high time complexities for allocation and deallocation (Table 1) and may not be suitable for higher dimensional hypercubes. An efficient allocation scheme that is capable of handling both cubic and non-cubic allocation efficiently; while exhibiting low time complexities for both allocation and deallocation, low memory overhead, high processor utilization and low waiting times for incoming tasks is of practical interest. This paper presents two such schemes, possessing the lowest time complexities and lowest memory overhead among all existing schemes, while exhibiting superior performance.

The paper is organized as follows. The following section presents the pertinent preliminaries. A discussion of the existing allocation strategies is presented in Section 3. Our approach is delineated in Section 4, also compared against existing schemes. Section 5 provides simulation results comparing the performance of our strategy against some existing strategies. The conclusion appears in Section 6.

2 Preliminaries

We consider an n-cube where the individual processors or subcubes are represented by an n-bit string of ternary symbols from $\sigma = \{0, 1, x\}$, where x denotes a 'don't care'. For example, in a 2-dimensional hypercube, $1x$ denotes the processors 10 and 11, whereas xx denotes all the four processors in the hypercube.

Allocation and deallocation of processors may result in 'physical fragmentation' of the hypercube, where an incomplete subcube can not be formed even with a sufficient number of free processors. This may occur due to the sequence of incoming and outgoing tasks or simply a 'bad' allocation strategy [1].

In addition to the physical fragmentation described

---

*This work was supported in part by AFOSR.

†This author is a Ph.D. degree candidate at the Dept. of ECE, Univ. of Mass., Amherst.

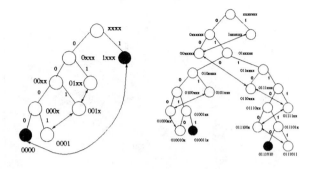

Figure 1: A 4-D Hypercube   Figure 2: ISC representation of a 7-D Hypercube

above, 'virtual fragmentation' of the hypercube may occur due to the allocation policy implemented. An allocation policy may not recognize an existing incomplete subcube; this forces an incoming task to wait.

2.1 Definitions and Notations

Definition 1 The **Hamming Distance** [6] between two subcubes $a = a_1 a_2 ... a_n$ and $b = b_1 b_2 ... b_n$; where $a_i \epsilon \sigma$ and $b_i \epsilon \sigma$ for $i \epsilon [1, n]$; can be defined as $H(a, b) = \Sigma_{i=1}^{n} h(a_i, b_i)$; where $h(a_i, b_i) = 1$ if $a_i \neq b_i$ and $a_i, b_i \epsilon \{0, 1\}$ and 0 otherwise. For example, $H(00x, 1xx) = 1$ and $H(1x0, x0x) = 0$.

Definition 2 The **Exact Distance** [6] between the subcubes a and b is $E(a, b) = \Sigma_{i=1}^{n} e(a_i, b_i)$; where $e(a_i, b_i) = 0$ if $a_i = b_i$ and 1 otherwise. For example $E(0x1, 10x) = 3$ where as $H(0x1, 10x) = 1$.

Definition 3 An **Incomplete Subcube** (ISC) S can be defined as follows:
(i) It consists of a group of disjoint subcubes $\{S_1, S_2, ..., S_m\}$, $(1 \leq m \leq n)$, with dimensions $d_1, d_2, ..., d_m$ respectively $(d_1 > d_2 > ... > d_m)$ [1].
(ii) $H(S_i, S_j) = 1$ for all $1 \leq i, j \leq m, i \neq j$.
(iii) $E(S_i, S_j) = d_i - d_j + 1$ for all $1 \leq i \leq j \leq n$.
d_1 is the **dimension** and $d = \sum_{i=0}^{n} 2^{d_i}$ is the **size** of the ISC S. S_1 is called the **head** of the ISC S.
Example 1 The subcubes $\{1xxx, 00xx, 010x\}$ form a 11-node ISC (i.e. size=11) of dimension 3 and $1xxx$ is the head. But subcubes $\{00xx, x100\}$ do not form an ISC as the exact distance between them is 3 though the hamming distance is 1.

Definition 4 A **binary representation** of a hypercube is a binary tree, where nodes in level i denote a subcube of dimension $n - i$ (Fig. 1). For example, the root node (level 0) represents the entire hypercube, its left and right children denote the subcubes $0xxx$ and $1xxx$ respectively and so on.

Definition 5 A **Sibling generated Incomplete Sub-**

Cube (SISC) is an incomplete subcube consisting of the subcubes $\{S_1, S_2, ..., S_m\}$ $(1 \leq m \leq n)$ such that the sibling of S_i is the common ancestor of all the subcubes $\{S_{i+1}, S_{i+2}, ..., S_m\}$, for all $i \epsilon [1, m - 1]$ in the binary tree representation.

Example 2 (Fig. 1) The free subcubes $\{01xx, 001x, 0001\}$ form a SISC of size 7 as the sibling of $01xx$ $(00xx)$ is the ancestor of $001x$ and 0001 and the sibling of $001x$ $(000x)$ is the ancestor of 0001. [2]

Definition 6 **Maximal Set of Incomplete Subcubes (MSIS)** is a set of free, disjoint ISCs [1] that is greater than or equal to all other sets of free disjoint ISCs of the same set of free nodes. For example, in a 4-D hypercube, if all nodes other than 0000, 1000, 1001, 1111 are free; $\{01xx, 00x1, 0010\}$, $\{110x, 1110\}$ and $\{101x\}$ form a MSIS whereas $\{x01x, 011x, 0101\}$, $\{110x, 1110\}$, $\{0001\}$ and $\{0100\}$ is not a MSIS as the former has more number of ISCs of size 2.

3   Existing Allocation Schemes

The existing allocation strategies may be broadly classified as the bit mapping strategies and the list type strategies. The bit-mapping strategies (e.g. Buddy [2], Gray Code [3], Modified Buddy [4] and Tree Collapsing [7]) have the disadvantage of suffering from physical as well as virtual fragmentation in addition to their exponential and even super-exponential time complexities [1]. However, the problem of internal fragmentation may be alleviated to some extent, by searching for the required number of processors instead of a full subcube. Thus these strategies may not be suitable for processor allocation.

The list type strategies (Free List [6], Prime Cube [8] and MSS [5]) also have the disadvantage of physical as well as virtual fragmentation due to their associated heuristics [1] in addition to their super-exponential time complexities. Recognizing the ISCs in list type strategies is non-trivial as the relationship among free subcubes in various lists has to be examined [1], causing an additional time overhead (Table 1). This time complexity will further increase if we try to allocate the 'best' ISC in order to reduce physical fragmentation. Table 1 provides a detailed description of the time and space complexities of various strategies. A detailed discussion on the various aspects of these strategies appears in [1].

We define the goal of an allocation (cubic and noncubic) strategy as the ability to maintain the Maximal Set of Incomplete Subcubes (MSIS) after every allocation and deallocation. We use a strategy that maintains a MSIS of ISCs of the type SISC. The allocation and deallocation time complexities of the proposed strategies are polynomial. Because they are

---

[1]For the rest of this paper it will be assumed that the subcubes are arranged in the decreasing order of dimensions, unless otherwise mentioned.

[2]Throughout the paper a shaded node indicates an allocated node.

best-fit strategies, the proposed allocation schemes do not cause significant physical fragmentation and exhaustively coalesces a released subcube; if it is recognizable by the strategy. In addition, the proposed schemes are capable of recognizing the adjacency of upto n subcubes in the tree, unlike free list and prime cube; the only two non-cubic allocation strategies in the literature. We also do not incur the penalty of higher job turn-around time due to reduced bisection bandwidth, unlike the prime-cube approach. This is because we recognize SISCs, which are essentially incomplete subcubes (Theorem 1 of [1]). Although our proposed strategies do not have complete ISC recognition capability, simulation results indicate that they exhibit better performance than all the previously proposed strategies due to their best-fit nature and low time overheads.

4 The Proposed Strategy

In the proposed strategy, a dynamic binary tree along with an array of free lists is used for processor allocation. The nodes in the tree represent various subcubes. If a subcube is free it does not have any descendants in the binary tree. Hence, initially, when the entire hypercube is free, the tree consists of the root node only. As requests are being honored, the tree branches out. A leaf node can represent either an allocated subcube or a free subcube, whereas a nonleaf node indicates a partially allocated subcube to task(s). In the rest of this Section 'ISC' refers to SISC only and 'MSIS' refers to MSIS of type SISC only.

Example 3 Consider a 4-D hypercube whose dynamic binary tree representation is shown in Fig. 1. The leaf nodes $1xxx$ and 0000 represent allocated subcubes whereas the leaf nodes $01xx$, $001x$ and 0001 represent the free subcubes. The non-leaf nodes $xxxx$, $0xxx$ etc. are partially allocated.

A higher dimensional free subcube can combine with a lower dimensional free subcube to form an ISC, if the former's sibling is an ancestor of the latter. This process can repeat and we may have a SISC type incomplete subcube of upto n subcubes. For example, in Fig. 2, the free subcubes $1xxxxxx$, $00xxxxx$, $0110xxx$, $01111xx$, $011100x$ and 0111011 together form an incomplete subcube of $64 + 32 + 8 + 4 + 2 + 1 = 111$ processors and the free subcubes $0101xxx$, $01000xx$ and 010010 form an incomplete subcube with 13 processors.

Incomplete subcubes are represented in the algorithm by a bidirectional link between two adjacent subcubes; the subcubes being arranged in the decreasing order of their dimensions. The highest dimensional subcube is the head of the ISC. Free ISCs are kept in an array of lists ('isc'). The list isc has $n + 1$ entries in it. The ISCs in these lists are all mutually disjoint.

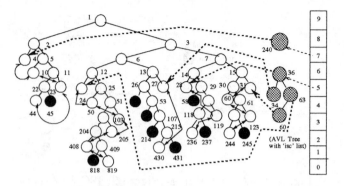

Figure 3: Representation of the 'isc' list in a 9-D Hypercube. The lightly shaded nodes represent the entries for various ISCs in the isc list arranged as an AVL tree.

The ISCs of dimension i are arranged as a height-balanced AVL tree [15], the key being the number of processors in the ISC. $isc[i]$ points to the root of AVL tree associated with dimension i. Fig. 3 illustrates the free ISCs associated with a hypercube of dimension 9. Free ISCs belonging to the AVL tree associated with any isc list will be represented by dotted lines as shown in Fig. 3 where nodes in the AVL tree point to the head of the ISC they represent.

In Fig 3 the ISC formed by nodes $\{4, 10, 23, 44\}$ has 240 processors and is of dimension 7. Hence it has a corresponding entry in $isc[7]$ as the head (node 4) is of dimension 7. There are four ISCs of dimension 5 in the system and their corresponding entries are arranged according to the number of processors in them in the form of an AVL tree (Fig. 3).

Whenever a node (or a set of nodes) becomes free it combines with its sibling, if free, to form the next higher dimension subcube. The parent is marked free and both the siblings are removed from the tree. This process continues and the corresponding free list(s) (in the form of AVL tree(s)) is updated.

Example 4 Fig. 1 represents a 4-D hypercube. Subcubes $1xxx$ and 0000 are allocated to a task I_j and the rest of free subcubes $01xx$, $001x$ and 0001 form a free ISC with 7 nodes. When I_j releases the subcubes $1xxx$ and 0000, they combine with the existing free subcubes and the entire hypercube becomes free.

4.1 Noncubic Allocation

In this subsection we discuss algorithms for noncubic allocation that work efficiently for cubic allocation as well. The algorithms proposed in this subsection will use the basic data structure described above.

4.1.1 Algorithm 1

In this approach each node N in the dynamic binary tree maintains a pointer to the head of the largest ISC ('iscptr') beneath N (i.e., all subcubes in this ISC are descendants of N) along with the number of processors in the ISC ('iscnodes'). For instance, in Fig. 3, the iscptr entry of node 3 will point to the ISC headed by 24 with 63 processors. Thus its iscnodes entry is 63. A completely allocated node, a free node (i.e. a leaf node in the tree) and a nonleaf node, all of whose descendants are allocated, will have its iscptr set to NULL and iscnodes set to 0. Free ISCs are kept in the form of an AVL tree associated with a list isc as described earlier. During allocation, the ISC with the minimum number of nodes, that can satisfy the request, is chosen. The required subcubes are extracted from this ISC and the rest are put in the form of one (or two) free ISC(s). The corresponding entry (entries) in the isc list is updated and so are the iscptr and iscnodes of the ancestors of the head(s) of the ISC(s). A formal description of the allocation algorithm follows. In this algorithm allocatedlist list maintains the list of subcubes to be allocated to the task after execution of the algorithm; newlist maintains the list of higher dimensional subcubes that are not required for the current allocation in progress; brokenlist maintains the list of free subcubes (not to be allocated) generated if we have to fragment a higher dimensional subcube to satisfy the request.

Allocation

Step 1 Form the dimensions $D_1, D_2, ..., D_m$ of the subcubes required to satisfy the task I_j (requiring D processors) arranged in the decreasing order.

Step 2 Search the isc lists from $isc[D_1]$ onwards and choose the ISC $S = \{S_1, S_2, ..., S_t\}$ with the minimum number of processors that can satisfy the request I_j (dimensions of S are $d_1, d_2, .., d_t$ respectively). If no such ISC is found, keep the request in waiting queue. Skip the remaining steps.

Step 3 $i = j = 1$. $allocatedlist = newlist = brokenlist = NULL$. (Initialization)

(d_i is the highest dimension subcube in S that may be allocated to I_j. D_j is the highest dimension subcube in I_j that is being considered for allocation. The following Steps illustrate the various scenarios that may arise based on the relationship between d_i and $D_j, D_{j+1}, ..., D_m$.)

Step 4 If $d_i > D_j$ and the available subcubes $\{S_{i+1}, S_{i+2}, .., S_t\}$ of S can satisfy the request for subcubes of dimensions $D_j, D_{j+1}, ..., D_m$ yet to be allocated : Append S_i to newlist; $i = i + 1$. Go To Step 4. (Ex. 5) (S_i need not be allocated since lower dimensional subcubes can satisfy the request.)

Step 5 If $d_i = D_j$: (i.e. S_i has to be allocated)
(a) Append S_i to allocatedlist. (Example 5)
(b) If $j = m$ (all subcubes allocated) : $list = \{S_{i+1}, ..., S_t\}$. Go To Step 7. Allocate the subcubes in allocatedlist to I_j. Skip the remaining Steps.
(c) If the sibling S_i' of S_i has enough number of pro-

cessors in the maximal ISC $\{\bar{S}_1, \bar{S}_2, ..., \bar{S}_l\}$ beneath it to satisfy the request for subcubes of dimensions $D_{j+1}, D_{j+2}, ..., D_m$ do the following : (Example 6) (trying to see if lower dimensional ISCs can be used instead of the remaining subcubes in S)
(i) Remove the ISC headed by $\bar{S}_1$ from the isc list and update its ancestors.
(ii) $list = \{S_{i+1}, S_{i+2}, .., S_t\}. S = \{\bar{S}_1, \bar{S}_2, .., \bar{S}_l\}. i = 1. j = j + 1$. Go to Step 7. (Example 7)
(Step 7 tries to form maximal ISC(s) out of unused subcubes of previous ISC S.)
(d) Go to Step 4.

Step 6 If $d_i > D_j$: (Now subcube S_i can be used to satisfy the remaining requests.)
Break the subcube S_i to allocate the remaining subcubes of dimensions $D_j, D_{j+1}, ..., D_m$ as follows : (Example 8)
(a) Form the two children of S_i in the dynamic binary tree (of 1 dimension less).
(b) If the dimension of the sibling $= D_j$: (i) Allocate the right sibling to I_j by appending it to allocatedlist. (ii) $S_i =$ left sibling of S_i. (iii) $j = j + 1$.
Else (i) Append the left sibling to brokenlist (to be released). (ii) $S_i =$ right sibling of S_i.
(c) If $j \leq m$ Go To Step 6a. (Continue allocation.)
(d) If the number of processors in brokenlist is greater than that in $\{S_{i+1}, S_{i+2}, ...\}$: • Insert $\{S_{i+1}, S_{i+2}, ...\}$ to the appropriate isc list and update the ancestors (i.e. their iscptr and iscnodes entries) of S_{i+1}. $list = brokenlist$.
Else: • Insert brokenlist to the appropriate isc list and update the ancestors. $list = \{S_{i+1}, S_{i+2}, ...\}$.
(e) Go to Step 7.
(f) Allocate the subcubes in allocatedlist to the task I_j. Skip the remaining steps.
(This Step is used by Steps 5 and 6 only. It forms maximal ISCs out of the free subcubes in list and newlist and returns to the point from which it was invoked.)

Step 7 (a) If newlist is empty : Insert the subcubes in list to the appropriate isc list and update the ancestors of its head and RETURN .
(b) Let L be the lowest dimension subcube of newlist and L_s be its sibling. Remove L from newlist.
(c) If the maximal ISC beneath L_s has more number of nodes than list :
• Insert the subcubes in list to the appropriate isc list and update the ancestors of its head.
• Remove the maximal ISC beneath L_s (I_s) from the isc list and update the ancestors of its head.
• $list = \{L\}$ append $\{ I_s \}$.
Else : Add L to head of list (Example 12).
(d) Go to Step 7a.

Example 5 Consider the scenario of Fig. 4. Suppose we have a request for 149 processors (dimensions

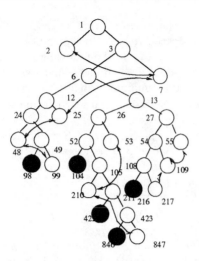

Figure 4: Representation of ISCs in a 9-D hypercube

7, 4, 2, 0) (Step 1). The ISC headed by node no 2 $S = \{2, 7, 25, 48, 99\}$ is selected from $isc[9]$ as $isc[7]$ and $isc[8]$ are empty (Step 2). Subcube 2 of dimension 9 need not be allocated as the rest of the subcubes can satisfy the request. Node 7, however, needs to be allocated and hence it is kept in *allocatedlist* whereas 2 goes to *newlist*.

Example 6 (Continuing from Ex.5) The sibling of the allocated node 7 (node 6) is checked to see if the maximal ISC beneath it (ISC $\{55, 109, 217\}$ with 28 processors) can satisfy the remaining dimensions 4, 2 and 0. Since, the ISC headed by 55 can do so, we use that instead of the subcubes $\{25, 48, 99\}$ in an attempt to preserve higher ISCs. However, the choice of the ISC $\{53, 210, 847\}$ would have maintained the MSIS after allocation. This drawback arises as nodes do not store all the ISCs beneath them to make the 'best' choice. The second algorithm overcomes this shortcoming by maintaining all the ISCs beneath a node in the form of an AVL tree.

Example 7 (Continued from Ex. 6) After the ISC headed by 55 is chosen for allocation, the free subcubes of the current ISC (headed by 2) are to be deallocated. Nodes 25, 48, 99 are in *list* and 2 is in *newlist*. After removing the ISC headed by 55, the sibling of node 2 (which is the lowest dimension subcube in *newlist*) is checked for its maximal ISC. In this case the *list* has more processors than the maximal ISC beneath 3 (headed by 53 with 21 processors). Thus the node 2 is removed from *newlist* and is added to the head of *list*. Since *newlist* becomes empty, we include this new ISC $\{2, 25, 48, 99\}$ in the free list of isc and update the ancestors of 2.

Example 8 Suppose we have the scenario of Fig. 4 and a request of size 57 (dimensions 5, 4, 3, 0). We have to choose the ISC headed by 2. Subcubes 2 and 7 go to *newlist*. Node 25 has to be allocated

as the remaining subcubes in the list cannot meet the demand. Thus node 25 (with 64 processors) has to be fragmented to allocate 57 processors from it. Node 25 branches out (not shown in the figure) and nodes $\{51, 101, 201, 1607\}$ will join *allocatedlist* and *brokenlist* will contain subcubes not allocated, i.e., $\{400, 802, 1606\}$ consisting of 7 processors. Since this is less in number than the remaining processors of S (48, 49) (Step 6d), the subcubes $\{400, 802, 1606\}$ is inserted in $isc[2]$ and the parents of 400 updated (i.e. nodes 200, 100, 50, 25 point to 400 as the highest ISC beneath them with 7 nodes). Hence $list = \{48, 99\}$ and $newlist = \{2, 7\}$.

Example 9 (Continuing from Ex. 8) Following the procedures mentioned in Step 7 $\{48, 99\}$ join $isc[5]$ as $\{2, 7\}$ combine with a higher ISC $\{55, 109, 217\}$ to form an ISC to join $isc[8]$.

Deallocation

The deallocation procedure maintains the maximal set of subcubes of type SISC (Theorem 2 in [1]). The algorithm is described below, followed by examples, illustrating the different steps of the algorithm.

Step 1 Let $L = \{S_1, S_2, ..., S_m\}$ be the subcubes of the ISC (to be deallocated) arranged in the *increasing* order of their dimensions $d_1, d_2, .., d_m$ respectively.

(*Steps 2-4 try to combine free sibling nodes to form higher dimensional free subcubes as much as possible.*)

Step 2 $i = 1$. Remove S_1 from L.

Step 3 If the sibling S_i^s of S_i is free : (i.e. S_i^s has no subcubes beneath it in its ISC now)
- If S_i^s is the head of the ISC : Remove the ISC from the isc list and update the ancestors of S_i^s.
- Combine the two siblings S_i^s and S_i by removing them from the dynamic binary tree and mark their parent as S_i. Go To Step 3.

Step 4 If $S_i^s = S_{i+1}$ (*i.e. the sibling is itself among the released subcubes.*)
- Remove S_{i+1} from L. Eliminate S_i and S_{i+1} from the tree and mark their parent as S_{i+1}.
- $i = i + 1$. Go To Step 3.

(*Combining free sibling nodes to form the free parent node ends here. Now we have to form maximal ISCs out of the released subcubes by traversing up the tree. S_i is the lowest dimensional free subcube. S denotes the ancestor of S_i whose sibling S^s is examined to form the maximal ISCs.*)

Step 5 If S_i is the root of the dynamic binary tree (i.e. the entire hypercube is free) : insert S_i in $isc[CUBESIZE]$ and skip the remaining steps.

Step 6 $S = S_i$. $S_i^s =$ Sibling of S_i. $I_i^s =$ maximal ISC under S_i^s. $list = \{S_i\}$. $L = L - \{S\}$.

(*S_i being the lowest dimensional free subcube in L, I_i^s denotes a lower dimensional ISC that can be readily appended to L to form a higher ISC. list maintains the free subcubes*) *that will form an ISC.)*

Step 7 S = Parent of S. If S is not root of the tree S^s = Sibling of S.

Step 8 If S is the root of the dynamic binary tree : (search over)

(a) If I_i^s is not NULL : Remove I_i^s from the corresponding *isc* list; Update ancestors of the head of I_i^s and *append* I_i^s to *list*.

(b) Insert *list* in the appropriate *isc* list and update the ancestors of its head.

(c) Skip the remaining Steps.

(Step 9 tries to combine the sibling S^s of the ancestor S of the released subcube S_i.)

Step 9 If S^s is free do the following Steps :

(a) If S^s has more number of nodes beneath it in its ISC than that in *list* and I_i^s combined :

(i.e., it will yield an ISC of higher size if we choose the ISC to which S^s belongs to; instead of the current list)

• If I_i^s is not NULL remove the ISC I_i^s from the *isc* list and update the ancestors of its head. Append I_i^s to *list*. I_i^s = NULL.

• Insert the ISC formed by *list* in the appropriate *isc* list and update the ancestors of the head of *list*.

• If L is empty skip the remaining Steps. (No higher ISCs possible.)

• If S^s is the head of an ISC : Remove the ISC headed by S^s from the *isc* list and update the ancestors of S^s and name them as *list*. Else : Remove the subcubes from S^s onwards from the ISC and name them as *list*.

• $S = S^s$. Go To Step 6.

(b) *(S^s has less number of nodes than list and I_i^s. Hence, the ISC to which S^s belongs is altered. The subcubes of dimensions lower than S^s are discarded and subcubes in list is combined with S^s instead. The discarded subcubes form a separate ISC.)*

• If S^s is head of an ISC : Remove the ISC from the corresponding *isc* list and update the ancestors of S^s.

• Remove the subcubes beneath S^s in the ISC and form a new ISC (call it N).

• If the subcubes in N are descendants of S_i^s and $S_i \epsilon list$: Append N to *list*. I_i^s = NULL.

Else If I_i^s is not NULL :

- Remove I_i^s from the *isc* list and update the ancestors of its head.

- Append I_i^s to *list*. Set I_i^s = NULL.

- Insert N to the appropriate *isc*. Update ancestors of its head.

- *list* = S^s append *list*. Go To Step 6.

Step 10 If $S^s = S_{i+1}$: Remove S_{i+1} from L; Add S_{i+1} to the head of *list*, $i = i + 1$ and Go To Step 6.

(The sibling node S^s turned out to be a higher dimensional released subcube.)

(Since S^s is not free we check if any of its highest descen-

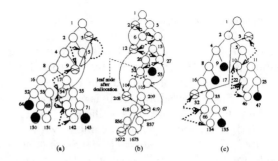

Figure 5: Examples of deallocation of ISCs in a 9-D hypercube. The unshaded crossed nodes represent the ISC being deallocated. The resulting ISCs after deallocation are indicated by a dark dotted bidirectional link

dant ISC (iscptr), if any, has more number of processors than what we have accumulated so far from the released nodes. If so, this ISC will combine with the higher dimensional released node to form a higher ISC.)

Step 11 If S^s is not free (partially allocated) and has more number of nodes in the maximal ISC beneath it than that in *list* and I_i^s combined do the following steps. (Since the maximal ISC beneath S^s has more number of nodes than our current *list* we choose the former to combine with higher dimensional subcubes.)

(a) If I_i^s is not NULL : Remove I_i^s from the corresponding *isc* list and update the ancestors of its head. Append I_i^s to *list*.

(b) Insert the ISC *list* in the appropriate *isc* list. Update the ancestors of its head.

(c) If S = NULL : Skip the remaining Steps.

(d) Remove the highest ISC beneath S^s from the *isc* list; update its ancestors. *list* = this new ISC.

Step 12 Go to Step 6.

Initially we try to combine the released nodes with the free subcubes (released as well as existing ones). For example, in Fig. 5b, the subcubes denoted by node no.s 2, 12, 208, 419 and 1672 are released (each denoted by a unshaded circle that is crossed). Subcube 1672 combines with subcube 1673 (which is free) both are eliminated and their parent 836 is marked free. This process continues and we have node 52 free and all its children in the tree are removed. At this point $L = \{2, 12, 52\}$ and we enter Step 6 as the entire hypercube is not free (Step 5). The process of forming ISC(s) starts with the lowest dimensional node 52. It is removed from L and *list* = $\{52\}$. The sibling of node 52 (node 53) has no ISC beneath it and $I_i^s = NULL$ (Step 6). Since the node 27 (S^s, as the sibling of ancestor node $S = 26$) is free, it joins the *list*. Thus, *list* = $\{27, 52\}$, $S = 13$ and $S^s = 12$ (Step 7). Since S^s is in L it is added to *list* after being removed from L (Step 10). Thus $L = \{2\}$, $list = \{12, 27, 52\}$. In the next iteration $list = \{2, 7, 12, 27, 52\}$ and $L = \{\}$. The algorithm terminates after inserting the ISC formed by $list$ to $isc[8]$.

In the scenario of Fig. 5a the released ISC $L = \{3, 9\}$. Since the sibling of 9 is not free, we come to Step 6 without any tree collapsing. Here, $I_i^s = \{33, 65\}$, which can potentially combine with 9 to form an ISC. $L = \{3\}$, $list = \{9\}$. Since 5 is a free node we can combine 5 with 9 as the lower dimensional subcubes ($\{34, 70, 142\}$) in the ISC associated with 5 have lower number of processors than what $list$ and I_i^s can potentially offer (Step 9b). The subcubes $\{34, 70, 142\}$ are the descendants of 8 (sibling of 9) and have more number of nodes than I_i^s (Step 9b) we form the $list = \{5, 9, 34, 70, 142\}$ which further combines with node 3 to form the ISC $\{3, 5, 9, 34, 70, 142\}$ which goes to $isc[8]$ before the algorithm terminates.

Figure 5c depicts another scenario during deallocation. Here the released subcubes $\{3, 32\}$ form the ISCs $\{32, 66, 134\}$ and $\{2, 10, 22, 46\}$. The formation of two ISCs is to maintain MSIS after deallocation by using Step 11 [1].

4.2 Algorithm 2

The allocation strategy proposed in algorithm 1 does not maintain the MSIS after every allocation, as illustrated in Example 6. In this approach, each node maintains a list of **all** ISCs beneath it in the form of an AVL tree [15], the key being the number of processors in the corresponding ISCs. The AVL tree helps to maintain the polynomial time complexities [1]. Updating the ancestors will involve updating the AVL trees associated with them. The deallocation and allocation procedures remain basically the same as in Algorithm 1 with the following modification to the allocation procedure in Step 5c.

Step 5 c (i) Remove the minimal ISC $\{\bar{S}_1, \bar{S}_2, ..., \bar{S}_l\}$ beneath $\bar{S}_i$ (found by searching the AVL tree associated with $\bar{S}_i$) from the corresponding isc list. Update all the ancestors of $\bar{S}_1$.

Examples of the use of this algorithm appears in [1].

4.3 Noncubic and Cubic Allocation : An Integrated Approach

The proposed non-cubic algorithms perform extremely well for cubic allocation too as observed by simulations. Although the objectives of maintaining MSIS and MSS [5] may seem to be contradictory in many cases; maintaining the MSIS often acts as a 'good' look-ahead allocation scheme for cubic allocation and outperforms the MSS-based strategies [1].

4.4 Analysis of the Proposed Strategies

The allocation and deallocation time complexities of Algorithm 1 are $O(n^2)$ each and that for Algorithm

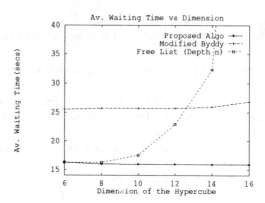

Figure 6: Av. Waiting Times for Uniform Distribution

2 are $O(n^3)$ each. The space complexity of Algorithms 1 is $\Omega(n)$ and $O(2^n)$ and Algorithm 2 is $\Omega(n)$ and $O(n2^n)$. A step by step derivation of these time and space complexities appears in [1]. The O, Θ, Ω notations used here are the same as in [15].

The time complexities of allocation and deallocation along with the space complexity of our algorithm is compared with the existing algorithms in Table 1. Depth indicates the number of adjacent subcubes that can form an ISC. The time and space complexities of these algorithms have been derived in [1].

5 Simulation Results and Performance Analysis

The performance of the two proposed strategies is compared to that of free-list[3] and the modified buddy strategy as they are the representative of the existing strategies [1]. The simulator is written in C and is run on a DEC-Station 5000. The simulation is event-driven, the events being allocation and deallocation of tasks and the actual CPU time for allocation/deallocation was also taken into account. The parameters of interest are (i) the average waiting time of a job, (ii) the average allocation time, (iii) the average deallocation time and (iv) the memory required. The following assumptions are made: (i) *Exponential* interarrival time. (ii) *Exponential* service time. (iii) FCFS Scheduling. and (iv) *uniform* and *pseudo-normal* ISC sizes.

Our interest is primarily in the steady-state behavior of the system under different allocation strategies.

One set of simulations was performed by changing the dimension of the hypercube and keeping all the other parameters constant. The average interarrival time and the average task completion time were assumed to be 11 seconds and 10 seconds, respectively to have a 'balanced' workload. The ISC size distribu-

---

[3]The procedure for free list was provided by J. Kim and C. R. Das

Strategy	Depth	No ISCs	Allocation	Deallocation	Memory	Type
Buddy	n	$(n-2).2^{n-1}+3$	$O(2^n)$	$O(2^n)$	$\Theta(2^n)$	first-fit
Gray	n	$n.2^{n+1}-3.2^n+1$	$O(2^n)$	$O(2^n)$	$\Theta(2^n)$	first-fit
M. Gray	n	3^n+2^{2n} $-5.2^{n-1}-1$	$O(C^n_{\lfloor\frac{n}{2}\rfloor}.2^n)$	$O(C^n_{\lfloor\frac{n}{2}\rfloor}.2^n)$	$\Theta(C^n_{\lfloor\frac{n}{2}\rfloor}.2^n)$	first-fit
Mod. Buddy	n	$n^2.2^n-2^{n+1}+1$	$O(C^n_{\lfloor\frac{n}{2}\rfloor}.2^n)$	$O(2^n)$	$\Theta(2^n)$	first-fit
TC	n	3^n+2^{2n} $-5.2^{n-1}-1$	$O(C^n_{\lfloor\frac{n}{2}\rfloor}.2^n)$	$O(2^n)$	$\Theta(2^n)$	first-fit
PC Graph	2	$n.2^{2n-1}-n.2^n$ $-(n-1).3^n+3n$	$O(n^{-2}.3^{3n})$	$O(n^{-1}.3^{2n})$	$O(n^{-1}.3^{2n})$	first-fit
Free List (Depth 2)	2	$n.2^{2n-1}-n.2^n$ $-(n-1).3^n+3n$	$O(n^3.2^n)$	$O(n^3.2^n)$	$O(n.2^n)$	first-fit
Free List (Depth n)	n	$O((n!)^2)$	$O(n^{-n}.2^{n^2}$ $+n^3.2^{2n})$	$O(n^3.2^{2n})$	$O(n.2^n)$	first-fit
Proposed-1	n	2.4^n	$O(n^2)$	$O(n^2)$	$O(2^n)$	**best-fit**
Proposed-2	n	2.4^n	$O(n^3)$	$O(n^3)$	$O(n.2^n)$	**optimal**

Table 1: **Analysis of Various Strategies**

tion was assumed to be uniform. It can be seen that the proposed strategies perform the best in terms of the average waiting time of a job (Fig. 6) The modified buddy strategy performs poorly even for lower dimensions of the hypercube, presumably due to its first-fit nature and the lower ISC recognition capability. The free list strategy does not perform well when the search depth is 2. However, for the depth of n, the free list strategy is comparable to the proposed strategies [1] for lower dimensions. However, as the dimension of the hypercube increases, the performance of the free list with depth n starts degrading to the extent that it has a delay of about 4.5 times more than that of modified buddy and about 7 times worse than the proposed strategies for a 16-dimensional hypercube. This can be attributed to the heuristics used in deallocation [1] which do not perform well when the dimension increases and to the associated time overheads in allocation and deallocation. The variance of the modified buddy scheme is much higher than the proposed strategies (Fig. 7). The variance of FL is comparable for lower dimensions, but for higher dimensions it performs poorly with variance being approximately 100 times worse than ours for a 16-D hypercube. This suggests a higher predictability in waiting times in addition to the lower waiting times for our strategies when compared to the other strategies. The arrival rate was also varied keeping the dimension of the hypercube constant. Results for various dimensions indicate that our strategy outperforms the other two under all traffic conditions due to reasons mentioned above. Figure 9 shows the response of average delay for various arrival rates for a 12-D hypercube.

A second set of simulations were run assuming the ISC size distribution to be $p_i = \frac{(i+1)mod2^{n-1}}{\sum_{i=0}^{2^n}(i+1)mod2^{n-1}}$. This distribution mimics the normal distribution. The trends of the various performance parameters are exactly similar to that of the first set of simulations [1].

A third set of simulations were run assuming a job-mix of both cubic and non-cubic requests. The probability of having a cubic request was varied for a 12-dimensional hypercube (since the FL exhibits poor performance for higher dimensions). The average waiting delays are shown in Fig. 8. It can be observed that our scheme outperforms the FL strategy even when all the requests are just cubic. Although the FL strategy tries to maintain the MSS (as opposed to our MSIS) and has full subcube recognition capability, our strategy outperforms the FL even for just cubic allocation, possibly due to the reasons mentioned in Section 4.3. Since the Modified Buddy exhibited poor performance its graph is deleted from Fig. 8 to reduce the scaling factor.

In all the three sets of simulations our strategy had the least execution times for both allocation and deallocation. Figure 10 presents results for the first simulation set. The free list performs the worst, which is understandable, given the time complexities in Table 1. The modified buddy also has a higher average allocation time than the proposed strategies as expected. The free list had a large average deallocation time (e.g.

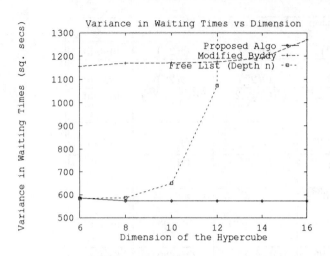

Figure 7: Variance in Waiting Time vs Dimension

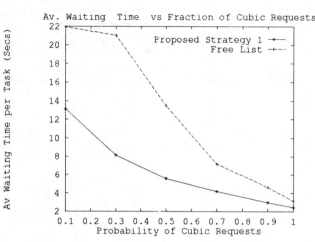

Figure 8: Av. Waiting Times for various job-mixes

more than 2000 times than ours for a 16-dimensional hypercube), which probably is a contributing factor to the high waiting times. A typical scenario would be a job waiting to get its turn for allocation while the host is busy deallocating other ISCs. The modified buddy also has a higher average deallocation time than the proposed strategies. The allocation and deallocation times for both the proposed strategies are almost identical. The variance in allocation and deallocation times were also least for our strategies [1] which suggests a higher predictability in these times.

In all the three sets of simulations both our strategies performed identical. The difference in time complexities arise because the second algorithm keeps track of all the ISCs beneath a node whereas the first algorithm just keeps the maximal ISC. Possibly, for the sizes of hypercubes we have considered, not many nodes in the dynamic binary tree have many ISCs beneath them to make an impact in the performance and allocation/deallocation time. It may be possible that if we further increase the dimension of the hypercube we may see some difference in their performance under high arrival rates.

A fourth set of simulations were run to determine the maximum amount of memory required by various strategies (Fig. 11). The modified buddy requires fixed amount of memory throughout as it is a bit-mapping strategy. For lower dimensions (Fig. 11), the modified buddy has the lowest memory requirement as the overhead of maintaining the free lists in terms of pointers and other information per node in the dynamic binary tree dominates. However, as the dimension of the hypercube increases, our strategy has the least memory requirement, as the dynamic binary tree saves us the space by pruning the nodes that are not needed. The Free-list has the worst memory requirement as it needs to form all possible subcubes from a released subcube.

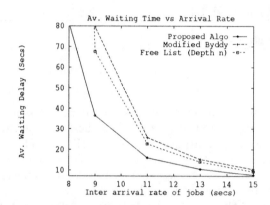

Figure 9: Uniform Distribution for a 12-D Hypercube

This scenario does not change even under extremely high traffic rates [1].

From the above four sets of simulations it can be said that our strategy performs the best in terms of all the four parameters of interest for both cubic and non-cubic requests. Though our ISC recognition capability could be better, the low time complexities and our capability to effectively maintain the MSIS of ISCs of type SISC after every allocation and deallocation along with our lower memory requirement for higher dimensions makes it extremely effective. This effect becomes more prominent as the dimension of the hypercube increases.

6 Conclusions

The sizes of the manufactured hypercubes are increasing and expected to keep increasing. Currently, NCUBE manufactures upto 8192 node hypercube systems and is targeting for 65536 node hypercube systems for tera-flops performance by 1995 [14]. As this size grows, it will become virtually impossible to use

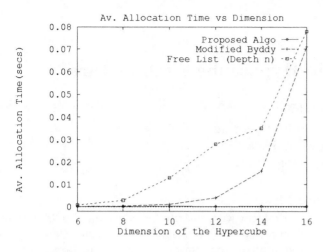

Figure 10: Av. Allocation Time vs Dimension

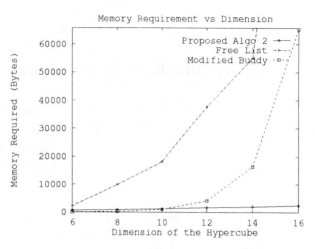

Figure 11: Memory Requirement of Various Strategies

near optimal algorithms with exponential and super-exponential time complexities given the limitations on processor speed imposed by the technology. Using another multiprocessor system to alleviate this problem may prove to be rather expensive and most of the existing algorithms may not be parallizable. It will, thus, become desirable to implement efficient allocation and deallocation algorithms with very low time and space complexities, so that host computers do not just perform allocations and deallocations. This will ensure lower turn-around times for individual tasks too. Because the proposed scheme is polynomial in both allocation and deallocation, has a low memory overhead and has the lowest average waiting times for jobs among all existing strategies, it can indeed be a practical solution.

References

[1] D. Das Sharma and D. K. Pradhan, "Novel Strategies for Cubic and Noncubic Allocation in Hypercube Multiprocessors", Tech. Rep. TR-93-023, Dept. of Computer Science, Texas A&M University.

[2] K. C. Knowlton, "A fast storage allocator", *Communications of the ACM*, Vol. 8, No. 10, Oct 1965, pp. 623-625.

[3] M. S. Chen and K. G. Shin, "Processor Allocation in an N-Cube Multiprocessor Using Gray Codes", *IEEE Transactions on Computers*, Vol. C-36, No. 12, Dec 1987, pp. 1396-1407.

[4] A. Al-Dhelaan and B. Bose, "A new strategy for processor allocation in an N-cube multiprocessor", *Proc. Int. Phoenix Conf. Comput. Commun.*, Mar 1989, pp. 114-118.

[5] S. Dutt and J. P. Hayes, "On Allocating Subcubes in a Hypercube Multiprocessor", *Proc. Third Conf. on Hypercube Computers and Applications*, Jan 1988,

pp. 801-810.

[6] J. Kim, C. R. Das and W. Lin, "A Top-Down Processor Allocation Scheme for Hypercube Computers", *IEEE Transactions on Parallel and Distributed Systems*, Vol 2, No 1, Jan 1991, pp. 20-30.

[7] P. J. Chuang and N.-F. Tzeng, "Dynamic Processor Allocation in Hypercube Computers", *Proc. 17th Annual Int'l Symp. on Comp. Arch.*, May 1990.

[8] H. Wang and Q. Yang, "Prime Cube Graph Approach for Processor Allocation in Hypercube Multiprocessors", *Proc. 1991 Int'l Conf. Parallel Processing*, Vol I, pp. 25-32.

[9] D. Das Sharma and D. K. Pradhan, "A Novel Approach for Subcube Allocation in Hypercube Multiprocessors", *Proc. 4th IEEE Symp. on Parallel and Distributed Processing*, 1992, pp.336-345.

[10] H. P. Katseff, "Incomplete Hypercubes", *IEEE Transactions on Computers*, vol. 37, pp. 604-608, May 1988.

[11] N.-F. Tzeng, H. L. Chen and P. J. Chuang, "Embeddings in Incomplete Hypercube", *Proc. 1990 Int'l Conf. on Parallel Processing*, Aug 1990.

[12] C. Hu, M. Bayoumi, B. Kearfott and Q. yang, "A parallelized algorithm for the the preconditioned interval newton method", *Proc. 5th SIAM Conf. on Parallel Processing for Sci. Comp.*, March 1991.

[13] N.-F. Tzeng, "Structural properties of Incomplete Hypercubes", *Proc. 1990 Int'l Conf. on Dist. Comp. Sys.*, May 1990.

[14] nCUBE 2 Systems : Technical Overview, nCUBE Corp., Foster City, CA, 1992.

[15] D. Knuth, *"The Art of Computer Programming: Sorting and Searching"*, Addison-Wesley, Reading, Massachusetts, 1973.

Random Routing of Tasks in Hypercube Architectures

Arif Ghafoor

School of Electrical Engineering

Purdue University

West Lafayette, IN, 47907

Abstract

We propose a general technique to study the performance of random routing in a hypercube system. The proposed method is a based on association schemes instead of using representation theory.

1   Introduction

A distributed computing system comprises a number of autonomous computers connected together through communication channels. The analogy of a distributed system with a graph where the vertices represent the processing nodes and the edges represent the links between them, helps us envisage the network interconnections structure. Distributed database systems, local area networks and loosely coupled multiprocessor systems are typical examples of a distributed computing environment. The most desired feature of a distributed system is efficient resource sharing among concurrent jobs [1, 2, 3]. Load balancing is supposed to achieve a system wide objective of enhanced performance subject to the constraint of reduced communication overhead incurred by information exchange and job transfers. The reason that there is no "optimal" load balancing strategy is because this problem has been dealt with various perspectives. One can name a wide range of resource scheduling algorithms from the load balancing point of view [2]. These distributed scheduling algorithms work on two principles. The first approach which is an extension of centralized control utilizes decomposition technique that involves parsing of a large problem into submodules to meet certain criterions. The other approach employs queuing model in which individual task scheduling is done independently. [4] describes a heuristic approach based on Bayesian theory with decentralized control. A taxonomy of different algorithms and compared their relative merits and demerits with varying degree of information dependency is given in [2].

Problem with distributed algorithms is that they require too much information collection. Contrary to the intuitive assumption that gathering large amount of data about the state of the system could yield better results, simple schemes have been reported to give performance not far from what one can expect from very complex schemes [2]. Moreover excessive amount of information cause extra overhead and decisions have to be postponed until all data becomes available.

It is known that network topology and the node architecture play an important role in the performance of these algorithms. Diameter of the network, hop distances and connectivity provide a heavily loaded noded with variety of choices for transferring the load. The architecture of the node determines the way incoming and outgoing traffic is handled.

In this paper we propose a general analytical model to evaluate any distributed load balancing algorithm which is based on randomly selecting the neighbors for job migration. In a steady state homogeneous environment, most of the distributed scheduling algorithms behave as random selection of neighbors [2]. We restrict ourselves to homogeneous networks where all nodes have identical configuration. We assume that the nodes communicate on a "virtual" hypercube structure, on which the migration of a task is modeled as a homogeneous Morkov process. The hypercube structure determines the pattern of communication between the scheduling nodes and needs not to be the subgraph of the actual network. We analyze the performance of the random scheduling on such a structure.

2   Combinatorics of Hypercube Structure

In this section, we briefly describe various combinatorial properties of an n-cube (hypercube) topology, represented as Q_n. These properties are elaborated in [5]. Specifically, we know that a hypercube is a *distance-regular* graph, that is,

$$\forall (i,j,k) \in [0..d]^3 \forall (x,y) \in U, L_{xy} = k \Leftrightarrow$$

$$\mid \{ z \in U, L_{xz} = i \& L_{yz} = j \} \mid = p_{ij}^k$$

where p_{ij}^k are constants and for a Q_n these are given as [10]:

$$p_{ij}^k = \begin{cases} \binom{\frac{k}{i-j+k}}{2}\binom{n-k}{\frac{i+j-k}{2}}, & \text{if } i+j-k \text{ is even} \\ 0, & \text{if } i+j-k \text{ is odd} \end{cases}$$

It is also known that Q_n is a *distance-transitive* graph as well [5]. For such a graph, the number of nodes at distance i from a given node is a constant called valency and denoted by v_i. Every distance-transitive graph is edge-transitive and distance-regular with valencies $v_i = p_{ii}^0$ [7] which is given by $\binom{n}{i}$ for Q_n.

Distance-transitivity is a highly desirable property for a graph, in order to design symmetric distributed algorithms as well as providing maximal fault-tolerance in the network [5].

Definition: Let X be the set of d x n matrices over $GF(q)$, where q is a power of prime and $(d \le n)$. Define the $i-th$ relation R_i on X by:

$$(x,y) \in R_i \Leftrightarrow rank(x-y) = i$$

Algebraically, Q_n can be characterized using the so called *Hamming or hypercubic association scheme* [7][10]. A *symmetric association scheme* consists of a finite set X consisting of v elements along with a set of partitioning relations $R = (R_0, R_1, \ldots, R_n)$ of the cartesian product $X.X$. These relations satisfy the following axioms:

- $R_0 = \{(x,x)/x \in X\}$

- Every R_i is symmetric.

- for a given ordered pair of elements (x,y) related by R_k, the number of elements z such that $(x,z) \in R_i$ and $(y,z) \in R_j$ only depends on k to which the ordered pair (x,y) belongs. This number is denoted by p_{ij}^k, as also mentioned above.

For Q_n, X is the vertex set of all binary codewords of length n and the Hamming distances[1] between codewords constitute the relation set R. For example, if Hamming distance (H_{xy}) between vertices x and y in Q_n is i, the $(x,y) \in R_i$. This scheme is called the Hamming scheme and is denoted as $H(n,2)$ [7][10]. If the graphical distance between any two vertices, x and y, is L_{xy}, the for Q_n, $L_{xy} = H_{xy}$.

---

[1] The Hamming distance between two codewords is the number of positions at which they disagree.

The following discussion is necessary for analyzing the performance of the hypercube structure. Let D_i the adjacency matrix of Q_n, for the relation R_i. For example, D_1 is the adjacency matrix of the network Q_n. It can be shown that these matrices span an algebra of dimension $n+1$ over the complex field, called the Bose-Mesner (BM) algebra of the scheme $H(n,2)$. This algebra admits a basis of idempotents J_i, such that:

$$D_k = \sum_{j=0}^{n} p_k(j) J_j,$$

where the $p_k(j)$ are the first eigenvalues. Let P be the $n+1$ by $n+1$ matrix with (i,j) entry $p_j(i)$. Let matrix Q be such that $PQ = vI$. The (i,j) entry of Q is denoted by $q_j(i)$. $p_j(i)$ and $q_j(i)$ will be used to evaluate the performance of random migration process of tasks.

3 Random Task Migration on Hypercube as a Markov Process

We assume that the number of nodes in the distributed system are $N = 2^n$ where each node can directly or virtually communicate with n neighbors and the overall communication structure is topological equivalent to a Q_n graph. Each node, therefore, has a unique binary address. Each node is assumed to have an I/O port, through which tasks enter into the node, a CPU, and $n+1$ queues for holding tasks. One of the queues is assigned to the local CPU which holds the tasks ultimately scheduled to be executed locally. The rest of the n queues are for the n virtual links to carry the outgoing traffic to n neighbors of the Q_n network. The reason to maintain a queue for each link is because different tasks have different sizes and consequently variable communication times.

The primary component of a node is a task manager which is responsible for deciding whether an incoming task is to be serviced locally or to be entered into one of the outgoing queue. The task manager and the local CPU are assumed to operate concurrently. For task scheduling, the Task Manager handles each task in the same manner. *Upon receiving a task, it chooses one of the queues out of $n+1$ queues, according to some predetermined rule.* After selecting a queue, the task is entered at the end of the queue. As mentioned above, in a steady state homogeneous environment, this task migration scheme is similar to *random routing* [12].

Therefore, the migration of the task according to a random scheduling algorithm can be modeled as a Markov process with each node of the Q_n representing a state for the Markov chain with as the migration of task through the network according to the following rule: if the task is on node x at time t, the position at time $t+1$ is chosen with an equal probability amongst the neighbors of x including the local CPU.

The performance parameter we are concerned is to find average number of hops a task migrates before it gets executed. Let T the random variable which counts the number of steps (hops) to reach a fixed point say x_0 of X. We are also interested in finding the first moments and limit law (for large n) of T.

An interesting feature of the association scheme structure is that lumping of states [13] of the Markov chain is possible and we can obtain a Markov chain which is still homogeneous, i.e. with transition probabilities independent of time. This is essentially due to the *distance transitive* property of the Hamming scheme. Suppose, x_0 be the vertex with all 0's codeword. We can lump all the states which are at graphical distance of i from x_0 into a single state, E_i (we are essentially lumping all the vertices of Q_n which are at the surface of distance i from the origin $x_0=0$). The resulting lumped states can be defined as follows:

$$E_i = \Gamma_i(x_0), \; i = 0, 1, \ldots, n$$

where Γ_i corresponds to the relation R_i.

The resulting transition probabilities (from state i to j) can then be given as:

$$P(i/j) = p^i_{j1}/v_1$$

We can define some new random variables T_i as the number of steps to reach the state 0 (x_0), starting from the state E_i. These variables are important in order to calculate the average value of T.

4 Performance Analysis of Random Task Migration in Q_n

The desired averages can be found by using the probability generating function. The results are given by the following Theorems. The derivations of theorems 1 and 2 are given in [6].

Theorem 1: Let $f(z)$ be the probability generating function for the random variable T. Then we have:

$$f(z) = (z + 2^n(1 - z)(1 + z\phi_1(z)))^{-1}. \quad (1)$$

where

$$\phi_1(z) = (\tfrac{1}{2^n}) \sum_{j=1}^n q_j(1)/(1 - z(p_1(j)/n))$$
(2)

We give the following alternative expression for $\phi_i(z)$, for our future reference.

Theorem 2:

$$\phi_k(z) = (1/vv_k) \sum_{j=1}^n \mu_j p_k(j)/(1 - zp_1(j)/v_1)$$

The desired average values $E(T)$ and $E(T_k)$ as well as the second moments of these variables can now immediately be obtained from thr derived generating functions.

Theorem 3: The mean and variance of T are:

$$E(T) = v(1 + \phi_1(1)) - 1,$$

$$var(T) = v^2(1 + \phi_1(1))^2 + v(2\phi'_1(1) + \phi_1(1) - 1)$$

Proof: As is well-known: $f'(1) = E(T), f''(1) = E(T^2) - E(T), var(T) = E(T^2) - E^2(T)$. We use Taylor formula of function f at order 2, in (1) to compute $f'(1)$ and $f''(1)$. Setting $z = 1 + h$, $a = \phi_1(1)$, $b = \phi'_1(1)$, in (1) we obtain immediately:

$$f(1 + h) = 1 + ((a + 1)v - 1)h +$$

$$((a + 1)^2v^2 + (b - a - 2)v + 1)h^2 + O(h^3).$$

The result follows. Q.E.D.

Theorem 4: The mean of T_k is:

$$E(T_k) = (\phi_1(1) - \phi_k(1) + 1)v - 1$$

Proof: Setting $z = 1 + h$, $a = \phi_k(1)$, $b = \phi'_k(1)$, $c = \phi_1(1)$, $d = \phi'_1(1)$, we obtain immediately: $f_k(1 + h) = 1 + ((c - a + 1)v - 1)h + O(h^2)$. The same techniques as in preceding theorem still apply here, and the result follows. Q.E.D.

We know that for a Hamming scheme (i.e. for a Q_n) we have $\mu_j = v_j = \binom{n}{j}$, as well as $v = 2^n, p_1(j) = n - 2j$ ([10]). We need the following lemmas, (see proofs in [6]):

Lemma 1: For large n, we have:

$$\phi_1(1) = -1 + (1/2 Log(2)) + o(1).$$

Lemma 2:

$$\phi_1'(1) = O(n^2)$$

The main result is:

Theorem 5: In a random walk on a Q_n, for large n:

$$lim(E(T)) = 2^{n-1}/Log(2), \qquad (3)$$

$$lim(var(T)) = 2^{2n-2}/(Log(2))^2, \qquad (4)$$

$$lim(P(T > t2^n) = e^{-t/2Log(2)}. \qquad (5)$$

Proof: Equation 3 and 4 follows directly from the lemmata and Theorem 3. Equation 5 follows from the Continuity Theorem for generating functions ([8]) applied to the equation

$$lim(E(e^{-\lambda T/2^n}) = 1/(1 + \lambda/2Log(2))$$

$$= \int_0^\infty e^{-t} exp(-\lambda t/2Log(2))dt$$

which comes from:

$$\phi_1(1) - \phi_1(e^{-\lambda/2^n}) = O(n(e^{\lambda/2^n} - 1)) = o(1)$$

as a tedious, but straightforward calculation shows. Q.E.D.

Since Q_n is a distance regular graph the number n_{ij} only depends on the distance from node i to node j, and it is easy to see that indeed:

$$N = E(T)$$

In the general case, a similar statement holds: n_{ij} only depends on the class to which the pair (i, j) belongs, and $N = E(T)$. This value of N is remarkably low as compared to the worst possible value which is of the order of V^3, where V is the number of nodes in the network [9].

5   Conclusion

In this paper we have developed a general technique to study the performance of random routing in a hypercube system. Our method is a direct generalization of [11]. We use association schemes instead of using representation theory.

6   Acknowledgement

The author wish to acknowledge the helpful discussion provided by P. Sole. This research is partially funded by a grant provided by the National Science Foundation, # 8913133A1-ECD, through SERC.

References

[1] D.L. Eager, E.D. Lazowska and J. Zahorjan, "Adaptive Load Sharing in Homogeneous Distributed Systems," IEEE Trans. on Soft. Engr., May 1986, pp. 662-675.

[2] I. Ahmad, A. Ghafoor, and K. Mehrotra, "Performance Prediction for Distributed Load Balancing on Multicomputer Systems," Proc. of Supercomputing'91, Nov. 1991, pp: 898-907.

[3] Y-T. Wang and R.J.T. Morris, "Load Sharing in Distributed Systems," IEEE Trans. on Computers, Mar. 1985, pp. 204-217.

[4] J.A. Stankovic, "An Application of Bayesian Decision Theory to Decentralized Control of Job Scheduling," IEEE Trans. on Comp., Feb. 1985, pp. 117-130.

[5] A. Ghafoor, and P. Sole, "Performance of Fault-Tolerant Diagnostics in the Hypercube Systems," IEEE Trans. on Comp., Aug. 1989, pp. 1164-1172.

[6] A. Ghafoor, "Performance of Distributed Task Scheduling in Distance-Transitive Networks," Tech. Report, 1993, School of Electrical Engineering, Purdue University, W. Lafayette, IN.

[7] E. Bannai, and T. Ito, *Algebraic Combinatorics I: Association Schemes.* Benjamin-Cummings (1984).

[8] W. Feller, *An Introduction to Probability Theory and its Applications.* John Wiley (1970).

[9] J.E. Mazo, "Some Extremal Markov Chains," Bell System Technical Jour., Vol. 61, 8, 1982.

[10] F.J. MacWilliams, and N.J.A. Sloane, *The Theory of Error-Correcting Codes*, North-Holland (1981).

[11] L. Flatto, A.M. Odlyzko, and D.B. Wales, "Random Shuffles and Group Representations," Ann. of Prob., Vol.13, 1, pp.154-178 (1985).

[12] L. Kleinrock, *Communication Nets*, McGraw Hill (1964).

[13] J.G. Kemeny, and J.L. Snell, *Finite Markov Chains*, Van Nostrand (1963).

Fault Tolerant Subcube Allocation in Hypercubes

Yeimkuan Chang and *Laxmi N. Bhuyan*
Department of Computer Science, Texas A&M University
College Station, Texas 77843-3112

Abstract – *The subcube allocation problem in faulty hypercubes is studied in this paper. An efficient method for forming the set of regular subcubes is proposed. A concept of irregular subcubes is then introduced to take advantage of the advanced switching techniques such as wormhole routing to increase the size of available subcubes. In this paper, a two-phase fault tolerant subcube allocation strategy is proposed. The first phase is the reconfiguration process based on a modified subcube partitioning technique which finds the set of disjoint subcubes in the faulty hypercube. The second phase is to apply an existing fault-free subcube allocation strategy such as Buddy strategy to each disjoint subcube for assigning the fault-free available subcubes to the incoming tasks. The simulation results using Buddy strategy are also given.*

1 Introduction

Hypercube architectures have received much attention due to their attractive properties, such as regularity, scalibility, fault tolerance, and multitasking capability. As the size of the system grows, the probability of some processors or links failing in the system becomes larger. In a normal situation, many parallel programs are executed concurrently in different subcubes allocated by the host system. In this paper, we study how to allocate subcubes in a faulty hypercube and thus maintain the multitasking capability of the system.

The main idea in assigning subcubes to incoming tasks is to avoid the interference among different tasks. This allocation problem of hypercubes has been studied in the literature. It was introduced by Chen and Shin [1]. Many other researchers use different approaches to tackle the problem[2, 3, 4]. Parallel algorithms for hypercube allocation have been developed in [5, 6]. The full subcube recognition ability and the efficiency of the allocation algorithms are the two main issues in hypercube allocation. In this paper, we address this allocation problem for a system with faults.

The straightforward method to accomplish fault tolerant subcube allocation is to use the concept of *exclusion*. In other words, the faulty processors are treated as being allocated permanently. To improve on the above approach, Jokanovic et al. [7] propose a two-phase fault tolerant gray code (FTGC) strategy. FTGC is based on the observations that in a fault-free hypercube, the left half of the system becomes more fragmented than the right half since the GC strategy takes a first-fit approach. Therefore in FTGC, the first phase is to find a parameter sequence [1, 7] such that the distribution of the faulty processors is concentrated to the left as much as possible. The second phase is to apply the same allocation and deallocation procedures of the GC strategy.

We consider wormhole routing in addition to store-and-forward routing. The wormhole routing employed in current commercial n-cube systems [8, 9] has the property that communication delay is insensitive to the physical distance provided that there is no interference on the communicating links. We provide results both with and without wormhole routing property. It is assumed that all the faults are static and are detected before the reconfiguration algorithm starts. We assume only node faults and as a result, the links incident on the faulty nodes are also faulty. The faulty nodes can not perform either computation or communication.

We present a modified subcube partitioning strategy and discuss how to construct the subcubes in a hypercube with two faults, since in the worst case, two faults in an n-cube are sufficient to destroy every possible regular $(n-1)$-cubes. The technique first selects a regular $(n-1)$-cube and then the faulty processor in the selected $(n-1)$-cube is replaced with its corresponding healthy processor in the other $(n-1)$-cube. We then extend the results to more than 2 faults and show that at least $\lceil \frac{n}{2} \rceil$ faults can be tolerated while maintaining a fault-free $(n-1)$-cube [10]. We carry out extensive simulation to find the probability of a healthy $(n-1)$-cube in an n-cube system as a function of the number of faults. It is shown that the modified subcube partitioning technique performs better than the method which only searches for the regular subcubes.

Finally, we use the above modified subcube partitioning technique to develop a two-phase fault-tolerant subcube allocation strategy. The first phase is to construct the set of disjoint subcubes in a faulty hypercube according to our modified subcube partitioning technique. The second phase is to apply an existing fault-free subcube allocation strategy such as Buddy strategy to each disjoint subcube. We simulate our fault tolerant allocation strategy based on Buddy strategy. The simulation results show that the fault tolerant subcube allocation strategy based on the modified subcube partitioning is always better than the others.

The rest of the paper is organized as follows. An efficient construction method of regular subcubes is proposed in section 2. In section 3, the concept of irregular subcube is introduced. The modified subcube partitioning technique is proposed to take advantage of wormhole routing technique. In section 4, a two-phase fault tolerant subcube allocation strategy is presented. Simulation results are given in section 5. Finally, the concluding remarks are given in the last section.

2 Construction of Regular Subcubes

We first present an efficient method to construct the set of disjoint subcubes in a faulty hypercube. The construction of the set of disjoint regular subcubes (SDS) is described in the procedure Form_SDS as follows.

> **Procedure** Form_SDS(n, F)
> /* n: size of hypercube and F: set of faults */
> **begin**
> $d :=$ Form_Regular_n_1_cube(n, F, F_0, F_1);
>   **if** $|F_0| = 0$ **then**
>   SDS $:=$ SDS $\cup *^{n-i-1}0*^i$;
>   **else** Form_SDS$(n - 1, F_0)$;
>   **if** $|F_1| = 0$ **then**
>   SDS $:=$ SDS $\cup *^{n-i-1}1*^i$;
>   **else** Form_SDS$(n - 1, F_1)$;
> **end**

The procedure Form_Regular_n_1_cube finds a dimension d at which the n-cube is divided into two $(n - 1)$-cubes such that the difference of the numbers of the faulty nodes in these two $(n - 1)$-cubes is the maximum. If there is a tie, we will solve it as follows. Let Q_d be the $(n - 1)$-cube in which the number of faults is less than that in the other $(n - 1)$-cube. We define the *weight* of a faulty processor in a hypercube as the number of neighbors which are also faulty and the *system weight* of the hypercube as the sum of the weights of all the faulty processors in the same hypercube. Let the system weight of Q_i be W_i. The dimension i is selected if $W_i \geq W_j$ for all $j \in \{0, .., n - 1\}$. If there is a tie for W_i again, we arbitrarily select one dimension.

Example 1: Fig.1 shows a 4-cube containing 5 faulty nodes marked as shaded circles. Fig.1(a) and (b) illustrates that the 4-cube is split into two 3-cubes at dimension 3 and 1, respectively. Thus dimension 3 is selected for split instead of dimension 1 since the system weight of Q_3 is 2 which is greater than the system weight, 0, of Q_1. After the dimension d is selected, the procedure Form_SDS is continued for each $(n - 1)$-cube until the subcube contains no faults or all the processors in the subcube are faulty.

The time complexity of Form_SDS is $O(|F| \times n^3)$ since Form_Regular_n_1_cube takes $O(|F| \times n^2)$ time units for each dimension and there are n dimensions. We shall see that the split and select process does not guarantee to find the *maximum set of subcubes* (MSS) introduced in [2]. However, our method shown in procedure Form_SDS is efficient whereas finding the MSS is an NP-hard problem.

Let us analyze the worst case number of faults that procedure Form_SDS can tolerate in order to maintain a fault-free $(n - m - 1)$-cube. After calling Form_SDS once, there exists an $(n - 1)$-cube containing at most $\lfloor \frac{|F|}{2} \rfloor$ faulty nodes. Thus Form_SDS can continue m times until the finally selected $(n - m)$-cube contains only one fault, when $|F| \leq 2^m$. Obviously, a fault-free $(n - m - 1)$-cube is available in an n-cube containing $|F| \leq 2^m$ faults, for $n \geq m + 1$.

3 Construction of irregular subcubes

We introduce the concept of *irregular subcubes* to take advantage of the advanced routing technology such as wormhole routing currently used in the commercial hypercube multiprocessors [8, 9]. We define an *irregular* subcube as a hypercube in which there exists one or more pairs of logically adjacent nodes such that the physical distance between the adjacent nodes is more than one.

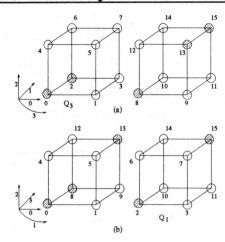

Figure 1: Construction of regular subcubes.

Due to wormhole routing, the communication speed between the logically adjacent nodes is the same as that between a set of physically adjacent nodes [8, 9]. A regular subcube can be embedded on an irregular subcube such that the load of the embedding is one and dilation of the embedding is equal to the maximum physical distance between the adjacent nodes.

We will briefly review the *modified*[1] subcube partitioning technique which constructs the irregular subcubes.

Modified Subcube Partitioning

The idea of modified subcube partitioning comes from how to utilize the unused links to achieve fault tolerance in the faulty hypercube. We shall see that the modified subcube partitioning technique tolerates more faults than only allocating regular subcubes.

The procedure of constructing an irregular subcube based on modified subcube partitioning is as follows. An $(n - 1)$-cube is first selected in the faulty hypercube. Then the faulty processor in the selected $(n - 1)$-cube is replaced by the corresponding processor in the adjacent $(n - 1)$-cube. The replacing processor and its neighbors must work since the links between it and its neighbors are utilized in the reconfiguration process. We shall see that this reconfiguration process works as long as the nodes which are two or fewer hops from any faulty node of the selected $(n - 1)$-cube, are healthy. The worst case of two faults being at antipodal positions can be solved by the modified subcube partitioning technique. The following example shows how the technique is applied to the 2-fault case.

Example 2: Assume there are two faulty nodes at antipodal positions of a 4-cube, i.e. (0000) and (1111) as shown in Fig.2. First, the node (1111) is replaced by node (0111). An irregular 3-cube is formed by the links shown as bold lines and the nodes marked with label 3 in parentheses. We call nodes 0011, 0101, and 0110 as the intermediate nodes of the irregular 3-cube since they reside in between node 0111 and its logical neighbors of the irregular 3-cube. If the intermediate nodes can be used for forming other subcubes, then one 2-cube consisting of nodes 0011, 0100, 0101, and 0110

---

[1] called *modified* in order to distinguish the subcube partitioning technique proposed in [11]

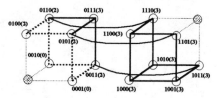

Figure 2: Subcube allocations of 4-cube with 2 faults.

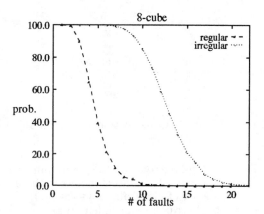

Figure 3: Probabilities of fault-free 7-cubes in an 8-cube with faults.

and two 0-cubes, 0001 and 0010, can be constructed as shown in the figure. However, if the intermediate nodes can not be used, the nodes 0001, 0010, and 0100 and the intermediate nodes can be used only as 0-cubes.

We call the assumption that the intermediate processors can not be used to form other disjoint subcubes to be *pessimistic*. However, in all the commercial hypercubes[8, 9], there are two kinds of processing units in a processor node, namely, a communication unit and a computation unit. Inside the communication unit, there is an $(n + 1) \times (n + 1)$ crossbar switch. For a message, there will be no delay in the switch if no other message competes for the same output. Thus the intermediate nodes can still be used for other tasks as long as there are no shared links with other tasks.

The time complexity of the modified subcube partitioning technique to find an $(n - 1)$-cube is given as follows. Let f be number of faults. For any dimension i, the faulty nodes are divided into two groups. The faulty node whose i^{th} bit value is zero is assigned to group 1, otherwise it is assigned to group 2. Next, the Hamming distances of any fault in group 1 and any fault in group 2 are computed. Each Hamming distance takes n time units. There are at most $O(f^2)$ Hamming distances. Thus the time complexity leads to $O(f^2 \times n)$.

Simulation Results

Finding a fault-free $(n - 1)$-cube in an n-cube containing a certain number of faulty nodes is the objective of our experiments. We carry out the simulation experiments by randomly generating faults in an n-cube. The probabilities of recognizing regular $(n - 1)$-cubes (section 2) are compared with the probabilities of recognizing irregular and regular $(n - 1)$-cubes based on the modified subcube partitioning technique.

Fig.3 shows the comparison of probabilities of recog-

nizing fault-free 7-cubes in an 8-cube with 2 to 22 faults. We can see that when the number of faulty nodes exceeds 5, the probability of successful recognition of regular 7-cubes drops under 50%. The probability of successful recognition of regular 7-cubes drops to 0 when the number of faulty nodes reaches 10. However, the probability of successful recognition of fault-free irregular 7-cubes drops to 0 when the number of faulty nodes reaches 20.

4 Subcube Allocation

We have seen that $\lceil \frac{n}{2} \rceil$ faults can be tolerated in the worst case by the modified subcube partitioning technique while maintaining a fault-free $(n - 1)$-cube. However, the simulation results show that in the average case, more faults can be tolerated in larger systems. In this section, we propose a two-phase fault tolerant subcube allocation strategy which allocates the regular and irregular subcubes. Since the faults do not occur as frequently as the processors are allocated in the system, we assume that the faults are detected and located before the construction starts.

The first phase is the construction of the set of disjoint subcubes by the modified subcube partitioning technique. The second phase is to sort the disjoint subcubes by their sizes in an increasing order and apply the existing fault-free subcube allocation strategy such as Buddy strategy etc. to each disjoint subcube. When an incoming task requesting a d-cube arrives in the system, starting from d-cube of the sorted disjoint subcubes, an available d-cube is searched and assigned to the incoming task.

Subcube Construction Phase

The algorithm Find_Disjoint_Subcubes for constructing the set of disjoint regular and irregular subcubes in a faulty hypercube is presented as follows. If the number of faults, $|F|$, is equal to the number of processors in the system then the process stops. Otherwise the procedure Find_n_1_cube described later is called to find an available $(n - 1)$-cube. If an $(n - 1)$-cube exists, the set F of faulty processors is updated. The number of faults remains the same and the fault-free $(n - 1)$-cube is put in Disjoint_Set. Finally, Find_Disjoint_Subcubes is called recursively with the dimension decreased by one. If an $(n - 1)$-cube does not exist, then Split_n_cube is called to find a dimension at which the n-cube is split into two $(n - 1)$-cubes such that the difference between the numbers of faulty processors in the two $(n-1)$-cubes is maximum. If the sets of faults in the two $(n-1)$-cubes are in F_0 and F_1, then two Find_Disjoint_Subcubes are called recursively for F_0 and F_1.

```
Procedure Find_Disjoint_Subcubes(n, F)
/* n: size of hypercube and F: set of faults */
begin
    if |F| = 2^n then return;
    Q_{n-1} := Find_n_1_cube(n, F);
    if an (n − 1)-cube Q_{n-1} exists then
        Disjoint_Set := Disjoint_Set ∪ Q_{n-1};
        Find_Disjoint_Subcubes(n − 1, F);
    else
        Split_n_cube(n, F, F_0, F_1);
        Find_Disjoint_Subcubes(n − 1, F_0);
        Find_Disjoint_Subcubes(n − 1, F_1);
    endif
end
```

Find_n_1_cube first checks if there exists a regular $(n-1)$-cube, i.e. if there exists a dimension i such that the ith bit values of the addresses of faulty nodes are all 0's or all 1's. If there exists an $(n-1)$-cube, the dimension i is put in P which is the set of dimensions that have been processed. The found $(n-1)$-cube is then returned. Otherwise, the irregular $(n-1)$-cube is searched according to the modified subcube partitioning technique. Hamming distances of addresses of any two faulty nodes are computed. For each pair of faulty nodes f_i and f_j whose Hamming distance is 1 or 2, we computes the dimensions spanning f_i and f_j and puts them into Dim_Set. If there exists a dimension i which does not belong to P and Dim_Set, then i is put in P and the $(n-1)$-cube is constructed by the procedure construct_n_1_cube(n, F, i) as follows. The n-cube is split into two $(n-1)$-cubes along the dimension i. One $(n-1)$-cube is selected and the faults in the selected $(n-1)$-cube is replaced with the healthy processors in the other $(n-1)$-cube. To update the set of faults, F, the ith bits of the addresses of faults in the selected $(n-1)$-cube is complemented. The following example illustrates the construction process of procedure Find_n_1_cube.

Example 3: Based on Fig.2, there are two faults, 0 and 15 in the 4-cube. There is no regular 3-cube available since the two faults are at antipodal positions. Since the Hamming distance between the two faults is 4, any dimension from 0 to 3 can be selected for constructing the irregular 3-cube by procedure construct_n_1_cube.

When Find_Disjoint_Subcubes finishes, Disjoint_Set contains all the fault-free disjoint subcubes. The last step is to sort the Disjoint_Set by the size of the subcubes.

Subcube Allocation Phase
We gives the procedure FT_Subcube_Allocation, for fault tolerant subcube allocation as follows. The procedure checks each subcube in the set of disjoint subcubes, Disjoint_Set, from the smallest subcube to the largest subcube and uses the existing fault-free subcube allocation strategy to allocate an available subcube from the set of disjoint subcubes. The deallocation procedure does not change except that the subcubes are released to the disjoint subcube from which they were allocated.

```
Procedure FT_Subcube_Allocation(n, d)
/* d: size of requested subcube*/
begin
    for all Q_i ∈ Disjoint_Set, from i = 1 to
        |Disjoint_Set| do
    if |Q_i| ≥ d then
        Q_d := Fault_Free_Subcube_Alloc(Q_i, d);
        if Q_d ≠ Null then return (Q_d);
    endif
    endfor
    return (Null);
end
```

Example 4: Assume there are two faults in a 4-cube as in Fig.2. Procedure Find_Disjoint_Subcubes returns Disjoint_Set = $\{Q_0, Q_0, Q_2, Q_3\}$ or $\{Q_0, Q_0, Q_0, Q_0, Q_1, Q_3\}$ if the intermediate nodes such as nodes 0011, 0101, and 0110 can not be used for forming subcubes except being used as 0-cubes. Take Buddy strategy as the dedicated allocation strategy. Assume that the incoming task sequence is $\{1,1,0,3\}$. For former case, two 1-cubes in Q_2,

a 0-cube in Q_0, and a 3-cube in Q_3 will be granted. However, for the later case, a 1-cube in Q_2, a 1-cube in Q_3, and a 0-cube in Q_0 are granted. The incoming task requesting a 3-cube needs to wait for the other tasks to finish.

5 Subcube Allocation Simulation

Performance based on the Buddy and single GC strategies is studied. We select the Buddy and single GC strategies because Buddy strategy is currently employed in the commercial hypercube multiprocessors and the single GC strategy is simple to implement. The following allocation strategies are adopted in the simulation.

1. the straightforward *regular* allocation strategy which treats the faulty processors as being allocated permanently,
2. the *modified regular* subcube allocation strategy which is based on the procedure Form_SDS proposed in section 2.
3. *irregular* subcube allocation strategy which is based on our two-phase fault tolerant subcube allocation strategy in section 4,
4. *pessimistic* irregular subcube allocation strategy which is same as the *irregular* except that the intermediate nodes can not be used to form other disjoint subcubes. However, the intermediate nodes can be assigned as individual 0-cubes.

The simulations on the two systems without any fault are also conducted for comparison. One follows the existing allocation strategy (called *no fault*) and the other is the two-phase fault-free allocation strategy called *no fault split*. The latter is motivated by the observation [7] that the left half of the system is more fragmented than the right half in the first-fit allocation approaches. The first phase is to split the n-cube into a set of subcubes, one k-cube for $1 \le k \le n-1$ and two 0-cubes. The second phase is the same as the one for irregular subcubes. Since there is no way we can service an incoming task which requests an n-cube or even an $(n-1)$-cube in a faulty hypercube, we will double (or quadruple) the residence time of the incoming task and reduce the subcube size by one dimension (or by two). This gives a fair comparison and indicates the impact of faults on the performance of the subcube allocation strategy.

In the experiments, K faults are generated 20 times with $K \ge 2$. For each K faults, 100 incoming tasks are generated and queued. The dimensions of the requested size by the incoming tasks are assumed to follow a given distribution such as uniform and normal distributions. The residence time of the allocated subcube is assumed to be uniformly distributed. In our experiments, the residence time is kept as $uniform(5, 11)$. Let p_i be the probability that an incoming task requests a subcube of size i, for $0 \le i \le D$ where D is the size of the system. Thus we have $\sum p_i = 1$. The p_i's$(p_0..p_8)$ for normal distribution used in our experiments are (.0228, .044, .919, .1498, .383, .1498, .919, .044, .0228).

The service discipline of the system is assumed to be first come first served (FCFS). At each time unit, the system attempts to find a fault-free subcube of the requested size to the first task in the queue and the assigned task is removed from the queue. After an incoming task in the system finishes, the subcube assigned to it is released. The process continues until all the 100 tasks

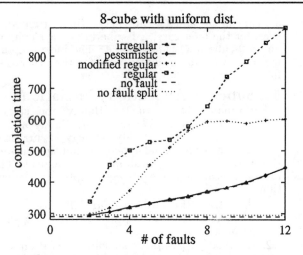

Figure 4: Completion time of Buddy strategy.

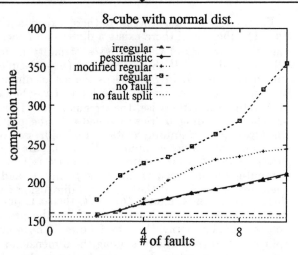

Figure 5: Completion time of Buddy strategy.

are finished. Under the above simulation model, the performance is measured in terms of completion time. For each K faults, 20 independent runs are performed. The average of these parameters for 20 runs are computed and are used in the plots.

Based on Buddy strategy, Fig.4 shows the completion time in the 8-cube with 2 to 12 faults and with uniform distribution of requested sizes. We can see that the performance of allocating irregular subcubes is always better than allocating regular subcubes. The *pessimistic* irregular approach is slightly worse than the *irregular* approach. This indicates that bigger subcubes play an important role in the allocation process since the *pessimistic* approach does not lose much bigger subcubes, especially the biggest possible available subcube, $(n-1)$-cube. Notice that *no fault split* approach does not work better than *no fault* approach since there is equal chance that an incoming task requests n-cube and *no fault split* approach assigns an $(n-1)$-cube with double the residence time. Fig.5 shows similar results.

6 Final Remarks

Wormhole routing, currently employed in commercial hypercube multiprocessors, is considered in constructing subcubes in the presence of faults. As long as the links of an irregular subcube assigned to one task are not shared by other subcubes, the performance of the task execution will be the same as the tasks running in the regular subcubes. This is because the subcubes generated by the modified subcube partitioning technique are disjoint. We developed a two-phase fault tolerant subcube allocation strategy which is general enough to apply to any existing fault-free subcube allocation strategy. The two-phase approach also improves the performance of larger fault-free hypercube systems. The extensive simulation results show that our approach always performs better than others.

References

[1] M. S. Chen and K. G. Shin, "Processor Allocation in an N-Cube Multiprocessor Using Gray Codes," *IEEE Transactions on Computers*, pp. 1396–1407, Dec. 1987.

[2] S. Dutt and J. P. Hayes, "Subcube Allocation in Hypercube Computers," *IEEE Transactions on Computers*, pp. 341–351, March 1991.

[3] J. Kim, C. R. Das, and W. Lin, "A Top-down processor Allocation scheme for hypercube Computers," *IEEE Transactions on Parallel and Distributed Systems*, pp. 20–30, January 1991.

[4] P. J. Chuang and N. F. Tzeng, "A Fast Recognition-Complete Processor Allocation Strategy for Hypercube Computers," *IEEE Transactions on Computers*, pp. 467–479, April 1992.

[5] Y. Chang and L. N. Bhuyan, "Parallel Algorithms for Hypercube Allocation," In *International Parallel Processing Symposium(IPPS)*, April 1993.

[6] M. Livingston and Q. F. Stout, "Parallel Allocation Algorithms for Hypercubes and Meshes," In *Proceedings of 4th Hypercube Concurrent Computers and Applications*, 1989.

[7] D. Jokanovic, N. Shiratori, and S. Noguchi, "Fault Tolerant Processor Allocation in Hypercube Multiprocessors," *Japan's IEICE Transactions*, vol. E 74, no. 10, pp. 3492–3505, Oct. 1991.

[8] nCUBE Corporation, *nCUBE 2 Processor Manual*, nCUBE Corporation, Dec. 1990.

[9] Intel, *Intel iPSC/2*, Intel Scientific Computers, 1988.

[10] Y. Chang and L. N. Bhuyan, "Constructing Subcubes in Faulty Hypercubes," Technical report, Texas A&M University, 1992.

[11] J. Bruck, R. Cypher, and D. Soroker, "Tolerating Faults in Hypercubes Using Subcube Partitioning," *IEEE Transactions on Computers*, pp. 599–605, May 1992.

SESSION 6A

ARCHITECTURE (I)

Performance of Redundant Disk Array Organizations in Transaction Processing Environments*

Antoine N. Mourad   W. Kent Fuchs   Daniel G. Saab

Center for Reliable and High-Performance Computing

Coordinated Science Laboratory

University of Illinois

Urbana, Illinois 61801

Abstract   *We study the performance of two redundant disk array organizations in a transaction processing environment and compare it to that of mirrored disk organizations and organizations using no striping and no redundancy. Redundant disk arrays and mirrored disks are used for providing rapid recovery from media failures in systems requiring high availability. The disk array organizations examined are: data striping with rotated parity (RAID5) and Parity Striping. RAID5 provides high data transfer rates by striping the data over multiple disks. It also provides better load balancing over the disks in the array. At the same time, data striping increases disk arm use which can lead to longer queuing delays. In transaction processing environments, because of the nature of I/O requests, namely a large number of small size requests, disk arms are a more valuable resource than data transfer bandwidth. Hence, parity striping was proposed as an alternative to RAID5. It provides rapid recovery from failure at the same low storage cost without interleaving the data over multiple disks. In this study, we use data from a large scale commercial transaction processing site to evaluate and compare the performance of the above organizations. We consider both non-cached systems as well as systems using a non-volatile cache in the controller.*

1   Introduction

Reliable storage is a necessary feature in transaction processing systems requiring high availability. Media failure in such systems is traditionally dealt with by periodically generating archive copies of the database and by logging updates to the database performed by committed transactions between archive copies into a redo log file. When a media failure occurs, the database is reconstructed from the last copy and the log file is used to apply all updates performed by transactions that committed after the last copy was generated. In such a case, a media failure causes significant down time and the overhead for recovery is quite high. For large systems, e.g., with over 150 disks, the mean time to failure (MTTF) of the permanent storage subsystem can be less than 28 days*. Mirrored disks have been employed to provide rapid media recovery [1]. However, disk mirroring incurs a 100% storage overhead which is prohibitive in many cases. Redundant disk array organizations [2,3] provide an alternative for maintaining reliable storage. Their storage overhead is only 10% for a 10-disk array. However, even when disk mirroring or redundant disk arrays are used, archiving and redo logging may still be necessary to protect the database against operator errors or system software design errors. Redundant disk arrays can also be used in conjunction with a twin-page parity storage scheme to reduce the overhead of undo recovery in transaction processing systems [7].

Chen *et al.* [4] compared the performance of RAID0[†], RAID1[‡], and RAID5[§] systems. They used a synthetic trace made of a distribution of small requests representing transaction processing workloads combined with a distribution of large requests representing scientific workloads. Disk measurements were made on an Amdahl 5890. Gray *et al.* [2] proposed the Parity Striping organization and used analytical models to derive the minimum (zero load) response time and the throughput at 50% utilization for fixed size requests. Their results suggest that Parity Striping is more appropriate than RAID5 for database and transaction processing systems. However, their model ignores the effect of skew in the distribution of accesses to disks which turns out to be an important element in favor of RAID5. Menon and Mattson [5] analyzed the performance of RAID5 systems in the transaction processing environment using analytical models. They compared the performance of arrays made of different size building blocks and studied the effect of caching. Reddy [6] analyzed the effect of various parameters and policies in the design of a non-volatile I/O cache for systems where the cost of writes is higher than the cost of reads. He did not assume any particular array organization and the effect of the parity update traffic on read miss access time was not modeled. In our evaluation of cached systems, we concentrate on comparing the behavior of the various array organizations when an I/O cache is used. Both read miss accesses and write (destage) accesses to disk are simulated.

---

*This research was supported in part by the National Aeronautics and Space Administration (NASA) under Contract NAG 1-613 and in part by the Department of the Navy and managed by the Office of the Chief of Naval Research under Grant N00014-91-J-1283.

*Assuming an MTTF of 100,000 hours for each disk.
[†]Data striping without redundancy.
[‡]Data striping with mirroring.
[§]Data striping with rotated parity.

Figure 1: RAID with rotated parity on four disks.

2 Redundant Disk Arrays

2.1 Data Striping

Striped disk arrays have been proposed and implemented for increasing the transfer bandwidth in high performance I/O subsystems [8–11]. In order to allow the use of a large number of disks in such arrays without compromising the reliability of the I/O subsystem, redundancy is included in the form of parity information. Patterson *et al.* [3] have presented several possible organizations for Redundant Arrays of Inexpensive Disks (RAID). One interesting organization for transaction processing environments is RAID with rotated parity (RAID5) in which blocks of data are interleaved across N disks while the parity of the N blocks is written on the $(N + 1)$st disk. The parity is rotated over the set of disks in order to avoid contention on the parity disk. Figure 1 shows the data and parity layout in a RAID5 organization with four disks. This pattern is repeated for the next set of blocks. An important parameter in the RAID5 organization is the striping unit. The striping unit can be defined as the "maximum amount of logically contiguous data stored on a single disk" [12].

The RAID5 organization allows both large (full stripe) concurrent accesses or small (individual disk) accesses. For a small write access, the data block is read from the relevant disk and modified. To compute the new parity, the old parity has to be read, XORed with the new data and XORed with the old data. Then the new data and new parity can be written back to the corresponding disks.

2.2 Parity Striping

Gray *et al.* [2] studied ways of using an architecture such as RAID in transaction processing systems. They argued that because of the nature of I/O requests in OLTP systems, namely a large number of small accesses, it is not convenient to have several disks servicing the same request. In other words, since in transaction processing systems I/O time is dominated by seek time and rotational latencies rather than by transfer time, it is not advantageous to have a request spread over multiple disks because that will make all those disks spend a significant amount of time seeking and rotating in order to decrease an already small transfer time. Hence, the Parity Striping organization shown in Figure 2 was proposed. The shading in the figure indicates the areas that belong to the same parity group. The organization consists of reserving an area for parity on each disk and writing data sequentially on each disk without interleaving. For a group of $N + 1$ disks, each disk is divided into $N + 1$ areas. One of these areas on each disk is reserved for parity and the other areas contain data. N data areas from N different disks are grouped together in a *parity group* and their parity is written on the parity area of the $(N + 1)$st disk.

 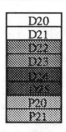 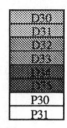

Figure 2: Parity striping of disk arrays.

Table 1: Disk and channel parameters.

Rotation speed	5400 rpm
Average seek	11.2 ms
Maximal seek	28 ms
Tracks per platter	1260
Sectors per track	48
Bytes per sector	512
Number of platters	15
Channel transfer rate	10 MB/s

3 Workload and System Model

To evaluate the different redundant disk array organizations, we have used data from an operational transaction processing system, from an IBM customer site. We use a very large trace containing over 3.3 million I/O requests accessing a database of more than 110 GBytes. The trace was collected using a low overhead tracing facility at an installation running the DB2 database management system. The trace entries contain the absolute address of the block accessed, the type of access (read, write or prefetch) and the time since the previous request. The time field is set to zero when both accesses are part of the same multiblock request.

Using this data, we simulate the behavior of the I/O subsystem. We account for all channel and disk related effects but we ignore cpu and controller processing times. The disk parameters used in the simulations are shown in Table 1. To compute the seek time as a function of the seek distance we use a non linear function of the form $a\sqrt{x - 1} + b(x - 1) + c$, x denoting the seek distance. Table 2 shows the characteristics of the trace used.

Disks are grouped into arrays containing N data disks. Each array has one controller and one independent path to the host. In the *Base* organization, an array consists of N data disks. In the mirrored disk organization, each array contains $2N$ disks consisting of N mirrored pairs. In the case of RAID5 and Parity Striping, the number of disks in the array is $N + 1$ in order to provide space for the parity information. No spindle synchronization is assumed. Block size is 4 KBytes.

In non-cached organizations we assume that a number of track buffers is used in the controller to reduce the effects of channel contention on performance. Write data are transferred on the channel to the buffers and when the disk head arrives to the appropriate location they are written to

Table 2: Trace characteristics.

Duration	3hr 3min
Database size	111 GB
# of I/O accesses	3,362,505
# of blocks transferred	4,467,719
# of single block reads	2,977,914
# of single block writes	312,961
# of multiblock reads	47,324
# of multiblock writes	24,306

Table 3: Default parameters.

Array size	$N = 10$
Block size	4KB
Striping Unit for RAID5	1 block
Parity placement for ParStrip	middle cylinders
For cached organizations:	
Cache size	16MB
Destage period	100 sec

the disk surface. Similarly reads are transferred from the disk to the buffers and when the channel is available they are sent to the host. This avoids having to wait an extra rotation if the disk head is at the appropriate location but the channel is busy. The buffers are also used to hold the old data and old parity that are read from disk in order to compute the new parity. The number of track buffers in the controller is proportional to the number of disks in the array attached to the controller. In our simulations we use 5 track buffers per disk.

In cached organizations, non volatile memory should be used to protect against the loss of write data in the event of a power failure. If volatile memory is used, then the cache should be flushed frequently to reduce the extent of data loss when a failure occurs [13]. There is one cache per array. Read hits are satisfied from the cache. The response time for a read hit is equal to the response time (waiting time and transfer time) at the channel. On a read miss the block is fetched from disk. If the replaced block is dirty, it has to be written to disk. The cache replacement policy is LRU. On a write hit, the block is simply modified in the cache. In organizations using parity (RAID5 and Parity Striping), when a block is modified the old data is kept in the cache in order to save the extra rotation needed to read the old data when writing the block back to disk. The old parity still has to be read and an extra rotation is still required at the disk holding the parity. On a write miss, the block is written to the cache and the block at the head of the LRU chain is replaced. A background *destage* process groups consecutive blocks and writes them back to disk in an asynchronous fashion. By using such a process, dirty blocks are destaged to disk before they reach the head of the LRU chain. Hence, write misses typically do not incur the cost of a disk access to write back a dirty replaced block. Only read misses have to wait for the block to be fetched from disk. The overall I/O response time is mainly determined by the read miss access time. The *destage* process turns small random synchronous writes into large sequential asynchronous writes. In our simulations the destage process is initiated at regular intervals. The time between two initiations of the process is the *destage period*. The write accesses issued by the destage process are scheduled progressively so that they will cause minimal interference with the read traffic.

For organizations using parity, the destage process accomplishes two purposes: it groups several dirty blocks together to perform a single multiblock I/O and it frees up space in the cache by getting rid of blocks holding old data.

Decreasing the destage period increases the write traffic seen by the disk. Increasing it reduces the hit ratio and increases the likelihood that the block at the head of the LRU chain is dirty which may cause a miss to wait for the replaced block to be written to disk.

One might wonder whether the destage policy used is better than the basic LRU policy in which dirty blocks are written back only when they get to the head of the LRU chain and a miss occurs. The question is even more relevant in the case of the Base and Mirror organizations in which old data blocks are not kept in the cache. We have compared the two policies for various cache sizes and we found that the periodic destage policy always performs better for all organizations. Reddy [6] uses a background process to write dirty blocks from the head of the LRU chain along with other dirty blocks in the cache that belong to the same track. In organizations using parity, there is a need for freeing old data blocks periodically even if the corresponding dirty block is not at the head of the LRU chain. It might be useful though to decouple the two issues by using the destage process that writes dirty blocks from the head of the LRU chain more frequently while a flushing process is only initiated from time to time to scan the entire cache and free old data blocks.

4 Experiments

Unless otherwise specified, the parameters shown in Table 3 are used by default in the following experiments.

4.1 Uncached Arrays

In a first experiment, we looked at non-cached organizations and we measured the performance of all four organizations for different array sizes. Figure 6 shows the response time in milliseconds for values of N from 5 to 20. In the Parity Striping organization, the parity areas were placed on the middle cylinders.

For mirrored disks, the response time for writes is the largest of the response times at the two disks in the mirrored pair. Reads, however, encounter less queuing since both disks of the pair can service reads in parallel. Moreover the shortest seek optimization[¶] is used to further reduce read response time. Since there are many more reads than writes,

---

[¶] A read is directed to the disk that has its arm nearest to request's location.

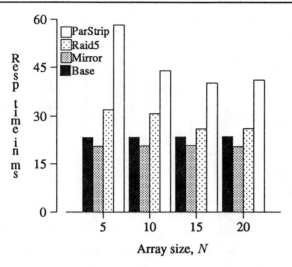

Figure 3: Response time vs. array size.

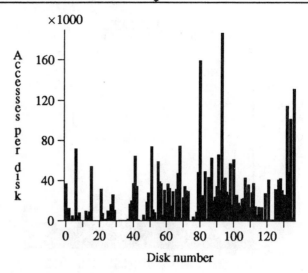

Figure 4: Distribution of accesses to disks in the Base organization.

the overall performance of mirrors is better than the Base organization. For $N = 10$, the response time of mirrors is about 12% shorter than that of the Base organization.

Comparing RAID5 to the basic organization, we notice that there is a significant decrease in performance associated with RAID5. Given that the fraction of large requests is small, the advantage of RAID5 in terms of high transfer rates cannot be fully exploited. One cause of the performance degradation encountered in the RAID5 scheme is the cost of small write requests. To service a single block write request, the data and parity disks have to be read to get the old data/parity then the data and parity blocks have to be written to the disk. Reading the old data/parity adds an extra rotation time to the response time of the request at each of the two disks involved. However, the response time of the parity update can be affected by queuing delays at the data disk since the parity write cannot be initiated until the old data has been read. The increased cost of write requests affects also read requests since it increases queuing for the disks. Another parameter that affects both read and write requests is *seek affinity*. Seek affinity is a measure of the spatial locality that may exist among disk accesses. The higher the seek affinity, the smaller the disk arm movements. Data striping decreases seek affinity and hence increases average seek distance and seek time.

The difference in performance between Parity Striping and RAID5 is mainly due to the ability of RAID5 to balance the load over all the disks in the array. For single block accesses, the service time at the disk (seek + latency + transfer) is higher in the case of RAID5 because of a decrease in seek affinity but RAID5 more than makes up for it by reducing queuing delays. The main argument used by Gray *et al.* [2] against RAID5 is the increased disk utilization due to having many arms service a single request. This does not happen, however, if the striping unit is chosen appropriately. In transaction processing workloads most requests are for single blocks. If the striping unit is a multiple of the block size then most small requests are serviced by a single disk.

Note that tuning the placement of the data on disk can reduce the skew in disk accesses and hence reduce the gap in performance between RAID5 and Parity Striping. How-

ever, RAID5 provides a way to balance the load automatically. Figures 4 and 5 illustrate this effect. In Figure 4 the total number of accesses to each disk is plotted for the Base organization while Figure 5 plots the distribution in the case of the RAID5 organization with 4KB striping unit. Figure 4 shows that there is a significant amount of skew in the disk access rate. Most of the skew within the array is smoothed out in the RAID5 organization.

4.1.1 Array Size

Changing the array size does not significantly affect the performance of the Base and Mirror organizations. There is only a very small increase in response time as the array size increases due to added channel contention since the same channel is servicing more disks. In the case of RAID5, the performance is affected by the fact that smaller arrays use more disks (for $N = 5$, there is one extra disk for every five data disks while, for $N = 10$, there is one extra disk for every ten disks). The other effect is that for large arrays the load is balanced over more disks which means that the risk of encountering large queuing delays is further reduced. $N = 20$ has the lowest cost and the best performance. However, large arrays are less reliable and have worse performance during reconstruction following a disk failure.

Figure 3 shows that the Parity Striping performance is worse for small arrays. One cause of this behavior is the fact that the parity area becomes larger for small arrays which increases the seek distance of reads and data writes since the parity area is in the middle of the disk. The effect of the placement of the parity is analyzed in more detail in Section 4.1.3. Another cause is that this organization aggravates the skew problem because when a hot spot appears on one disk and the disk becomes a bottleneck, the disk holding the corresponding parity area also experiences increased load and possibly long queues which in turn affects the performance of other disks in the array. This phenomenon is more severe for small array sizes. One possible solution to this problem would be to use a finer grain in striping the

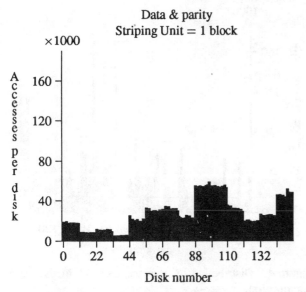

Figure 5: Distribution of accesses to disks in the RAID5 organization.

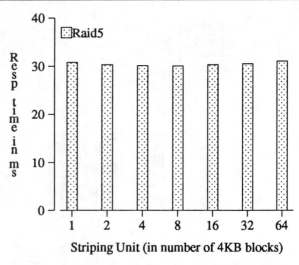

Figure 6: Response time vs. striping unit for RAID5.

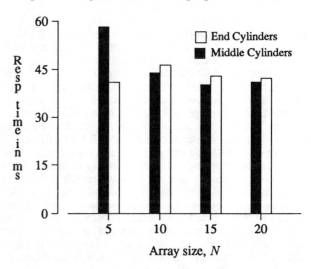

Figure 7: Comparison of parity placements for Parity Striping.

parity so that the parity update load is more balanced over the disks. Such an organization would preserve seek affinity for read accesses and writes to the data. It also preserves other useful properties of parity striping, such as better fault containment than RAID5, control over the distribution of data over the disks and various software benefits.

4.1.2 Striping Unit in RAID5 Organizations

The striping unit for RAID5 was varied from 1 block to 64 blocks with $N = 10$. The tradeoff between small and large striping sizes is similar to the tradeoff between RAID5 and Parity Striping. Large striping sizes provide better seek affinity and reduce total disk utilization by avoiding that multiple arms move to service a small multiblock request. However, they do not balance the load as well as with small striping units. Figure 6 shows the results. The optimal striping unit is 8 blocks or 32KBytes. There is little difference in performance, however, between values from 1 to 64 blocks. For striping units larger than 64 blocks the performance starts degrading significantly and, for very large striping units, it approaches that of Parity Striping.

4.1.3 Parity Placement in Parity Striping Organizations

Gray *et al.* [2] suggested that since the parity area is accessed frequently, it should be placed on the cylinders at the center of the disk. We found that this does not always improve performance especially for small arrays where the parity areas are quite big and when the workload has a high read-write ratio. A simple model can be used to explain this effect. Assuming that accesses are uniformly distributed over all disks in the array and that accesses to a given disk are uniformly distributed over all data areas on the disk, the access rate to any one of the N data areas on each disk is equal to $1/N^2$ times the total access rate to the array while the access rate to any given parity area is w/N times the total

access rate to the array, where w is the fraction of accesses that are writes. Hence, the parity areas are accessed more often than the data areas if and only if $w > 1/N$. In the workload considered, we have $w = 0.1$. Hence, according to the model, for $N > 10$, the parity area should be placed in the middle of the disk while for $N < 10$, it should be placed at the end of the disk. In Figure 7 the results for the two placements are show for various values of N. From the figure we observe that the cutoff point occurs somewhere between $N = 5$ and $N = 10$ (probably closer to 10 than to 5 given the large difference in performance seen for $N = 5$).

4.1.4 Modifying Trace Speed

In order to examine the performance of the various organizations at higher or lighter loads, we have conducted an experiment in which the trace was speeded up by a factor of 2 and another one where the trace was slowed down by a factor of two. Note that the workloads obtained by speed-

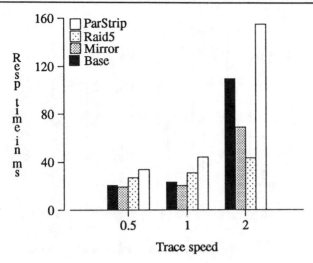

Figure 8: Response time vs. trace speed.

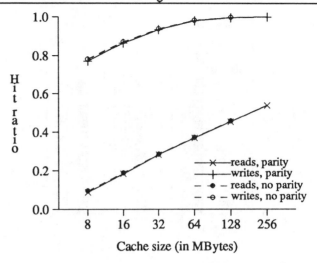

Figure 9: Hit ratio vs. cache size.

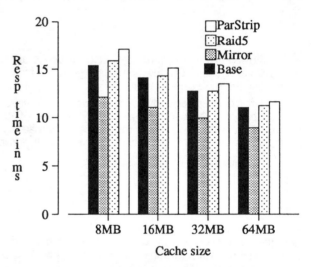

Figure 10: Response time vs. cache size.

ing up the trace do not reflect the characteristics of any real system. Doubling the processor speed does not imply that I/O's will be issued twice as fast since transactions may have to wait for one I/O to finish before issuing another one. RAID5 response time degrades gracefully as the load increases. RAID5 does even better than mirrors when the load doubles. The response time for Parity Striping and to a lesser degree that of the Base organization degrades severely as trace speed is doubled. Figure 8 shows the results.

4.2 Performance of Cached Organizations

4.2.1 Cache Size

All organizations benefit from larger cache sizes. The hit ratio is slightly lower for RAID5 and Parity Striping because of the space held by the old blocks in the cache. The read and write hit ratios are plotted in Figure 9 both for organizations using parity (RAID5, Parity Striping) and for those not using parity (Mirror, Base). Multiblock accesses are counted as hits only if all the blocks requested are in the cache. The write hit ratio is almost one for large caches because blocks are usually read by the transaction before being updated. The read hit ratio is relatively low for a small cache size (< 10% for an 8MB cache) but it increases to about 54% for a cache size of 256MB. The effect on hit ratio of keeping the old blocks in the cache is minimal. The difference between the parity and the no parity organizations is always less than 1% for writes. For reads, the hit ratio of the parity organizations is 6% lower for an 8MB cache but the difference goes down to 2% for a 16 MB cache.

The response time results are shown in Figure 10. The performance of mirrors relative to the Base organization improves because in this case the I/O response time depends mainly on read miss access time. Since each of the disks in the mirrored pair sees the same destage traffic as the corresponding disk in the base organization, the contribution of the destage traffic to read miss access time is the same in both organizations. In addition mirrors service reads faster because the read load is shared between the two disks in the

mirrored pair and because of the shorter seek optimization. For a 16MB cache size, mirrors perform 22% better than the Base organization.

The gap in performance between RAID5 and the Base organization reduces considerably because the larger cost of writes in RAID5 does not affect the overall response time directly but only through its contribution to read miss waiting time. Write costs are also reduced by the fact that old blocks are kept in the cache. As cache size increases, the gap gets even smaller in relative terms because the difference in miss ratios becomes smaller, the likelihood of having the old block in the cache when the write is destaged becomes higher and the probability (which should be already very small) of having to wait for a replaced page to be written to the disk on a cache miss becomes even smaller thus further reducing the contribution of the higher costs of write in RAID5 to the response time. For a cache size of 16MB, RAID5's performance is only about 1% worse than that of the Base organization. These results are in agreement with Menon and Mattson's analysis [5]

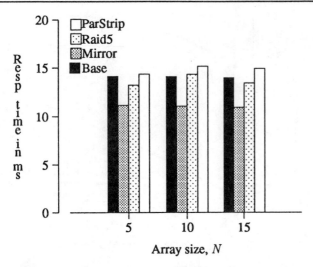

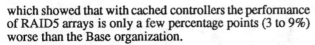

Figure 11: Response time vs. array size for cached organizations.

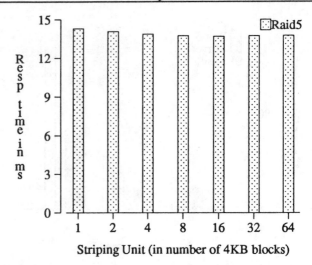

Figure 12: Response time vs. striping unit for RAID5 with cache.

which showed that with cached controllers the performance of RAID5 arrays is only a few percentage points (3 to 9%) worse than the Base organization.

RAID5 still does better than Parity Striping. However, the gap between the two narrows mainly because of the reduced load at the disks which makes RAID5's load balancing advantage less important in determining response time.

4.2.2 Array Size

In Figure 11, we compare three organizations with different array sizes but with the same total cache size. For $N = 5$, the cache size in each array is 8MB while for $N = 10$, the cache size is 16MB and for $N = 15$, the cache size is 24MB. We see that for the Base and Mirror organizations, the performance improves slightly in the larger array in spite of the higher channel contention. This implies that a big shared cache for 10 disks is better than two partitioned caches for every five disks. In the case of RAID5 and Parity Striping, the number of disk arms and the load balancing issue have more effect on performance than the difference in hit ratio between a global and a partitioned cache.

4.2.3 Striping Unit

The response time of the cached RAID5 organization is plotted in Figure 12 as a function of the striping unit. The optimal striping unit in this case is 16 blocks or 64 KBytes compared with 4 blocks for the non-cached organization. The reason for the difference is that the load on the array is much lighter in a cached organization and therefore the need for load balancing is not as high. This makes larger striping units more efficient because they increase seek affinity and reduce disk utilization on multiblock accesses.

5 Conclusions

We have used a trace from a large commercial transaction processing system to evaluate the performance of two redundant disk array organizations and compare them to mirrored disks and to non-redundant non-arrayed systems (Base organization). The I/O workload is dominated by single block I/O's and contains a significant amount of skew in the distribution of accesses to disks. We evaluated both cached and non-cached organizations. We found that RAID5 outperforms Parity Striping in all cases mainly because of its load balancing capabilities.

In non-cached organizations, RAID5 and Parity Striping perform significantly worse than the Mirror and Base organizations. For an organization with 10 data disks per array, RAID5's response time is 32% worse than the Base organization. It was also found that, for a workload with a large amount of skew in disk accesses, large RAID5 arrays perform better than smaller arrays by balancing the load more evenly over the disks. By speeding up the trace, it was shown than RAID5 behaves better than the other organizations under very high loads. For Parity Striping organizations, we found that placement of the parity area on the disk can affect performance significantly and we derived a rule for placing the parity in a way that minimizes seek times.

For cached organizations, we found that all organizations benefit from higher cache sizes. The write hit ratio is close to 1 for cache sizes over 32 MBytes. The read hit ratio on the other hand keeps increasing almost linearly as cache size increases. RAID5 performance is very close to the Base organization's performance in cached organizations. For a 16MB cache, RAID5 performance in only about 1% worse than the Base organization. Mirrors, however, do even better in cached organizations because of their efficiency in servicing reads. Their response time is about 22% shorter than the Base organization. The optimal striping unit in RAID5 organizations is smaller for non-cached organizations (32 KB) than for cached organizations (64 KB) due to the stronger need for load balancing in non-cached organizations. The response time curve is, however, relatively flat around the optimal value.

There are several issues that we have touched upon that warrant further investigation. Two of these issues are the optimal destage policy when old data blocks are kept in the cache and the use of a smaller striping unit for the parity in order to balance the parity update load in the Parity Striping organization.

Acknowledgements

The authors would like to thank Kent Treiber, Ted Messinger, Honesty Young and Robert Morris of IBM Almaden Research Center for providing the traces and for answering numerous questions.

References

[1] D. Bitton and J. Gray, "Disk shadowing," in *Proceedings of the 14th International Conference on Very Large Data Bases*, pp. 331–338, Sept. 1988.

[2] J. Gray, B. Horst, and M. Walker, "Parity striping of disk arrays: Low-cost reliable storage with acceptable throughput," in *Proceedings of the 16th International Conference on Very Large Data Bases*, pp. 148–161, Aug. 1990.

[3] D. Patterson, G. Gibson, and R. Katz, "A case for redundant arrays of inexpensive disks (RAID)," in *Proceedings of the ACM SIGMOD Conference*, pp. 109–116, June 1988.

[4] P. M. Chen, G. A. Gibson, R. H. Katz, and D. A. Patterson, "An evaluation of redundant arrays of disks using an Amdahl 5890," in *Proceedings of the ACM Sigmetrics Conference on Measurement and Modeling of Computer Systems*, pp. 74–85, May 1990.

[5] J. M. Menon and R. L. Mattson, "Performance of disk arrays in transaction processing environments," in *Proceedings of the 12th International Conference on Distributed Computing Systems*, June 1992.

[6] A. L. N. Reddy, "A study of I/O system organizations," in *Proceedings of the International Symposium on Computer Architecture*, pp. 308–317, 1992.

[7] A. N. Mourad, W. K. Fuchs, and D. G. Saab, "Recovery issues in databases using redundant disk arrays," *Journal of Parallel and Distributed Computing*, vol. 17, pp. 75–89, Jan. 1993.

[8] M. Y. Kim, "Synchronized disk interleaving," *IEEE Trans. Computers*, vol. C-35, pp. 978–988, Nov. 1986.

[9] M. Livny, S. Khoshafian, and H. Boral, "Multi-disk management algorithms," in *Proceedings of the ACM Sigmetrics Conference on Measurement and Modeling of Computer Systems*, pp. 69–77, May 1987.

[10] K. Salem and H. Garcia-Molina, "Disk striping," in *Proceedings of the IEEE International Conference on Data Engineering*, pp. 336–342, Feb. 1986.

[11] A. Reddy and P. Banerjee, "An evaluation of multiple-disk I/O systems," *IEEE Trans. Computers*, vol. 38, pp. 1680–1690, Dec. 1989.

[12] P. M. Chen and D. A. Patterson, "Maximizing performance in a striped disk array," in *Proceedings of the International Symposium on Computer Architecture*, pp. 322–331, May 1990.

[13] A. L. N. Reddy, "Reads and writes: When I/O's aren't quite the same," in *Proceedings of the Hawaii International Conference on System Science*, pp. 84–92, 1992.

THE CHUTED-BANYAN (CANYAN) NETWORK: AN EFFICIENT DISTRIBUTION NETWORK FOR GROWABLE PACKET SWITCHING BASED ON FREE-SPACE DIGITAL OPTICS

Thomas J. Cloonan & Gaylord W. Richards

AT&T Bell Laboratories

263 Shuman Blvd.

Naperville, IL. 60566

cloonan@iexist.att.com

Abstract- This paper studies the architectural trade-offs found in a novel ATM packet switch that offers several benefits for applications based on free-space digital optics. The architecture is based on a banyan network, but it uses a combination of deflection routing and output buffering to solve the contention resolution problem. It also makes novel use of chutes to remove packets from the banyan network when they have arrived in their desired output row.

INTRODUCTION

Free-space digital optics (FSDO) is a new packaging and interconnection technology which permits signals to be routed between digital integrated circuit chips as beams of light propagating orthogonal to the plane of the device substrates [1]. This approach to device connectivity may offer several system-level benefits, including high bandwidth, high density connectivity (parallelism), low signal skew, low channel crosstalk, and lower overall system power dissipation. A very simple arrangement of optical components that can provide many useful interconnection patterns between device substrates is shown in Fig. 1. This arrangement is based on a simple 2-lens system with a grating (diffractive hologram). Changes in the periodic structure of the grating can diffract different amounts of optical power from the original beam into different spots in the output image plane. The grating in Fig. 1 routes each input beam to three spots in the output image plane. If some of the beams in Fig. 1 are blocked, then the connections for the Banyan network [2] can be provided.

Recent research efforts on networks have yielded a novel packet switch architecture: the general growable packet switch architecture (Fig. 2). It contains two basic sub-systems: the distribution network and output packet modules. The distribution network routes each packet to the output packet module that is connected to the packet's desired output port. The distribution network must support

N input ports, and it must also support (Nm/n) output lines (chutes) that are routed to the mxn output packet modules, where N and n must both be a power of 2. Each of the output packet modules has access to n output ports, so a total of K=N/n output packet modules are required. Each output packet module stores the arriving packets in n FIFO queues, and it then routes the packets at the front of each of the FIFO queues to their desired output ports when the output ports are available. The overflow problems associated with the finite-length FIFO queues can be analyzed to determine satisfactory queue lengths based on the traffic load. According to the generalized Knockout Principle [2], if the arriving packets on the N input lines (N large) are statistically independent and if p is the switch loading, then the probability that more than m packets will be simultaneously destined for the n output ports in a single output packet module (the cell loss probability) is given by

$$P(loss) = \left[1 - \frac{m}{np}\right]\left[1 - \sum_{k=0}^{m}(np)^k\frac{e^{-np}}{k!}\right] + (np)^m\frac{e^{-np}}{m!} \quad (1)$$

A large, high-speed electronic implementation of the growable packet switch may encounter packaging problems due to the large number of connections, so it may be beneficial to explore the feasibility of implementing the distribution network using FSDO. However, networks based on FSDO must have simple nodes to minimize the size of device arrays and decrease the required lens field of view. In addition, the inter-stage interconnections must require only a few distinct beam-steering operations, because excessive optical power losses result in long device switching times. Finally, the number of stages must be minimized to reduce the cost of the system optics.

Many types of network topologies have been proposed for the distribution network, but most of these topologies tend to violate some of the aforementioned optical constraints. In this paper, we will explore a new network topology that may be better suited for the growable packet switch distribution network given the design constraints of FSDO. This photonic packet switch will be called the Chuted-Banyan network, or the Canyan network. It is a self-routing packet switch that has very simple nodes and inter-stage interconnections that require only three beam-steering operations. It will also be shown that a Canyan network with very low blocking probabilities can be implemented with a small number of device arrays.

THE CANYAN NETWORK

The Canyan network is a synchronous packet switch network based on deflection routing. The actual structure of the Canyan distribution network is shown in Fig. 3 within a growable packet switch architecture containing N=16 input ports, N=16 output ports, and output packet modules with m=8 inputs and n=4 output ports. Thus, the growable packet switch must provide K=N/n=4 output packet modules. The Canyan distribution network (to the left of the output packet modules) must support N=16 inputs and Nm/n=32 outputs, so the network provides an effective fanout of m/n=2. The particular Canyan distribution network in Fig. 3 contains S=7 node-stages, but it will be shown that the value of S can be modified to yield any desired throughput or packet loss probability within the network.

The Canyan distribution network can be viewed as two distinct sub-networks (the switching sub-network and the chute-multiplexing sub-network) overlayed on top of one another. As a result, the nodes in Fig. 3 are effectively split into two distinct parts: the switching portion (the white portion of each node) and the chute-multiplexing portion (the shaded portion of each node). The switching portion of each node must provide the functionality required for the routing of each packet to an appropriate output chute in the chute-multiplexing sub-network, while the chute-multiplexing portion of each node merely transports the packets toward the output packet modules. Because of similarities between the beam-steering operations required for their link-stage interconnections, both the switching sub-network and the chute-multiplexing sub-network can share the same imaging optics to provide the connections between adjacent node-stages.

The switching sub-network. The switching sub-network is a standard Banyan network containing 2x2 self-routing switching nodes (the white portion of the large nodes in Fig. 3). Two packets arriving at the node input ports can be routed on straight paths or on crossed paths to the two node output links. The dark-colored links within the link-stages of Fig. 3 provide the Banyan interconnection pattern connecting the outputs from the switching portion of the nodes in one node-stage to the switching portion of the nodes in the next node-stage. The Banyan pattern in an N-input Canyan network is repeated after every set of $\log_2(N)-1$ link-stages. To identify the link-stage type within a Canyan network, a new network parameter

known as the connection variable (V) can be defined. In any link-stage i of an N-input Canyan network, the connection variable for that particular link-stage is defined to be $V = \log_2(N)-1-i \mod [\log_2(N)-1]$. .If a particular packet is destined for an output port emanating from output packet module j, then the goal of the switching sub-network is to route that packet on the Banyan connections to ANY node in ANY node-stage that is in ANY horizontal row connected via the straight output chutes to output packet module j. Once the packet has arrived at any of these desired nodes, then the packet is transferred within the node to the chute-multiplexing portion of the switching node (the shaded portion of the large nodes in Fig. 3)

The chute-multiplexing sub-network. The entire chute-multiplexing sub-network is a simple network with Nm/n straight output chutes (the light-colored links in Fig. 3) connecting the chute-multiplexing portion of nodes (the shaded portion of the large nodes in Fig. 3). The 2m/n output chutes leaving the chute-multiplexing portion of node U in node-stage i are connected directly to the 2m/n inputs on the chute-multiplexing portion of node U in node-stage i+1, and the 2m/n output chutes leaving the chute-multiplexing portion of node U in the last node-stage are connected directly to the output packet module. Once a packet is routed to an idle output chute, it is transported straight down that output chute from node to node until it arrives at the output packet module. In effect, a packet that is routed to an output chute takes posession of that chute for the entire packet slot, and no other packets are permitted to use that chute during the entire packet slot.

A routing example. An example of a packet being routed through the Canyan network is shown by the bold, dark links in Fig. 3. This packet is to be routed from input port 11 to output port 5 within the growable packet switch. Since output port 5 is attached to output packet module 1, the packet can be routed down any one of the eight output chutes directed at output packet module 1. Another path connecting input 11 to output 5 is also shown by the bold, dashed links in Fig. 3. This alternate path would be used by the packet if it were deflected (blocked by another packet) from its desired path in node-stage 0. Once a packet is deflected, it must propagate through a minimum of $\log_2(N/n)+1$ link-stages before it can be transferred to the chute-multiplexing logic. However, if the Canyan network is designed with a sufficient number of link-stages, then the alternate paths will typically permit most of the deflected packets to be routed to their desired destinations.

Packet routing. Each packet must have a header prepended to the raw data that can be used by the nodes to identify the output port to which the packet is destined. For a growable packet switch with N output ports, this header contains $\log_2(N)+1$ bits, where the first bit of the header field is the activity bit and the remaining $\log_2(N)$ bits specify the destination address for the packet. To route a packet through the node-stages of a Banyan network, a particular node must first determine if the incoming packet is active- i.e., if the activity bit is set to a logic one. If so, then the node only needs to compare one bit of the binary destination address to one bit of the node's own binary address (defined by the connection variable V) to calculate the appropriate output link to which the packet should be routed. A node can determine when a packet should be transferred from the switching sub-network to the chute-multiplexing sub-network by examining the most-significant $\log_2(N/n)$ bits of the destination address. If these bits match the binary address of the output packet module to which the node is connected, then they will also match the most-significant $\log_2(N/n)$ bits of the node address, and the node will transfer the packet to an available chute within the chute-multiplexing sub-network (if an available chute exists).

Contention resolution. Packet deflection can occur if two packets arrive at a node and attempt to use the same link emanating from that node, because only one packet can be routed on the desired link, and the other packet is deflected to the second undesired link. Packet deflection can also occur if a packet arrives at a node connected to its desired output packet module and finds that all of the $2m/n$ output chutes for that node are already occupied by other packets.

PERFORMANCE ANALYSIS

The delay that a packet encounters as it passes through the Canyan distribution network is dependent on the number of node-stage in the network. The total delay within an S-stage network is given by $(S)[\log_2(N/n)+4]$ flip-flop delays. The number of node-stages (S) required for a particular application will typically be determined by the required packet loss probability. Two different analyses were used to study this network characteristic: a computer simulation and an analytical model based on a Markov chain. Both analyses produced similar results. These analyses were used to study the operating characteristics for Canyan distribution networks with various sizes (N), dimensionalities (S, m, and n), and offered loads (p). The plots in Fig. 4 indicate how the packet loss probability varies as a function of network size (N). As N is increased, the Canyan network requires more node-stages to produce a desired packet loss probability.However, all curves approach an asymptotic limit. This limit is actually the growable packet switch limit of Eq. 1. The plateaus in Fig. 4 indicate that some stages in the Canyan network can be removed to produce the Compressed-Canyan network, whose performance is shown in Fig. 5.

CONCLUSIONS

In this paper, we have analyzed the hardware requirements of an optical Chuted-Banyan (Canyan) distribution network. The hardware requirements were shown to depend significantly on the number of input ports (N). In an attempt to minimize the hardware within each Canyan node (and the associated size of the device array), buffering of entire packets within each node-stage was not allowed. Instead, routing decisions are made as soon as sufficient header bits have been received, and the packets are then routed to the next stage whether their desired paths are available or not. To solve the contention problems that often occur within a self-routing packet switch, a deflection routing scheme is used that merely sends a blocked packet out of a node on any available path. To return the deflected packet to its desired path, additional stages are added to the network. It was shown that this type of self-routing network satisfies most of the constraints placed on systems by optics.

REFERENCES

[1] F. B. McCormick, F. A. P. Tooley, J. L. Brubaker, J. M. Sasian, T. J. Cloonan, A. L. Lentine, R. L. Morrison, R. J. Crisci, S. L. Walker, S. J. Hinterlong, and M. J. Herron, "Design and tolerancoing comparisons for S-SEED-based free-space switching fabrics," Optical Engineering 31(21), (1992) pp. 2697-2711.

[2] R. Goke and G. J. Lipovski, "Banyan networks for partitioning multiprocessor systems," Proc. of the 1st Annual Symposium on Computer Architecture, (1973) pp. 21-28.

[3] K. Y. Eng, M. J. Karol, and Y.-S. Yeh, "A growable packet (ATM) switch architecture: design principles and applications," in Proc. GlobeCom'89, (IEEE, New York, 1989), pp. 1489-1499.

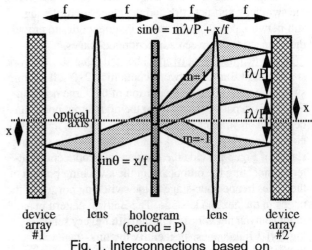

Fig. 1. Interconnections based on diffractive holograms.

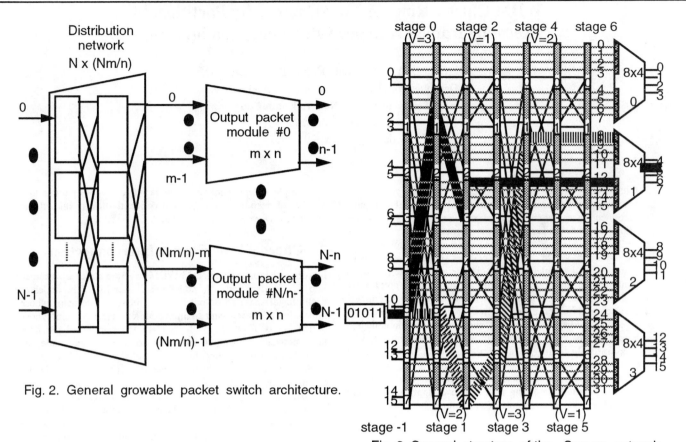

Fig. 2. General growable packet switch architecture.

Fig. 3. General structure of the Canyan network.

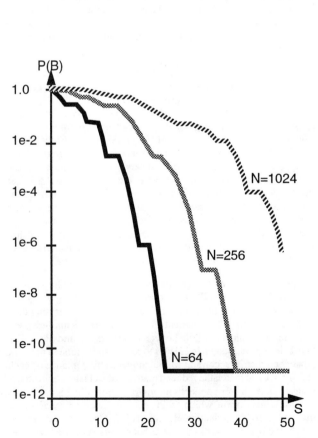

Fig. 4. Packet loss probability (P(B)) vs. # stages (S) for Canyan networks (m=32, n=8, p=1.0).

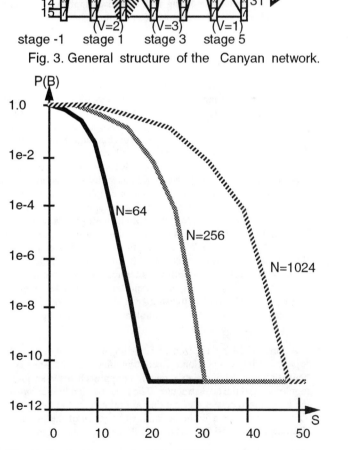

Fig. 5. Packet loss probability (P(B)) vs. # stages (S) for Compressed-Canyan networks (m=32, n=8, p=1.0).

WDM Cluster Ring: A Low-Complexity Partitionable Reconfigurable Processor Interconnection Structure

KHALED A. ALY AND PATRICK W. DOWD*

Department of Electrical and Computer Engineering,
State University of New York at Buffalo
Buffalo, NY 14260
dowd@eng.buffalo.edu

Abstract – Wavelength selectivity of light sources and filters enable reconfigurable and partitionable parallel computer interconnection at low space complexity via wavelength-division multiplexing (WDM). Large scale processor networks can be built by mixing wavelength-division and passive space-division. In a multi-domain WDM network clusters of processors can be assigned ordered wavelength channel sets such that interprocessor communication takes place on a conflict-free point-to-point basis. This paper presents a multi-domain WDM processor interconnection structure based on a ring of processor clusters. The cluster ring has a limited space complexity and is easily partitionable into sub-networks, each can be reconfigured to multiple connectivities by appropriately tuning node transceivers. Reconfiguration of arbitrarily-sized sub-networks into grids, binary trees, and multi-stage connectivities is demonstrated. The objective is supporting multi-user parallel computing with a relaxed application dependence on the physical interconnection topology.

1 Introduction

WDM networks have the potential of achieving passive reconfiguration through tunable light sources and filters and passive coupling devices [1, 2]. Reconfiguration to multiple point-to-point connectivities enables efficient execution of synchronous parallel tasks with various interprocessor communication patterns. Partitioning into multiple independent sub-networks enables independent concurrent execution of multiple tasks exploiting different modes of parallelism [3]. Network scalability is constrained by limitations on optical power budget (star coupler fanout in the order of 10's to a few 100's of ports) and on the optical devices tuning range (typically 10's to a few 100's of channels are available) [4]. Overcoming this limitation is possible through wavelength *spatial re-use*. Multi-domain WDM structures (M-WDM) employing discrete broadcast and select coupling domains were introduced to achieve scalable conflict-free interprocessor communication [5]. Clusters of processors, interconnected according to a regular topology via fiber links, transmit over distinct ordered channel sets through *broadcast* output couplers and receive *conflict-free* through *select* input couplers.

A form of the generalized chordal ring was considered in [2] for cluster interconnection. Reconfiguration was studied via a heuristic connectivity matrix matching approach between guest topologies and the host structure. The potential and limitations of the generalized cluster chordal ring are summarized in Section 2. The cluster ring is a limited-complexity version of the previous approach, described in Section 3. Partitioning the cluster ring

---

*This work was supported by the National Science Foundation under Grants CCR-9010774 and ECS-9112435.

to multiple independent sub-networks, each reconfigurable to a number of desirable point-to-point connectivities, is studied and an illustrative example is presented.

2 Chordal Ring Cluster Interconnection

In a discrete broadcast-select M-WDM structure nodes are grouped into M_1 clusters with M_0 nodes per cluster [5]. The nodes of each cluster share an output coupler (broadcast domain) whose dimension is $M_0 \times F$ and an input coupler (select domain) whose dimension is $F \times M_0$. Clusters are interconnected via fiber links between the output and input couplers. Each node has a single fixed transmitter and a multi-channel tunable receiver. Nodes within the same cluster transmit over distinct wavelength channels. Each cluster P_k, represented as a set of processors $\{(i, 0), \ldots, (i, M_0 - 1)\}$, transmits over an ordered channel set $\Lambda_i = (\lambda_{i0}, \lambda_{i1}, \ldots, \lambda_{i(M_0-1)})$. Transmit channel sets are assigned to clusters such that no conflicts occur at any input coupler, i.e any two clusters fanned out to the same input coupler are assigned disjoint channel sets.

The chordal ring cluster interconnection considered in [2] is defined by the set of clusters $P = \{0, 1, \ldots, M_1 - 1\}$ and the set of links $E = \{\{i, i \pm k\}, k \in [0, (F-1)/2] \ \forall \ i \in [0, M_1 - 1]\}$, where F represents the cluster degree. The necessary number of disjoint channel sets to guarantee conflict free communication is given by $X = F + (M_1 - F\lfloor M_1/F \rfloor)$ ($X = F$ if F divides the number of clusters M_1). X channel sets each has M_0 channels are assigned to clusters in a way that preserves conflict-free reception. If a total of C channels are available, then $XM_0 \leq C$. A larger cluster degree results in better physical connectivity but requires more channels. The chordal ring reduces to a ring of clusters with self links when the cluster degree $F = 3$, which is shown in Figure 1. Besides reducing the space complexity, a lower F has the advantage of reducing the required number of channel sets X, which implies more nodes can be packed in each cluster.

The principle of reconfiguration to an arbitrary topology can be explained as follows [2]. The adjacency matrix of a given guest topology is divided into square sub-matrices, each corresponding to a cluster. Any "1" element in a submatrix ij implies there need to be a physical link between clusters P_i and P_j. The heuristic approach taken in [2] was to reduce the emulated network adjacency matrix according to the connectivity demands of each sub-matrix and match the reduced matrix to the cluster connectivity matrix. The cluster connectivity matrix, shown in Figure 2, has a modulo band form which facilitates the matching by applying matrix bandwidth reduction algorithms [6]. One limitation is that emulating larger networks demands better cluster connectivity, i.e. larger F. This implies more channels are required for conflict free communication. In turn, this poses a limit on the cluster size, and therefore the network size. The heuristic matrix transformation

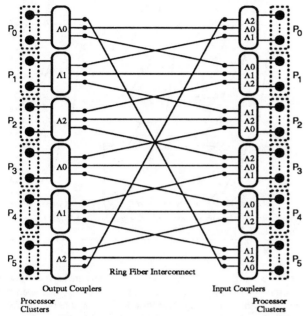

Figure 1: Cluster ring with $M_1 = 6$ clusters; $F = 3$ (conflict-free channel set assignment shown)

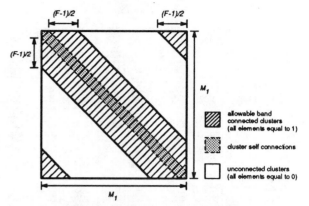

Figure 2: *Modulo band-form* connectivity matrix; general F

approach allows emulating arbitrary connectivities. However, the procedure must be carried out for each allowable connectivity depending on the available portions of the network.

3 WDM Cluster Ring

This paper considers a top-down approach. That is, given the described chordal ring structure with $F = 3$, i.e a cluster ring, we study the reconfiguration (method and constraints) to each of the following desirable connectivities: n-dimensional grid, binary tree, and multistage networks. It is shown that these connectivities can be systematically emulated by arbitrary portions of the network, thus enabling flexible partitioning. Size limitations are identified for grids with larger dimension and multi-stage connectivities. The cluster ring M-WDM network is shown in Figure 1 with $M_1 = 6$ clusters, M_0 nodes per cluster, and $X = 3$ channel sets whose conflict-free assignment is illustrated.

3.1 Grid Emulation

The adjacency matrix of a 64-node (8×8) torus is shown in Figure 3, divided into submatrices corresponding to an 8-cluster ring. An inherit correspondence exists between the grid adjacency matrix and the band-form connectivity matrix of the cluster ring. This can be observed by substituting every non-zero submatrix by a "1" element and matching against the physical connectivity matrix of Figure 2 when $F = 3$. All node adjacencies are confined to a cluster within a unity distance from the subject cluster. This correspondence allows direct mapping of n-dimensional grids ($n \leq 3$) with a size constraint that increases with larger n. The grid can be emulated also using a fraction of the nodes at each cluster provided that all involved clusters are successive ones (enabling flexible partitioning at node boundaries, rather than cluster boundaries).

In general, a symmetric n-dimensional grid with m nodes per axis, denoted as $\mathcal{GR}_{n,m}$, can be constructed by connecting m

$\mathcal{GR}_{n-1,m}$ levels. If the size of each $\mathcal{GR}_{n-1,m}$ does not exceed the cluster size, M_0, and if successive clusters are employed in the emulation, then the connectivity of $\mathcal{GR}_{n,m}$ is confined to clusters within a unity distance. This is since any node of a $\mathcal{GR}_{n-1,m}$ level connects only to a node from the following or preceding levels, whose sizes do not exceed the cluster size. In the case of torus emulation, however, the last level is connected to the first. This implies that the complete cluster ring has to be utilized in order to involve the wrap-around cluster links. This does not mean that the emulated torus size is restricted to the total network size. Fractions of clusters can be used as long as all included clusters are successive. The following are valid grid emulations using a full cluster size of $M_0 = 64$ nodes:

▷ A linear array whose size is a multiple of 64.

▷ A square (64×64) 2-D grid (using 64 clusters) or a rectangular grid consisting of a multiple of 64 rows of nodes.

▷ A symmetric ($8 \times 8 \times 8$) 3-D grid (using 8 clusters each with 64 nodes) or an asymmetric grid with multiple 8×8 grids.

▷ A symmetric ($4 \times 4 \times 4 \times 4$) 4-D grid (using 4 clusters each emulating a 64-node 3-D grid).

The example shows the potential for grid emulation by the cluster ring, as well as the size limitation for symmetric grids with a large dimension. Practically, dimensions of up to 3 are beneficial. If large symmetric processor networks are required, each node may consist itself of a cluster of processors, operating at a lower rate than the optical interface rate and thus can be time-multiplexed. Further scalability and cost effectiveness are possible by letting every node emulate a number of nodes, as was shown in [7].

3.2 Complete Binary Tree Emulation

Embedding a complete binary tree of depth k, $\mathcal{T}_k$, in m_1 successive clusters, $1 \leq m_1 \leq M_1$, is considered. The employed number of nodes per cluster is $m_0 = 2^{k+1}/m_1$. Consider values of m_1 that are powers of two, since nodes are then evenly distributed among clusters (one node is left out). The embedding is achieved by labeling the nodes of $\mathcal{T}_k$ using the set of labels (used clusters) $\{0, 1, \ldots, m_1 - 1\}$. Every label occurs exactly m_0 times, except for one label that occurs only $m_0 - 1$ times, which corresponds to the node that makes the difference between 2^{k+1} and the complete binary tree size. Labeling is restricted by one rule: labels x and y assigned to any two adjacent nodes in $\mathcal{T}_k$ are such that $x = y$

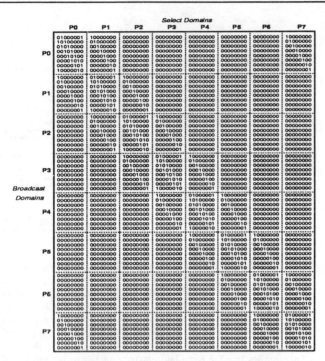

Figure 3: Adjacency matrix of 64-node square grid divided into 64 (8 × 8) submatrices corresponding to 8 clusters with 8 nodes each

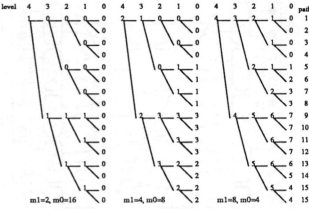

Figure 4: The labeling associated with emulating a 4-level binary tree; $m_1 = \{2, 4, 8\}$ and $m_0 = 32/m_1$

or $x = y \pm 1$. This process corresponds to assigning adjacent tree nodes to clusters that are physically connected. Configuration takes place by properly re-tuning receivers, as directly derived from the labeling.

The emulation uses at most $m_{max} = 2k$ successive clusters, each with $2^{k+1}/m_1$ nodes, starting with any arbitrary cluster and leaving only one node in one cluster free. T_k is labeled with cluster identifiers such that any two adjacent nodes carry identifiers of connected clusters and each identifier does not occur more than M_0 times in the entire tree labeling. To allow using the *maximum* number of clusters in the labeling process, the root is labeled $m_1/2$, where m is the number of clusters to be used ($m_1 m_0$ should match the tree size). Hence the condition that $(m_1/2) + 1 \leq k + 1$, i.e. $m_1 \leq 2k$. Since m_1 is a power of two, it can take on the values $\{1, 2, \ldots, \lfloor \log_2 2k \rfloor\}$.

The node labeling algorithm is given below. Labeling is carried on in order for all root-leaf paths of the tree, with nodes labeled by previous paths being skipped. The jth path is identical to the $(j-1)^{th}$ path in all nodes starting from the root until the first node $(i, \lceil j/2^i \rceil)$ is encountered, where 2^i divides $j - 1$, then it becomes disjoint. The algorithm proceeds by labeling both the root node $(j, 1)$ and one of its children, arbitrarily $(k - 1, 2)$, with $m_1/2$. Labeling continue in the left sub-tree with a value that decrements by 1 each level down until either 0 is reached or the label has occurred M_0 times. The policy then is to try to exhaust lower-valued labels first along each path and then increment again by 1 until $m_1/2 - 1$ is reached. The same procedure is carried on with the right sub-tree, except that the increment direction is first pursued and then the decrement direction. It can be seen that the left sub-tree, of size $2^k - 1$, is covered by $m_1 m_0/2 = 2^k$ labels, with one of the $m_1/2 - 1$ labels left out. The right sub-tree is also covered by $(m_1 m_0/2) - 1 = 2^k - 1$, since one of the $m_1/2$

labels is assigned to the root. The way labels are assigned always preserves a label within a distance of 1 for the next unlabeled node in any path. An example labeling for a complete binary tree of depth 4 is shown in Figure 4 for the valid values of m_1 ($m_1 = 1$, not shown, corresponds to the trivial case of assigning all 32 nodes to a single cluster).

```
begin
label[(k, 1)] = label[(k − 1, 2)] = m_1/2
for each root-leaf path j, 1 ≤ j ≤ 2^{k−1}
    for each node (i, ⌈j/2^i⌉) in path j, 0 ≤ i ≤ k
        if node is not labeled by path j − 1
            candidate_label = label[(i − 1, ⌈j/2^{i−1}⌉)] − 1
            assign (i, j, candidate_label)
            while reject assign (i, j, candidate_label + 1)
        end if
    end for
end for
for each root-leaf path k, 2^{k−1} + 1 ≤ j ≤ 2^k
    for each node (i, ⌈j/2^i⌉) in path j, 0 ≤ i ≤ k
        if node is not labeled by path j − 1
            candidate_label = label[(i − 1, ⌈j/2^{i−1}⌉)] + 1
            assign (i, j, candidate_label)
            while reject assign (i, j, candidate_label − 1)
        end if
    end for
end for
end

assign (i, j, candidate_label)
if ((((j ≤ 2^{k−1} and candidate_label > 0) or
    (j > 2^{k−1} and candidate_label < m_1/2)) and
    count[candidate_label] < m_0)
    then label[(i, ⌈j/2^i⌉)] = candidate_label
        increment count [candidate_label]
    else reject
end if
```

3.3 Multi-stage Network Emulation

A multi-stage interconnection network (MIN) consists of $k = \log_2 M$ interconnection stages linking $k + 1$ levels of nodes, with M nodes in each level. Two successive stages are linked according to a given interconnection function. Indirect cube multistage networks efficiently simulate other networks/algorithms. Bidirectional multistage networks with arbitrary stage permutations can be embedded in a cluster ring where the stage size does not exceed $M_0 \leq C/3$, proven as follows.

Any two clusters, P_i and $P_{i \pm 1 \mod(M_1)}$ can be configured as a stage with an arbitrary permutation in both directions with $F = 3$. To support larger level sizes, say kM_0 for some integer $k > 1$, it is necessary to have $F = 4k - 1$ ($2k - 1$ links at each side plus the self-cluster link). Recalling that $M_0 F \leq C$, the maximum level size is $kC/(4k - 1)$, with a peak at $k = 5/4$. Since k is an integer by assumption, and also F has to be odd, the maximum level size is $M_0 = C/3$ with $F = 3$ ($k = 1$). In other words increasing k beyond unity results in a stronger reduction of the maximum M_0. For example, let $C = 60$, $F = 3$, so $M_0 = 20$ nodes. Using $k = 2$, requires $F = 7$ and hence $M_0 = 8$. The resulting $kM_0 = 16$ nodes is less than the 20 nodes per level with $k = 1$.

3.4 Example: A Partitioned Cluster Ring

A detailed example of the cluster ring partitioning and reconfiguration is presented in Fig. 5. The number of clusters, M_1, is scalable to any desired value. It is assumed that M_1 is a multiple of 3 and therefore $X = 3$ to maximize the cluster size, taken to be $M_0 = 4$ for the purpose of illustration. The considered portion of the network, of size 10 clusters, is partitioned into a 16-node square mesh, a 15-node binary tree, and a two-stage binary shuffle sub-networks. The figure shows the wavelength assignment to the fixed transmitters which guarantees conflict free reception at all nodes, and the channels to which the receiver/s need to tune to for each embedding. The lines represent the virtual (wavelength) *point-to-point* connections between the processors of each sub-network.

The square grid $\mathcal{GR}_{2,4}$ is embedded into $\{P_0, P_1, P_2, P_3\}$. Receive channels are given according to the adjacency order (north, west, south, east). The binary tree $\mathcal{T}_3$ is embedded into a set of $m = 4$ clusters $\{P_4, P_5, P_6, P_7\}$. The wavelength realization of the downward links is only shown. Node labeling is done precisely according to the given algorithm, with the cluster numbers shifted by 4 since the first cluster is P_4. A bidirectional binary shuffle function is embedded into clusters $\{P_8, P_9\}$. This example illustrates the flexible partitioning and the principles of wavelength spatial re-use and conflict-free point-to-point communication.

4 Summary

This paper introduced a low complexity multi-domain WDM processor network structure based on a cluster ring. The network is partitionable into independent sub-networks, each can be reconfigured to a number of point-to-point connectivities. Partitioning and reconfiguration are achieved via a distributed *tuning* and *labeling* process. The considered connectivities are the n-dimensional grid, binary tree, and multi-stage networks. Size limitations were identified for some emulation cases. However, the number of clusters in the network does not suffer any limitation. Unlimited scalability is due to wavelength spatial re-use. Advantage is taken from the sparsity of interconnection topologies with limited node degree. Conflict-free point-to-point communication is enabled via a discrete broadcast and select approach. This approach overcomes the physical power budget and tuning range constraints. The cluster-based structure results in a modular architecture with limited space complexity thanks to the constant cluster degree. The described interconnection network has a potential for supporting multi-user/multi-tasking massive parallelism and alleviating the application dependence on the underlying physical topology.

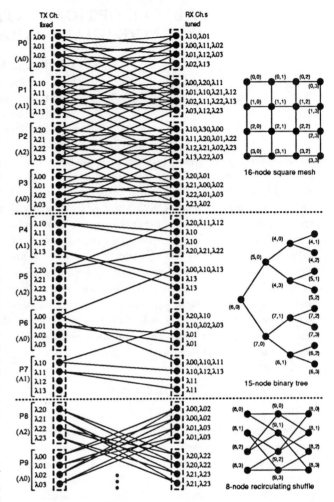

Figure 5: Upper portion of cluster ring partitioned into square grid, binary tree, and shuffle sub-networks

References

[1] B. Li and A. Ganz, "Virtual topologies for WDM star LANs: The regular structures approach," in *Proc. IEEE INFOCOM'92*, 1992.

[2] K. A. Aly and P. W. Dowd, "Parallel computer reconfigurability through optical interconnects," in *Proceedings of the 21st International Conference on Parallel Processing*, pp. I105 – I108, August 1992.

[3] H. J. Siegel *et al.*, "PASM: A partitionable SIMD/MIMD system for image processing and pattern recognition," *IEEE Transactions on Computers*, vol. c-30, pp. 934–946, Dec. 1981.

[4] C. A. Brackett, "Dense wavelength division multiplexing networks: Principles and applications," *IEEE Journal on Selected Areas of Communications*, vol. 8, pp. 948–964, Aug. 1990.

[5] K. A. Aly and P. W. Dowd, "A multi-domain WDM network structure for reconfigurable partitionable massively parallel computer architecture," *SPIE Proceedings (High-Speed Fiber Networks and Channels)*, vol. 1784, Sept. 1992.

[6] R. P. Tewarson, *Sparse Matrices*. Academic Press, 1973.

[7] J. P. Fishburn and R. A. Finkel, "Quotient networks," *IEEE Transactions on Computers*, vol. C-31, pp. 288–295, Apr. 1982.

A SCALABLE OPTICAL INTERCONNECTION NETWORK FOR FINE-GRAIN PARALLEL ARCHITECTURES

D. Scott Wills and Matthias Grossglauser
School of Electrical Engineering
Georgia Institute of Technology
Atlanta, Georgia 30332-0250
scotty@ee.gatech.edu

Abstract: *This paper introduces a scalable interconnection network based on integrated optical devices developed at Georgia Tech. It incorporates arrays of optical transmitters and detectors placed on top of silicon chips which are then arranged on a silicon substrate. Staggering these substrates to overlap chips in different planes forms an offset cube topology. This paper examines this novel optoelectric communication network. It explains the physical network architecture and the offset cube topology, and summarizes its performance. Combining this optoelectronic communication network with new fine-grain machine architectures (multi-node per chip) can lead to extremely dense, high performance parallel systems.*

INTRODUCTION

Fine-grain, message-passing architectures offer high performance for high-throughput parallel applications such as image processing (e.g., filtering, edge detection, convolution), object recognition, and data compression. The Pica group at Georgia Tech aims to produce a minimal parallel node which makes efficient use of the required resources (i.e., high performance per transistors required). The design includes a minimal sequential core architecture. Support for essential parallel operations is provided: communication, synchronization, naming, and fine-grain task and storage management. The high-throughput nature of the targeted applications allows reasonable processor utilization with a small amount of local memory. A network interface/router supports communication. (See [7].)

The simple machine architecture and reduced local memory requirements of target applications allow multiple high performance (50 MIPS/node) processing nodes to be integrated on a single chip. Future technology improvements in transistor density and speed are translated directly into more nodes per chip. However, before this goal can be realized, a major obstacle must be overcome: limited off-chip bandwidth.

In order to meet the I/O requirements of several nodes per chip, a high-bandwidth, three-dimensional optical network is being developed. Existing optical interconnect techniques such as index-guided and free-space networks are difficult to incorporate in three-dimensional interconnection networks.

This paper describes a three-dimensional optoelectrical network using hybrid integrated optoelectronic devices [2] and through-wafer transmissions [1]. The network is designed to provide 3.2 Gbits/sec off-chip bandwidth using vertical optical channels to eight neighboring chips. This interconnection forms an *offset cube* topology which is analyzed later in this paper.

PHYSICAL ARCHITECTURE

Silicon continues to be the material of choice for digital integrated circuits. However, because of its poor optoelectronic properties, efforts to incorporate integrated optical emitters and detectors have met with limited success. More effective optoelectronic devices have been demonstrated using GaAs and InP. However, GaAs and InP have been unable to match Si in density or cost for VLSI applications.

A new technique for combining GaAs and InP-based optoelectrical devices with Si VLSI has been developed in the Microelectronics Research Center at Georgia Tech. Using epitaxial liftoff, small (50 x 50 μm) GaAs and InP-based light emitting diodes and lasers have successfully been removed from an InGaAsP substrate, aligned, and selectively deposited on a silicon host substrate. This technique has been refined to allow the deposition of arrays of GaAs and InP-based emitters and detectors in parallel.

The GaAs and InP-based devices are deposited on an insulating planarizing layer over the Si substrate, so optoelectronic devices can be placed on top of electronic circuitry. Emission wavelengths for InGaAs and InGaAsP are selected for which Si is transparent, allowing through-wafer transmission.

This network exploits this technology to form a three-dimensional interconnect. Processing elements are designed and fabricated using existing Si VLSI techniques. Bonding pads are provided only for power contacts. (Probe pads are provided for testing.) Holes in the overglass allow GaAs and InP-based emitters and detectors to be attached in a post-processing step. Silicon-based driver circuits connect parallel buses from routers on the chip to the faster optical devices (0.1 – 1.0 Gbits/second).

After the Si chips containing the optical devices are fabricated and tested, they are attached to a larger

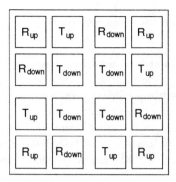

Figure 1: Chip Device Pattern

silicon substrate. This provides physical support and also participates in power distribution and cooling. A fully populated substrate forms one plane of the system. Completed substrates are then stacked to create the third dimension. To facilitate manufacturing, all chips and substrates are identical.

Each chip is broken into four quadrants which overlap with eight neighboring chips (four in the plane above, four in the plane below). Although any number of transmitter and receivers can be included in each quadrant, the minimum case requires two transmitter/receiver pairs, shown in Figure 1.

In order for transmitter/receiver pairs to be correctly aligned, the chips must be offset. To achieve this using identical plane substrates, a spacing equal to one half the chip width plus inter-chip spacing is added to two non-opposing sides of the substrate. Then alternating planes are rotated 180^0 during assembly providing the correct device alignment. Figure 2 shows the substrate pattern.

System I/O is provided by I/O layers which inject data directly into the network on the top and bottom surfaces of the stack. These layers can interface to high-bandwidth electrical or optical connections. Alternatively, the input surface can be covered by visible light detectors for direct focal-plane imaging. The output surface can incorporate a frame buffer for direct video output.

The sides of the stack are used for power distribution and cooling connections. Due to the dense packaging, liquid cooling between layers is anticipated.

Optical clock distribution is also incorporated in this system. In-line detectors on each layer detect broadcast clock signals from the bottom of the stack via the through-wafer transmission technique employed by the network.

TOPOLOGY

Since each chip in the system contains several nodes, the communication network is divided into two

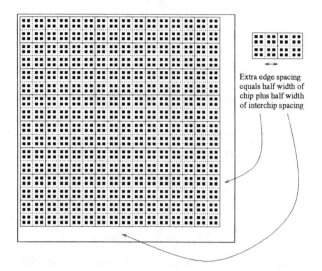

Extra edge spacing equals half width of chip plus half width of interchip spacing

Figure 2: Substrate Pattern for 8 x 8 Chip Layer

parts. The inter-chip network delivers messages between the chips containing the source and destination nodes. The intra-chip network is used to communicate messages within the source and destination chips. Messages sent between nodes within a chip are routed entirely via the intra-chip network.

Communication within a chip is accomplished using a wire-only network. This type of network is well-studied [3, 6]. Since the on-chip wire density is high and the number of nodes per chip is initially low (the first Pica prototype will contain four nodes), the focus of this paper is placed on the inter-chip network.

Each chip connects to eight neighboring chips: four in the plane above and four in the plane below. There are no inter-chip connections within a substrate plane. This interconnection forms an *offset cube* topology which bears similarities to the well-studied *k*-ary 3-cube. While the physical architecture of the inter-chip network is not symmetric in the vertical and horizontal dimensions, the offset cube topology implemented by this network is isotropic. The topology, shown in Figure 3, is formed by vertices that have all even or all odd coordinates.

In describing offset cubes, L represents the number of layers, and K represents the number of chips in either dimension of a layer. For a symmetric cube, $L = 2K + 1$ creating a *L-ary offset cube*.

Each chip is designated by a 3-tuple (X, Y, Z), where Z is the layer number between 0 and $L - 1$. Chips on even numbered layers are assigned even X and Y coordinates between 0 and $2K - 2$. Chips on odd numbered layers are assigned odd X and Y coordinates between 1 and $2K - 1$.

For example, layers $Z = 0, 2, 4, \ldots, L-2$ contain the chips $(0, 0, Z), (0, 2, Z), \ldots, (2, 0, Z), \ldots, (2K-2, 2K-$

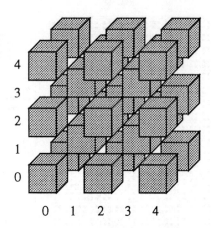

Figure 3: Offset Cube Topology

$2, Z$). Layers $Z = 1, 3, \ldots, L - 1$ contain the chips $(1, 1, Z), \ldots, (2K - 1, 2K - 1, Z)$. Each chip (X, Y, Z) (except for border chips) connects to $(X \pm 1, Y \pm 1, Z \pm 1)$. Border chips have $Z \in \{0, L - 1\}$, or $X = 0$ or $Y = 0$ for even layers and $X = 2K - 1$ or $Y = 2K - 1$ for odd layers. The total number of chips is LK^2.

In a k-ary 3-cube, a shortest path routing algorithm uses channel routes to reduce the difference in coordinates of the source and destination chips. A well-performing deterministic shortest path algorithm reduces the coordinate differences one at a time (i.e., x, then y, then z, or x-y-z routing).

A shortest path routing algorithm for an offset cube is more complicated since each channel hop changes all three coordinates. If one coordinate difference reaches zero, it must be increased in the next hop to reduce differences in other dimensions. This creates "bouncing" routes where a message alternates between neighboring planes as it moves through one dimension.

Therefore the minimum path cost of routing a message between chips A and B on an offset cube is:

$$hops_{min} = \max(|X_B - X_A|, |Y_B - Y_A|, |Z_B - Z_A|)$$

The analysis of this network is similar to that of wire-only networks [3], except that the optical channels require a different set of assumptions.

1. Unlike the wire density limits for wire networks, optical channels are limited entirely by characteristics of the transmitter and receiver. The communication medium (i.e., the space through which light travels) does not enforce any fundamental density or bandwidth limits.

2. The channel delay is not determined by channel length. Since the messages travels fast (near c), and the channel length is small (around $1mm$), the time of flight in a channel is around three femtoseconds. The channel delay is dominated by transmitter and receiver response time (around $1nS$).

3. Power and heat dissipation are the dominant issues in communication costs. GaAs LEDs have a low quantum efficiency ($< 3\%$). GaAs lasers are are more efficient (75%), but still dissipate several milliwatts during operation. The cumulative effect of thousands of lasers operating in a few hundred cm^3 will require special consideration.

These assumptions affect the performance aspects of a network. However, many existing techniques for multiprocessor interconnection networks (e.g., deadlock avoidance [4]) are applicable to this network.

NETWORK PERFORMANCE

Performance of the offset cube topology has been analyzed through simulation. This section presents a summary of the results of this analysis. The complete details are available in [5].

To compare the k-ary 3-cube and offset cube topologies, a simulator has been constructed which analyzes network performance under similar conditions and loads [5]. In these experiments, a 16-ary 3-cube (4096 nodes) is compared with a 25-ary offset cube (3925 nodes). A random traffic load is simulated. Wormhole routing is employed with eight virtual channels for each physical channel. All message lengths are 25 flits.

Figure 4 compares the performance of deterministic and adaptive routing algorithms on the two topologies. For the deterministic case, the 16-ary 3-cube routes messages x-y-z. The offset cube routes along a diagonal shortest path. The offset cube utilizes less than half of the capacity of the 3-cube.

The capacity loss is due to loading in the center of the network. For random traffic, deterministic routing on a 3-cube does a reasonable job of distributing traffic across the entire network. The offset cube's diagonal routing creates much higher traffic in the center of the network. This has been verified by comparing cross-sectional loading profiles for a 3-cube with x-y-z and diagonal routing algorithms.

To better utilize the capacity of an offset cube network, a simple adaptive routing algorithm is employed. Local information (the number of virtual channels in use) is used to select between shortest paths at each vertex. Figure 4 compares each topology for both deterministic and adaptive routing strategies. The adaptive offset cube performance exceeds that of the best 3-cube strategy. The adaptive 3-cube performance is *lower* than the deterministic case because of increased diagonal routing.

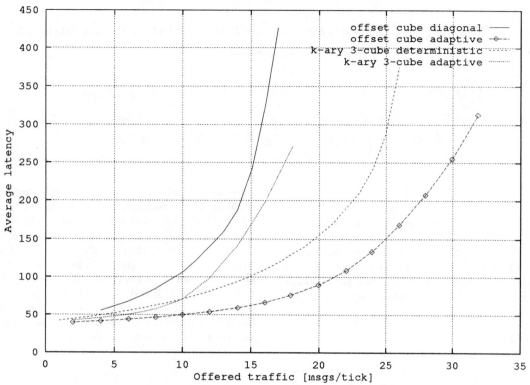

Figure 4: Performance of Offset Cube, 3-Cube

STATUS

Currently, a test chip is being built containing parts of the offset cube router and optical channel. A proposed full-scale prototype will contain 32 100 Mbit/sec transmitter/receiver pairs. Four 50 MIPS processing elements are included on each chip. A processing plane contains 64 chips (256 nodes, 12,800 MIPS) and measures approximately 10 cm by 10 cm. 16 planes contain 1024 chips (4096 nodes, 204,800 MIPS) and fit inside a cube 10 cm on a side. 819.6 Gbits/sec of I/O bandwidth is available from the top and bottom surfaces of the cube. Combining this optoelectronic communication network with new fine-grain machine architectures can lead to extremely dense, high performance parallel systems.

REFERENCES

[1] K. H. Calhoun, C. B. Camperi-Ginestet, and N. M. Jokerst. Vertical Optical Communication Through Stacked Silicon Wafers Using Hybrid Monolithic Thin Film InGaAsP Emitters and Detectors. *IEEE Photonics Technology Letters*, 5(2):254–257, February 1993.

[2] C. Camperi-Ginstet, M. Hargis, N. Jokerst, and M. Allen. Alignable Epitaxial Liftoff of GaAs Material with Selective Deposition Using Polyimide Diaphragms. *IEEE Transactions Photonics Technology Letters*, 3(12):1123–1126, December 1991.

[3] William J. Dally. Performance Analysis of *k*-ary *n*-cube Interconnection Networks. *IEEE Transactions on Computers*, C-39(6):775–785, June 1990.

[4] William J. Dally and Charles L. Seitz. Deadlock-Free Message Routing in Multiprocessor Interconnected Networks. *IEEE Transactions on Computers*, C-36(5):547–553, May 1987.

[5] Matthias Grossglauser and D. Scott Wills. Performance Analysis of an Offset Cube Network using Optical Interconnect. VLSI Architectures Group Technical Report, January 1993.

[6] Daniel A. Reed and Richard M. Fujimoto. *Multicomputer Networks: Message-Based Parallel Processing*. Scientific Computation Series. MIT Press, 1987.

[7] D. Scott Wills, W. Stephen Lacy, Huy Cat, Michael A. Hopper, Ashutosh Razdan, and Sek M. Chai. Pica: An Ultra-Light Processor for High-Throughput Applications. *submitted to* ICCD '93 International Conference on Computer Design, VLSI in Computers & Processors, October 1993.

Bus-Based Tree Structures for Efficient Parallel Computation

(*Extended Abstract*)

O. M. Dighe   R. Vaidyanathan*   S. Q. Zheng

Dept. of Electrical & Computer Eng.   *Dept. of Computer Science*
Louisiana State University, Baton Rouge, LA 70803

Abstract

We propose a class of new multiprocessor structures called bus based trees (BBTs) that are based on multiple buses. We show that a BBT can simulate a tree machine optimally. We also discuss optimal VLSI layouts for the BBT and show that the BBT can be used as a building block to construct new powerful parallel computing structures.

1 Introduction

A *tree machine* is a multiprocessor parallel computer architecture in which the processors are connected as a complete binary tree. As binary trees capture the essence of divide-and-conquer strategies, tree machines support several useful logarithmic-time functions such as broadcast, search, census, and select-and-rotate, assuming that data communication along a link takes constant time. Designing efficient parallel algorithms for tree machines has received extensive attention (see [1, 2, 3, 5, 10] for examples). Several tree-based parallel computing structures like the mesh of trees and pyramid networks, have been proposed and shown to be very useful.

In [13], a simplified tree machine called the compressed tree machine, is proposed. A compressed tree machine has fewer processors and links than the corresponding tree machine. It has been shown that the compressed tree machine has a more compact VLSI layout, and for a class of applications, the compressed tree machine can simulate the tree machine without any slowdown. In [12], a multiple bus based multiprocessor structure is proposed. This structure also consists of fewer processors than the tree machine, and it has been shown that for a class of algorithms, this structure can perfectly simulate the tree machine. This bus based structure is generalized in [4] to construct other useful structures. However, none of the methods proposed in [4, 12, 13] can simulate all tree machine algorithms with a small constant slowdown factor.

In this paper, we propose a structure called the *bus based tree* (BBT) that can simulate each computation and communication step of a tree machine in two and three

*Supported in part by the Summer Stipend Program Grant from the Council on Research, Louisiana State University, Baton Rouge, LA 70803.

substeps, respectively. This simulation is shown to be optimal under the assumptions laid out in the next section. The BBT also has other advantages over the tree machine like ease of broadcasting and more flexibility of design parameters. Other work on multiple bus based structures have been done before (see [6, 9] for more references). We also propose an extension of the BBT to the *bus based mesh of trees* (BBMOT), and show it to be a powerful parallel computing structure. We believe that the BBT and the techniques presented in this paper can be used to design new parallel computing structures.

2 Simulation Lower Bound

A tree machine $T(p)$ consists of $2^{p+1} - 1$ nodes (processors) connected as a complete binary tree. We use the terms "node" and "edge" to refer to the processors and communication links of $T(p)$; the terms "processor" and "bus" will be used in the context of multiple bus based networks (MBNs).

In this section we derive a lower bound on the time needed to simulate each step of $T(p)$ on $M(p, b)$, an arbitrary MBN with 2^p processors and 2^b buses, where $2 \leq b < p$. Since $T(p)$ has $2^{p+1} - 1$ nodes, it is clear that all but one processor of $M(p, b)$ must simulate two nodes of $T(p)$. Therefore, each computation step of $T(p)$ needs at least 2 substeps on $M(p, b)$. Normally, the leaves of $T(p)$ are more powerful than the internal nodes, which are responsible primarily for routing information in the tree. Therefore, we require each processor of $M(p, b)$ to simulate at most one leaf of $T(p)$. Each edge of $T(p)$ represents a possible communication. We assume that each node of $T(p)$ may communicate with all of its neighbors in one step. Therefore, each of the $2^{p+1} - 2$ edges of $T(p)$ may be involved in a communication during a step. We refer to such a communication step as a *maximal communication step*. Since $2^{p+1} - 1$ nodes are mapped to 2^p processors, it is possible for both nodes associated with an edge of $T(p)$ to be mapped to the same processor of $M(p, b)$. In this case, the edge does not represent a communication in $M(p, b)$. Since each processor of $M(p, b)$ simulates one leaf, it is easy to verify that there can be at most 2^{p-1} edges in $T(p)$ that do not constitute communications in $M(p, b)$. In other words, each step of $T(p)$ involves at least $2^{p+1} - 2 - 2^{p-1} = 3 \times 2^{p-1} - 2$ communications in

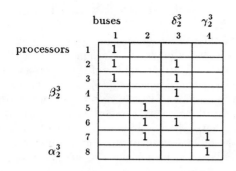

Figure 1: $B(3,2)$; a 1 denotes a connection

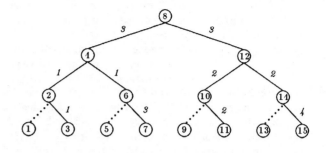

Figure 2: Simulation of $T(3)$ by $B(3,2)$

$M(p,b)$. Since $M(p,b)$ has 2^b buses, the simulation requires at least $\lceil \frac{3\times 2^{p-1}-2}{2^b} \rceil = 3 \times 2^{p-b-1}$ communication substeps, for $p > b \geq 2$.

Theorem 1 *Simulation of a maximal communication step of $T(p)$ on an MBN with 2^p processors and 2^b buses (where $2 \leq b < p$) requires at least $3 \times 2^{p-b-1}$ communication substeps, if each computation step is simulated in 2 substeps and if each processor of the MBN simulates exactly one leaf of $T(p)$.* ∎

<u>Remarks:</u> The above lower bound is independent of the processor fan-out of the MBN.

3 Construction of Bus Based Trees

We denote a bus based tree (BBT) with 2^p processors and 2^b buses by $B(p,b)$. Consider the following constructions for $p > b \geq 2$.

Construction 1: Given $B(p,2)$, construct $B(p+1,2)$.

Construction 2: Given $B(p,b)$, construct $B(p+1,b+1)$.

Given $B(3,2)$ (see Figure 1), for all $2 \leq b < p$, $B(p,b)$ is constructed by first applying Construction 1, $p - b - 1$ times followed by Construction 2, $b - 2$ times.

We denote processors 2^p and 2^{p-1} of $B(p,b)$ by α_b^p and β_b^p, respectively. Two, not necessarily distinct, buses of $B(p,b)$ are designated as special buses and denoted by γ_b^p and δ_b^p. Figure 1 shows α_2^3, β_2^3, γ_2^3 and δ_2^3.

Construction 1 starts with two $B(p,2)$'s placed one above the other (i.e., they share the same 4 buses). For convenience, let the upper and lower $B(p,2)$'s be $B_1(p,2)$ and $B_2(p,2)$, respectively. Let their special processors and buses be $\alpha_{1,2}^p$, $\alpha_{2,2}^p$, $\beta_{1,2}^p$, $\beta_{2,2}^p$, $\gamma_{1,2}^p = \gamma_{2,2}^p$, and $\delta_{1,2}^p = \delta_{2,2}^p$. In Construction 1, we connect processors $\beta_{1,2}^p$ and $\beta_{2,2}^p$ to bus $\gamma_{1,2}^p = \gamma_{2,2}^p$. The resulting MBN is $B(p+1,2)$. Clearly, $\alpha_2^{p+1} = \alpha_{2,2}^p$ and $\beta_2^{p+1} = \alpha_{1,2}^p$. We designate both γ_2^{p+1}, and δ_2^{p+1} to be $\gamma_{1,2}^p = \gamma_{2,2}^p$.

Construction 2 starts with two $B(p,b)$'s that represent the second and fourth quadrants of a $2^{p+1} \times 2^{b+1}$ Boolean matrix whose first and third quadrants are all initially 0's. Let the $B(p,b)$'s in the second and fourth quadrants be

$B_1(p,b)$ and $B_2(p,b)$, respectively. As before, we have $\alpha_{1,b}^p$, $\alpha_{2,b}^p$, $\beta_{1,b}^p$, $\beta_{2,b}^p$, $\gamma_{1,b}^p$, $\gamma_{2,b}^p$, $\delta_{1,b}^p$, and $\delta_{2,b}^p$. Note here that $B_1(p,b)$ and $B_2(p,b)$ have distinct sets of buses. In Construction 2, we connect processors $\beta_{1,b}^p$ and $\beta_{2,b}^p$ to bus $\gamma_{1,b}^p$. The resulting MBN is $B(p+1,b+1)$. Clearly, $\alpha_{b+1}^{p+1} = \alpha_{2,b}^p$ and $\beta_{b+1}^{p+1} = \alpha_{1,b}^p$. We designate γ_{b+1}^{p+1} and δ_{b+1}^{p+1} to be $\gamma_{2,b}^p$, and $\gamma_{1,b}^p$, respectively.

The following lemma can be proved by induction on p.

Lemma 1 *For $p > b \geq 2$, $B(p,b)$ satisfies the following:*
(i) α_b^p is connected only to γ_b^p.
(ii) β_b^p is connected only to δ_b^p.
(iii) Each processor is connected to at most 2 buses. ∎

4 Simulation of a Tree Machine

In this section, we outline a method for optimally simulating each step of $T(p)$ by $B(p,b)$ in $3 \times 2^{p-b-1}$ communication substeps and 2 computation substeps. We will refer to the mapping of the nodes and edges of $T(p)$ on the processors and buses, respectively, of $B(p,b)$ as the mapping of $T(p)$ on $B(p,b)$. Let $N(x) = \{1, 2, \cdots, x\}$. Let the nodes of $T(p)$ be indexed with indices from $N(2^{p+1} - 1)$ so that an in-order traversal of $T(p)$ enumerates these indices in ascending order. We assign the nodes of $T(p)$ to processors of $B(p,b)$ by a function $f : N(2^{p+1} - 1) \longrightarrow N(2^p)$ defined as, $\forall u \in N(2^{p+1} - 1)$, $f(u) = \lceil \frac{u}{2} \rceil$. It is easy to verify that at most two nodes are mapped to the same processor and each leaf is mapped to a different processor. Therefore, each computation step of $T(p)$ needs 2 substeps on $B(p,b)$. It can also be verified that for any two leaves ℓ_1 and ℓ_2 that have a common parent u, either $f(\ell_1) = f(u)$ or $f(\ell_2) = f(u)$. In other words, only communications through $3 \times 2^{p-1} - 2$ of the $2^{p+1} - 2$ edges of $T(p)$ need be simulated by $B(p,b)$.

What remains is mapping the $3 \times 2^{p-1} - 2$ edges of $T(p)$ to the 2^b buses of $B(p,b)$. Figure 2 shows the mapping of the nodes and edges of $T(3)$ to the processors and buses of $B(3,2)$. The indices of the nodes of $T(3)$ are shown enclosed in circles. For a node indexed u, $f(u)$ is shown in bold. Edges that do not count as communications in

$B(3,2)$ are shown dotted. For each of the remaining edges of $T(3)$, the italicized number next to it gives the bus of $B(3,2)$ to which it is mapped.

For $p \geq 3$, $T(p+1)$ can be recursively expressed as two $T(p)$'s, $T_1(p)$ and $T_2(p)$, whose roots are connected to the root of $T(p+1)$. Given the mapping of $T(p)$ on $B(p,2)$, consider the mapping of $T(p+1)$ on $B(p+1,2)$. The nodes are mapped by the function f described above. Let the two $B(p,2)$'s that make up $B(p+1,2)$ be $B_1(p,2)$ and $B_2(p,2)$ (as in Construction 1). The corresponding edges of $T_1(p)$ and $T_2(p)$ are mapped on identical buses as $B_1(p,2)$ and $B_2(p,2)$ share their buses. The edges between the roots of the subtrees and the root of $T(p+1)$ are both mapped to $\gamma_{1,2}^{p} = \gamma_{2,2}^{p} = \gamma_2^{p+1} = \delta_2^{p+1}$.

Consider now the mapping of $T(p+1)$ on $B(p+1,b+1)$ given the mapping of $T(p)$ on $B(p,b)$, where $p > b \geq 2$. Once again, the nodes are mapped by the function f. As before, we consider $T_1(p)$, $T_2(p)$, $B_1(p,b)$ and $B_2(p,b)$ (as in Construction 2). The corresponding edges of $T_1(p)$ and $T_2(p)$ are mapped on corresponding buses of $B_1(p,b)$ and $B_2(p,b)$. The edges between the roots of the subtrees and the root of $T(p+1)$ are both mapped to $\delta_{b+1}^{p+1} = \gamma_{1,b}^{p}$. For brevity, we state the following lemma without proof.

Lemma 2 *For $p > b \geq 2$, the mapping of $T(p)$ on $B(p,b)$ satisfies the following:*
(i) Each processor simulates at most 2 nodes.
(ii) Each bus simulates at most $3 \times 2^{p-b-1}$ communications.
(iii) γ_b^p simulates at most $3 \times 2^{p-b-1} - 2$ communications.
(iv) If nodes u and v are adjacent in $T(p)$ and the edge $<u,v>$ is mapped to bus j, then $f(u)$ and $f(v)$ are both connected to bus j. ∎

A direct consequence of parts (i), (ii) and (iv) of Lemma 2 is the following theorem.

Theorem 2 *For $p > b \geq 2$, each step of $T(p)$ can be simulated by $B(p,b)$ in 2 computation substeps and $3 \times 2^{p-b-1}$ communication substeps.* ∎

Remarks: $B(p,b)$ can simulate $T(p)$ optimally (by Theorem 1). In particular, if $b = p - 1$, then 3 substeps are sufficient for $B(p,p-1)$ to simulate any communication step of $T(p)$. In the tree machine the number of communication links (edges) is a function of the number of leaves. The BBT, on the other hand, allows p and b to be relatively independent of each other, thus permitting a trade-off between the time and the number of communication elements in the structure. Each processor of $B(p,b)$ is connected to at most two buses (from Lemma 1 part (iii)), whereas a node in $T(p)$ can have three neighbors. The processor fan-out of 2 for $B(p,b)$ is the best possible; if the processor fan-out is 1, then either all processors must be connected to the same bus, or the MBN would not be connected.

So far, we have restricted b to be at least 2. If $b < 2$, then all processors of $B(p,b)$ are connected to all bus(es). This configuration, though dense, still restricts the processor fan-out to at most 2. It is also easy to verify that it can simulate each step of $T(p)$ optimally.

5 VLSI Layouts for the BBT

In this section we discuss optimal layouts for $B(p,b)$ in two conducting layers. Each processor of $B(p,b)$ is connected to a constant number of buses (and simulates a constant number of nodes of $T(p)$); therefore, each processor can be assumed to occupy unit area in the word model. Besides the area, two important aspects of a VLSI layout are its aspect ratio (width/height) and the placement of the processors relative to the perimeter of the layout. For a layout to be implemented effectively, its aspect ratio should be close to 1. To avoid complex input/output interfaces for a chip, it is important to place processors close to the periphery of the layout. A layout in which all processors are at the periphery is called a *perimeter layout*. Usually, perimeter layouts result in large wastage of area. Therefore, we refer to a layout in which the processors are not necessarily at the perimeter as a *dense layout*. We refer to a layout whose width and height are W and H, respectively, as a $W \times H$ layout. For brevity, we present the following theorems without proofs.

Theorem 3 *For $p > b \geq 0$, $B(p,b)$ has a $O(2^p) \times O(b)$ perimeter layout.* ∎

Theorem 4 *For $p > b \geq 0$, $B(p,b)$ has a $O(2^{p/2}) \times O(2^{p/2})$ dense layout.* ∎

Remarks: The tree bus network (TBN) proposed in [4] also has similar layouts; however, the TBN cannot simulate an arbitrary step of the tree machine optimally. Using arguments similar to the proof of the area lower bound in [11] for a perimeter layout of $T(p)$, it can be shown that any perimeter layout of $B(p,b)$ requires $\Omega(b2^p)$ area. The dense layout clearly has optimal area.

6 Constructing New Multiprocessor Structures Using BBT's

We have shown in previous sections that the BBT is a versatile structure with a compact VLSI layout. The BBT can be used as a building block for constructing more complex multiprocessor parallel computing structures, that have compact VLSI layouts. To illustrate this point, we use the BBT to construct a new interconnection structure called the bus based mesh of trees (BBMOT). The BBMOT is based on the well known mesh of trees (MOT) topology.

The MOT consists of $2^{p_r + p_c}$ (primary) processors arranged in a $2^{p_r} \times 2^{p_c}$ grid. Each row and column has a binary tree, whose leaves are the elements of the row or column in question. In the MOT, besides the $2^{p_r + p_c}$ processors described above, the binary trees require an additional $2^{p_r}(2^{p_c} - 1) + 2^{p_c}(2^{p_r} - 1)$ (secondary) processors, whose function is usually to route information in the tree. The MOT is a very versatile parallel computing structure and algorithms for several applications have been proposed for it. A comprehensive discussion of the MOT appears in [7]. An MOT with $2^{p_r+p_c}$ primary processors is denoted by $MOT(2^{p_r}, 2^{p_c})$. An $MOT(2^{p_r}, 2^{p_c})$ has an $O(p_r p_c 2^{p_r+p_c})$ area layout.

The BBMOT is based on the MOT. It has $2^{p_r+p_c}$ processors arranged in a $2^{p_r} \times 2^{p_c}$ grid. Each row and column of the BBMOT has a $B(p, b)$, for some $p > b \geq 0$ and $p = p_r$ or p_c. A BBMOT with $2^{p_r+p_c}$ processors, with 2^{p_r} $B(p_c, b_c)$'s along the rows and 2^{p_c} $B(p_r, b_r)$'s along the columns is denoted by $BBMOT(2^{p_r}, 2^{p_c}, 2^{b_r}, 2^{b_c})$. It can be shown that $BBMOT(2^{p_r}, 2^{p_c}, 2^{b_r}, 2^{b_c})$ has a $O(b_c 2^{p_r}) \times O(b_r 2^{p_c})$ layout.

A $BBMOT(2^{p_r}, 2^{p_c}, 2^{b_r}, 2^{b_c})$ is functionally very similar to the $MOT(2^{p_r}, 2^{p_c})$. By the results presented in previous sections, every algorithm for $MOT(2^{p_r}, 2^{p_c})$ can be implemented on $BBMOT(2^{p_r}, 2^{p_c}, 2^{p_r-1}, 2^{p_c-1})$ with the same time complexity. In addition, the BBMOT can use the buses to broadcast information in constant time, which is not possible on the MOT. The following theorem, though presented in [4] in the context of the tree bus network, also holds for the BBMOT.

Theorem 5 (*Dighe et al.* [4]) *For* $\log N \leq T \leq \frac{N}{2}$, *a* $BBMOT(N, \frac{N}{T}, 1, \frac{N}{2T})$ *can sort* N *numbers in* $O(T)$ *time.* ∎

Remarks: This theorem leads to a sorting circuit with an AT^2 complexity of $O(N^2 T \log(\frac{N}{T}))$. Though a dense layout of the BBMOT is used, all inputs and outputs of the sorting algorithm are through processors placed on the perimeter of the layout. For $T = \log N$, the AT^2 complexity of the circuit is $O(N^2 \log^2 N)$. The corresponding MOT algorithm has an AT^2 complexity of $O(N^2 \log^3 N)$. The primary reason for this difference is due to the fact that the above algorithm only uses broadcasting along the columns. While the MOT requires a binary tree of depth $\log N$ for this, the BBMOT requires only one bus. It can also be shown that using Leighton's column sort [8] and the BBMOT, N numbers can be sorted in $\log N \leq T \leq \sqrt{N}$ word steps with an optimal AT^2 complexity of $O(N^2)$. Though the same (asymptotic) result can be achieved with the MOT, our method uses a smaller overall area and fewer processors.

With small modifications, the BBT can also simulate a ring with a constant factor of slowdown, while maintaining all properties discussed so far. Using such modified BBTs, we can construct an improved BBMOT structure that has advantages over the MOT. This improved BBMOT can also simulate a 2-dimensional torus with a small constant slowdown factor; no such method is known for the MOT.

7 Concluding Remarks

We have proposed in this paper a new multiprocessor parallel computing structure called the bus based tree (BBT). In addition to being able to simulate a tree machine optimally, the BBT can out-perform the tree ma-

chine in carrying out certain operations like constant time broadcast and ring simulation (by the leaf processors). We have also discussed optimal VLSI layouts for the BBT. BBTs can be used as building blocks for more complex structures. To illustrate this point, we have proposed a multiprocessor structure called the bus based mesh of trees (BBMOT), that has advantages over the mesh of trees. The BBT can also be used to construct new parallel computing structures that correspond to the mesh with diagonal trees, X-tree, shuffle-tree, and the pyramid.

References

[1] S.G. Akl, *The Design and Analysis of Parallel Algorithms*, Prentice-Hall, Englewood Cliffs, NJ, 1989.

[2] J.L. Bentley and H. T. Kung, "A Tree Machine for Searching Problems", *Proc. Int. Conf. Parallel Processing*, 1979, pp. 257–266.

[3] E. Dekel and S. Sahni, "Binary Trees and Parallel Scheduling", *IEEE Trans. Comput.*, **32**, 1983, pp. 307–315.

[4] O. M. Dighe *et. al.*, "VLSI Parallel Architecture Based on Multiple Buses", *Proc. ISCA Int. Conf. on Computer Applications in Design, Simulation and Analysis*, 1993, pp. 52–55.

[5] E. Horowize and A. Zorat, "Divide-And-Conquer for Parallel Processing" *IEEE Trans. Comput.*, 1983, **32**, 1983, pp. 582–585.

[6] J. Kilian *et al*, "The Organization of Permutation Architectures with Bussed Interconnections", *IEEE Trans. Comput.*, **39**, 1990, pp. 1346–1358.

[7] T. Leighton, *Introduction to Parallel Algorithms and Architectures: Arrays, Trees, Hypercubes*, Morgan Kaufmann Publishers, San Mateo, CA, 1992.

[8] T. Leighton, "Tight Bounds on the Complexity of Parallel Sorting," *IEEE Trans. Comput.*, **34**, 1985, pp. 344–354.

[9] T. N. Mudge *et al*, "Multiple Bus Architectures", *IEEE Computer*, 1987, pp. 42–48.

[10] Q.F. Stout, "Sorting, Merging, Selecting and Filtering on Tree and Pyramid Machines" *Proc. Int. Conf. Parallel Processing*, 1983, pp. 214–221.

[11] J. D. Ullman, *Computational Aspects of VLSI*, Computer Science Press, Rockville, MD, 1984.

[12] R. Vaidyanathan, "Design of Multiple Bus Interconnection Networks for Fan-in Computations," *Proc. 29th Annual Allerton Conf. on Comm, Control & Comput.*, 1991, pp. 1093–1102.

[13] S.Q. Zheng, "Compressed Tree Machines", to appear in *IEEE Trans. Comput.*

SESSION 7A

CACHE MEMORY (II)

Efficient Stack Simulation for Shared Memory Set-Associative Multiprocessor Caches

C. Eric Wu, Yarsun Hsu, and Yew-Huey Liu

IBM T. J. Watson Research Center
P.O. Box 218
Yorktown Heights, NY 10598

Abstract

We propose efficient stack simulation algorithms for shared memory multiprocessor (MP) caches. A stack simulation algorithm for write-updated MP caches is first presented. It produces the number of write-updates as well as misses for all cache configurations in a single run. We then devise a new stack simulation algorithm for write-invalidate MP caches. Our algorithm takes into account cross-invalidation among processors, and generates the number of invalidations as well as misses for all cache configurations in a single run. A cache simulator based on our algorithms for MP caches is developed and the results on sample traces are reported. Our results show that efficient stack simulation is a powerful technique for multiprocessor cache analysis.

1. Introduction

Cache has been a common approach to bridge the gap between fast processors and slow main memories since it was first introduced by IBM on the System/360 Model 85 [7]. In a shared-memory MP system, high-speed caches connected to each processor maintain local copies of memory locations and supply instructions and operands at the rate required by each processor. If such a system is to correctly execute computations, all references to a given location, no matter from which processor they originate, should reference the same value; i.e., the contents of the cache memories must be consistent.

There are two policies for maintaining cache consistency: *write-update* and *write-invalidate*. In a write-update MP cache, read requests are carried out locally if a copy of the block exists. When a processor updates a block, all other copies in private caches are also updated, thus prevents invalidating other copies and subsequently forcing misses on next references in other caches. The Firefly multiprocessor at DEC and the Dragon multiprocessor at Xerox PARC [1], for example, use the write-update policy.

The write-invalidate policy maintains cache consistency of multiple copies differently. Instead of updating all other copies, it invalidates them so that subsequent updates by the same processor can be performed locally in the cache, since other copies of the block are no longer valid. The Berkeley [6] and the Illinois [9] protocols and those used in many cache-coherent network architectures, such as the Wilson's hierarchical bus [15], the Wisconsin Multicube [3], and the Data Diffusion Machine [4], adopt the write-invalidate policy.

Trace-driven cache simulation has been used as a standard technique to evaluate cache performance in computer systems. Since such trace-driven simulation are usually repeated for many alternative caches with many reference traces, the whole process can be tedious and the CPU time required for simulation can be enormous.

Mattson and his colleagues [8] proposed an efficient trace-driven simulation method for storage hierarchies. Their approach, called *stack algorithm*, allows miss ratios for all cache sizes to be computed in a single pass over the reference trace. Traiger and Slutz [13] extended the stack algorithm to compute miss ratios for variable line sizes and variable associativity in a single pass.

Stack algorithm works because the replacement policy guarantees the inclusion property, i.e., the contents of any size cache is a superset of those in any smaller cache. Thus the cache at any time can be represented as a stack, with the upper d elements of the stack representing the lines present in a cache of size d. For a reference at stack distance d from top of the stack (where stack distance is one), all caches of size d or more would have a hit.

Since multiple caches with different number of sets are considered during the stack simulation, stack distances of a reference may be different in different caches. For caches with up to 2^U sets and set-associativity up to S, the common tag stack is searched for each reference until it is found or the end of the stack is reached. During the search each element in the stack is compared with the reference tag, and a vector of Right-Matched counters $RM(i)$, $0 \le i \le U$, is used to record the number of stack elements with exactly i right-matched bits. If a stack element has more than U right-matched bits when comparing to the reference, $RM(U)$ is incremented.

Elements in the stack with exactly n right-matched bits are in the same set in a 2^i-set cache, where $i \le n$. When a reference X is found in the stack, the right-matched counters $RM(i)$, $0 \leq i \leq U$, have been incremented for all elements ahead of the line. Therefore, the stack distance $d(n)$ of X in a 2^n-set cache is one plus the number of elements ahead of X with at least n right-matched bits, i.e.,

$$d(n) = 1 + \sum_{i=n}^{U} RM(i) \,,$$

where $0 \leq n \leq U$. For an s-way 2^n-set cache, the current reference X is a hit if $s \geq d(n)$, otherwise a miss.

To generate the number of misses for various cache configurations, we can use a distance counter, $DC(n, s)$, where $0 \leq n \leq U$ and $1 \leq s \leq S$, to record the number of references that have stack distance $d(n) = s$. Assuming the total number of references is N, the number of misses $miss(n, s)$ for an s-way 2^n-set cache is given by

$$miss(n, s) = N - \sum_{i=1}^{s} DC(n, i) \,.$$

Although a single common tag stack is used in the original stack algorithm, we may use multiple stacks to improve the speed of stack simulation. In this paper we are interested in caches with 2^L to 2^U sets, where $0 < L \leq U$. Thus, we can use 2^L stacks instead of just one common stack.

There have been a number of papers in the literature dealing with efficient trace-driven cache simulation based on the stack algorithm. Gecsei [2] showed how the stack algorithm could be generalized to multiple levels with different block sizes for LRU and related replacement policies. Hill [5] generalized the algorithm to an arbitrary number of sets instead of being a power of two. Thompson and Smith [12] proposed an efficient stack algorithm to obtain the number of write backs as well as the number of misses for fully-associative caches by attaching a dirty level to each element in the stack. A dirty level is a kind of stack distance associated with an element in a stack to indicate how far the dirty line has been pushed into the stack. Since a line may have different stack distances in caches with different numbers of sets, it may have different dirty levels as well in these caches. Thus, Wang and Baer [14] extended the approach by attaching a vector of dirty level to each element in a stack to calculate the number of write backs.

For virtual address caches with real tages, Wu et. al. [16] proposed stack simulation techniques to handle synonyms and calculate pseudonym frequencies as well as miss and write back ratios in a single run. Thompson [11] proposed several stack simulation techniques for fully-associative shared-memory MP caches. Wang and Baer [14] discussed the feasibility of extending their stack algorithm for write-update MP caches. While their contribution is by no means small, extensions of the stack algorithm are required for simulating set-associative shared-memory MP caches. Our motivation, therefore, is to find efficient stack algorithms for both write-update and write-invalidate MP caches. For a shared-memory MP system with 2^P-byte page size and 2^B-byte line size, configurations under our investigation are caches with 2^L to 2^U sets (for simplicity, a power of two), where $0 < L \leq U$ and set-associativity from 1 (direct-mapped) to S.

This paper is organized as follows. Section 2 describes our stack simulation technique for write-update caches, and Section 3 describes our approach for write-invalidate caches. Experimental results on sample MP traces are reported for both write-update and write-invalidate caches, and concluding remarks are given in Section 4.

2. Write-Update MP Caches

Performance metrics for write-update MP caches include miss ratio, defined as the number of local reference misses over the total number of local references, and update frequency, defined as the number of updates in a cache over the total number of local references. The operation of the protocol can be specified by the actions taken on processor reads and writes. Read hits can always be performed locally in the cache. On a read miss, if no dirty copy exists in other caches, then memory has a consistent copy and supplies a copy to the cache. If a dirty copy exists, then the corresponding cache inhibits memory and sends a copy to the requesting cache. On a write hit, if the copy is shared, all other copies are updated, otherwise the write can be carried out locally. On a write miss, if a dirty copy exists, the corresponding cache sends a copy to the requesting cache, otherwise the requesting cache gets a copy from memory.

On a miss (read or write) when a dirty copy exists or on a write hit, whether the memory copy is updated or not depends on how the protocol is implemented. In either case the miss ratio and update frequency won't change. Therefore, we are interested in miss ratio and write-update frequency, while the write-back ratio depends on specific implementation.

To generate the number of updates for various cache configurations, an Update Counter $UC(n, s)$ is used to record the number of remote writes from other CPUs which update shared lines in the local cache with stack distance $d(n) = s$. Thus, the number of updates for an s-way 2^n-set cache is given by

$$update(n, s) = \sum_{i=1}^{s} UC(n, i) \,.$$

Algorithm 1 shows a stack simulation algorithm for write-update MP caches.

The algorithm for write-update MP caches works as follows. For each local reference or remote write reference X, find the stack(a) to which it is mapped. A search is then performed through the stack, the number of right-matched bits b is recorded, and the histogram is accumulated by incrementing $RM(b)$ for later producing stack distances $d(n)$, $L \leq n \leq U$. If the reference is found in the stack, stack distances of the line for caches with different numbers of sets are accumulated, starting from caches with 2^U sets, as indicated at lines 8 and 9. The stack distances are then used to increment distance counters $DC(n, i)$ if it is a local reference, or increment update counters $UC(n, i)$ if it is a remote write.

When a remote write updates a local line, whether the updated line is moved to the top of the stack depends on how the protocol is implemented. At line 12 of Algorithm 1, the line is moved to the top of the stack only if it is a local reference. If it is a remote write reference, the line is not moved. However, the algorithm could be easily modified to reflect protocol operations. When a local reference is not found in the stack, the line is brought in, as indicated at line 13.

```
    FOR each reference X DO {
1.    IF local reference THEN N++
2.    IF local reference or remote write THEN {
3.      a = X mod (2 ** L)
4.      RM(i) = 0 for all i, L <= i <= U
        search stack(a) until found or EOS {
          /* for each stack element Xi */
5.        b = # of right-matched bits (X vs Xi)
6.        RM(min(b, U))++
        }
        IF found THEN {
7.        c = 1
8.        FOR (i = U; i >= L; i--) DO {
9.          c = c + RM(i)
10.         IF local THEN DC(i, c)++
11.         ELSE /*remote write */ UC(i, c)++
        }
12.       IF local THEN move it to top of stack(a)
        } ELSE IF local reference THEN {
13.       bring new tag X to top of stack(a)
    } } }
14. Gather statistics
```

Algorithm 1. Simulation for write-update MP caches

Note that no action is needed for a remote read reference. This is because a remote read won't change the stacks for the local cache, thus will change neither miss ratios nor update frequencies.

An IBM System 370 MP trace is used in our experimental study. It was derived from a dual-CPU instruction trace of IMS (Information Management System) running on MVS (Multiple Virtual Storage) operating system. It is a mixed operating system and application trace of an interactive commercial workload. The trace file is around 25,315 Kbytes in size. We use combined I/D caches, 4-KByte page size, and 64-byte line size. Configurations under our investigation are caches with 32 to 2048 sets, and set-associativity from 1 to 8.

A cache simulator based on Algorithm 1 is developed on an IBM RS/6000 model 530. Figure 1 and 2 show the results in miss ratios and update frequencies for Cache 0 and Cache 1, respectively. As expected, miss ratio decreases as the cache size increases when set-associativity is fixed. For update frequency, it increases as cache size increases when set-associativity is fixed.

For a fixed cache associativity the update frequency increases as the cache size increases, indicating a potential conflict in optimizing both the cache miss ratio and the update frequency.

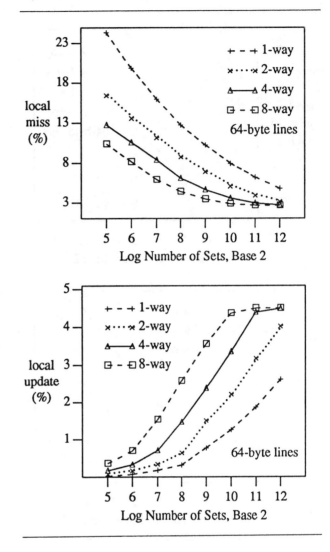

Figure 1. Write-update MP cache 0

3. Write-Invalidate MP Caches

Instead of updating all other copies, the write-invalidate policy invalidates all other copies when a processor writes a cache line. Performance metrics for write-invalidate MP caches include miss ratio and invalidate frequency, which is defined as the number of invalidations in a cache over the total number of local references. The operation of the protocol can be specified by the actions taken on processor reads and writes.

Read hits can always be performed locally in the cache. On a read miss, if no dirty copy exists, then memory has a consistent copy and supplies a copy to the cache. If a dirty copy exists, then the corresponding cache inhibits memory and sends a copy to the requesting cache. On a write hit, if the copy is shared, all other copies are invalidated, otherwise the write can be carried

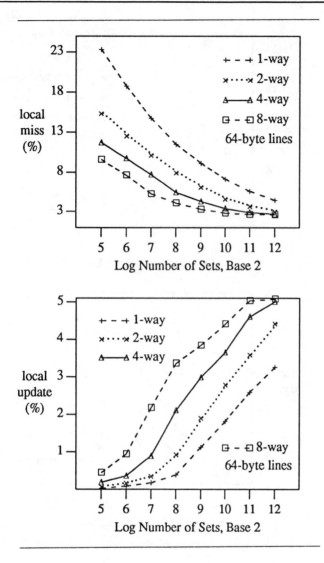

Figure 2. Write-update MP cache 1

out locally. On a write miss, if a dirty copy exists, the corresponding cache sends a copy to the requesting cache, otherwise the requesting cache gets a copy from memory.

On a miss (read or write) when a dirty copy exists or on a write hit, whether the memory copy is updated or not depends on how the protocol is implemented. In either case the miss ratio and invalidation frequency won't change. Therefore, we are interested in the miss ratio and write-invalidate frequency, while the write-back ratio depends on specific implementation.

3.1. Stack Distances And Marker Handling

Contrary to write-update MP caches, cache lines could be deleted in write-invalidate MP caches. A deleted line is a hole in the stack, represented by a special entry called

a *marker*. For fully-associative caches, a marker will remain at the same position until another line below the marker is accessed. In this case the marker jumps to the position of the newly accessed line while that line is moved to the top of the stack.

For a set-associative MP cache a deletion also leaves a marker in the stack. Since the contents of a stack include references to more than one set, a marker may need to move for some caches, but need to stay in others. Thus, a marker can split into multiple markers. In other words, one need more than just a marker for a deleted line in the stack to do stack simulation for set-associative write-invalidate MP caches.

Our solution of manipulating markers is to have a valid range associated with each marker. The valid range $V = (rmmin, rmmax)$ of a marker is a set of consecutive integers, from $rmmin + 1$ to $rmmax$. The upper inclusive limit $rmmax$ and the lower exclusive limit $rmmin$ in a marker indicates the logarithmic maximum and minimum number of sets for the hole to exist. When a line is deleted from the stack due to cross-invalidation, a marker is created with its valid range $V = \{L, ..., U\}$, i.e., $rmmin = L - 1$ and $rmmax = U$, to indicate that all caches having the shared line would have a hole.

When a marker is encountered during a stack search for a newly accessed line, the number of right-matched bits b between the current reference and the marker tag is recorded. If $b \leq rmmin$, the marker is not observed, because the hole does not exist in 2^n-set caches, where $L \leq n \leq rmmin$. If $b > rmmin$, on the other hand, the hole may be observed in a 2^n-set cache, where $rmmin < n \leq \min(b, rmmax)$. In case of $b \leq rmmax$, the hole is not observed in caches with more than 2^b sets, because the reference and the hole are in different sets. Thus, the set of consecutive integers $G = \{rmmin + 1, ..., \min(b, rmmax)\}$, also represented by the pair $(rmmin, \min(b, rmmax))$, is called the observed range of the marker.

With the presence of markers, it is not sufficient to have only one vector of $RM(i)$ for keeping track of stack distances. If we use $RMM(j, k)$ to record the number of markers with $rmmin = j$ and $\min(b, rmmax) = k$, then the number of holes $h(n)$ ahead of the currently referenced line in a 2^n-set cache is given by

$$h(n) = \sum_{j=L-1}^{n-1} \left[\sum_{k=n}^{U} RMM(j, k) \right],$$

and the number of elements $e(\overline{n})$ (excluding markers) ahead of the line is

$$e(n) = \sum_{i=n}^{U} RM(i).$$

Thus, its stack distance $d(n)$ in a 2^n-set cache is given by

$$d(n) = 1 + e(n) + h(n).$$

Figure 3 shows the stack distances of a reference in caches with different numbers of sets ($2 \leq n \leq 5$). A pair of $(rmmin, rmmax)$ is associated with each marker, and both markers have $rmmax = U = 5$. A marker in the stack is indicated by an M in Figure 3.

The referenced line is found in the stack with two elements and two markers ahead of it in Figure 3. In caches

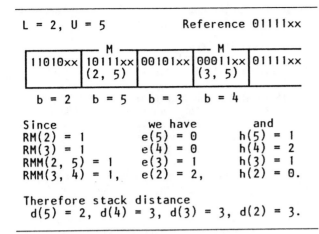

Figure 3. Stack distances

with 4 sets ($n = 2$), the reference would have stack distance $d(2) = 3$ because both markers have $rmmin \geq 2$, indicating that there would be no holes in a 4-set cache. In caches with 8 sets, the stack would split into 2 substacks, and one of the two stack elements is still ahead of the line while the other is in the other sub-stack. In addition, the marker with a tag 10111xx is observed in caches with 8, 16, or 32 sets, therefore $d(3) = 3$. In caches with 16 sets, the stack would split into 4 sub-stacks, with no stack element and two holes (from both markers) ahead of the line. Thus, we have $d(4) = 3$. In caches with 32 sets ($n = 5$), the stack would split into 8 sub-stacks, and we have only one hole ahead of the line in the same set. Thus, we have $d(5) = 2$.

When a line below a marker is accessed, it is moved to the top of the stack and may fill up the hole and leave another hole at its own location. If the marker is the first marker encountered in the stack, the hole in a 2^n-set cache, where $rmmin < n \leq \min(rmmax, b)$, jumps to the position of the newly accessed line. In case of $rmmax > b$, the hole remains at its location in 2^n-set caches, where $b < n \leq rmmax$, because the newly accessed line and the hole are in different sets. Thus, a marker may split into multiple disjoint ones.

Given a cache configuration, only the first hole observed by the reference would be filled. Other holes remain unfilled and observable. When a marker is observed and filled in all caches which have the hole, i.e., $rmmax \leq b$, the marker is moved to the location of the newly accessed line. If the reference is not found in the stack, the marker is then deleted. Figure 4 shows examples of marker splitting and moving.

In Figure 4, the marker splits into two disjoint markers. It is observed and filled in 16-set caches, but not in 32-set caches. Note that one split marker moves to the location of the newly accessed line.

Since a reference may fill only one hole, a Boolean variable for each number of sets is required during a stack search to indicate if it has filled a hole. Thus, a

Boolean vector $FILL(i)$, $L \leq i \leq U$, initialized to all false, is used for marker manipulation. For a marker observed by the current reference, its observed range $G = (rmmin, \min(b, rmmax))$ is checked against the fill vector to determine if the hole could be filled. Figure 5 shows an example of such marker handling.

Before the marker is observed in Figure 5 the FILL vector contains two filled regions, $(L - 1, \ldots, X)$ and $(Y, \ldots, Z)$, indicating that the current reference has filled a hole in a 2^n-set cache, where $L \leq n \leq X$ or $Y + 1 \leq n \leq Z$. When the marker is observed in the range $(rmmin, b)$ with $L - 1 < rmmin < X$ and $Y < b < Z$, the split marker 1 having valid range (X, Y) is created due to the hole fillings. It is then moved to the location of the newly accessed line in the stack, or discarded eventually if the line is not found. The other two markers (2 and 3), having valid range $(rmmin, X)$ and $(Y, rmmax)$, stay at the original marker location, because the hole is not filled in 2^n-set caches, where $rmmin + 1 \leq n \leq X$ or $Y + 1 \leq n \leq rmmax$.

If we use E to represent the set of integers in the range $(L - 1, U)$ of the fill vector, i.e., $E = \{L, \ldots, U\}$, and F to represent the union of filled regions in the fill vector, then $E - F$ defines the caches in which hole fillings have not occurred thus far. For example, we have

$$F = \{L, \ldots, X, Y + 1, \ldots, Z\}$$

in Figure 5 before the marker is observed. The observed range G of a marker for a reference is given by

$$G = \{rmmin + 1, \ldots, \min(b, rmmax)\} \, .$$

The intersect of G and $E - F$, $G \bigcap (E - F)$, defines cache sizes (logarithmic number of sets) in which a hole may be filled by the reference. The union of $V - G$ and $V \bigcap F$, $(V - G) \bigcup (V \bigcap F)$, on the other hand, defines cache sizes in which a hole has been filled before the marker is encountered, and therefore the hole cannot be filled by the reference. If we define a *maximum disjoint subset* to be a maximum subset of consecutive integers in $(V - G) \bigcup (V \bigcap F)$ or $G \bigcap (E - F)$, then each maximum disjoint subset in $(V - G) \bigcup (V \bigcap F)$ needs a split marker at current marker location, and each maximum disjoint

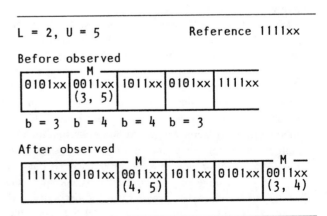

Figure 4. Marker splitting and moving

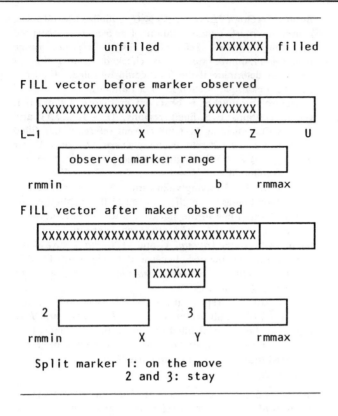

Figure 5. Observed range vs fill vector

subset in $G\bigcap(E - F)$ indicates a split marker on the move. In Figure 5, for example, we have

$$G\bigcap(E - F) = \{\{X + 1, ..., Y\}\},$$

and

$$(V \overset{.}{-} G) \bigcup (V\bigcap F)$$
$$= \{\{rmmin + 1, ..., X\}, \{Y + 1, ..., rmmax\}\}.$$

If $(V - G) \bigcup (V\bigcap F) = \phi$, i.e., if the hole can be filled in all caches having the hole, the marker is then deleted from the stack. On the other hand, if $G\bigcap(E - F) = \phi$, there would be no need for split markers.

Routine 1 is a stack searching routine for a given reference X. For a local reference the split marker list is initialized to empty, and the fill vector $FILL(i)$, $L \leq i \leq U$, is set to false. In cases of either a remote write or a local reference, $RM(\min(b, U))$ is incremented for each stack element ahead of the line. For each marker ahead of the line $RMM(rmmin, \min(b, rmmax))$ is incremented if the observed range G is not empty. At line 9, we split the original markers for each maximum disjoint subset in $(V - G) \bigcup (V\bigcap F)$ in case of a local reference. If there is only one consecutive subset, the marker is not split, but just updated with a new valid range, if necessary. If $(V - G) \bigcup (V\bigcap F) = \phi$, the marker is deleted from the stack. At line 10, split markers are added to the split marker list for each maximum disjoint subset in $G\bigcap(E - F)$. The split marker list is eventually added to

the stack at the position of the newly accessed line, or discarded if the line is not found in the stack.

```
   IF local reference THEN {
1.    MarkerList = empty
2.    FILL(i) = FALSE for all i, L <= i <= U
   }
3. IF local reference or remote write THEN {
4.   search stack for X until found or EOS {
         /* stack element Xi */
5.      b = number of right-matched bits (X vs Xi)
        IF not a marker THEN
6.         RM(min(b, U))++
        ELSE IF min(b, rmmax) > rmmin THEN {
8.         RMM(rmmin, min(b, rmmax))++
           IF local reference THEN {
9.            update/split marker for each maximum
              disjoint subset in (V - G) U (V ∩ F),
              and the marker is deleted if
              (V - G) U (V ∩ F) = φ
10.           add a split marker to MarkerList for
              each maximum disjoint subset in
              G ∩ (E - F) and set FILL(i) = TRUE
              for all i, i ∈ G ∩ (E - F)
} } } }
```

Routine 1. Search routine for each reference

For a remote write, we search the stack and calculate $RM(i)$ and $RMM(j, k)$ as indicated in the search routine, without dealing with the split marker list. This is because remote writes may only invalidate local lines, but cannot fill holes.

3.2. Miss Ratios and Invalidate Frequencies

Algorithm 2 outlines the approach for generating number of invalidations as well as number of misses for various write-invalidate MP cache configurations in a single run. The algorithm works as follows. For each local or remote write reference X, find the stack(a) to which it is mapped. A search is then performed through the stack and the number of right-matched bits b is checked. For a stack element rather than a marker, the element histogram is accumulated by incrementing $RM(\min(b, U))$. For an observed marker with valid range $V = (rmmin, rmmax)$ and observed range $G = (rmmin, \min(b, rmmax))$, the hole histogram is accumulated by incrementing $RMM(rmmin, \min(b, rmmax))$. In case of a local reference a split marker list is maintained during the search, as described in the stack searching routine, until the line is found or the end of stack is reached.

For a local reference found in the stack, the split marker list is added into the stack at where the line was found. The line is then moved to the top of the stack. If the line is not found in the stack, both element and hole histograms are not used, and the split marker list is discarded in the case of a local reference. If it is a local reference, the line is then brought in at the top of stack.

```
    FOR each reference X DO {
1.    IF local THEN N++
      IF local reference or remote write THEN {
2.      RM(i) = 0 for all i, L <= i <= U
3.      RMM(j, k) = 0, L - 1 <= j < U, L <= k <= U
4.      a = X mod (2 ** L)
5.      search stack(a) using the search routine
          to get RM(i), RMM(j, k), and MarkerList
          for a local reference, or get RM(i) and
          RMM(j, k) for a remote write
        IF found THEN {
6.        c = 1
7.        FOR (i = U; i >= L; i--) DO {
8.          c = c + RM(i)
9.          compute h(i)
10.         d(i) = c + h(i)
11          IF local reference THEN DC(i, d(i))++
12.         ELSE /* remote write */ IC(i, d(i))++
          }
          IF local reference THEN {
13.         add MarkerList right ahead of the line
14.         move the line to top of stack(a)
          } ELSE /* remote write */
15.         make it a marker with V = (L-1, U)
        } ELSE IF local reference THEN {
16.       bring new tag X to top of stack(a)
    } } }
17. Gather statistics
```

Algorithm 2. Simulation for write-invalidate MP caches

For a remote write the stack is also searched as in cases of local references. However, the stack distances are used to increment Invalidation Counters rather than distance counters. An Invalidation Counter $IC(i, s)$ is used to record the number of invalidations which have stack distance $d(n) = s$ for 2^n-set caches. Thus, the number of invalidations for an s-way 2^n-set cache is

$$invalidate(n, s) = \sum_{i=1}^{s} IC(n, i) .$$

After the MP address trace has been read through, miss ratios and invalidation frequencies are then calculated for all cache configurations.

Figure 6 and Figure 7 show miss ratios and invalidate frequencies of the first (cache 0) and second (cache 1) cache in various cache configurations using the write-invalidate protocol. Miss ratio decreases as either the number of sets or set-associativity increases.

Invalidate frequency results from remote references rather than local ones. It increases as the cache size increases, indicating a potential conflict in optimizing both cache miss ratio and invalidation frequency.

We expect caches using the write-invalidate policy have higher miss ratios than those using the write-update policy. It is interesting to note that the miss ratio of a write-invalidate cache is slightly higher than that of its counterpart using the write-update policy. For example, a 64-KByte 4-way write-invalidate cache (256 sets) has a miss ratio of 6.14%, while a write-update cache of the same configuration has 6.05%.

The major difference in performance for the sample trace comes from the comparison between update ratios

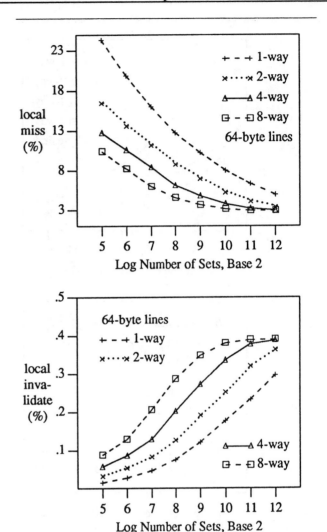

Figure 6. Write-invalidate MP cache 0

and invalidate ratios. Our simulation shows that the invalidate ratio of a write-invalidate cache is roughly one order of magnitude smaller than the update ratio of a write-update cache. Taking the cost of updates due to remote writes into consideration, we conclude that the write-invalidate policy is better for our sample trace.

Since stack simulation techniques are used to run trace-driven simulation efficiently, the speed of the simulator is important. Our experiments show that our stack algorithms may be up to 2.5 times slower than a naive one-cache-per-run approach. On the other hand, our techniques generate performance metrics for 64 cache configurations (number of sets from 32 to 4096, set-associativity from 1 to 8) in a single run. Future research directions may include efficient implementation techniques, such as using high-balanced trees instead of linked lists for stack construction and simulation.

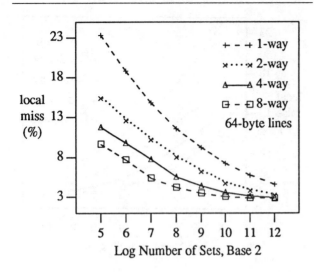

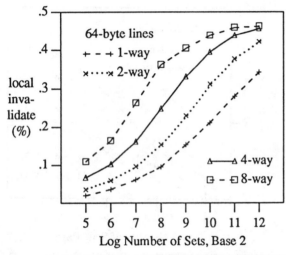

Figure 7. Write-invalidate MP cache 1

4. Summary

Stack simulation is a powerful tool for cache analysis. In this paper, we first extended stack simulation for write-update MP caches. A stack simulation algorithm for write-invalidate MP caches was then devised to generate the number of invalidations as well as misses. Our approach carefully handles markers so that stack distances and thus performance metrics are calculated in a single pass. Experimental results show that our stack simulation approaches are practical and efficient for simulating shared-memory MP caches.

Acknowledgements

We would like to thank W.-H. Wang for discussions on stack simulation and J. Knight for generating the address trace from an instruction trace provided by the System Performance Group of IBM DSD Poughkeepsie.

References

[1] Archibald, J. and J.-L. Baer, Cache Coherence Protocols: Evaluation Using a Multiprocessor Simulation Model, *ACM Trans. on Computer Systems*, 4, 4, Nov. 1986, pp. 273 - 298.

[2] Gecsei, J., Determining Hit Ratios in Multilevel Hierarchies, *IBM Journal of Research and Development*, 18, 4, July 1974, pp. 316 - 327.

[3] Goodman, J. and P. Woest, The Wisconsin Multicube: A New Large-Scale Cache-Coherent Multiprocessor, *Proc. 15th ISCA*, 1988, pp. 422 - 431.

[4] Haridi, S. and E. Hagersten, The Cache Coherence Protocol of the Data Diffusion Machine, *Proc. PARLE'89*, Vol. 1, 1989, pp. 1 - 18.

[5] Hill, M., Aspects of cache memory and instruction buffer performance, PhD dissertation, Tech. Report 87/381, Department of EECS, CS Division, U. C. Berkeley, California, Nov. 1987.

[6] Katz, R., S. Eggers, D. Wood, C. Perkins, and R. Sheldon, Implementing a Cache Consistency Protocol, *Proc. of 12th ISCA*, 1985, pp. 276 - 283.

[7] Liptay, J. S., Model 85 Cache, *IBM Systems Journal*, 7, 1, 1968, pp. 15 - 21.

[8] Mattson, R., J. Gecsei, D. Slutz, and I. Traiger, Evaluation techniques for storage hierarchies, *IBM Systems Journal*, 9, 2, 1970, pp. 78 - 117.

[9] Papamarcos, M. and J. Patel, A Low Overhead Coherence Solution for Multiprocessors with Private Cache Memories, *Proc. of 11th ISCA*, 1984, pp. 348 - 354.

[10] Sweazey, P. and A. Smith, A Class of Compatible Cache Consistency Protocols and Their Support by the IEEE Futurebus, *Proc. of 13th ISCA*, 1986, pp. 414 - 423.

[11] Thompson, J., Efficient Analysis of Caching Systems, *Ph.D. Thesis*, Technical Report UCB/CSD 87/374, UC Berkeley, Oct. 1987.

[12] Thompson, J. and A. Smith, Efficient (Stack) algorithms for analysis of write-back and sector memories, *ACM Trans. on Computer Systems*, 7, 1, Feb. 1989, pp. 78 - 117.

[13] Traiger, I. and D. Slutz, One pass technique for the evaluation of memory hierarchies, *IBM Research Report RJ 892 (#15563)*, July 1971.

[14] Wang, W.-H. and J.-L. Baer, Efficient Trace-Driven Simulation Methods for Cache Performance Analysis, *ACM Trans. on Computer Systems*, 9, 3, Aug. 1991, pp. 222 - 241.

[15] Wilson Jr., A. W., Hierarchical Cache/Bus Architecture for Shared Memory Multiprocessors, *Proc. 14th ISCA*, pp. 244 - 252, 1987.

[16] Wu, C. E., Y. Hsu, and Y.-H. Liu, Stack Simulation for Set-Associative V/R-Type Caches, Proc. of the 16th Computer Software and Applications Conference, pp. 332 - 339, Sep. 1992.

PARALLEL CACHE SIMULATION ON MULTIPROCESSOR WORKSTATIONS

Luis Barriga and Rassul Ayani

Royal Institute of Technology, Dept. of Teleinformatics/Computer Systems Division

KTH-Electrum/204, S-164 40 Kista, Sweden

{luis,rassul}@it.kth.se

Abstract -- *Trace-driven simulation is the most widely used method to evaluate caches. This demands large amounts of storage and computer time. Several techniques have been proposed to reduce the simulation time of sequential trace-driven simulation. However, little has been done to exploit parallelism. In this paper, we present some efficient parallel simulation techniques that exploit set-partitioning as the main source of parallelism. We show that a straightforward implementation does not give much speedup as one might expect. We develop more efficient parallel simulation techniques by introducing more knowledge into the cache simulator. The techniques presented here can be efficiently used on multiprocessor workstations.*

1. INTRODUCTION

To produce statistically reliable results, trace-driven cache simulation requires huge traces that demand lots of storage and computer time. Therefore it is important to develop techniques and algorithms that reduce these requirements. All techniques that reduce simulation time can be divided into techniques that: (i) reduce sequential simulation time [4]; (ii) evaluate several cache configurations in a single pass of the trace [5][6]; (iii) exploit parallelism [1][2].

One source of parallelism is set partitioning. For a given cache configuration, an address reference can be mapped to one single cache set. Thus, each set can be simulated independently by a process, and the degree of parallelism is equal to the number of sets. Another source of parallelism proposed by Heildelberger et.al.[1] is called time-partitioning, where the trace is partitioned into a finite number of sequential subtraces of equal length. Each subtrace is used to simulate the same cache configuration in parallel with the other subtraces. Partial metrics are obtained for each sub-simulation. To compute the final metrics, a number of re-simulations are necessary, where the partial metrics are merged and corrected. The authors show that time partitioning is statistically more efficient than set-partitioning under certain circumstances, and that the speedup could be as much as the number of available processors. However, this scheme requires preprocessing the main trace to divide it into subtraces, and it is dependent on the number of available processors. During simulation, it is necessary to save references that create misses. The resimulation phase demands extra overhead that increases with the number of processors. Another parallel algorithm based on stack distances was proposed by Nicol et.al [2]. The number of misses can be easily computed from the stack distances, which in turn can be computed via some well-known parallel algorithms such as merging, sorting or 2D ranking, suitable for SIMD architectures.

2. A PARALLEL SIMULATION SCHEME

During trace-driven cache simulation the following operations are performed on an address: (1) fetch address from the trace (2) break up address into tag, block number, block offset; (3) compute set number; (4) search block in corresponding set; (5) update set status and metrics. The main objective of cache simulation is to measure the miss rate.

A whole trace cannot be loaded into memory since that would require hundreds of megabytes of core memory. Therefore, during simulation, the trace is read into memory by parts. Although a trace can be processed in parallel with the fetching process, references to the same set must preserve the relative order as in the original trace, otherwise the following unfortunate scenario can occur: a process A fetches the n-th reference, and a process B fetches the m-th reference (m>n), and both references map to the same set. If ordering is not preserved B might deliver its m-th reference before A leading to wrong metrics. Obviously, synchronization is needed to make insertions to the same set.

Let one process, the parser, handle the parsing of the trace, and let each set be simulated by a dedicated process. To make it more efficient, the parser fetches a window of N references at a time. Each address is broken up, and a record is created containing the set number. The parser inserts the record into the corresponding set list. After all N records have been inserted, the parser hands over the accumulated set lists to the corresponding set processes. For each reference from the received list, the set process updates its set data structures and metrics. Synchronization is mandatory among the parser and set-processes every time an insertion is done, and whenever a handing over is performed. We assume that locks are used to synchronize access to common lists, having one lock per set. For a window size N=1 we would need 2 pairs of lock/unlock operations per reference which is a high overhead to be paid. If N=1000 the impact of locking is reduced to 2 pairs of lock/unlock operations per thousand references. For N equal to the trace length, the execution of the simulation would be sequential, apart from the huge demands in memory.

A practical estimation of speedup

Assuming that the time to generate a reference is negligible, we show that the maximum speedup (the number of sets) is not achievable due to program locality and cache configuration. Let:

L - the length of the trace.

T - the time to process the trace by a sequential simulator in absence of input latency.

n - the number of sets.

l_i - the total number of references to set i in the trace.

l_{max} - the maximum value among all l_i ($i = 1, n$).

To avoid causality errors the evaluation of one set must be done sequentially. The total simulation time will be equal to the time to process the set with the highest number of references $t_{max}=l_{max}*(T/L)$. The potential speedup that can be obtained is:

$$potential\ speedup=T/t_{max}=\ L/l_{max} \quad (1)$$

For an uniformly distributed trace, ie a trace with equal number of references to all sets, $l_i\ =\ l_{max}\ =\ L/n$, the speedup is n. For a trace that tends to use a group of sets more than others the speedup would diminish. The value of l_{max} is minimal for a direct-mapped cache, and equal to L for a fully associative cache.

Considering the input latency, the simulation process with respect to each reference consists of 2 strictly sequential phases: read reference and process reference. Let:

t_g - the time to read/generate a reference

t_p - the time to process a reference

$k_{pg}\ =\ t_p/t_g$

For N references the simulation time is $N(t_g+t_p)$. If t_p is completely hidden by t_g, then the total simulation time would be equal to $N*t_g$. An upper bound for the speedup would then be:

$$U_{speedup}=(t_g+t_p)/t_g=1+t_p/t_g=1+k_{pg} \quad (2)$$

As can be seen from equation (2), the speedup is limited by k_{pg}. If the trace resides on disk, k_{pg} is dependent on I/O bandwidth. The parameter t_g is also affected by the techniques used to store and retrieve the trace. Other factors that affect the speedup are the granularity of reference processing, the cache configuration and the trace characteristics. If the amount of computation to be done for a reference increases, then we get higher speedup. To such cases relate: (i) evaluating different cache configurations for a single pass of the trace [5]; (ii) simulating caches with large associativities; (iii) analyzing transient states at the beginning of simulation or after a cache flush; and (iv) investigating cache performance under multiprogramming.

3. OTHER TECHNIQUES

As a target for parallelization the DineroIII [5] sequential uniprocessor cache simulator was selected. Our experiments were mainly conducted on the shared memory multiprocessor Sequent Symmetry S81. For some experiments we used a Sun SPARCstation 10/30. As the runtime system we chose the FastThread (FT) lightweight processes package developed by Anderson [7].

We experimented with two cache configurations whose parameters are given in table 1.

TABLE 1. Cache configurations

conf	degree of associativity	block size	i-instr, d-data size	nr of sets
conf1	4	32	i-32k, d-32k	512
conf2	8	64	i-128k, d-128k	512

As input traces we selected 3 traces. Their characteristics are given in table 2. The first two traces were obtained by running the respective program under the Abstract Execution environment AE[a] [9]. The last trace was obtained from a synthetic trace generator as explained in [11]. For each trace we give the potential speedup and, in parenthesis, the largest subtrace of common references.

TABLE 2. Sample Traces

trace	trace size	potential speedup (largest set subtrace)	
		conf1	conf2
tex	832k	8 (104k)	7 (120k)
cc1	1000k	112 (9k)	74 (14k)
pascal	5000k	192 (26k)	152 (33k)

Benchmarking the sequential simulator

We simulated the cache configuration conf2 using the original DineroIII cache simulator. Timing results of our experiments are shown in table 3. For each trace we also

TABLE 3. Sequential simulation benchmarks for Sequent Symmetry (Sym) and Sun Sparc 10 (Sun)

trace	Sim. time sec.		Input Latency		Coeff. k_{pg}	
	Sym	Sun	Sym	Sun	Sym	Sun
tex	166	24	127	19	0.2	0.2
cc1	201	29	151	23	0.3	0.26
pascal	1010	109	667	83	0.5	0.3

computed the coefficient k_{pg}. We notice that around 70% of the time is spent reading the reference and 30% processing it.

The impact of locking

To measure the impact of locking, we tested our parallel scheme for a window size of 1. The results showed that the total parallel execution time was of the order of the sequential simulation time. The main source of overhead was the high contention on the set lists. Besides, many processes were idle most of the time waiting for references to be processed, since the parser could not catch up with the set processes. Even parallel trace parsing could not reduce the input latency due to high list contention.

Synthetic traces.

Synthetic traces are attractive since their space require-

(a) The AE has been extended for Parallel Programs. See the SPAE trace generator [10]

ment is small. Thiebaut et.al. [11] describe an algorithm that generates traces that behave much like real traces. Address generation follows a hyperbolic distribution controlled by locality, working set, address space size, and trace length. The algorithm generates a trace by using a distribution that produces stack indexes. Each new index is saved into an LRU stack that expands with each new index, and can be as large as the number of unique references ever generated. Indexes that were encountered before are used to address the LRU stack and to generate an old reference that is moved to the top of the stack becoming the Most Recently Used (MRU) address. The "hole" that is left is filled by "pushing down" the stack. We implemented a synthetic trace generator and used it to produce a synthetic trace for a pascal compiler with an 20MB address space. The trace was 10 millions long, had 44000 unique references, and contained 90% of read references.

In the proposed scheme each new reference is associated with a `read/write` tag bit which is computed using a predefined probability obtained from the corresponding real trace. This scheme lacks the distinction between instruction and data references so that an instruction reference can become a `write` one. Moreover, the tag bit is hardwired to the reference, so that if it was a `read` tag the reference can never become a `write` one. An enhanced scheme could use the knowledge about the division of the address space into text, data and stack segments, and hardwire a `read` tag only to instruction references.

Although computing an index into the LRU stack and generating a new reference are fast operations, by contrast, an old reference is very expensive to re-generate since a search in the LRU stack is needed, followed by an update of the stack. Pushing down the stack is very expensive for large arrays. If the stack were implemented as a double-linked list, then the update operation would be fast, but a linear search would be costly. Our experiments showed that for both cases, generating a trace of one million references is 15 times slower than reading it from disk. The longer the traces were the higher the slowdown was. We opted for not using synthetic traces.

Raw Buffered Data.

To avoid slow scanning of formatted data, we converted the traces to binary form. We also rewrote the input process so that it could buffer blocks references crossing procedure calls and the file system as less as possible. For optimal I/O transfer, the buffer size was selected as a multiple of the disk sector size. In table 4 we show the new input latencies for both the Sequent Symmetry and the Sun SPARC.

TABLE 4. Raw buffered input

trace	Symmetry		Sun Sparc10/30	
	latency sec.	Coeff. k_{pg}	latency sec	Coeff. k_{pg}
TeX	45	0.9	14	0.3
cc1	53	1.0	16	0.4
pascal	280	1.3	60	0.5

The coefficient k_{pg} has been averaged using the results from table 3. We can see that the input overhead was reduced by 50-60% for the Symmetry and by around 30% for the Sun.

Increasing granularity and exploiting locality.

To increase granularity that can reduce synchronization latencies, we let the parser collect a group of references in a exclusive local list, and then hand it over to the set process. For this, the parser needs to extract a window of references from the trace. Thus, only one lock/unlock operation would be needed for a group of references. This technique is efficient since it relies on the principle of locality, a property that most programs enjoy. This also implies that for programs with high temporal locality only a subset of the set processes will be active at a time. The others will be blocked not consuming processor time. In small multiprocessors with a low number of processors such as multiprocessor workstations this method is feasible. We implemented the windowing technique for our cache configurations and traces. The results are illustrated in figure 1. For each trace we show the actual speedup that was achieved and the "clean" speedup, i.e. the speedup in absence of input latency.

The results show that for slow trace generation rates our techniques can reduce the simulation time at most by 40-50%. Using more than 3 processors does not give better speedups. Furthermore, the results reveal that if the input latency could be hidden by reference processing, then higher speedups can be obtained for up to 5-6 processors.

4. CONCLUSIONS

Set-partitioning exploits parallelism by assigning one process per set. Although the speedup could be equal to the number of sets, we showed that for extremely fast generation rates, the speedup is limited by the cache and trace characteristics. For slow generation rates, the speedup is proportional to the ratio between reference processing time and reference generation time. To get a fast reference generation, we tested 3 types of traces: formatted disk-based, unformatted disk-based, and synthetic traces. The worst generation rate was exhibited by synthetic traces, followed by formatted traces. Unformatted raw buffered traces provided a rate of the order of the processing rate. To reduce the latency effects during reference processing, we proposed a windowing technique that exploited locality and handed over bulks of references to set processes. Note that we simulated one cache at a time using non-preprocessed traces. There exists several common cases in cache simulation where the amount of work to be performed for each reference is higher than in our experiments. For such situations the reference processing time is slower implying better speedups. Such is the case when: simulating several cache configurations in one pass of the trace; simulating with stripped traces, where each reference causes a miss; simulating large caches or caches with large associativities; and investigating the effects of multiprogramming, where a new jobs cause flushes or high percentage of miss rates. Based on our results, we claim that our techniques are suit-

I-173

able for parallel cache simulation on multiprocessor workstations.

The following techniques deserve further investigation: (i) parallel parsing of the trace; (ii) test our techniques in a execution- and program-driven environment that provides high generation rates.; (iii) clustering set lists to increase granularity; (iv) impact of the window size on performance.

Acknowledgments. Thanks to D. Thiebaut for his remarks on synthetic traces, M. Hill for hints on the DineroIII cache simulator, D. Wagner and J. Larus for providing information about Abstract Execution environments. This research is a part of the parallel simulation project financed by the Swedish National Board for Industrial and Technical Development (NUTEK) under contract 90-01773p.

REFERENCES

[1]   P. Heildelberger, H. Stone, *"Parallel Trace-Driven Simulation by time Partitioning"*, Proceedings of the Winter Simulation Conference (1990), pp. 734-737.

[2]   D. Nicol, A.G. Greenberg, B.D. Lubachevsky, *"Massively Parallel Algorithms for Trace-Driven Cache Simulation"*, Proceedings of the 6th workshop on Parallel and Distributed Simulation (1992) pp. 3-11.

[3]   R. Mattson, J. Gecsei, D. Slutz, and I. Traiger, *"Evaluation Techniques for Storage Hierarchies"*, IBM Systems Journal, Vol. 9, (12(2),1970) pp. 78-117.

[4]   T.R. Puzak., *Cache-Memory Design*, Ph. D. thesis, ECE Dept., Univ. of Massachusetts, 1985.

[5]   M. D. Hill and A. J. Smith, *"Evaluating Associativity in CPU Caches"*, IEEE Transactions on Computers, (December, 1989), pp.1612-1630.

[6]   J.L. Baer, W.H. Wang, *"On the inclusion property for multi-level cache hierarchies"*, Proc. of 15th Annual International Symposium on Computer Architectures, Honolulu, (May,1988), pp. 73-80.

[7]   T.E. Anderson, *FastThreads User's Manual*, Dept. of Computer Science and Engineering, Univ. of Washington, (Jan,1990), 10 pp.

[8]   Mark D. Hill, *The DineroIII Cache Simulator*, Univ. of Wisconsin, Comp. Sciences Dept. 1991.

[9]   J. R. Larus, *"Abstract Execution: A Technique for Efficiently Tracing Programs"*, Software Practices & Experience, (December, 1990), pp 1241-1258.

[10]   D. Grunwald, G.J. Nutt, A.M. Sloane, D. Wagner, W. Waite, B. Zorn, *"Executuion architecture independent program tracing"*, Dept. of Comp. Science, Univ. of Colorado, Tech. Rep. CU-CS-525-91, (April,1991)..

[11]   D. Thiebaut, J.L. Wolf, and H. Stone, *"Synthetic Traces for Trace-Driven Simulation of Cache Memories"*, IEEE Transactions on Computers, (April, 1992), pp. 388-410.

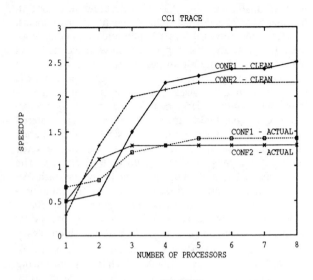

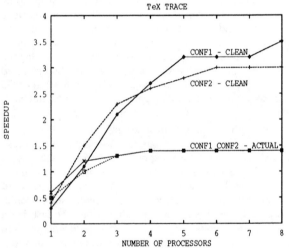

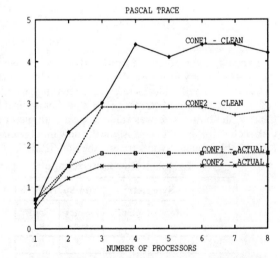

FIGURE 1. Speedups for cc1, TeX and pascal traces. Clean speedup means no input latency.

A CHAINED-DIRECTORY CACHE COHERENCE PROTOCOL FOR MULTIPROCESSORS

Soon M. Chung † and Longxue Li
Department of Computer Science and Engineering
Wright State University
Dayton, Ohio 45435, U.S.A.

Abstract — *In this paper, a chained-directory cache coherence protocol for a shared-memory multiprocessor system is proposed, which uses linked lists to store the information for the coherence maintenance. The proposed scheme uses small memory space and has high scalability, and it requires less memory accesses by distributing the coherence maintenance to caches. We evaluated the performance of the proposed scheme, and compared it with those of two other directory-based schemes; the singly-linked list and the full-map schemes. The analysis shows that the proposed scheme provides better processor utilization and results in less memory contention.*

1. Introduction

In a shared-memory multiprocessor system, as shown in Figure 1, a private cache is attached to each processor to balance the speed gap between main memory and processors. Each memory block may have more than one copy in different caches belonging to different processors, and different processors may access the shared copies simultaneously and modify them locally without knowing what is happening to the other copies in the other caches. Therefore, the cached copies in different caches may be inconsistent and a processor accessing such shared data may not be able to obtain the latest value of the data. A multiprocessor system is said to be cache consistent if a read access to any block always returns the most recently written value of that block, and various cache coherence protocols have been proposed for cache consistence [5]:

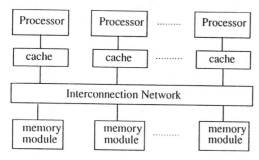

Figure 1. Shared-Memory Multiprocessor Architecture

In this paper, we propose a chained directory cache coherence protocol which is similar to the singly-linked list approach [6], but has the following differences:

(i)   As in the case of the ownership protocol [2], the head of the sharing list is the owner of the shared data.

(ii)   Each cached block has a head pointer as well as a forward pointer. This difference results in better perfor-

† This research was partially supported by National Science Foundation under grant No. IRI-9008694.

mance in maintaining cache consistency.

2. Protocol Description

The proposed scheme uses a singly linked list to specify which cache has a copy of shared block, and each cached copy maintains a head pointer to the current head of the sharing list. For each block in main memory, there is only one head cache field associated with it pointing to the head cache if there exists any shared copy in any cache. Otherwise, that field is set to nil. Each block in a cache is in one of the following five states: *Invalid, Unowned, Read, Ex–Read,* and *Write. Invalid* means the cached block is not a valid copy. *Unowned* means that the cache has a valid shared copy, but the cache is not the owner of the block (head in the list). Thus it refers to the head (owner) cache. *Read* means the cache has a valid copy and some other caches also have valid copies. *Ex–Read* means the cached copy is the only one copy in the system and it is clean. *Write* means the copy is a head copy which is dirty and there may exist some other invalid copies in the system. In each cache, there is a cache directory containing an entry for each cached block, which stores the state information of the block for cache consistency maintenance.

In order for the cache controller to keep track of the state of each block, the state field is composed of some flag bits: A Valid bit (V) indicates whether the block is valid or not; a Modified bit (M) indicates whether the block is dirty (modified) or not (clean), which is used for writing back when cache block replacement occurs; A Owner bit (O) indicates whether the cache is the owner or not. Also there are two pointer fields: one is the Head Pointer (HP) pointing to the current head of the sharing list (owner), the other is the Forward Pointer (FP) pointing to the next cache in the sharing list. When the cache is the last one in the sharing list, its FP is TAIL. With this state information the cache controller can take appropriate actions when a shared data is accessed, and hence maintains cache coherence.

In the following, we describe different actions to be taken by our protocol when a shared copy is accessed, which can be divided into the following four classes. By hit here we mean that there is a valid copy in the cache, and by miss there is an invalid copy or no copy in the cache.

1. Read Hit

The copy is consistent, and the read operation is carried out locally.

2. Read Miss

(a)   The requesting cache has an invalid copy.

A load request is forwarded directly to the head, and the head sends only the requested datum to the requester. The requesting cache loads the datum, then carries out the read operation. Its state is not changed.

(b)   The requesting data is non-existent in the cache.

A load request is forwarded to the memory module, the memory controller takes an action depending on whether the requested data block is cached or not by checking its head pointer field:

i. Uncached

The memory controller sets a pointer to the cache. Then the data block is loaded into the cache from the memory. The state of this block in the cache is set to *Ex−Read*, its HP is left nil, O=1 and FP=TAIL.

ii. Cached

In this case the memory controller directs the request to the head cache together with the requester's identifier. After receiving the requesting message, the head cache sets its FP to the requester, then sends its previous FP value and the block to the requester which finally sets its HP and FP fields. Then, the read operation is done locally. The final states of the head copy and the requester's copy are *Read* and *Unowned*, respectively. As a result, the requesting cache is inserted next to the head cache in the sharing list. The reason why it is put next to the head, instead of becoming the head as in the singly-linked list scheme [6], is to avoid the changing of HPs in other caches in the list. A block replacement may be needed at the requesting cache, and the corresponding actions to be taken are explained later in this section.

3. Write Hit

Depending on the state of the cached copy, there are four possible cases:

(a)   *Write*

The write operation is done locally.

(b)   *Ex−Read*

The write operation is carried out locally, and cache copy becomes dirty (M=1), and its state is changed to *Write*.

(c)   *Read*

The head sends out an invalidation message down to the sharing list. The second cache invalidates its copy (V=0) and transmits the invalidation message to cache next to it, each cache in the list repeat the same actions until the last cache which, after invalidating its copy, sends an invalidation-reply message to its head. After receiving invalidation-reply message, the head changes its state to *Write*, and sets the dirty bit (M=1), and carries out the write operation. In this way, the sequential consistency is guaranteed.

(d)   *Unowned*

A write-request message is sent to the memory controller, which then changes its head pointer to the requesting cache and directs the write request to the old head. The old head sets its HP to the new head, invalidates its copy, and sends the new head identifier and invalidation message down to the sharing list. The second cache in the list does the same thing and so on. The cache whose FP is the new head sets its FP to TAIL to become the last cache in the new list. This process repeats until the last cache in the old list receives the message. It updates the HP, sets the FP to the old head, and then sends an invalidation-reply message to the new head. Then the new head clears its HP, sets dirty bit (M=1), changes its state to *write*, and finally carries out the write operation.

4. Write Miss

When a write miss occurs, a load request message is sent to the memory module. There are three possible cases listed below. In any case, the final state of requesting cache is *Write*.

(a)   The requested data block is uncached.

The memory controller sets the block's head pointer to the requester and sends the block to the cache, which then sets its FP=TAIL, M=1, O=1, and carries out the write operation.

(b)   The requested data block is cached elsewhere but non-existent in the requesting cache.

The memory controller changes the block's head pointer to the requester and directs the request to the old head. An invalidation message and new head identifier are sent down to the sharing list, and an invalidation-reply message is returned to the old head by the tail cache. Then the old head sends a copy and write-grant message to the new head. Now for the new head, M=1, its FP is set to the old head, and the write operation is carried out.

(c)   The requested data block is cached, and the requesting cache has an invalid copy.

The memory controller changes the block's head pointer to the requester and sends the identifier of the requesting cache and a write-invalidation message to the old head. The old head sets its HP to the new head, invalidates its copy, and sends the new head identifier and invalidation message down to the sharing list. The second cache in the list does the same thing and so on. The cache whose FP is the new head sets its FP to TAIL. This process continues until the last cache receives the message, updates the HP, changes the FP to the old head, and sends an invalidation-reply message to the old head. Then the old head sends a copy and write-grant message to the new head, which then sets its state field and carries out the write operation.

5. Block Replacement

Depending on the state of the block to be replaced, there are four possible cases:

(a)   *Invalid* or *Unowned*

The cache sends a message to its head, then the head cache sends a message down to the sharing list to eliminate it. The cache whose FP is the requesting cache updates its FP to the FP of the requesting cache.

(b)   *Read*

A message is sent to the memory module to change the block's head pointer. As a result, the second cache in the list becomes the new head, and each cache copy in the list changes its HP.

(c)   *Ex−Read*

A message is sent to the memory module to clear the block's head pointer field.

(d) *Write*

In this case, the block is written back to the memory module. All the invalid copies in the sharing list are purged by clearing the state field of the copies. If a head copy is replaced, the next copy in the list becomes the head copy after the necessary changes.

3. Performance Analysis

3.1. Modeling

Processor utilization reflects the performance degradation due to coherence maintenance, and the average number of memory references shows the degree of memory contention for each scheme. We use the method introduced in [4] to evaluate the processor utilization and the average number of memory references.

(a) System Parameters

Following parameters are used to model the system performance:

N---number of processors.
w---memory reference rates for writes.
r ---memory reference rates for reads.
m--- cache miss ratio.
s ---fraction of memory references to shared blocks.
s' ---sharing degree among processors, that is, fraction of processors having a copy of a shared block.
u ---probability that a shared block is unmodified.
h ---rate of write hit to unmodified shared data. $h = w(1-m)su$
i ---probability that a read miss is due to an invalid copy.
t_n ---average message transmit time through the interconnection network.
t_c ---the number of processor cycles required for a cache reference.
t_m---the number of processor cycles required for a memory reference.
t_M---the number of processor cycles required for the memory module to access the directory.
t_{invld}---the number of processor cycles required for a cache to invalidate a block.

(b) Processor Utilization

According to [4], processor utilization is defined as:

$$U = \frac{1}{1 + T_{idle}} \tag{1}$$

where T_{idle} is the processor idle time in terms of processor cycle time. It is the delay due to the required message passing through the interconnection network, memory and/or cache reference, and the interference from other caches. Thus, we have

$$T_{idle} = T_{msg} + T_{int} \tag{2}$$

T_{msg} represents the delay due to message passing through the network and the memory/cache accesses, and T_{int} results from the interference (invalidation requests and the data transfer requests) from other caches.

Here, we assume that the interconnection network is a buffered Omega network implemented with $k \times k$ switches. By using the model developed in [3], a message composed of B packets is transmitted through the $\log_k N$ stages of the network

within

$$t_n = \log_k N \left[t_t + \frac{t_s(1-1/k)B^2 p}{2(1-Bp)} \right] + (B-1)t_s \tag{3}$$

where N is the number of processors, t_t is the transit time of a packet from one stage to next one when the buffer is empty; t_s is the network switch cycle time; p is the probability that a packet arrives on each input at each network cycle. Generally, t_t is smaller than t_s. However, in our analysis we assume that $t_t = t_s$. We can easily obtain p from the number of interconnection network requests issued by each processor during a processor cycle as:

$$p = (m + h) U \tag{4}$$

For a processor cycle, a message is generated with the probability $m + h$: the first term corresponds to the message transmit caused by cache misses, and the second term represents the invalidation messages due to a write to a block in *Read* state.

In the following derivations, we denote t_{nd} as the average message transmit time through the network when the message includes a datum, and t_{nb} as the time when the message includes a data block, and t_{nm} as the time when the message is an invalidation message, respectively.

In the following, we give the derivation of the processor utilization and the average number of memory references for the proposed protocol. For the proposed scheme, the average delay for each processor due to message passing and memory/cache reference is as follows:

$$T_{msg} = T_{miss} + T_{inv} \tag{5}$$

Here, T_{miss} is the overhead caused by all the read and write misses, and T_{inv} is the overhead due to invalidations for cache coherence. Thus,

$$
\begin{aligned}
T_{miss} = {}& rmi(t_{nd} + t_c) + rm(1-i)s(t_{nb} + t_{nm} + t_M + t_c) \\
& + rm(1-i)(1-s)(t_{nb} + t_m) + wm(1-s)(t_{nb} + t_m) \\
& + wms(t_{nb} + t_c)
\end{aligned} \tag{6}
$$

In the above equation, the first term is the delay due to reads of invalid data, in which case only the requested data need to be transmitted; the second term represents the delay due to read miss to non-existent shared data, in which case the cache owning the data block forwards the requested block; the third term is due to read miss to nonshared data, and remaining terms are due to write miss to nonshared and shared blocks, respectively. T_{inv} is obtained as

$$
\begin{aligned}
T_{inv} = {}& h(s'N-1)(t_{nm} + t_{invld}) \\
& + wmsu[(1-i)(t_{nm} + t_M + (s'N-1)(t_{nm} + t_{invld})) \\
& \qquad + i(s'N-1)(t_{nm} + t_{invld})] \\
& + wms(1-u)[(1-i)(t_{nm} + t_M + t_{nm} + t_{invld}) + i(t_{nm} + t_{invld})] \\
& + (r + w)ms(t_{nm} + t_{invld})Ns'/2
\end{aligned} \tag{7}
$$

Here, the first term is due to write hit to shared unmodified data. The second and third terms represent the overhead resulting from write miss to unmodified and modified data, respectively. The forth term accounts for the overhead caused by updating the sharing list when a shared block is replaced. We assume that the updating message goes through half of the sharing list on the average.

Interferences from other caches are composed of two parts: one is the invalidation requests from other caches due to write, and the other is the data transfer request when the other caches have reads to invalid block. Thus, we have

$$T_{int} = w(1-m)sut_{invld} + wms(t_{invld} + t_c/(s'N))$$
$$+ rms(it_{invld} + (1-i)t_c)/(s'N) \qquad (8)$$

By using the above equations developed, we can evaluate the processor utilization U, and from [4], the average number of memory references, which can be used to analyze the memory contention, is given as:

$$R = NT_m(1-U) \qquad (9)$$

where T_m is the delay due to memory references, which can be easily determined from the previous analysis.

3.2. Result of Analysis and Comparison

In this section, we discuss the results of performance analysis based on the equations developed in Section 3.1, and we compare the performance of the proposed scheme with those of the singly-linked list scheme [6] and the full-map scheme [1]. Table 1 shows the system parameter values used to evaluate the performance of the three schemes.

Table 1. System Parameter Values Used for Analysis

Parameter	Value
switch size	4×4
packet size	12 bytes
switch cycle time	$t_s = 1$
processor cycle time	t_s
cache cycle time	t_s
cache block size	32 bytes
memory cycle time	$10t_s$
memory address size	4 bytes

Figure 2 shows the effect of memory request rate (m) on the processor utilization for the proposed scheme. As the m increases, the processor idle time increases, so the processor utilization decreases. For the same value of m, the processor utilization decreases as the number of processors increases due to higher network and memory contentions. Other two schemes showed the similar results.

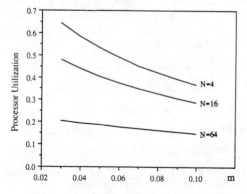

Figure 2. Effect of m on the Processor Utilization
(w=0.3, r=0.7, i=0.3, u=0.7, s=0.2, s'=0.3)

Figure 3 shows the effect of sharing degree among processors (s') on the processor utilization for the proposed scheme. As more processors share a data block, the bigger the overhead for consistence maintenance. Therefore, as s' increases, the processor utilization decreases. For the same value of s', as the number of processors increases, the length of the sharing list increases and it takes more time to update the list. That is why the processor utilization decreases as the number of processors increases for the same value of s'.

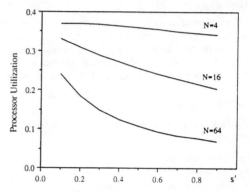

Figure 3. Effect of s' on the Processor Utilization
(w=0.3, r=0.7, i=0.3, u=0.7, m=0.1, s=0.2)

Figure 4 shows the difference in time required for handling read miss and write miss for the three schemes. For the proposed scheme, a read miss to an invalid block is directly handled by the head cache without a memory access, whereas the read miss results in an additional memory access for the case of the singly-linked list scheme. Thus, the proposed scheme saves one memory access for a read miss to an invalid block. Furthermore, for the proposed scheme, the sharing list can be restructured without intervention of the memory, which results in a better processor utilization. The advantage of the proposed scheme is at the cost of the head pointers at each cache block. The full-map scheme incurs the most delay since each miss to a shared block has to go through the memory.

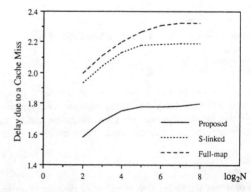

Figure 4. Comparison of the Delay due to a Cache Miss
(w=0.3, r=0.7, i=0.3, u=0.7, m=0.1, s=0.2, s'=0.3)

Figure 5 compares the processor utilizations of the three schemes. The processor utilization of the proposed scheme decreases as the number of processors N increases because the length of the sharing list becomes longer and it takes more time to update. For example, the proposed scheme may have to invalidate each cache in the sharing list serially and hence result in more processor idle time, while the full-map scheme can broadcast an invalidation message to the caches. As a result, the full-map scheme has better performance than the proposed scheme when the number of processors is larger than a certain value. The processor utilization of the full-map scheme decreases slowly as N increases for the same reason. However, the full-map scheme requires much more memory space to store the state information. The proposed scheme has a slightly better processor utilization than the singly-linked list scheme when N is small since it saves some network traffic and memory access with the help of the head pointer for each cached block for the case of read misses.

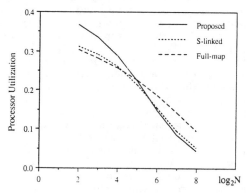

Figure 5. Comparison of the Processor Utilization
(w=0.3, r=0.7, i=0.3, u=0.7, m=0.1, s=0.2, s'=0.3)

Figure 6 compares the average number of memory references of the three schemes for cache coherence. The full-map scheme has more memory contention as N becomes larger because every transaction regarding the coherence should go through the memory. On the other hand, the proposed scheme has less memory contention compared to others because a significant part of protocol-related transactions are distributed to the caches.

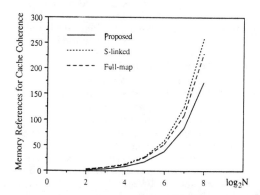

Figure 6. Comparison of the Memory Contention
(w=0.3, r=0.7, i=0.3, u=0.7, m=0.1, s=0.2, s'=0.3)

4. Conclusion

The proposed chained-directory scheme is a merging of the singly-linked list and the ownership protocols. Each cached copy in a list maintains a pointer to the head of the sharing list. With this linkage, most of the consistency-related transactions are carried out among the caches without involving the memory. Therefore, the memory contention due to cache consistency is reduced to a great extent compared to other directory-based protocols. The memory space required for the proposed scheme is proportional to the logarithm of the number of caches, whereas the full-map scheme needs a memory space proportional to the number of caches. Thus, the proposed scheme has a better scalability than the others.

References

[1] D. Chaiken, C. Fields, K. Kurihara, and A. Agarwal, "Directory-Based Cache Coherence in Large-Scale Multiprocessors," Computer, Vol. 23, No. 6, 1990, pp.49-58.

[2] R. H. Katz, S. J. Eggers, et al., "Implementing a Cache Consistency Protocol," Proc. Int'l Symp. on Computer Architecture, 1985, pp-276-183.

[3] C. P. Kruskal and M. Snir, "The Performance of Multistage Interconnection Networks for Multiprocessors," IEEE Trans. on Computers, Vol. 32, No. 12, 1983, pp.1091-1098.

[4] J. H. Patel, "Analysis of Multiprocessors with Private Cache Memories," IEEE Trans. on Computers, Vol. 31, No. 4, 1982, pp.296-304.

[5] P. Stenstrom, "A Survey of Cache Coherence Schemes for Multiprocessors," Computer, 1990, Vol. 23, No. 6, pp.12-24.

[6] M. Thapar and B. Delagi, "Stanford Distributed-Directory Protocol," Computer, Vol. 23, No. 6, 1990, pp.78-80.

Evaluating the impact of cache interferences on numerical codes *

O. Temam, C. Fricker, W. Jalby[§]

HPC Division at University of Leiden, INRIA, IRISA

Abstract

In numerical codes, the regular interleaved accesses that occur within do-loop nests induce cache interference phenomena that can severely degrade program performance. Cache interferences can significantly increase the volume of memory traffic and the amount of communication in uniprocessors and multiprocessors. In this paper, we identify cache interference phenomena, determine their causes and the conditions under which they occur. Based on these results, we derive a methodology for computing an analytical expression of cache misses for most classic loop nests, which can be used for precise performance analysis and prediction. We show that cache performance is unstable, because some unexpected parameters such as arrays base address can play a significant role in interference phenomena. We also show that the impact of cache interferences can be so high, that the benefits of current data locality optimization techniques can be partially, if not totally, eradicated.

Keywords: memory reference patterns, software optimization, data locality, numerical codes, modeling cache interferences, performance analysis, performance prediction.

1 Introduction

As CPU cycle time decreases, main memory and network latencies rapidly increase and cache misses become very costly. Furthermore, the increasing issue rate of processors worsen the burden on caches. Moreover, most CPU chips are now being designed for integration into a massively parallel supercomputer or a parallel workstation, and therefore minimizing memory traffic, i.e. optimizing memory hierarchy utilization, is becoming critical. For all these reasons, optimizing the cache behavior has become a major issue.

An important step towards accurate evaluation of cache behavior and cache interferences has been made in [1], where blocked Matrix-Matrix multiply is carefully studied. Cross and self-interference misses are evaluated, and a model for this algorithm is provided. This paper clearly unveils that interferences can severely alter locality exploitation.

Indeed, many powerful software optimization techniques for exploiting numerical codes locality have now been designed [5, 6, 7]. Although they stress the possible impact of cache interferences, no real evaluation of these phenomena nor a study of their frequency of

occurrence have been performed yet. In the next sections, we will show that cache interferences can have a strong impact on performance and occur frequently.

To address the issues related to cache interferences, we are currently developing a model named *NUMODE* (NUmerical MODEl) [8]. The goal of NUMODE is to provide a framework for modeling cache interferences of a given loop nest, and then deriving an analytical expression for the number of cache misses. Therefore, NUMODE is both a model and a methodology.

Using such techniques, it is possible to precisely quantify the interference phenomena that occur in real caches, determine their causes and conditions of occurrence. Many such phenomena can be identified. Moreover, the cache performance is shown to be actually unstable in many situations. It namely appears that software optimization techniques lack accuracy and can consequently lose their efficiency. It is also shown that the number of additional memory references due to cache interferences can be obtained as an analytical function of problem parameters. This result allows a precise analysis of the behavior of most classic algorithms on caches, and can then be used to design new optimization techniques or tune existing ones. Making such a function available to compilers can be profitable to code restructuring techniques. *NUMODE* is currently under development and will be implemented to perform extensive testings of its scope and accuracy.

In section 2 the problem is defined and hypotheses are given. In section 3, the general method for computing the number of misses is indicated. In section 4, conclusions are drawn and further work is discussed.

2 Problem statement

The goal of this paper is to introduce a method for estimating cache interferences. General principles and main steps of the technique are indicated while details of the technique can be found in [10].

Cache architecture Direct-mapped caches have been chosen for several reasons:

- Direct-mapped caches are more sensitive to interferences than w-way associative caches. Therefore, they are more likely to benefit from studies and optimizations on that matter.

- Since the replacement policy of direct-mapped caches is straightforward, computing interferences is easier in direct-mapped caches. Though, we strongly believe the technique can be extended to w-way associative caches with moderate modifications.

- Among the three newest processor chips (DEC Alpha, MIPS R4000, SuperSPARC), two chips (DEC Alpha, MIPS R4000) include a small (8kbytes) direct-mapped on-chip data cache. Since the frequency of

* This work was funded by the BRA Esprit III European Project APPARC, European Agency DGXIII.

[†] HPC Division, University of Leiden, Niels Bohrweg 1, 2333 CA Leiden

[‡] INRIA, Domaine de Voluceau, BP 105, 78153 Le Chesnay Cedex, France.

[§] IRISA/INRIA, Campus de Beaulieu, 35042 Rennes Cedex, France.

such processors is very high, the cost of a cache miss is huge, making it critical to reduce the amount of interferences.

• The placement policy in the DEC Alpha, for example, is such that, a data cache location can generally be determined from the data virtual address. Therefore, a study based on virtual addresses would accurately describe real cache behavior.

In the remainder of the paper, the cache size is indicated by C_S and the line size by L_S. The unit size is 8 bytes, i.e. the size of a double-precision floating point data. In all experiments, cache size is equal to 8-kbyte and line size is equal to 32-byte (the characteristics of the DEC Alpha data cache), so $C_S = 1024$ and $L_S = 4$.

Codes Let us now discuss which types of codes are considered. In numerical codes most data traffic occurs in do-loops, only these code constructs are examined. Only array references are considered because it is probable that other variables would be stored in registers if they are frequently used, and otherwise they would induce minimal perturbations of cache behavior.

Loop Nests A loop nest is composed of n distinct loops, j_i being the loop index of the ith loop, and j_n being the loop index of the innermost loop. Column-major storage is assumed, so, for example, the virtual address of array reference $A(j_1, j_2)$ is $a_0 + N j_1 + j_2$ where N is the leading dimension of array A and of the starting address of array A (cf figure 1).

3 Modeling numerical codes behavior
3.1 Restrictive hypotheses

Figure 1: *An example of loop nest considered.*

A number of restrictions are imposed on the loop nests considered. First, the array subscripts must all be of the form $A(\alpha_1^A j_{k_1} + \beta_1^A, \ldots, \alpha_p^A j_{k_p} + \beta_p^A)$ where $(j_{k_i})_{1 \le i \le p}$ (with $p \le n$) is any subset of loop indices, and $(\alpha_i^A)_{1 \le i \le p}$ and $(\beta_i^A)_{1 \le i \le p}$ are constants. For example, subscripts such as $A(j_1 + j_2)$ are not considered. A close analysis of benchmark suites such as the *Perfect Club* [2] or *NAS* [3] and studies such as the one done by Yew and al. [4] show that such hypotheses encompass the subscripts found within most numerical loop nests. Furthermore, in dusty-deck codes, "irregular" subscripts often correspond to linearization, and therefore, in terms of memory reference patterns, are equivalent to those considered within the scope of the model. For all these reasons, these restrictions on array subscripts are considered to be reasonable constraints. Moreover, further developments of the present model may include more complex subscripts.

3.2 General principles

If cache was fully-associative and replacement was optimal, cache misses would occur in two cases only. First, when data are loaded in cache for the first time; such misses are called *compulsory misses*. Second, when cache space is too small to store all loop nest data. Then, an element is flushed from cache each time a new element needs to be loaded; such misses are called *capacity misses*.

However, in direct-mapped caches (and in set-associative caches as well), cache misses can occur though cache space is sufficient, because a data element can only be mapped into one specific cache location (or w locations in w-way associative caches). Therefore, such cache misses do not occur because of *capacity* conflicts, but because of *mapping* conflicts; they are called *mapping misses* [9].

The main effect of such unexpected misses is to degrade the spatial and temporal reuse of data. Interferences can either correspond to interferences of an array with itself (*self-interferences*), or with another array (*cross-interferences*). In the remainder of the paper, these two types of interferences are analyzed.

For self-interferences the principle is to study the mapping of the set of elements of an array to be reused, and check whether these elements overlap with themselves. If so, self-interferences occur, and estimating the degree of overlapping yields the number of additional memory accesses brought by self-interferences. In general, self-interferences mostly correspond to temporal interferences.

For cross-interferences, once the set of elements of an array to be reused is identified (taking into account self-interferences), the overlapping between these elements and elements of another array is determined. Knowing the number of times the two sets of elements overlap, and the amount of overlapping each time, is sufficient to compute the number of additional memory accesses brought by cross-interferences. Cross-interferences can correspond to either temporal or spatial interferences.

3.3 Self-interferences

Theoretical reuse set Thanks to the subscript types considered, it is easy to identify where reuse due to self-dependences occurs. If coefficient $a_i = 0$ in the virtual address expression $a_0 + a_1 j_1 + \ldots + a_n j_n$ of a reference to array A, then loop i carries reuse. Let us define l as the lowest loop level where reuse occurs. On all loops k with $k > l$, no reuse occurs. So, on each iteration of these loops, the array elements referenced are all distinct. This set of elements is called the reuse set. During each execution of the sub loop nest $j_n, \ldots, j_{l+1}$, all elements of the reuse set are referenced.

Definition For array reference $a_0 + a_1 j_1 + \ldots + a_n j_n$, the reuse set can be defined if there exists a_k such that $a_k = 0$. The loop level of a reuse set is $l = \max\{k / a_k = 0\}$. The theoretical reuse set is equal to $RS(A)_l = \{a_0 + a_1j_1 + \ldots + a_nj_n, (0 \le j_i \le N_i - 1)_{i>l}\}$.

Reuse can occur on different loop levels. However, reuse that is carried on loop levels higher than l is at least one order of magnitude less important than the reuse on loop l (see [10]).

Size of the theoretical reuse set Let us compute the number of cache lines corresponding to the theoretical reuse set assuming no self-interferences. First, let us determine the cache line stride of the reuse set of a reference A (where reuse occurs on loop l). It is equal to $\min(1, \frac{\alpha}{L_s})$, where $\alpha = \min_{l+1 \le k \le n}(a_k)$. This term is the ratio of the smallest coefficient (only considering loop levels defining the reuse set) to the line size. If this ratio is greater than 1, then it is necessarily equal to 1, since at most one new cache line is referenced on each iteration. For example, the access stride of reference $A(3j_3)$ is $\min(1, \frac{3}{L_s})$. Let us consider another example. The virtual address of reference $A(j_3. B*j_2)$ is $a_0 + Nj_3 + Bj_2$ where N is the leading dimension of A (here reuse occurs on loop level 1). Then the access stride of the reuse set is $\min(1, \frac{\min(N,B)}{L_s})$. Then, the number of cache lines in the theoretical reuse set is equal to $N_n \times \ldots \times N_{l+1} \times acces\ stride$.

Actual reuse set Because self-interferences occur, not all elements of the reuse set can actually be reused. In direct-mapped caches, as soon as two elements of the reuse set compete for the same cache line, none of the two elements can be reused: they are victim of *self-interferences*. The elements of the reuse set not victim of self-interferences belong the *actual reuse set*. **The actual reuse set** *is the set of cache lines of the theoretical reuse set where no self-interferences occur.* Characterizing self-interferences is equivalent to determining the actual reuse set. And determining the actual reuse set is equivalent to studying the mapping of the theoretical reuse set in cache.

To compute the overlapping within the theoretical reuse set of a reference A, the loops n to $l + 1$ are successively executed, starting with loop n. On each loop level k, the cache lines used by loops $n, \ldots, k+1$, which are not victim of interferences, form a temporary reuse set called $RS(A)_l^k$. On loop level k, the interferences between N_k such temporary reuse sets $RS(A)_l^{k+1}$ are determined, and the cache lines still not victim of interferences form the new temporary reuse set $RS(A)_l^k$.

Proposition 3.1 *The total number of additional memory requests is equal to $N_1 \times \ldots \times N_l \times$ (Number of cache lines of the theoretical reuse set −Number of cache lines of the actual reuse set) (see [10]).*

3.4 Cross-interferences

Two main cases of cross-interferences can occur between two references: either the difference between the corresponding two virtual addresses is constant (independent of loop indices), or it varies with the loop indices. These two cases must be distinguished because such cross-interferences are very different. In the first case, the two references are in translation,

they always overlap and the amount of overlapping is constant, while in the second case the two references overlap only periodically and the amount of overlapping varies. So, detecting and estimating cross-interferences in the first case basically amounts to comparing the constant parameter of the two virtual addresses (it generally depends on arrays dimensions and base address), while in the second case, the relative movement of the two references must be analyzed.

The first type of interferences is called *internal cross-interferences* (because a set of references in translation constitutes a kind of *class* of references, and such cross-interferences then occur within a class), and the second type of interferences is called *external cross-interferences* (interferences among two references not belonging to the same class).

3.4.1 Estimating cross-interferences

Computing the impact of cross-interferences between two references amounts to estimating how much of the reuse of one reference is lost because of cross-interferences with the other reference. So, let us consider two references R^1, R^2, and compute the impact of R^2 on the reuse of R^1.

First, the set of elements R^1 can reuse must be estimated; it is the reuse set defined in section 3.3. Second, the set of elements of R^2 that can interfere with R^1 must be estimated as well; this set is called the interference set. Because the reuse set is computed on a given loop level l, the interference set should be computed on the same loop level. Recall, that below loop l, no reuse can occur for R^1. Therefore, the number of additional memory requests for R^1 due to cross-interferences with R^2, on each reutilization of the reuse set (i.e. on each iteration of loop l), is exactly equal to the number of cache lines used by both the reuse set and the interference set. This notion is fundamental in the computation of cross-interferences. Computing interferences this way allows to make abstraction of time considerations, i.e. *when* interferences occur. It is sufficient to estimate the intersection between the set of cache lines corresponding to the reuse set and the interference set. It is important to note that, in the following sections, the reuse set considered is the actual reuse set, otherwise cross-interferences would be counted where self-interferences already occur, resulting in an overestimate of additional memory requests.

The interference set The definition of the theoretical interference set is the same as that of the theoretical reuse set.

Definition For array reference $a_0 + a_1j_1 + \ldots + a_nj_n$, the **theoretical interference set**, on loop level l (this loop level is determined by the victim reuse set), is equal to $IS(A)_l = \{a_0 + a_1j_1 + \ldots + a_nj_n, (0 \le j_i \le N_i - 1)_{i>l}\}$.

Moreover, determining the actual interference set is done much the same way as for the actual reuse set. The actual reuse set is the subset of cache lines of the theoretical reuse set where no self-interferences occur, while the **actual interference set** is simply the set of cache lines used by the theoretical interference set. So if a cache line of the theoretical interference set is victim of self-interferences, this cache line is still counted in the actual interference set (while such a cache line is rejected from the actual reuse set). Intuitively, the actual interference set corresponds to the cache surface (the number of cache lines) used by the theoretical interference set. The amount of cross-interferences is directly correlated to the size of the actual interference set (the larger the set, the higher the probability of overlapping with the reuse set).

3.4.2 Internal cross-interferences

Internal cross-interferences occur between two references in translation. Thanks to that property the relative cache position between the reuse set of the *victim* reference and the interference set of the *interfering* reference is always the same. Therefore, if the reuse set is defined on loop level l (reuse occurs on loop l), the total number of additional memory requests due to cross-interferences between the two references is equal to the total number of iterations of loop l times the number of cache lines used by both the actual interference set and the actual reuse set $CL(R^1, R^2)$ on one iteration of loop l.

Proposition 3.2 *The total number of additional memory requests due to internal cross-interferences between R^1 and R^2 is equal to $N_1 \times \ldots \times N_l \times CL(R^1, R^2)$ (see [10]).*

3.4.3 External cross-interferences

If two references R^1, R^2 are not in translation, the cross-interferences on R^1 due to R^2 are called external cross-interferences. Computing such interferences amounts to estimating the relative position of R^1 and R^2. Each time the reuse set of R^1 and the interference set of R^2 overlap, the amount of overlapping is estimated (the time unit here is one iteration of the loop where reuse occurs for R^1, i.e. loop l).

The virtual addresses of references of R^1, R^2 are $r_0^1 + r_1^1 j_1 + \ldots + r_n^1 j_n$ and $r_0^2 + r_1^2 j_1 + \ldots + r_n^2 j_n$. The positions of the interference set and the reuse set are determined by loop indices j_i with $i \le l$ (loop indices j_i with $i > l$ determine the sets themselves). Therefore, the relative address distance of the two sets is $(r_0^1 - r_0^2) + (r_1^1 - r_1^2)j_1 + \ldots + (r_l^1 - r_l^2)j_l$. Let $r_i = r_i^1 - r_i^2$ for $i > 0$, $r_0 = r_0^1 - r_0^2 \bmod C_S$, and $D = gcd_{1 \le i \le l}(r^i)$. Then for any value of $(j_i)_{1 \le i \le l}$, there exists $\lambda \in \mathbf{Z}$ such that $r_0 + r_1 j_1 + \ldots + r_l j_l = r_0 + \lambda D$. Let $d = gcd(D, C_S)$. The possible relative cache positions of the two references are necessarily of the form $r_0 + \lambda d \bmod C_S$. Therefore, $RS(R^1)_l$ and $IS(R^2)_l$ have only $\frac{C_S}{d}$ possible relative cache positions. Approximately on each iteration of reuse loop l, a new relative position is reached. Therefore, after $\frac{C_S}{d}$ iterations of loop l, all possible relative positions have been reached. Consequently, the relative "movement" of the two references is periodic, of period $\frac{C_S}{d}$. The total number of iterations of loop l executed is $N_1 \times \ldots \times N_l$, so there are $\frac{N_1 \times \ldots \times N_l}{\frac{C_S}{d}}$ periods.

Since the period is $\frac{C_S}{d}$, for any interval of $\frac{C_S}{d}$ values of λ, $r_0 + \lambda d \bmod C_S$ describes all possible relative cache positions. Let I_λ be one such interval.

Then for any value of λ, the number of cache lines used by both the reuse set and the interference set, i.e. $CL(R^1, R^2; \lambda)$, is computed the same way internal cross-interferences are evaluated, i.e. by computing the beginning and the end of each set and then deducing their overlap.

Proposition 3.3 *The total number of additional memory requests for one period is $\sum_{\lambda \in I_\lambda} CL(R^1, R^2; \lambda)$ and the total number of additional memory requests is $\frac{N_1 \times \ldots \times N_l}{\frac{C_S}{d}} \times \sum_{\lambda \in I_\lambda} CL(R^1, R^2; \lambda)$ (see [10]).*

4 Conclusions and Further work

The importance and frequency of occurrence of cache interferences are generally considered to be very irregular. These interference phenomena prevent optimum utilization of cache memory, and render cache performance unstable. The purpose of NUMODE is to understand, detect and quantify cache interferences. NUMODE is basically a framework and a method for computing the number of cache interferences within a given numerical loop nest. The different types of cache interferences are identified and separated into distinct classes. Then, for each class, it is shown how to evaluate the number of additional memory requests due to such interferences.

Software optimization for cache interferences is one of our main research goals. Main goals include determining the accurate value of the optimal block size in blocked algorithms and developing of a strategy for determining when and where to apply data copying.

References

[1] M. S. Lam, E. E. Rothberg, M. E. Wolf: *The cache performance and optimizations of blocked algorithms*, Proceedings of 4th ASPLOS, 1991.

[2] L. Pointer: *Perfect Club Report*, July 1989, CSRD Report No. 896.

[3] D. Bayley, J. Barton, T. Lasinsky and H. Simon: *The NAS parallel benchmarks*, Technical Report RNR-91-002, NASA AMes Research Center, August 1991.

[4] She, Zhiyu, Zhiyuan Li and P-C. Yew: *An empirical study on array Subscripts and Data Dependencies*, August 1989, CSRD Report No. 840.

[5] M. E. Wolf and M. Lam: *A Data Locality Optimizing Algorithm*, Proc. of PLDI.

[6] C. Eisenbeis, W. Jalby, D. Windheiser, F. Bodin: *A strategy for array management in local memory*, Advances in Languages and Compilers for Parallel Processing, MIT Press, 1991.

[7] K. Kennedy, K. McKinley: *Optimizing for parallelism and data locality*, Proceedings of ICS'92.

[8] O. Temam: *Study and optimization of numerical codes cache behavior*, PhD Thesis, Univeristy of Rennes, 1993.

[9] R.A.Sugumar and S.G.Abraham: *Efficient Simulation of Caches under Optimal Replacement with Applications to Miss Characterization*, to appear in the 1993 ACM SIGMETRICS Conference.

[10] O. Temam, C. Fricker, W. Jalby: *Impact of cache interferences on usual dense numerical loop nests*, University of Leiden Technical Report, Leiden, The Netherlands, March 1993.

Performance Evaluation of Memory Caches in Multiprocessors *

Yung-Chin Chen

MIPS Technologies, Inc.
Silicon Graphics, Inc.
Mountain View, CA 94039
ychen@mti.sgi.com

Alexander V. Veidenbaum

Center for Supercomputing Research and Development
University of Illinois at Urbana-Champaign
Urbana, IL 61801
alexv@csrd.uiuc.edu

Large-scale MIN-based shared-memory multiprocessor systems have long shared memory latency. Private caches can improve memory access latency but they may suffer from the cache coherence problem and potentially lower data locality due to data sharing and multiprocessor scheduling. These two problems also increase shared memory load and may result in frequent memory stalls. In this paper, we evaluate the performance of memory caches, a cache memory placed in front of shared memory, in a large-scale multiprocessor system in the presence of processor caches. The memory cache is shown to have good performance and scalability.

1 Introduction

Large-scale multiprocessor systems in which processors and memory modules are linked by interconnection networks have been proposed and designed in recent years. One of the major problems associated with this type of architecture is the speed of shared memory access due to long network latency and slow shared memory. Private caches and local memory that are associated with each processor are used to reduce the frequency of shared memory accesses. The use of private caches in multiprocessor systems introduces the cache coherence problem, and caches may not perform as effectively as those in uniprocessor systems.

The cache coherence problem may also increase the shared memory load because coherence maintenance can cause additional locality loss and memory updates. For software coherence schemes, there could be frequent cache invalidations. For directory schemes, the false sharing effect may induce unnecessary cache misses. Parallel processing can also cause spatial and temporal locality losses due to data sharing and multiprocessor scheduling. Our previous study [4] showed that a memory stall is still one of the major performance bottlenecks even when processor caches are present.

In addition, the speed gap between processors and memory is increasing. The problem will become even worse in the future. Faster memory can serve memory requests more efficiently but at a higher cost. More interleaving can increase memory bandwidth but it is expensive and is not always effective, e.g. hot spots. A cost-effective approach is to place a cache memory in front of a shared memory module. The memory cache with high hit ratios turns a system having a slow shared memory access latency into one having a fast shared memory access latency.

The memory cache is not a new idea. It was proposed [9] as an alternative way to use cache memory without the cache coherence problem. It was also implemented in the Alliant FX/2800 machine. However, the improvement in performance due to memory caches is not as large as for processor caches because a memory request still has to traverse the interconnection networks to access data.

It is possible to have both types of caches in the system. The processor cache can eliminate most of the requests from a processor to shared memory, and the memory cache can reduce the access latency for these requests. In this paper, we evaluate the performance of memory caches in a large-scale multiprocessor system with processor caches. We study the performance improvement from the use of the memory cache, the effects of its configurations and its scalability.

2 Memory Caches

The use of a memory cache has several advantages. First, the memory cache does not induce any coherence problem, and its design is transparent to processor cache coherence protocols and architectures. Second, the implementation of memory caches can utilize the latest technology such as low-cost high-speed CDRAMs (Cache DRAM) [6]. A CDRAM integrates a cache into a large DRAM with fast block transfer and options of set associativity. Third, the memory cache has very high hit ratios, as will be shown in this paper.

We assume that the shared memory is interleaved in such a way that a cache line belongs to a single memory module. If we use a memory cache line size equal to that

*This work was supported in part by the National Science Foundation under Grant No. NSF 89-20891, NASA Ames Research Center under Grant No. NASA NCC 2-559, and MIPS company.

of the processor cache, the memory cache uses mainly the temporal locality of each memory line. It could utilize the spatial locality if it had a larger line size. The temporal locality depends mainly on the degree of sharing, either true sharing or false sharing. Therefore, it is affected by the number of processors, the line size, scheduling algorithm, and program behavior.

The temporal locality is high for a multiprocessor system due to data sharing and the cache coherence problem, the two major problems that reduce the hit ratios of processor caches. Part of the performance losses can be compensated by the memory cache. The sharing of lines causes unavoidable cold misses for each processor. When the memory cache is used, the first cold miss can bring the shared data into the memory cache, and subsequent cold misses by other processors can hit in the memory cache.

The improvement on performance by the memory cache can be thought of as increase of the processor cache hit ratio. Without considering the effects of overlapped memory accesses and resource conflicts, such an increase can be expressed by

$$\delta = (1 - h_{pc}) \times h_{mc} \times \frac{t_{mem} - t_{mc}}{t_{mem} - t_{pc}}. \tag{1}$$

where t_{pc} is the access time for a processor cache hit, t_{mc} is the access time for a memory cache hit, t_{mem} is the access time to shared memory, h_{pc} is the processor cache hit ratio, and h_{mc} is the memory cache hit ratio.

The improvement is proportional to the memory cache hit ratio and the processor cache miss ratio. In addition, it is related to the access times of the shared memory and caches. When the shared memory becomes slower or the memory cache becomes faster, the memory cache is more effective. In practice, the actual improvement in shared memory latency is hard to estimate without doing simulation.

3   System Architecture

The system organization that we simulate consists of an equal number of 32-bit RISC processors and shared memory modules interconnected by unidirectional, multistage Omega networks. The weak consistency model [7] is used. The processor data cache is a 64-KB direct-mapped cache, and the line size is 32 bytes.

We use the simple software scheme [8] to maintain processor cache coherence in most of our experiments. We also use the directory scheme in one experiment. The directory scheme in this study is based on Censier and Feautrier's distributed full-map directory scheme [2]. For the software scheme, we use a write-through write-allocate cache with a small write-back cache as its write buffer [3]. This write policy was shown to have better performance than pure write-through or write-back caches for software schemes.

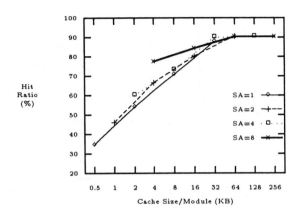

Figure 1: Average hit ratios of memory caches (SA: set associativity).

A memory cache is placed in front of each shared memory to handle all of the requests to shared memory. We assume that the memory cache uses the write-back policy. We observed that the write-through policy yields worse performance. The LRU (least-recently-used) replacement policy is used.

For more detailed description of these modules and their timing models, see [4].

Six Fortran programs are used as our benchmarks: *ARC3D*, *OCEAN*, *FLO52*, *TRFD*, *MDG*, and *MG3P*. Details about these benchmarks can be found in [1] [5]. Event-driven timing simulation is used in our study. More details about the trace generation, simulation methodology, and benchmark characteristics can refer to [4].

4   Performance Evaluation

In this section, we evaluate the performance of memory caches. Unless otherwise specified, we assume a 32-processor system; both processor and memory caches use 32-byte lines.

Figure 1 shows the overall hit ratios for the memory cache averaged over six benchmarks for different memory cache configurations. The set associativity is important only when the cache size is small. A direct-mapped 32KB cache is a cost-effective configuration, whose average hit ratio is approximately 88%, while the maximal average hit ratio is 91% for a larger cache or a higher set associativity. Most of the misses are read misses, and most of the read misses are cold start misses (see [4]).

The memory cache can have a line size larger than that of the processor cache and therefore capture spatial locality. A longer processor cache line will increase data trans-

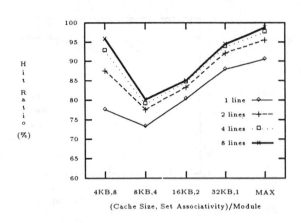

Figure 2: Hit ratios of memory caches with different line sizes (in number of processor cache lines).

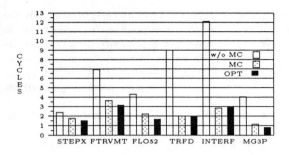

Figure 3: Comparison of average stall cycles for a read miss fetch between different memory systems.

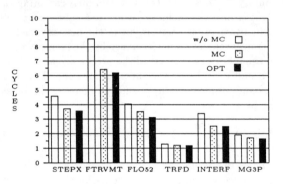

Figure 4: Comparison of average memory read latencies between different memory systems.

fer time and cause more contention for caches, networks, and memory. It also increases false sharing for directory schemes. Consequently, the processor cache line size cannot be made too long. On the other hand, there is not so much overhead for memory caches because the physical location of shared memory and memory caches are close, and the data transfer does not have to traverse the networks.

Figure 2 shows the average hit ratios for the memory cache using memory cache line sizes that are equal to 1, 2, 4 and 8 processor cache lines, respectively. A longer line has higher hit ratios due to higher spatial locality. Increasing the line size from 1 to 2 processor cache lines yields the largest improvement. Any further increase in line size is less important. Also, when a long cache line is used, a small cache with a high set associativity can achieve hit ratios equivalent to a large cache with a lower set associativity. This appears to be an economic choice for memory caches if the use of a long cache line does not incur a lot of overheads.

Hit ratios alone do not characterize performance of caches. Some overheads, such as write stalls and network and memory contention, are not reflected in the hit ratios. Therefore, we also measured average memory latency. The latency for a given memory access consists of the default minimal cycles required (22 cycles in our case) and extra cycles due to resource contention (called the *stall cycles*). The memory cache can reduce memory latencies for those accesses that have processor cache misses in two ways. One is to provide faster memory access (i.e., a shorter minimal latency, 15 cycles in our case) for requests having memory cache hits. The other is to reduce stalls on memory modules. The system with memory caches has fewer memory stalls than the system without memory caches because the cache memory can serve requests faster so that the average

queuing delays of these requests would be shorter, and the load is shared among shared memory and memory caches.

To demonstrate how memory caches reduce stalls, Figure 3 shows the average stall cycles for a read miss fetch for systems with and without memory caches. The OPT system assumes 100% hit ratios for memory caches, a performance that can be thought of as the optimal performance of memory caches as well as that of a system using shared memory which is as fast as cache memory. The system with the memory cache (MC) uses a direct-mapped 64KB cache for each memory module, which yields maximal hit ratios for most of these benchmarks. In general, MC performs nearly as well as OPT because of its high hit ratios. Both eliminate most of the memory stalls that occur in the system without memory caches. However, they will increase network contention at the same time due to a higher traffic rate. The reduction in stall cycles for a read miss fetch ranges from 26% to 76% with an average 58% for MC. Write accesses to shared memory modules also benefit from the memory cache in the same way.

Figure 4 shows the average memory read latencies for the same systems. The reduction in read latency ranges from 7% to 26% with an average 17%. The reduction is not as much as that in stall cycles because there is a minimal latency required for each read miss, and the processor cache has high hit ratios that mask off the improvement coming from a shorter processor cache miss service time.

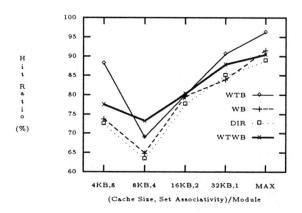

Figure 5: Comparison of hit ratios of the memory cache for different cache write policies and coherence schemes.

Figure 5 shows the average overall hit ratios of the memory cache on systems using the directory scheme (DIR) and the software scheme with write-back (WB), write-through (WT) and write-through with a write-back write-buffer (WTWB) caches, for several configurations of memory caches. In general, their hit ratios are not very different.

When the number of processors increases, the memory cache hit ratios increase for the same memory cache configuration (not shown here). When the system size grows, although the memory cache size for each module remains the same, the *combined* memory cache size becomes much larger. When the combined cache size is large enough to contain the problem size, further increase in the cache size will not improve hit ratios. We did observe that when two systems with different numbers of processors have memory caches with the same set associativity, cache line size and *combined* cache size, their memory cache hit ratios were about the same. The combined cache size that yields maximal hit ratios depends on individual programs rather than the system size. This argument is not always true for processor caches because in this case different caches may have a copy of the same memory line, whereas memory caches do not have sharing among each other.

5 Conclusions

We evaluated the use of memory cache in front of each memory module to further reduce shared memory access latency in the presence of processor caches. The memory cache does not have cache coherence problems. Its design is transparent to processor cache coherence protocols and architectures. We have shown that a reasonable size of memory cache can always achieve high hit ratios. Memory caches reduce average shared memory access latency in two ways. One is to reduce the minimal latency required for each processor cache miss when it has a memory cache hit. The other is to reduce memory stalls on shared memory modules. The memory cache can cut stall cycles by approximately a half.

We also evaluated the performance of memory caches for systems using directory schemes or software schemes with different processor cache write policies. They always achieve similar high hit ratios.

The memory cache hit ratios depend on the combined memory cache size, not the cache size per memory module. As the system size grows, we can reduce the memory cache size per module to maintain a constant combined memory cache size and still have similar performance.

References

[1] D. Bailey and H. Simon. "The NAS Parallel Benchmarks". Technical report, NASA Ames Research Center, 1991.

[2] L.M. Censier and P. Feautrier. A New Solution to Coherence Problems in Multicache Systems. *Transactions on Computers*, C-27:1112–1118, November 1978.

[3] Y.-C. Chen and A.V. Veidenbaum. An Effective Write Policy for Software Coherence Schemes. In *Supercomputing' 92*, 1992.

[4] Y.-C. Chen and A. Veidenbaum. "Performance Evaluation of Memory Caches in Multiprocessors". Technical Report 1266, CSRD, University of Illinois at Urbana-Champaign, 1993.

[5] G. Cybenko et al. "Supercomputer Performance Evaluation and the Perfect Benchmarks". In *Int. Conf. on Supercomputing*, 1990.

[6] K. Dosaka et al. "A 100MHz 4Mb Cache DRAM with Fast Copy-Back Scheme". In *IEEE Int. Solid-State Circuits Conf.*, pages 148–149, 1992.

[7] M. Dubois et al. "Memory Access Buffering in Multiprocessors". In *Int. Sym. on Computer Architecture*, pages 434–442, 1986.

[8] A.V. Veidenbaum. "A Compiler-assisted Cache Coherence Solution for Multiprocessors". In *Int. Conf. on Parallel Processing*, pages 1029–1035, August 1986.

[9] P.C. Yeh et al. "Performance of Shared Cache for Parallel-Pipelined Computer Systems". In *Int. Sym. on Computer Architecture*, pages 117–131, 1983.

SESSION 8A

PERFORMANCE EVALUATION

TRANSMISSION TIMES IN BUFFERED FULL-CROSSBAR COMMUNICATION NETWORKS WITH CYCLIC ARBITRATION

A.J. Field   P.G. Harrison

Dept. of Computing
Imperial College of Science, Technology and Medicine
180, Queen's Gate, London SW7 2BZ., U.K.
email {ajf,pgh}@doc.ic.ac.uk

Abstract—*In this paper we consider the distribution of message transmission times in buffered full cross-bar interconnection networks with cyclic arbitration in which the input buffers are serviced in a 'round robin' fashion. The system is modelled as an open queueing network in which the queues appear at the network outputs and with the cyclic arbiter being modelled by queue jumping. We obtain the Laplace Transform of the transmission time by deriving a conditional Laplace Transform and solving by the use of a generating function. The density function is then enumerated by numerical inversion and compared with similar results from a simulation model. The analysis is then extended to general service times by modelling each output as a LCFS queue with a suitably modified arrival rate. In the special case of exponential service times, this model is less versatile than the previous one since it only works in the case where the jump probability is fixed. In this case, however, it is shown to produce the same result as the original.*

1   Introduction

The objective of the work presented here is to quantify the behaviour of a 'cyclic' arbitration scheme within a full crossbar communication network with buffering. This work follows on naturally from [5] which considered a circuit-switched configuration in which transmitters wait until completion of their message transfer is complete. The cyclic arbitration system is conceptually more complicated than that of FCFS but has a particularly straightforward implementation in hardware using a multi-phase clocking arrangement as described in [4].

Whilst communication networks of the type we describe have been analysed to obtain both resource-based and task-based performance measures, see for example [6,9,11,2] and [10] respectively, these analyses have assumed FCFS arbitration, either in keeping with the physical system structure, or as an approximation to the actual discipline used. As expected, the resource-based measurements are the same in both cases (the corresponding queueing disciplines for each are work-conserving). However, as will be described in more detail in Section 2, the cyclic arbiter allows some tasks to jump the queue and obtain service ahead of other tasks already queueing for service.

The rest of the paper is organised as follows: In Section 2 we describe the architecture of the communication networks in question and show how the system can be modelled by an open single-server queue with overtaking. In Section 3 we derive the Laplace Transform of the transmission time density function for the cyclic arbitration scheme. This is obtained from a conditional Laplace Transform using a generating function together with the marginal equilibrium queue length probabilities. In Section 4 we use a numerical inversion method to obtain the distribution function of the transmission time through the network for varying network loads. These are compared with similar results from a simulation model of the system, and also with an equivalent system with FCFS arbitration. Section 5 shows how the anlysis can be extended to general service time distributions. This model is more versatile in this sense, but is more difficult to enhance than the previous model in the special case of exponential service times because of the inherent assumptions about independent and identically distributed delays in its delay cycle analysis. The conclusions from the paper are laid out in Section 6.

2   The System

We consider the well-known full crossbar interconnection network, with x input and output ports, in which messages are placed in FIFO buffers prior to being transmitted. The buffers enable the arbiter to serve messages in the order in which they arrive; at the same time they help to smooth the traffic through the network and hence improve the throughput. If starvation is to be avoided, however, a second arbitration scheme is required at the network inputs in the event that one of these buffers fills: messages blocked by the full buffer need to be admitted to the output queue fairly when buffer space is eventually released.

An alternative approach, which we explore here, uses a single arbiter and locates the message buffers at the *input* side of the network. Each such buffer is divided into x separate FIFO buffers—messages for output $j, 1 \leq j \leq x$, are then queued at the back of buffer j.

The operation of the arbiter is now similar to that of the well known token passing access method, see for example [8]: we envisage that associated with each network output (i.e. each transmitter) is a token, and that in order to transmit a message to that output, the input must first acquire the token. When there are no messages for an output, the corresponding token circulates around the inputs in the order 1,2,...,x,1,2,... and so on, until such a message arrives. When a message for the output appears at one of the inputs, that input waits for the token to arrive and then holds it for the duration of the transmission. During this time other messages for the same output may be loaded into their respective input buffers. Also, at the same time, messages originating from other inputs and destined for other outputs may be being transmitted.

When the message has been transferred to the receiver the input releases the token which then continues circulating the inputs, even though it may have a second message for that output in its buffer. In theory, an input may hold all x tokens, that is it has the capacity to service all of its input buffers simultaneously. This is in contrast to an unbuffered crossbar where each input can be connected to at most one output at any time.

The 'cyclic' arbitration scheme is starvation free since every transmitter has the opportunity to transmit a message on each cycle of the token.

2.1   Representation as an Open Queueing Model

An exact analysis of this system as a queueing network would be intractable since we would have to represent explicitly the state of every input queue in the system. We therefore choose an approximate analysis in which we model the system as though the queues were located at the outputs only.

To obtain the transmission times we follow the progress of a special 'monitored' task (a message) from its arrival at an output queue to its service (transmission) completion. We now note that once the task is queued, a newly arriving task may jump ahead of the monitored task by virtue of the cyclic arbiter. Our approximating assumption in this respect is that a 'jump' of this sort occurs with a probability, p, that remains fixed throughout the task's lifetime in the system. Note that we do not allow preemption since the start of a message transfer corresponds to the originating input holding the required output token. A discussion of the choice of value for p is given in Section 3.1 below.

Our model of the system is therefore an open queueing network comprising x independent and identical Poisson input streams, each with rate λ, and x single-server queueing centres, each with service rate μ (Figure 1). An arriving task is assumed to select one of the x service centres with equal probability (i.e. $1/x$). Because of the branching properties of the Poisson process, this enables us to decompose the system into x independent and identical single-server queueing systems, each with a Poisson arrival process with rate $x(\lambda/x) = \lambda$ and service rate μ. The analysis can then proceed by examining just one single-server queue, but with the unusual 'jump' queueing discipline.

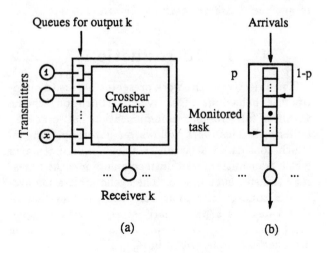

Figure 1: The Queueing Model

We consider first a general method which can be applied in the case of exponential service times. In Section 5 we present an alternative, but less versatile, formulation which handles general service times.

3 Response Times with Exponentially Distributed Message Lengths

In keeping with the above discussion we consider an open single-server queue with a Poisson arrival process with rate λ, and negative-exponentially distributed service times with mean $1/\mu$. We further identify a special task whose progression through the queue we shall monitor as the basis of our analysis. The general method is similar to that in [1].

Let T denote the response time random variable of a task, i.e. the sum of the tasks queueing time and service time, and define

$$F_n(t) = P(T \leq t \mid N = n)$$

where the random variable N is the number of tasks in the queue ahead of the monitored task at its arrival instant. Consider now the intervals $[0, h)$ and $[h, t+h)$ where time 0 is the arrival instant of the monitored task. During the initial interval $[0, h)$, as $h \to 0$, there are two possible events: firstly, an arrival with probability $\lambda h + o(h)$, and secondly a service completion with probability μh. Moreover, if an arrival occurs we make the approximation that the arriving task is serviced before the monitored task with a fixed probability $p = (x-1)/2x$.

A fuller justification of this approximation is deferred until Section 3.1 but it may be argued intuitively as follows: if the new arrival originates from the same input as the monitored task, M say, then it is certain to be be serviced after M because the input queues are FCFS. If it arrives on any other input (with probability $(x-1)/x$), there will be, on average, a 50% chance of it being served ahead of M and a 50% chance of it being served after M—an overall probability of $(x-1)/2x$. The approximation gets worse as $\lambda/\mu \to 1$, but is found to be accurate under light to moderate loading—see Section 3.1 for a more detailed explanation.

By the law of total probability, and using the fact that the forward recurrence time of an exponential random variable is an identical exponential random variable, we have

$$
\begin{aligned}
F_n(t+h) &= (1 - (\lambda + \mu)h)F_n(t) \\
&+ p\lambda h F_{n+1}(t)
\end{aligned}
$$

$$
\begin{aligned}
&+ (1-p)\lambda h F_n(t) \\
&+ \mu h F_{n-1}(t)
\end{aligned}
$$

for $n \geq 1$. If we now define $f_n(t) \equiv \frac{dF_n(t)}{dt}$ then we obtain

$$f_n(t) = -(p\lambda + \mu)F_n(t) + p\lambda F_{n+1}(t) + \mu F_{n-1}(t) \quad (1)$$

For n=0, the arriving task inevitably enters service immediately and so

$$F_0(t+h) = (1 - \mu h)F_0(t) + \mu h$$

since a new arrival cannot effect the response time and a service completion certainly results in $T \leq t$. Thus

$$f_0(t) = -\mu F_0(t) + \mu$$

This implies, as we already know, that $F_0(t) = 1 - e^{-\mu t}$. Now let $f_n(t)$ have Laplace Transform $L_n(s)$. Then, by a simple integration by parts, we can establish that $f'_n(t)$ has Laplace Transform $sL_n(s) + f_n(0)$. Moreover, for $n \geq 1$, the response time is a sum of two or more exponential random variables with parameter μ. Hence its distribution is a mixture of Erlang-k distributions with $k \geq 2$. But, the density function of an Erlang-k random variable vanishes at the origin and so $f_n(0) = 0$ for $n \geq 1$. We therefore have (after differentiating equation 1 above)

$$(s + p\lambda + \mu)L_n(s) = p\lambda L_{n+1}(s) + \mu L_{n-1}(s) \quad (n \geq 1)$$

with $L_0(s) = \mu/(s + \mu)$. To solve this recurrence, let $G(z) = \sum_{n=0}^{\infty} L_n(s)z^n$. Multiplying the recurrence throughout by z^n and summing over $n \geq 1$ now yields

$$
\begin{aligned}
&(s + p\lambda + \mu)[G(z) - L_0(s)] \\
&= p\lambda z^{-1}[G(z) - L_0(s) - zL_1(s)] + \mu z G(z)
\end{aligned}
$$

Thus,

$$G(z) = \frac{(-(s + p\lambda + \mu)z + p\lambda)L_0(s) + p\lambda z L_1(s)}{\mu z^2 - (s + p\lambda + \mu)z + p\lambda}$$

Now, we know $L_0(s)$, but must find one more equation in order to determine the value of $L_1(s)$ and hence $G(z)$. We therefore use the analyticity of $G(z)$ inside the unit disc: the denominator above vanishes at $z = r$ and $z = r'$, where r and r' are the roots of the quadratic equation $Q(x) = 0$, with

$$Q(x) = \mu x^2 - (s + p\lambda + \mu)x + p\lambda$$

Without loss of generality, let $r \leq r'$. Now, $Q(x)$ is positive as $x \to \infty$, and for all $x \leq 0$. But $Q(1) = -s$ so that $r \leq 1 \leq r'$ and for $s > 0$, $r < 1 < r'$.

A necessary condition for the analyticity of $G(z)$ for $|z| < 1$ is therefore that the numerator also vanishes at $z = r$. Hence, using the quadratic equation,

$$-\mu r^2 L_0(s) + p\lambda r L_1(s) = 0$$

so that $p\lambda L_1(s) = \mu r L_0(s)$. This serves to define $L_1(s)$ and hence $G(z)$.

Note, in the special case that p=0 (instead of $(x-1)/2x$), we obtain, since $L_0(s) = \mu/(s+\mu)$,

$$G(z) = \frac{\mu}{s + \mu - \mu z}$$

Thus, in the steady state, the response time density has Laplace Transform

$$L(s) = \sum_{n=0}^{\infty} p_n L_n(s)$$

where p_n is the stationary probability mass function of the queue length, by the random observer property, see for example [12], i.e. $p_n = (1 - \rho)\rho^n$, where $\rho = \lambda/\mu$. Hence,

$$
\begin{aligned}
L(s) &= (1 - \rho)G(\rho) \\
&= \frac{\mu - \lambda}{s + \mu - \lambda}
\end{aligned}
$$

as expected, since the queueing discipline is FCFS at $p = 0$. In the general case we have

$$L(s) = \sum_{n=1}^{\infty} P(N = n)\, L_n(s) + P(N = 0)\, L_0(s)$$

But, since a task in service cannot be preempted, $P(N = 0) = p_0 = 1 - \rho$, and we have

$$
\begin{aligned}
L(s) &= \sum_{n=1}^{\infty}\sum_{q=n}^{\infty} P(Q = q)\, P(N = n \mid Q = q)\, L_n(s) \\
&\quad + (1 - \rho)\, L_0(s)
\end{aligned}
$$

by the law of total probability, where Q is the equilibrium queue length random variable. A new arrival can arrive on *any* input to the crossbar and so we make the approximation that $P(N = n \mid Q = q) = 1/q$ ($n \le q$). For small to moderate values of Q (which occur with higher probability under light to medium loads) the approximation is good since the inputs will be relatively lightly populated with messages for the output in question, so that in the model, a new arrival will be roughly equally likely to jump into any position in the queue. Under heavy loads, the range of N and Q will be higher: although the equal probability assumption is less accurate in this case, $1/q$ will at least be small

for large q, so the probability of N being small when q is large will vanish in the limit, as we would want. Hence,

$$
\begin{aligned}
L(s) &= \sum_{n=1}^{\infty}\sum_{q=n}^{\infty}(1 - \rho)\frac{\rho^q}{q}L_n(s) + (1 - \rho)L_0(s) \\
&= (1 - \rho)\sum_{n=1}^{\infty}\int_0^\rho L_n(s)\sum_{q=n}^{\infty}z^{q-1}dz + \\
&\quad (1 - \rho)L_0(s) \\
&= (1 - \rho)\int_0^\rho \sum_{n=1}^{\infty} L_n(s)\frac{z^{n-1}}{1 - z}dz + \\
&\quad (1 - \rho)L_0(s) \\
&= (1 - \rho)\int_0^\rho \frac{G(z) - L_0}{z(1 - z)}dz + (1 - \rho)L_0(s)
\end{aligned}
$$

(Note that we are able to interchange the summation and integral because the series is absolutely convergent at $|z| < 1$.) We now simplify the integration term:-

$$
\begin{aligned}
\frac{G(z) - L_0(s)}{z} &= \frac{p\lambda L_1(s) - \mu z L_0(s)}{\mu(z - r)(z - r')} \\
&= \frac{(r - z)\mu L_0(s)}{\mu(z - r)(z - r')} \\
&= \frac{L_0(s)}{r' - z}
\end{aligned}
$$

The above integral is therefore

$$
\begin{aligned}
\frac{L_0(s)}{r' - 1}\int_0^\rho &\left\{\frac{1}{1 - z} - \frac{1}{r' - z}\right\} dz \\
&= \frac{L_0(s)}{r' - 1}\left[\log_e\frac{r' - z}{1 - z}\right]_0^\rho
\end{aligned}
$$

from which we obtain

$$L(s) = (1 - \rho)L_0(s)\left\{1 + \frac{1}{r' - 1}\log_e\frac{r' - \rho}{r'(1 - \rho)}\right\} \quad (2)$$

where

$$
\begin{aligned}
L_0(s) &= \frac{\mu}{(s + \mu)} \\
r' &= \frac{x + \sqrt{x^2 - 4p\mu\lambda}}{2\mu}
\end{aligned}
$$

and where $x = s + \mu + p\lambda$.

3.1 On the Calculation of p

In the above analysis, we have assumed that an arriving task is serviced ahead of the monitored task with a fixed probability $p = (x - 1)/2x$.

For lightly loaded systems, we would expect the approximation to be reasonably accurate: Consider the situation in which the monitored task, M say, is the only task queued for the (busy) output—a common siuation under light loading. Now consider a new arrival, A. For both M and A, the probability that they originate from input i is $1/x$ for all $1 \leq i \leq x$ by virtue of the random arrival process. Furthermore, if A originates from the same input as M, it will certainly be serviced after M, because the input queues are FCFS. Therefore, summing over all other possible inputs we obtain:

$$\sum_{n=1}^{x-1} \frac{1}{x}\left(\frac{n}{x}\right) = \frac{x-1}{2x}$$

which is the same as the intuitive expression for p we gave earlier.

In a heavily loaded system, the approximation will be good for the early part of the monitored tasks lifetime in the system, since by slicing off equal numbers of customers from each queue (assumed in equilibrium) we have a similar situation to that of the lightly loaded system. However, as the task nears the front of its input queue, and as further arrivals queue up behind it and in other queues, so new arrivals *should* be progressively less likely to be serviced ahead of it, hence the less precise the fixed p assumption becomes. We would therefore expect the model's accuracy to deteriorate as $\lambda/\mu \to 1$ giving pessimistic predictions; this effect can be seen in the density functions derived below.

4   Analysis

In the absence of an obvious analytical inversion of L_s we appeal to the numerical method of [3].

In order to validate these transmission times and so assess their accuracy under different loads, the density functions derived were compared with those from a simulation model of the actual network architecture under the same assumptions about the arrival process and message lengths. The results of these comparisons are shown in Figure 2 for $x = 4$. (The value of x was chosen simply to shorten the simulation times; similar curves are easily obtainable for $x > 4$.)

For light to moderate loads (Figures 2(a) and (b)) the derived and simulation-generated curves are almost indistinguishable. As the load increases beyond $\lambda/\mu = 0.5$, the number of arrivals experienced by the monitored task during its lifetime in the system increases accordingly, and this leads to a higher number of incoming tasks jumping the queue ahead of the monitored task. It is under these conditions that the

accuracy of the model gradually deteriorates, as is illustrated by the graph of Figure 2(c) which compares the densities for $\rho = 0.8$.

We find that the analytical and simulation models correspond closely for values of $\rho \leq 0.6$ approximately, but that the separation of the curves becomes visibly evident thereafter. Further experimentation with the value of p under heavy load also reveals that the curves become further coincident as p is lowered, as we would expect from the discussion above. However, the discrepency is still apparent even for very small values of p suggesting that even a rate-dependent (but still fixed) value of p is not sufficient to abtain accurate densitites under heavy loads. It would appear that the model can only be improved in these cases by making p state dependent in some way, possibly requiring additional conditioning parameters to that of n in $F_n(t)$. This is discussed further in Section 6 below.

A further comparison of some relevance is that between the cyclic and FCFS models under light to moderate loading. The original network with FCFS arbitration can be approximated by an open queueing network in which the output buffers are modelled by unbounded FCFS queues.

Figure 3 shows the derived density functions of the cyclic and FCFS queuing disciplines for $\lambda = 1$ and $\mu = 2$. The cyclic scheme promotes some messages in the queue at the expense of others when compared to FCFS and for this reason relatively more of the distribution lies in the tail. For example, approximately 0.9% of messages have transmission times longer than 5.0 in the FCFS scheme, compared to approximately 1.4% for the cyclic scheme. The net effect this has on the transmission time density is not so pronounced, however, and we may reasonably conclude that the cyclic scheme represents a viable and 'well behaved' alternative to FCFS in this respect.

5   General Message Lengths

We now consider an alternative formulation of the problem and relax our previous assumption of exponential service times. We observe that under the assumption that the jump probability if fixed, the queueing system we have considered can be regarded in the following more general form. A customer arrives at a queue in steady state at a random position in the queue. Susequent arrivals form two independent Poisson streams with rates $(1-p)\lambda$ of lower priority which can be ignored since the tagged customer will be served first and with rate $p\lambda$ of higher priority. In this section we write λ instead of $p\lambda$ to simplify

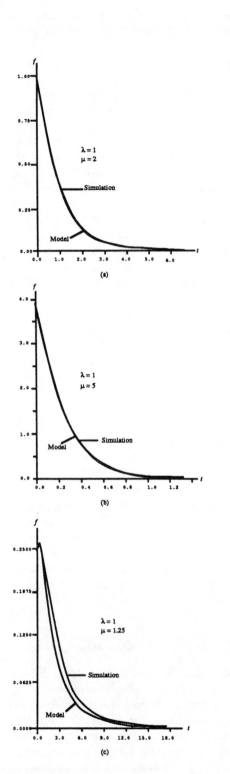

Figure 2: Comparison with Simulation for $\rho = 0.2, 0.5, 0.8$

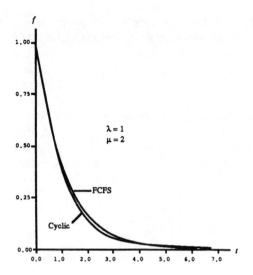

Figure 3: Comparison with FCFS $\lambda = 1, \mu = 2$

the notation. The customers ahead of the tagged customer at its arrival instant also have higher priority, i.e. will be served before the tagged customer. We define the following random variables:

S, S_i Service time random variable.

R Residual service time random variable.

A Number of customers in the queue immediately before the arrival instant of the tagged customer. For $A > 0$, the customer in service is numbered 0 and the others in first come first served order.

Z Number of arrivals during a service period.

Z' Number of arrivals during a residual service period.

M_i Sum of the service periods arising from the service period of customer i, i.e. comprising that of customer i, those of the arrivals during customer i's service period, those of the arrivals during their service periods and so on until there are no more arrivals. Thus the M_i are all independent and, for $i > 0$, identically distributed—customer 1 has a residual service time rather than a full one.

M, M_{ij} Generic random variables distributed as M_i for $i > 0$.

T Response time, i.e. the time elapsed between the tagged customer's arrival and completion of service.

Q Queueing time, i.e. the time elapsed between the tagged customer's arrival and beginning of service, so that $T = Q + S$. Thus $Q = M_0 + ... + M_{A-1}$.

We denote the probability generating function of a discrete random variable X by G_X and the Laplace transform of the distribution of a continuous random variable Y by Y^*. Thus:

$$
\begin{aligned}
Q^*(\theta) &= E[\, e^{-\theta Q}\,] \\
&= E[\, E[\, e^{-\theta Q} \mid A\,]\,]
\end{aligned}
$$

where

$$
\begin{aligned}
E[\, e^{-\theta Q} \mid A\,] &= 1 && \text{for } A = 0 \\
&= E[\, E[\, e^{-\theta Q} \mid R, A\,] \mid A\,] && \text{for } A > 0
\end{aligned}
$$

Now, labeling the service time of customer i in the queue at the arrival instant by S_i, we have

$$
\begin{aligned}
&E[\, e^{-\theta Q} \mid R, A\,] \\
&= E[\, E[\, e^{-\theta(M_0+...+M_{A-1})} \mid R, A, S_1, ..., S_{A-1}\,] \mid R, A\,]
\end{aligned}
$$

and

$$
\begin{aligned}
M_0 &= R + M_{01} + ... + M_{0Z'} \\
M_i &= S_i + M_{i1} + ... + M_{iZ} && \text{for } i > 0
\end{aligned}
$$

Thus,

$$
\begin{aligned}
&E[\, e^{-\theta Q} \mid R, A\,] \\
&= e^{-\theta R} E[\, e^{-\theta(S_1+...+S_{A-1})} \times \\
&\quad E[\, e^{-\theta(M_0-R)} \mid R\,] \times \\
&\quad \textstyle\prod_{i=1}^{A-1} E[\, e^{-\theta(M_i-S_i)} \mid S_i\,] \mid R, A\,]
\end{aligned}
$$

since the M_i are mutually independent and independent of A, M_i depends only on S_i ($i > 0$) and M_0 depends only on R. Thus we obtain, for $i > 0$,

$$
\begin{aligned}
&E[\, e^{-\theta(M_i-S_i)} \mid S_i\,] \\
&= E[\, E[\, e^{-\theta(M_{i1}+...+M_{iz})} \mid Z, S_i\,] \mid S_i\,] \\
&= E[\, M^*(\theta)^Z \mid S_i\,] \\
&= e^{-\lambda S_i(1-M^*(\theta))}
\end{aligned}
$$

since the M_{ij} are independent of S_i and since, given S_i, Z is Poisson with parameter λS_i. Similarly, $E[\, e^{-\theta(M_0-S_0)} \mid R\,] = e^{-\lambda R(1-M^*(\theta))}$. Thus we obtain

$$
\begin{aligned}
&E[\, e^{-\theta Q} \mid R, A\,] \\
&= e^{-(\theta+\lambda(1-M^*(\theta)))R} E\left[\, \prod_{i=1}^{A-1} e^{-(\theta+\lambda(1-M^*(\theta)))S_i}\,\right]
\end{aligned}
$$

and so, for $A > 0$,

$$
\begin{aligned}
&E[\, e^{-\theta Q} \mid A\,] \\
&= R^*(\theta + \lambda(1 - M^*(\theta)))S^*(\theta + \lambda(1 - M^*(\theta)))^{A-1}
\end{aligned}
$$

We obtain $M^*(\theta))$ by deconditioning the equation

$$
E[\, e^{-\theta(M-S)} \mid S\,] = e^{-\lambda S(1-M^*(\theta))}
$$

after multiplying by $e^{-\theta S}$ to get

$$
\begin{aligned}
M^*(\theta) \\
&= E[\, E[\, e^{-\theta M} \mid S\,]\,] \\
&= E[\, e^{-S(\theta+\lambda(1-M^*(\theta)))}\,] \\
&= S^*(\theta + \lambda(1 - M^*(\theta)))
\end{aligned}
$$

Finally, this yields the result

$$
\begin{aligned}
T^*(\theta) = S^*(\theta) \{\ &P(A=0) + \\
&\frac{R^*(\theta + \lambda(1 - M^*(\theta)))}{M^*(\theta)} \times \\
&[G_A(M^*(\theta)) - P(A=0)]\ \}
\end{aligned}
$$

For the case of exponential service times, we can recover the solution obtained before by noting that then $R^* = S^*$ so that $T^*(\theta) = S^*(\theta)G_A(M^*(\theta))$. The generating function $G_A(z)$ can be determined as follows:

$$
\begin{aligned}
G_A(z) &= \sum_{a=0}^{\infty} z^a P(A=a) \\
&= (1-\rho)\left\{1 + \sum_{q=1}^{\infty} \frac{\rho^q}{q} \sum_{a=1}^{q} z^a\right\} \\
&= (1-\rho)\left\{1 + \frac{z}{1-z} \sum_{q=1}^{\infty} \left[\frac{\rho^q}{q} - \frac{(\rho z)^q}{q}\right]\right\} \\
&= (1-\rho)\left\{1 + \frac{1}{z^{-1}-1} \log_e \frac{z^{-1}-\rho}{z^{-1}(1-\rho)}\right\}
\end{aligned}
$$

In the case of exponentially distributed service times, with parameter μ, the equation for $M^*(\theta)$ yields $M^*(\theta)$ as the smaller root of the quadratic in x:

$$
x = \frac{\mu}{\mu + \theta + \lambda(1-x)}
$$

so that $M^*(\theta)^{-1}$ is the larger root of the equation

$$
\mu z^2 - (\lambda + \mu + \theta)z + \lambda = 0
$$

i.e. $M^*(\theta) = r'^{-1}$. Thus, since in the exponential case, $S^*(\theta) = \frac{\mu}{\mu+\theta} = L_0(\theta)$, we have that $L = T^*$ with L as defined in equation 2. The models are therefore consistent in this case.

6   Summary and Conclusions

In this paper we have derived the Laplace transform of the transmission time density function of a buffered full crossbar communication network operating a cyclic arbitration scheme.

The cyclic scheme has the property that arriving messages can sometimes jump the queue and obtain service ahead of messages already queued in the system. We find that in this open model, the transmission time density function for the cyclic scheme is found to have a slightly longer tail than the equivalent FCFS scheme, although the mean transmission time for both is the same.

By validating the model with a simulation model, we find that it is accurate for light to moderate loads, but that the accuracy deteriorates as $\rho \to 1$. This is attributable to our approximating assumption that an arriving message jumps ahead of a given monitored message with a probability p which is independent of the current state of the queue. The longer the queue length on arrival, and the nearer the monitored message is to the head of the queue, so the jump probability *should* decrease in order to preserve an accurate representation of the scheme. A general analysis would therefore benefit from p being state dependent, requiring at least the solution of a differential equation for $G(z)$.

For the fixed p assumption, we noted that an alternative approach is to model the system as a LCFS queue with an arrival rate equal to λp. This enables us to handle general service time distributions, but the method is less versatile: the original approach is more easily extended to state-dependent jump probabilities than the adapted LCFS model.

A related area of interest is the analysis of unbuffered crossbar networks. In these systems, messages arriving at the network must wait for their requested output to become available before a second message can be accepted from the same transmitter. A model of this system is described in [5], under the assumption that the service times are exponentially distributed. Since the message population is fixed, this approach uses a closed queueing model and expresses the tagged customer's delay as the sum of an integral number of identical exponentially distributed service times. Curiously, in this situation the transmission time distribution is found to have a *shorter* tail than that of FCFS, which is the opposite of the open case considered here.

References

[1] E.G. Coffman, R.R. Muntz and H. Trotter. "Waiting Time Distributions for Processor Sharing Systems". *JACM* 17(1), pp. 123-130.

[2] D.M. Dias, J.R. Jump, "Analysis and Simulation of Buffered Delta Networks", *IEEE Trans on Computers*, Vol. C-30, No. 4, April 1981, pp. 273-282.

[3] H. Dubner and J. Abate. "Numerical Inversion of Laplace Transforms by Relating them to the Finite Fourier Cosine Transform". *JACM* 15(10), pp. 115-123.

[4] A.J. Field and M.D. Cripps. "Self-clocking Networks". *Proc. 1985 International Conference on Parallel Processing*, pp. 384-387.

[5] A.J. Field and P.G. Harrison. "Transmission Times in Unbuffered Crossbars with Cyclic Arbitration". *Proc. 1992 International Conference on Parallel Processing*, pp. 132-137 (Vol. 1).

[6] P.G. Harrison, N.M. Patel, "The Representation of Switching Networks in Queueing Models of Parallel Systems". *Proceedings of the 12th International Symposium on Computer Performance*, Editors P.-J. Courtois and G. Latouche, Brussels, December 1987, pp.497-512.

[7] P.G. Harrison, "On Non-uniform Packet Switched Delta Networks and the Hot-spot Effect". *IEE Proceedings*, Vol. 138, No. 3, May 1991, pp. 123-130.

[8] IEEE. "IEEE 802-4 Token Bus Access Method". *IEEE*, 1985.

[9] F.P. Kelly, "Blocking Probabilities in Large Circuit-switched Networks", *Advances in Applied Probability*, 18, pp. 473-505 (1986).

[10] C.P. Kruskal, M. Snir and A. Weiss. "The Distribution of Waiting Times in Clocked Multistage Interconnection Networks". *Proc. 1986 International Conference on Parallel Processing*, pp. 12-19.

[11] T. Lang, M. Valero, I. Alegre, "Bandwidth of Crossbar and Multi-bus Connections for Multiprocessors", IEEE Transactions on Computers, Vol C-31, December 1982, pp. 1227-1234.

[12] I. Mitrani, "Modelling of Computer and Communication Systems", Cambridge University Press, 1987.

Experimental Validation of a Performance Model for Simple Layered Task Systems[†]

Athar B. Tayyab and Jon G. Kuhl

Center for Computer Aided Design and Department of Electrical & Computer Engineering
University of Iowa, Iowa City, IA 52242

Abstract

Performance modeling for the class of parallel computations structured as simple layered task systems is considered. Specifically studied are the effects of using different forms of inter–layer sequencing mechanisms. The paper refines and generalizes an earlier analytical performance model and presents the results of experimental validation of the model on two different multiprocessor systems. The experimental validation results show that the model is quite accurate, with average prediction errors in the range of only a few percent. Finally, the paper uses the model to investigate performance tradeoff issues for layered task systems using barrier versus explicit intertask sequencing mechanisms and implemented on computing systems with differing architectural features. This study shows that the relationship between the type of sequencing mechanism, the nature of the computation, and the features of the underlying architecture, interact in complex ways to impact overall performance. Several interesting and non-intuitive results are shown.

1. Introduction

A number of past works have considered the performance analysis of general layered task systems [4,6,7,9]. Most of these models are based on queueing theory and are often too complex to provide a basic intuitive understanding of the relationship among various performance parameters. The accuracy of these models has usually been validated by means of simulation studies. Where experimental validation has been performed, reported results have generally been limited to measurements on a single system [6,12]. Furthermore, most of these studies do not consider the explicit cost of enforcing sequence control and the effect of this cost on overall application performance. It is often assumed that such costs are modest and therefore can be ignored. However, for medium-to-small grain computations, overheads associated with a particular sequence control mechanism can have significant effect on program performance. An analysis of these issues was recently reported in [10], where a simple analytical model was developed and compared with simulation results. The model to be developed and experimentally validated in this paper is based on a refinement and generalization of that model. The objective is to develop approximate, high-level, and tractable analytical performance models for layered task systems. Results of the experimental validation given later in this paper demonstrate both the qualitative and quantitative value of such models for predicting performance of a large class of parallel algorithms across a range of architectures.

The performance model developed in this paper is probabilistic in nature and uses concepts from order statistics. It compares performance of simple layered task systems using barrier-based sequencing to performance with less restrictive intertask sequencing. The modeling approach lumps complex architectural details into high level parameters which can be easily calibrated for a given system. A calibration-based approach has been previously used in [12] to predict performance of the Cm^* multiprocessor.

The performance prediction model presented in this paper has been experimentally validated against measurements on a 12 processor Alliant FX/2800 and a 12 processor Encore Multimax and the results show the model to be quite accurate. The calibration procedure used by the model requires a straight forward one time measurement of three variables on each machine. For the Alliant system, the model's average prediction error was less than 2.2% and maximum error was less than 10%. On the Multimax, average prediction error was less than 3.5% and maximum error was less than 15%. These results are described in more detail in section 4.

2. Performance Modeling

This paper is concerned with a large class of parallel computations which can be modeled as task systems with deterministic sequencing constraints expressed as Layered Directed Acyclic Graphs (L-DAG's). It is assumed that explicit synchronization mechanisms are used to enforce intertask precedence relations. That is, some tasks may have to wait for others to finish. Our architectural model is a simple shared memory multiprocessor system with n identical processors operating asynchronously and in parallel. Processors may contend for shared resources such as global memory or communication channels. We assume static construction of computational threads and therefore will not consider the effects of scheduling strategies in this work.

A probabilistic modeling approach is used to represent the complex behavior of parallel processing systems by means of stochastic processes. Task execution times are modeled as random variables, with known probability distribution and first two moments. The approach is to consider each layer in isolation, obtain an average case performance expression for that layer, and then combine the results to obtain an expression for the entire graph. In analyzing the execution of an individual layer, we will utilize concepts from the theory of order statistics.

2.1. Modeling Sequencing Overheads

As in [10], we consider two different sequencing mechanisms: explicit intertask sequencing and barrier-based sequencing. A single sequencing operation, whether intertask or barrier-based, involves a task incrementing, or decrementing, a shared counter while its successor task(s) "spinwait" for that counter to reach a certain value. This operation requires a fetch_and_ϕ primitive [5] which can read, modify, and write a memory location atomically. Such a primitive may require exclusive usage of critical system resources such as processor-

---

[†] This work was supported in part by the University of Iowa NSF/NASA, Industry/University Cooperative Research Center for Simulation and Design Optimization.

memory interconnection (IN). Contention among multiple tasks performing sequencing operations may result in some tasks waiting for system resources to become available. Let the expected task execution time be denoted by T_E, let T_W denote the expected waiting time for shared resources, and let T_S be the expected time to execute a single sequencing operation once the task has gained control of necessary resources. The total expected time for a task to execute and perform its sequencing operation is

$$T_{task} = T_E + T_W + T_S \qquad \qquad 1$$

We will now present a linear model that approximates T_W. Let μ be the mean task execution time and let σ^2 be its variance. For no variation in task execution time, i.e. $\sigma = 0$, all tasks attempt to perform sequencing operation simultaneously, resulting in maximum waiting time T_{Wmax}. Waiting time decreases gradually as variance increases, until it reaches zero or some minimum value T_{Wmin}. We model this behavior by the following two-piece linear function:

$$T_W(\sigma) = \begin{cases} T_{Wmax} - \left(\dfrac{T_{Wmax} - T_{Wmin}}{\sigma_{Wmin}} \right)\sigma & \sigma \le \sigma_{Wmin} \\[2ex] T_{Wmin} & \sigma > \sigma_{Wmin} \end{cases} \qquad 2$$

where σ_{Wmin} is defined as follows:

Let m be the *effective* number of tasks competing for sequencing resources and let $T_{first:m}$ be the shortest expected execution time among the m tasks and let $T_{last:m}$ be the longest expected execution time among the m tasks. We define σ_{Wmin} as the value of σ such that

$$T_{last:m} - T_{first:m} = T_{Wmax} \qquad \qquad 3$$

i.e. variance in task execution time is large enough to allow all but the last task to complete their sequencing operations.

The parameter m measures the effective number of tasks competing for a shared resource, e.g. shared memory. Its value depends on both application and system architecture. Let r be the throughput of the IN, i.e. r is the number of processors that can simultaneously access shared memory. For an n task system, r is defined as

$$r \equiv \frac{nt_1}{t_n} \qquad \qquad 4$$

where t_1 is the memory access time for a single processor, and t_n is the access time for single processor when total of n processors are competing for the IN. Given n and r, the value of m can be computed as follows

$$m = \frac{n}{r} \qquad \qquad 5$$

2.3. Modeling Barrier-Based Sequencing

In barrier-based sequencing of an L-DAG, a single barrier is used to synchronize all tasks within a layer. No task in the next layer can start execution until all tasks in the current layer have reached the barrier. Therefore the execution of a layer is constrained by the last of n tasks within that layer.

Let $T_{Barrier}^i$ be the expected execution time of the ith layer using barrier-based sequencing. We shall develop

our model for $T_{Barrier}^i$ from the perspective of the last task to reach the barrier. Let there be n tasks in a layer and let $T_{last:n}$ be the longest expected execution time among the n tasks. We have

$$T_{Barrier}^i = T_{last:n} + T_W + T_S \qquad 5$$

where T_W and T_S denote the waiting and sequencing time of the last task, respectively.

We will now apply this analysis to a specific example. Consider a single layer with n tasks. Let m be the effective number of tasks competing for barrier sequencing resources. If task execution times are i.i.d. random variables with gaussian distribution, mean μ, and variance σ^2, then from the results given in [8], we have

$$T_{first:m} = \mu - \sigma \sqrt{\log_e m} \qquad 6$$
$$T_{last:m} = \mu + \sigma \sqrt{\log_e m}$$

Assuming $T_S = h$, $T_{Wmax} = (m-1)h$, and $T_{Wmin} = 0$, we have from equations (2), (3) and (6)

$$\sigma_{Wmin} = (m-1)h \ / \ 2\sqrt{\log_e m}$$

and

$$T_W(\sigma) = \begin{cases} (m-1)h - 2\sqrt{\log_e m} & \sigma \le \sigma_{Wmin} \\[1ex] 0 & \sigma > \sigma_{Wmin} \end{cases}$$

Putting it all together, we have from (5)

$$T_{Barrier}^i = \begin{cases} \mu + \sigma\sqrt{\log_e n} - 2\sigma\sqrt{\log_e m} + mh & \sigma \le \sigma_{Wmin} \\[1ex] \mu + \sigma\sqrt{\log_e n} + h & \sigma > \sigma_{Wmin} \end{cases} \qquad 7$$

For shared memory systems using *CREW* or *EREW* model for shared memory updates [1], the nature of barrier sequencing makes $m \approx n$. For m = n, the above result simplifies to the one reported in [10].

2.4. Modeling Intertask Sequencing

Whereas in barrier-based sequencing all tasks in a layer have to wait for the completion of all tasks in the previous layer, the intertask sequencing requires only that tasks wait for their predecessors in the previous layer to finish execution. Clearly, this is a less restrictive scheme but the total number of sequencing operations per layer is greater than for barrier-based sequencing. This behavior introduces interesting tradeoffs. Contention for shared memory is now spread over n variables compared to a single variable in the barrier-based sequencing. Each of these n variables is accessed k times compared to n accesses for the single variable. This may result in reduced contention per variable, however one must consider the bandwidth of the processor-memory interconnection network for interlocked memory operations. For limited throughput networks, the waiting time or the delay experienced by intertask scheme may be proportional to $O(nk)$, while for high throughput networks, it will be proportional to $O(k)$.

Let us assume that task execution times are i.i.d. random variables and that each task in a layer performs exactly k sequencing operations. On the average, we can approximate the performance of a layer from the performance of one of its k-barriers as follows:

Let $T_{Intertask}^i$ be the expected execution time of the ith layer using intertask sequencing.

$$T_{Intertask}^{i} = T_{last:k} + T_W + kT_S \qquad \textbf{8}$$

where $T_{last:k}$ is the longest expected execution time among k tasks and T_W is given by equation (2).

We will apply this analysis to a specific case, as shown earlier for the barrier-based sequencing. Consider a single layer with n tasks and let task execution times be i.i.d. random variables with gaussian distribution, mean μ, and variance σ^2. Let m be the effective number of tasks competing for the sequencing resources. Assuming $T_S = h$, $T_{Wmax} = (m-1)kh$ and $T_{Wmin} = 0$, we have

$$\sigma_{Wmin} = (m-1)kh \,/\, 2\sqrt{log_e m}$$

and

$$T_W(\sigma) = \begin{cases} (m-1)kh - 2\sigma\sqrt{log_e m} & \sigma \le \sigma_{Wmin} \\ 0 & \sigma > \sigma_{Wmin} \end{cases}$$

Putting it together, we have from (6), (8) and above

$$T_{Intertask}^{i} = \begin{cases} \mu + \sigma\sqrt{log_e k} - 2\sigma\sqrt{log_e m} + mkh & \sigma \le \sigma_{Wmin} \\ \mu + \sigma\sqrt{log_e k} + kh & \sigma > \sigma_{Wmin} \end{cases} \qquad \textbf{9}$$

Equation (9) applies to layers second through (M-2). Layer M-1 is sequenced using an n-barrier so that model of equation (7) applies. Due to variance of the k-barriers in layers 2 through (M-2), the effective execution time variance of the tasks in (M-1)st layer will be greater than the expected variance σ^2. Effects of this increased variance are considered in [11] and will not be addressed in this paper.

3. Experimental Validation

The analytic performance models presented in this paper were validated with experiments on two commercial multiprocessor systems: a 12 processor Alliant FX/2800 [2] and a 12 processor Encore Multimax [3]. Experiments were run on both machines with identical application setup. Tasks were scheduled statically on individual processors. Once initiated, each processor runs a tight loop, executing the task assigned to it, performing the necessary sequencing operations, and waiting for any predecessor tasks to complete before proceeding to the next layer, until all layers are executed. Each task computes scalar sum of a vector of random length. The length of these vectors was generated using a random number generator which allowed control over both the distribution type (e.g. gaussian, uniform, and exponential) and distribution parameters (mean and variance). Results are reported here only for gaussian distribution. The Chi-Square statistics for all values of mean and variance reported in this paper are well within the threshold value for the usual 5% significance level.

Timing measurements were done on the initial or parent thread. Timing probes were placed before the first layer and after the barrier at the end of last layer. Execution time for initial and final task was set equal to zero. For each timing data reported, mean of a sample size of 5,000 was taken. For a 95% confidence interval, the width of the confidence interval for any sample mean was not more than 2% of the sample mean.

3.1. Calibrating the Model

Calibration of the model requires straightforward measurements of three parameters, namely T_S, t_1 and t_n. T_S can be calibrated by timing the function that implements the basic sequencing operation. Calibrating t_1 and t_n involves measuring the maximum and minimum amount of time taken by a processor to execute the sequencing function. Given t_1 and t_n for a system, its access throughput r can be computed using equation (4).

Results of the calibration experiments on Alliant and Encore are summarized below:

Parameter	Alliant	Encore
T_S (or h)	3.0 μsecs	9.0 μsecs
r	1.4	4.3

The values of r clearly indicates the significant difference between Alliant and Encore in terms of bandwidth of processor-memory interconnection for interlocked operations [2,3]. A higher value of r predicts that intertask sequencing should perform better on Encore compared to Alliant, especially for higher values of k.

4. Results

This section presents the results of the validation study. Measurements on both systems are compared with the predictions of the analytic model. Mean task execution time was chosen so that the ratio μ/h represents small-to-medium grain computations. For all test cases, a 7 layer simple L-DAG was used with execution time of the initial and final task set to zero. For each test, total execution tine of the task graph is plotted against its *coefficient of deviation*, which, for a fixed mean μ, is defined as follows:

$$\text{Coefficient of Deviation} \equiv \frac{2\sigma}{\mu} * 100\,\%$$

4.1. Barrier-Based Sequencing

Figures 1a and 1b present the results for tests on Alliant and Encore respectively. Both tests use similar values of μ/h, n and m (due to contention for single barrier variable, $m \approx n$). Both systems show similar behavior for low values of variance, where barrier-based sequencing takes advantage of increasing variance to reduce effective number of competing tasks to be less than n. For large values of variance, each layer is delayed by the slowest task in that layer and therefore overall execution time increases linearly with variance. These results confirm earlier findings reported in [10]. Maximum error for the prediction method occurs at the "knee" of the curve. This error is due to the two-piece linear approximation of $T_W(\sigma)$ whereas experimental data shows a more smoother behavior for $T_W(\sigma)$.

4.2. Intertask Sequencing

Figures 2a and 2b show the performance of intertask sequencing on Alliant and Encore respectively. The value of k is chosen to represent a moderate interlayer interaction. As predicted by the model in equation (9), performance of intertask sequencing is sensitive to the value of m for low values of σ, other parameters remaining constant. For the Encore, a lower value of m (or a higher value of r) results in reduced contention at low values of variance. On the other hand, for Alliant, a higher value of m results in relatively high contention at low values of variance and therefore the sequencing mechanism continues to experience reduced contention as variance increases. Consequently, the "knee" of the curve, which is proportional to m, is now at the far right side of the x-axis in Figure 2a.

4.3. Comparing the Sequencing Methods

Some interesting results are found when comparing the performance of the two sequencing methods across different architectures. This is shown in Figures 3a and 3b where the relative performance of the two methods is plotted against coefficient of deviation.

In order to better understand the results, we will develop an asymptotic expression for comparing the performance of the two sequencing methods. Assume that M is very large and for intertask sequencing, ignore the n-barrier in the $(M\text{-}1)$st layer. For barrier-based sequencing, we will assume that $m \approx n$, which is certainly true for the two systems under study. Based on equations (7) and (9) we can derive the following asymptotic expression:

$$\frac{T_{Intertask}}{T_{Barrier}} \approx \begin{cases} \dfrac{1+\beta mk+\alpha \sqrt{\log_e k} -2\alpha \sqrt{\log_e m}}{1+\beta n -\alpha \sqrt{\log_e n}} & \propto \; small \\[2ex] \dfrac{1+\beta k+\alpha \sqrt{\log_e k}}{1+\beta+\alpha \sqrt{\log_e n}} & \propto \; large \end{cases} \qquad 1 0$$

where $\alpha = \sigma /\mu$ is the normalized standard deviation, $\beta = h/\mu$ is the normalized cost of a sequencing operation, $T_{Intertask}$ is the expected execution of an L-DAG using intetask sequencing, and $T_{Barrier}$ is the expected execution of an L-DAG using barrier-based sequencing.

Experimental results show that intertask sequencing outperforms barrier-based sequencing only for low values of k/n. As predicted by equation (10), at low variance (i.e. low values of α), the performance ratio of intertask to barrier-based sequencing is $O(mk/n)$, or $O(k/r)$. Whereas intertask sequencing on Encore, a system with high throughput for interlocked operations, is able to

outperform barrier-based sequencing by about 10%, it is outperformed on Alliant by about 10%, a system with low throughput for interlocked operations.

For large variance, the asymptotic performance ratio is $O((k+\sqrt{\log_e k}) / \sqrt{\log_e n})$. Thus the restrictive nature of barriers cause the performance gap between the two methods to close as variance increases. This is true only for low values of k, since for large k, barriers still provide better performance by a ratio of $O(k/\sqrt{\log_e n})$.

5. Concluding Remarks

This paper has presented a high level analytical model to predict performance for a class of parallel computations that can be expressed as layered task systems. The model was experimentally validated against measurements on two commercial multiprocessor systems. The results show the model to be quite accurate. The results also demonstrate the effectiveness of approximate high-level models to provide both a qualitative and quantitative understanding of complex and interesting performance tradeoff issues for layered task systems under varying sequencing constraints and architectural features. These high level models were calibrated for specific architectures using simple measurements. The validation results demonstrate that a combination of high-level, architecture independent modeling approach, combined with simple calibration procedures on specific systems, can be an effective tool in performance prediction and tradeoff analysis. A major conclusion of this work is that choice of sequencing mechanisms for layered task systems can dramatically impact the performance of such systems. Although the choice of a particular sequencing mechanism is dependent on various parameters, a wide range of experiments on two different multiprocessor systems show that barriers can be an efficient sequencing mechanism for layered task systems with even small amount of inter-layer interaction. Future work in this area will consider more general layered task systems and additional sequencing mechanisms. Effectiveness of these models when applied to real applications also needs to be examined.

References

[1] G. Almasi and A. Gottlieb, *Highly Parallel Computing*, Bejamin/Cummings, Redwood City, CA, 1989.

[2] Alliant Computer Systems Corp., *FX/2800 Series System Description*, Littleton, MA, August 1991.

[3] Encore Computer Corporation, *Multimax Technical Summary*, Chicago, Illinois, January 1989.

[4] H. Jiang and L. Bhuyan, "Performance Analysis of Layered Task Graphs," *International Conference on Parallel Processing*, Vol. 3, August 1991, pp 275-279.

[5] C. P. Kruskal, L. Rudolph, and M. Snir, "Efficient Synchronization on Multiprocessors with Shared Memory," *5th ACM Symposium on Principles of Distributed Computing*, pp218-228, 1986.

[6] Victor W. Mak and Stephen F. Lundstrom, "Predicting Performance of Parallel Computations," *IEEE Transactions on Parallel and Distributed Systems*, Vol. 1, No. 3, July 1990, pp 257-270.

[7] Sridhar Madala and James B. Sinclair, "Performance of Synchronous Parallel Algorithms with Regular Structures" *IEEE Trans. on Parallel and Distributed Systems*, Vol. 2, No. 1, January 1991, pp 105-116.

[8] R.-D. Reiss, *Approximate Distributions of Order Statistics*, Springer-Verlag New York Inc., 1989.

[9] R. A. Sahner and K. S. Trivedi, "Performance and Reliability Analysis using Directed Acyclic Graphs," *IEEE Transaction on Software Engineering*, Vol. SE-8, no. 4, July 1982, pp. 319-331.

[10] Athar Tayyab and Jon Kuhl, "Analyzing Performance of Sequencing Mechanisms for Simple Layered Task Systems," Proceedings. of the *6th Int. Symposium on Parallel Processing*, March, 1992, pp 173-178.

[11] Athar Tayyab, *Performance Prediction for a Class of Parallel Computations*, PhD Thesis, Electrical and Computer Engineering, University of Iowa, Aug 1993.

[12] D. F. Vrsalovic, D. P. Siewiorek, Z. Z. Segall, E. F. Gehringer, "Performance Prediction and Calibration for a Class of Multiprocessors," *IEEE Tran. on Computers*, Vol. C-37, No. 11, Nov. 1988, pp. 1353-1365.

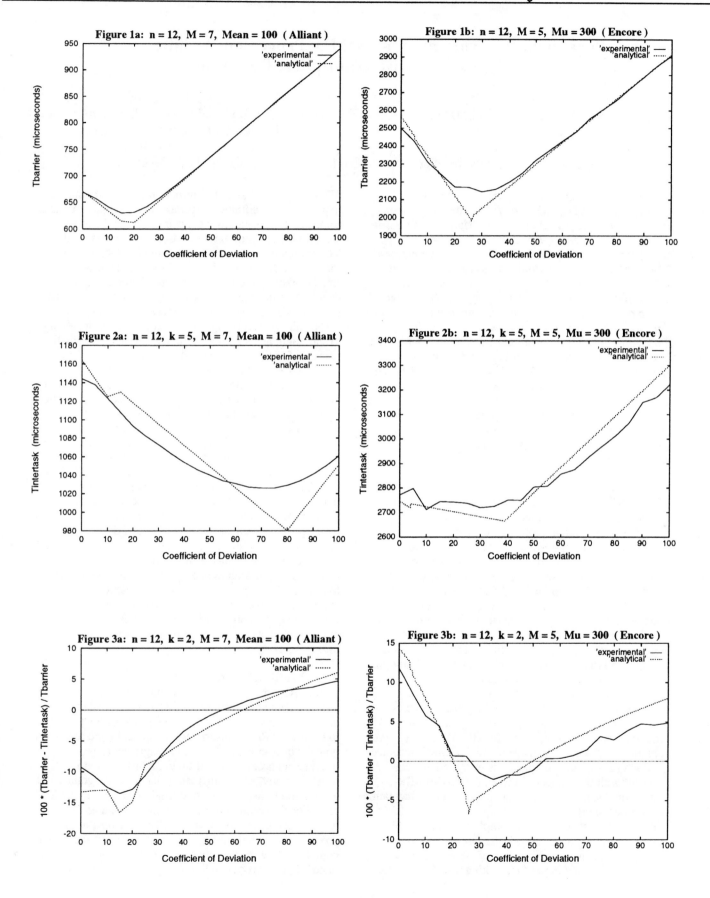

Figure 1a: n = 12, M = 7, Mean = 100 (Alliant)

Figure 1b: n = 12, M = 5, Mu = 300 (Encore)

Figure 2a: n = 12, k = 5, M = 7, Mean = 100 (Alliant)

Figure 2b: n = 12, k = 5, M = 5, Mu = 300 (Encore)

Figure 3a: n = 12, k = 2, M = 7, Mean = 100 (Alliant)

Figure 3b: n = 12, k = 2, M = 5, Mu = 300 (Encore)

Performance Evaluation of SIMD Processor Architectures Using Pairwise Multiplier Recoding

Todd C. Marek
2 Kimball Court #502
Woburn, MA 01801
(617) 981-3893
email: marek@ziggy.csl.ncsu.edu

Edward W. Davis
North Carolina State University
Department of Computer Science
P.O. Box 8206
Raleigh, NC 27695-8206

Abstract: Research in the area of massively parallel processing with a focus on processor architecture is presented. Multiplication using conventional techniques and a modified multiplier recoding scheme for 2's complement fixed-point multiplication is used to evaluate different SIMD architectures. Recoding is facilitated by a small amount of hardware which can be included at each processor even when the size of the individual processor is small. Performance figures are measured for 20-bit and 32-bit fixed-point multiplies, and for generation of Mandlebrot images using 20 bits and 32 bits of precision. Results indicate that although the 1 bit system has the highest cycle count for individual multiplies, this system also achieves the largest aggregate Mops rate when the systems are given equivalent processing power.

1. Introduction

This article presents research in the area of massively parallel processing with a focus on evaluating SIMD machines with different processing element (PE) architectures. Multiplication using both conventional techniques and a modified multiplier recoding scheme is used as a basis for the evaluations presented here. Booth's recoding scheme [1] and the modified Booth's algorithms [2],[3] are well-known and can be useful in uniprocessor systems, but are not particularly well-suited for the SIMD massively parallel environment. This is because these approaches seek to reduce the number of additions of the multiplicand to the partial product sum by identifying runs of 0s and 1s in the multiplier, but in the SIMD environment different processors will have different multipliers. The multiplication performance figures presented here are used to assess SIMD systems with processor widths of 1, 4, 8, and 16 bits. These "granularities" represent a large number of SIMD machines in which PE size and complexity is restricted in a cost trade-off with the number of PEs in the system. The architectural simulator used to conduct this experimentation is described in [4].

The recoding scheme for 2's complement fixed-point binary multiplication described here results in a reduction of execution time that is independent of the values of the multiplier presented in parallel, making it well-suited for SIMD systems. Pairs of multiplier bits are combined and require a single arithmetic operation for each pair of bits as compared to two additions per pair in shift-and-add approaches. The multiplier recoding method presented here has two features which make it unique. First, it is suited for 2's complement numbers, and second, it requires only a small amount of hardware at each processor and can therefore be included in a massively parallel system even when the size of the individual processors is very small.

The pair-wise recoding scheme requires only the ability to shift, add, and subtract. With the addition of a small amount of logic, these operations can be sequenced to efficiently implement multiplication using multiplier recoding. Other schemes which recode more than 2 bits of the multiplier require additional hardware such as adder trees or extra registers to maintain odd multiples of the multiplicand. Recoding of the 2 bits is implemented such that each pair of multiplier bits is treated separately without overlapping of bits, thus providing a potential halving of the number of operations required to form partial products. Recoding schemes based on accessing and positioning more than 2 bits, with overlapping of bits, require additional steps and do not reduce the number of addition iterations in the SIMD environment since this is a data-dependent approach. Multiplier recoding using pairs of bits is a well-known approach but has been impractical for massively parallel SIMD machines due to the hardware costs relative to the size of the PEs. A new implementation of this approach suitable for massively parallel systems of simple PEs, and an evaluation of PE granularities using multiplication as a measure of performance are the subjects of this paper.

2. Multiplication Support

Processor architectures investigated here possess register and data path widths corresponding to the granularities identified above. A full adder and logic block are provided to enable the PE to perform arithmetic and logic operations on pairs of operands. A variable length shift register provides an accumulator of up to 32 bits in length. Masked operation is implemented based on the contents of a single-bit register. A second single-bit register enables conditional execution of certain logic operations. In PEs where this register contains a "1" the logic operation proceeds as normal, and where it contains a "0" the logical complement of the operation is performed when the global instruction specifies conditional operation. Conditional operation can be used to simultaneously perform an addition in some PEs and subtraction in others, which is a key feature for the recoding scheme implemented here.

Multiplication in simple PEs may be achieved by a shift-and-add algorithm:

```
In all PEs{
  Clear shift register
  For i = 0 to N-1{
    Load mask register with multiplier bit i
    Perform masked add of shift register and multiplicand
    Shift right, storing a bit of the product
    Increment i
  }
  Store remaining product bits from shift register
}
```

The multiplication process described above is based on a sequential treatment of the individual bits of the multiplier. The multiplier recoding method is based on using a 2-bit multiplier recoder, and is shown to provide speedup approaching a factor of 2 since it enables multiplier bits to be processed in pairs rather than individually. For any given pair of multiplier bits, the desired action to be taken in the multiplication process is confined to one of only four possibilities. These possibilities correspond to the value of the pair of multiplier bits where the desired action is to form a partial product of the multiplicand and 0, 1, 2, or 3 (i.e. 00, 01, 10, or 11).

• Case 0: The partial product of the multiplicand and 0 is always 0, so these PEs are masked off, and remain masked off for the remaining steps of the iteration.
• Case 1: The partial product of the multiplicand and 1 always equals the multiplicand, which is therefore added to the partial product sum being collected in the accumulator.
• Case 2: The partial product of the multiplicand and 2 may be formed by doing a simple left shift of the multiplicand by one bit position, which is followed by adding the shifted multiplicand to the partial product sum in the shift register.
• Case 3: These partial products are formed by using the multiplier recoding principle that $X \cdot 3 = (X \cdot 4) - X$. This principle is implemented by incrementing the next pair of multiplier bits by 1, giving the effect of multiplying by 4, and subtracting the multiplicand from the partial product sum in the shift register.

These four cases appear to require two N-bit additions and an N-bit subtraction, representing an increase in operations as compared to the process described above where the multiplier bits are treated individually. It is possible to form all of the correct partial products for a pair of bits by handling all four possible cases simultaneously using masked operation and conditional operation. The SIMD implementation using the multiplier recoding technique is given by the following algorithm:

```
In all PEs{
  Clear shift register
  Load mask register with multiplier bit 0
  Preload shift register with multiplicand (masked)
  For i = 1 to N-1{
    Recode multiplier bits i and i+1
```

```
    Preshift and store bit in PEs with Case 2 (masked)
    Add shift register and multiplicand (masked, conditional)
    Shift right, storing a bit for Cases 0,1,3 (masked)
    Shift right, storing a product bit for all Cases
    Increment i by 2
  }
  Store remaining product bits from shift register
}
```

Although there appear to be more steps, the actual number of operations needed to implement this algorithm is less than that for the previous algorithm.

Hardware assistance for the multiplier recoding operation may be provided by including a simple circuit at each PE which can be loaded with the pair of multiplier bits. This circuit is designed to directly load the mask and conditional operation registers with the proper values. In order to take full advantage of such a circuit, a second mask register is also incorporated at each PE, providing one mask register, G1, that masks out PEs containing a 0 pair; one mask register, G2, that masks out PEs containing a 2 pair; and a conditional operation register, K, that allows subtraction in PEs containing a 3 pair. Table 1 gives the relationships between multiplier bits and the values that need to be loaded into G1, G2, and K. Since Case 3 produces a carry into the next pair, the values of G1, G2, and K must reflect the value of the carry in, C_i, and a value for C_{i+1} must be generated for the next pair. Logic equations for G1, G2, K, and C_{i+1} can be easily obtained in terms of M_i, M_{i+1}, and C_i. In 2's complement multiplication, it is also necessary to update the most significant bit of the current partial product sum based on whether or not the partial product is positive or negative.

Table 1. Control Bits for 2-bit Recoding.

M_{i+1}	M_i	C_i	G1	G2	K	C_{i+1}	Multiple of Multiplicand
0	0	0	0	0	0	0	+0D
0	0	1	1	0	0	0	+1D
0	1	0	1	0	0	0	+1D
0	1	1	1	1	0	0	+2D
1	0	0	1	1	0	0	+2D
1	0	1	1	0	1	1	-1D
1	1	0	1	0	1	1	-1D
1	1	1	0	0	0	1	+0D

Implementation of the multiplier recoding support and for updating the most significant bit is obtained at the expense of 5 additional flip-flops and 14 logic gates at each PE. This represents a negligible increase in hardware to 4, 8, and 16-bit processors, and even systems based on 1-bit processors can realistically incorporate this additional hardware. Several additional instructions (3 or 4 depending on granularity) must also be provided, and microcoded multiplication algorithms must be revised to take advantage of this feature.

3. Methodology and Experimental Results.

In order to maintain equality of computing resources between the granularities investigated in this research, the systems are simulated as indicated in Table 2. The product of granularity and number of PEs is constant, as is the product of number of PEs and memory per PE.

Table 2.   Experimental System Specifications.

Granularity	Number of PEs	Arrangement of PEs	On-chip PE Memory
1	16,384	128x128 X-grid	1,024 bits/PE
4	4,096	64x64 X-grid	4,096 bits/PE
8	2,048	32x64 X-grid	8,192 bits/PE
16	1,024	32x32 X-grid	16,384 bits/PE

Operand lengths in real applications may vary greatly. An operand length of 32 bits is common and may be considered to be a standard length, but non-standard lengths also warrant consideration. Performance of the SIMD systems is therefore determined for 20-bit and 32-bit fixed-point multiplies, and for generation of small Mandlebrot images using 20 bits and 32 bits of precision. Figures for the individual multiplies are given as a total number of clock cycles in Figure 1, and as an aggregate number of multiplies per second for each system in Figure 2. Mandlebrot set image generation performance figures are given in Figure 3 as a total number of clock cycles. The Mandlebrot images are produced at a resolution of 128x128 pixels using 128 iterations of the Mandlebrot algorithm.

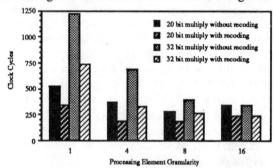

Figure 1. Cycle Counts for Individual Multiplies.

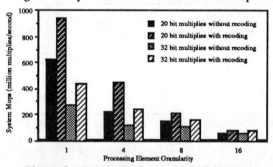

Figure 2. Aggregate System Multiply Rates.

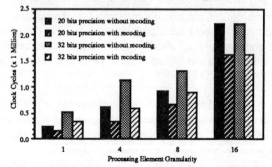

Figure 3. Cycle Counts for Mandlebrot Image Generation.

4.   Analysis of Performance

The preceding figures indicate that multiplier recoding provides a significant performance improvement over shift and add multiplication as expected. This experimentation has also provided several new insights. First, it has been demonstrated that a multiplier recoding approach can be implemented in SIMD systems with very small PEs, while maintaining the expected performance benefits. Second, quantitative information has been gathered related to SIMD systems of "equivalent processing potential" having different PE granularities. Third, it is observed that the performance improvements are not constant across the different processor granularities, and that the improvement is less than expected in the 1, 8, and 16-bit systems. These last two insights are now discussed.

Analysis of the multiplication algorithms reveals that larger PE granularities are able to save cycles during adds, logical operations, and data transfers between M-bit registers and memory locations, where M is the granularity. This is true only when the operands are at least as large as the granularity. These systems are only able to achieve their highest potential savings when the operand lengths are even multiples of the granularity. Multiplications involving 20-bit operands are expected to be somewhat less efficient on 8 and 16-bit systems where the closest multiple of the granularities are 24 and 32 bits respectively than on 1 and 4-bit systems where the closest multiple is exactly 20. Performance figures in Figure 1 for 20-bit and 32-bit multiplies using the 16-bit system are therefore identical since the 20-bit multiply must actually be implemented as a 32-bit multiply. This characteristic of the 16-bit system is also evident in the results shown in Figures 3. Performance benefits of the 1 and 4-bit granularities with regard to 20-bit operations are therefore partially a result of the length of the operands encountered.

Multibit systems are not able to save cycles during any operations involving single bits of information. Loading a single bit into a mask register takes 1 cycle regardless of the granularity of the PE. This type of single bit operation frequently occurs in both of the multiplication algorithms and accounts for most of the performance differentials observed in favor of smaller granularities. The analysis of the multiplication algorithms also shows that as the system granularity increases the number of adds, logical operations, and data transfers in multibit systems decreases relative to the total number of operations, while the number of single bit operations remains fixed. Thus there is increasingly less opportunity for the multibit systems to do what they do best as the granularity increases.

Closer inspection of the information presented in Figure 1 indicates that for 32-bit multiplies, with and without recoding, the observed cycle counts steadily decrease as granularity increases, reaching the smallest value at 16-bit granularity. The rate of decrease, however, is observed to significantly slow as granularity increases. Although not intuitively obvious, this can now be expected based on the preceding observations. The occurrence of different minima for 20-bit multiplies when recoding is used and when it is not used is explained by considering the source of savings gained by using recoding. Recoding reduces cycle counts because the number of adds are reduced. The 8-bit system beats the 4-bit system without recoding because it can do adds in fewer cycles, but loses when recoding is used because there are fewer adds required.

These observations are now formalized. Cycles for shift-and-add multiplication techniques are given by:

$$C = C_{init} + C_{load} + C_{add} + C_{sub} + C_{store} \qquad (4.1)$$

where C is the total number of cycles, C_{init} is initialization overhead, C_{load} is preloading the multiplicand in the 1-bit system and loading the blocks of the multiplier in the multibit systems, C_{add} is the additions required to form the partial product, C_{sub} is the final correcting subtraction for 2's complement multiplication, and C_{store} is storing the result from the shift register. Following the format of Equation 4.1, Equations 4.2 - 4.5 give the number of cycles required to perform a single multiplication in all PEs for 1, 4, 8, and 16-bit systems respectively, when multiplier recoding is not used. These expressions are given in terms of N, the length of the operands being multiplied.

$$C_1 = 3 + (N+1) + (N-2)(N+5) + (N+2) + (N+4) \qquad (4.2)$$

$$C_4 = 6 + \lceil \frac{N}{4} \rceil + (N-1)(\lceil \frac{N}{4} \rceil + 13) + (\lceil \frac{N}{4} \rceil + 4) + \frac{(N-4)}{4} \qquad (4.3)$$

$$C_8 = 10 + \lceil \frac{N}{8} \rceil + (N-1)(\lceil \frac{N}{8} \rceil + 9) + (\lceil \frac{N}{8} \rceil + 4) + \frac{(N-8)}{8} \qquad (4.4)$$

$$C_{16} = 18 + \lceil \frac{N}{16} \rceil + (N-1)(\lceil \frac{N}{16} \rceil + 9) + (\lceil \frac{N}{16} \rceil + 4) + \frac{(N-16)}{16} \qquad (4.5)$$

The preceding equations lead to the generalized expression of Equation 4.6, given in terms of N, the operand length, and M, the system granularity.

$$C_M \approx (M+2) + \lceil \frac{N}{M} \rceil + (N-1)(\lceil \frac{N}{M} \rceil + 10) + (\lceil \frac{N}{M} \rceil + 4) + \frac{(N-M)}{M} \qquad (4.6)$$

Note that as granularity and operand length vary, there is a significant portion of work that remains constant. As granularity increases, this constant term becomes the dominant part of the total for ordinary operand lengths.

When multiplier recoding is used, the expression of Equation 4.1 changes to:

$$C = C_{init} + C_{load} + C_{add} + C_{msb} + C_{store} \qquad (4.7)$$

where C, C_{init}, C_{load}, C_{add}, and C_{store} are as defined above and C_{msb} is necessary to allow for a possible carry out generated by the most significant pair of multiplier bits in multibit systems. Equations 4.8 - 4.11 give the number of cycles required to perform a single multiplication with multiplier recoding for 1, 4, 8, and 16-bit systems respectively.

$$C_{1R} = 5 + (N+1) + (\frac{N}{2}+1)(N+8) + (N-6) \qquad (4.8)$$

$$C_{4R} = 4 + \lceil \frac{N}{4} \rceil + \lceil \frac{N}{4} \rceil(2(\lceil \frac{N}{4} \rceil + 10)) + (\lceil \frac{N}{4} \rceil + 16) + \frac{(N-4)}{4} \qquad (4.9)$$

$$C_{8R} = 4 + \lceil \frac{N}{8} \rceil + \lceil \frac{N}{8} \rceil(4(\lceil \frac{N}{8} \rceil + 10)) + (\lceil \frac{N}{8} \rceil + 26) + \frac{(N-8)}{8} \qquad (4.10)$$

$$C_{16R} = 4 + \lceil \frac{N}{16} \rceil + \lceil \frac{N}{16} \rceil(8(\lceil \frac{N}{16} \rceil + 10)) + (\lceil \frac{N}{16} \rceil + 39) + \frac{(N-16)}{16} \qquad (4.11)$$

Equations 4.9 - 4.11 lead to the generalized expression of Equation 4.12. As in the case without multiplier recoding there is a significant portion of work that remains constant. As granularity increases, this constant term becomes the dominant part of the total for ordinary operand lengths.

$$C_{MR} \approx 4 + \lceil \frac{N}{M} \rceil + \lceil \frac{N}{M} \rceil(\frac{M}{2}(\lceil \frac{N}{M} \rceil + 10)) + (\lceil \frac{N}{M} \rceil + (2M+8)) + \frac{(N-M)}{M} \qquad (4.12)$$

5. Conclusions

Experimental results reported in this work support some prior conclusions regarding the massively parallel approach, but also provide new insights. When there are a large number of operations to be performed, the 1-bit granularity becomes the system of choice due to its high aggregate Mops rate. Conversely, when there are fewer operations to be performed the 8-bit or 16-bit granularities may be preferable due to their lower cycle counts for individual operations. The magnitude of the differences in performance reported here was unexpected. Comparative analysis of where computation time is actually spent allowed conclusive explanation of these differences.

One of the most significant issues motivating this work has been the relationship between machine structure and system performance. Machine structure in this sense includes PE granularity, PE complexity, and interconnection structure. Quantitative models of this relationship have not been found in the literature. The results presented here represent the first step towards developing models related to PE granularity and PE complexity. Other applications giving a more comprehensive view of granularity and performance, an in-depth analysis of these results, and new models of SIMD computation are reported in [5]. These models quantitatively reveal areas for improvement in various approaches to massively parallel computation.

References

[1] A.D. Booth, "A Signed Binary Multiplication Technique." *Quarterly Journal of Mechanics and Applied Mathematics*, 4(2), pp.236-240, 1951.
[2] O.L. MacSorley, "High Speed Arithmetic in Binary Computers." *Proc. of the IRE*, pp.67-91, January, 1961.
[3] L.P. Rubinfield, "A Proof of the Modified Booth's Algorithm for Multiplication." *IEEE Transactions on Computers*, 24(10), pp.1014-1015, October, 1975.
[4] T.C. Marek, "A New Simulator Workbench for Comparing SIMD Processing Element Architectures." *Proc. of the ACM Southeast Regional Conf.*, April, 1992.
[5] T.C. Marek. Comparative Evaluation of Processor Architectures for Massively Parallel Systems. Ph.D. Dissertation, Electrical and Computer Engineering, North Carolina State University, Raleigh, NC, April, 1992.

PERFORMANCE CONSIDERATIONS RELATING TO THE DESIGN OF INTERCONNECTION NETWORKS FOR MULTIPROCESSING SYSTEMS

Earl Hokens and Ahmed Louri
Department of Electrical and Computing Engineering
The University of Arizona
Tucson, AZ 85721

Abstract – *Choosing the best interconnection scheme for a multiprocessor is no easy task. This paper presents a two phase analysis to assist in this task. The two phases of the analysis are analytical, and simulation. The analytical phase introduces a new metric called the network bandwidth requirement, or nbr. The nbr is an estimate for the interconnecting network link speed of a multiprocessor. This result is used in the second phase of the analysis, different multiprocessor configurations are simulated using stochastic activity networks to verify the results of the analytical phase, and analyze the effects of contention and memory design on performance.*

INTRODUCTION

Processing need is surpassing current limits of single processor technology [1]. This fact is pushing the envelope of computer design deeper into the realm of parallel processing. The single most important issue of parallel processing is the interconnection network [1]. This paper will give a computer designer insights on how to select an interconnection network for use in a tightly coupled multiprocessor. The analysis however, can also be applied to Multicomputers and computer networks.

Our approach to this problem is from a different perspective than previous research [2]; our analysis centers on the individual interconnecting links as part of a complete multiprocessor. Rather than calculating a bandwidth for the entire network, we calculate a bandwidth requirement in words per cycle for the interconnecting links. Furthermore, we do not isolate the interconnection network for the purpose of analysis. Processor characteristics, memory issues, data consistency issues are all considered in addition to interconnection network issues.

We propose a two step method for the analysis of this problem that results in an estimate for the interconnection link speed. The first step is an analytical method that provides an initial estimate for the interconnecting link speed. We refer to this estimate

as a new metric called the network bandwidth requirement, or *nbr*. The units for the *nbr* are words per cpu cycle, or w/c. The second step is simulation.

Using the estimate for the interconnecting link speeds obtained in the analytical portion, models based on stochastic activity networks, or SANs [3] are used to simulate the multiprocessor. The processor, interconnection network and memory are modeled providing performance results for each part of these parts. The simulation is used to verify results obtained in the analytical step, and to explore contention and memory access characteristics.

EQUATION DEFINITIONS

The purpose of these analytical equations is to provide a speed estimate for the interconnection links. Four factors have been identified for this analysis: (1) cache line size, (2) memory cycle, (3) contention, and (4) data consistency issues. The analytical equations incorporate cache line size to account for transferring a cache line, and data consistency for the traffic generated to maintain the consistency of data in the system. Conspicuous in their absence are contention and memory cycle, both are deferred to the simulation phase. For equation development, the memory is assumed to have zero delay. To compensate for the effects of contention, the processors in the system are assumed to issue data requests 100% of the time, the actual probability of a data request should be somewhat less than one. In an actual network a percentage of these transactions will be rejected, but so long as the sum of the probability of rejecting a request and the actual probability of a request do not exceed one, the equations compensate for not factoring in contention and will provide an accurate estimate.

Overhead

Overhead represents the average number of transactions needed to maintain data consistency for each memory transaction initiated. A representative scheme was developed through a survey of directory based coherency schemes [4] and used to develop these equations. The analysis separates consistency transactions from data transfers to simplify equation development. Hierarchical interconnection network characteristics are incorporated into the equations, but will not be discussed. Units for all the overhead equations are words per memory request, or w/mr.

Overhead generated by reading dirty shared data. When a processor attempts to read dirty shared data the dirty item may be stored in the processor cache, or it may be stored elsewhere. In either case a dirty item is considered unusable. For the reading of dirty shared data the hit ratio is unimportant, because an attempt to read dirty data contained in a processor's cache is a forced miss. In the representative scheme an attempt to read dirty shared data is accomplished in *three* steps. First, the processor cache pair issues a read to main memory. Second, upon receipt of read request, main memory recognizes that the data is dirty, and forwards the request to the processor with the clean copy, which sends the cache line containing the data to the processor that initially requested the data. This is a net of *three* transactions. However, since data transfers are considered separately, overhead due to reading dirty shared data is(oh_{rds}):

$$oh_{rds} = 2 \times P_r \times P_s \times P_d \quad \text{w/mr.} \quad (1)$$

For each instruction issued, P_r is the probability an instruction is a read. Of the data being read, P_s is the probability it is shared and P_d is the probability that the shared data is dirty. For every read of this type, *two* one word transactions are generated.

Similarly, equations for overhead generated by: reading shared data (oh_{rs}), reading unshared data (oh_{ru}), writing shared data ($ohinv_i$), and writing unshared data (oh_{wu}) were developed.

$$oh_{rs} = P_r \times P_s \times (1 - P_d) \times (1 - h_i) \quad \text{w/mr.} \quad (2)$$

$$oh_{ru} = P_r \times (1 - P_s) \times (1 - h_i) \quad \text{w/mr.} \quad (3)$$

$$ohinv_i = (1-P_r) \times (2P_s(1+S*N)) \times (1 - \frac{C_i}{N}) \quad \text{w/mr.} \quad (4)$$

$$oh_{wu} = (1 - P_r) \times (1 - P_s) \times ((1 - h_i) + 2h_i) \quad \text{w/mr.} \quad (5)$$

h_i is the global hit ratio of the caches up to level i, S is the percentage of processors sharing data, N is the number of processors, and C_i is the number of processors serviced by a cache at level i.

By summing these equations a unique representation of the overhead at level i may be obtained. This results in

$$oh_i = oh_{rds} + oh_{rs} + oh_{ru} + ohinv_i + oh_{wu} \quad \text{w/mr.} \quad (6)$$

Data transfers. A data transfer must be initiated for every *miss* and also every *hit* on shared dirty data. For the shared dirty data case, *two* data transfers are initiated. First the processor sends a clean copy to the main memory, which forwards a copy to the requesting processor. Factoring in instruction fetches, which incur no overhead, and data transfers results in the following equation:

$$D_i = 2 \times ((1 - h_i) + h_i \times P_s \times P_d) \quad \text{t/mr.} \quad (7)$$

The units are data transfers per memory request, or t/mr.

Network bandwidth requirement - nbr.

We now introduce a new metric called the network bandwidth requirement, or nbr. The nbr is the average number of words per cpu cycle that each link should be capable of transferring to or from each source. For single level interconnection networks, the only source is the processor private cache pair. The equation for the nbr was derived through a realization of factors that will increase traffic on an interconnecting link. Each processor contributes $D_i L_i + oh_i$ transactions into the interconnection network where L_i is the cache line size in words. The overall network bandwidth requirement for each link is affected by: the number of sources per link (N_{L_i}), the average path length, P_{L_i}, that a transaction travels, the number of processors covered by the cache at this level, C_i, and the number of instructions each processor can execute per cycle (nbr_0). The P_{L_i} calculations are based on characteristics of the interconnection network [5]. This yields the following equation:

$$nbr_i = nbr_0 \times P_{L_i} \times N_{L_i} \times C_i \times (D_i \times L_i + oh_i) \quad \text{w/c.}$$

ANALYTICAL RESULTS

The results of the equations were generated using representative values for the different parameters found in literature [6, 7] and plotted. Networks evaluated include a bus, crossbar, Multistage Interconnection Network (MIN), and a 2D mesh. Analysis of the graphs showed that the curves maintain a consistent shape for each of the interconnection networks the nbr values are of course different. Figure 1 shows the typical shape of the curves for the nbr results. The influence each parameter has on interconnection network traffic relative to each other changes significantly as the multiprocessor size grows. The slopes of these lines increases and decreases with the number of processors. For small multiprocessors, the parameter P_r exerts the greatest influence on interconnection network performance. However, as the number of processors grows, the effect of P_s, and S on the interconnection network performance also grows, eventually becoming the dominant parameters. This information should be a clue to designers

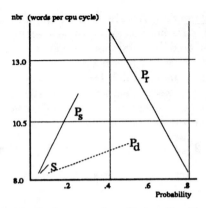

Figure 1: Typical curves for *nbr* vs. P_r, P_d, P_s, and S for a *64*-processor multiprocessing system.

as to which parameters need to be addressed in the design of multiprocessors.

The *nbr* results are graphed in Figures 2, and 3.

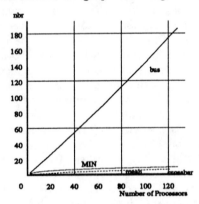

Figure 2: *nbr* vs. the number of processors. All network types are graphed.

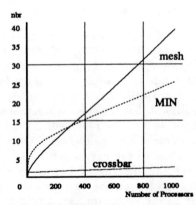

Figure 3: *nbr* vs. the number of processors. All network types are graphed.

As expected the bus is the worst performer, and the crossbar the best. The bus is the least capable interconnection strategy, but it is still a feasible choice for very small numbers of processors. The mesh is a better performer than the MIN up to about 300 processors, but the difference is not substantial. A

designer must remember that performance is not the only factor to consider in choosing an interconnection network.

The conversion of these *nbr* results to an interconnecting link speed is very simple. The *nbr* represents the number of one word transactions each link must be capable of handling in a cpu cycle to sustain processor execution. Therefore, if the cpu has a 20MHz clock cycle, the interconnecting links should have a cycle of $20 \times nbr$ MHz. Obviously, it is impossible to operate interconnecting links at some of the speeds indicated given current technological constraints, these configurations may be eliminated from further consideration. Once impossible configurations have been eliminated, simulation can be used to further investigate the remaining candidate networks.

SIMULATION RESULTS

To verify the results of the analytical section, the tool *UltraSAN* [3] was used. *UltraSAN* is a modeling and simulation tool based on stochastic activity networks, or SANs [3]. As a gauge for measuring the performance of the processors, a single processor bus was used. The interconnecting link speed used in this model evaluation was assumed to be equal to the processor speed. This uniprocessor system achieved a 66% processor utilization, a 12% memory utilization and a 3% link utilization. For all graphs of processor performance data a light line was included for the processor utilization of the uniprocessor bus model as a reference to this gauge. Furthermore, all results are graphed with their corresponding error bars. Some error bars may not be visible. This indicates that the simulation results are very accurate. Simulation models for all networks mentioned previously were developed and with the corresponding *nbr* values used to generate the results.

Performance results were obtained for processor, memory, and link utilizations. An accurate prediction of the *nbr* will result in a processor utilization that is equal to the gauge for all multiprocessor types and sizes.

Figures 4, 5, 6, 7, show the simulation results for the bus, crossbar, MIN, and mesh interconnection networks respectively. The *nbr* is a very good estimate, maintaining the processor utilization for all interconnection networks at a level close to the gauge. The memory utilization also remains at a fairly constant level of performance. The steady levels of performance of the processor and memory utilization demonstrate that the *nbr* value accounts for expected increases in the interconnection network traffic. If it had not, the utilization factor for both would drop since the interconnection network would necessarily have a higher utilization factor. The link utilizations for all models except the crossbar also remain essen-

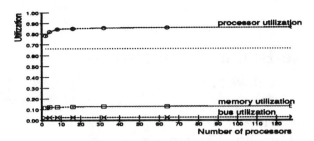

Figure 4: Simulation results for the bus multiprocessor with the capability of interleaving bus transactions.

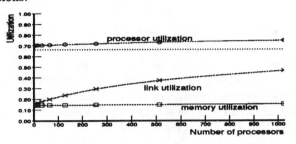

Figure 5: Simulation results for the crossbar multiprocessor model.

tially constant, which is in accordance with the previous statement. The link utilization in the crossbar increases with the number of processors, and is attributed to the invalidation/acknowledgement traffic and the method by which the link utilization is calculated.

CONCLUSIONS

The *nbr* calculations are shown to be accurate and flexible. Processors with different characteristics may be considered easily helping the designer to determine the best possible mating of processor to interconnection network strategy to obtain maximum performance within the current technological constraints. It can also be determined if the *nbr* of an interconnection network exceeds what is required by a given processing load.

Using the analytical and simulation phases in concert offers excellent aid to designers for the selection

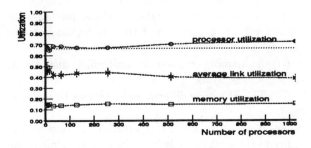

Figure 6: Simulation results for the MIN multiprocessor model.

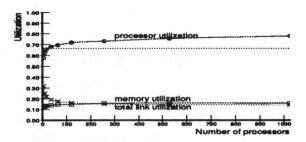

Figure 7: Simulation results of the mesh multiprocessor model.

of an interconnecting network and the determination of the interconnecting link speed requirements of a multiprocessing computer. The ease of use, the accuracy of the results, and the small time investment required to obtain the results makes this two-phase analysis an ideal starting block for multiprocessor design.

ACKNOWLEDGEMENT: We acknowledge the contributions of Dr. William Sanders and the *usan* group for assisting us in the use of *UltraSAN*.

References

[1] L. N. Bhuyan, Q. Yang, and D. Agrawal, "Performance of Multiprocessor Interconnection Networks," *IEEE Computer Magazine*, pp. 25–37, February 1989.

[2] A. Agrawal, "Limits on Interconnection Network Performance," *IEEE Transactions on Parallel and Distributed Systems*, vol. 2, pp. 398–412, October 1991.

[3] J. A. Couvillion, R. Freire, and et al., "Performability Modeling with UltraSAN," *IEEE Software Magazine*, pp. 69–80, Sept 1991.

[4] A. Agarwal, R. Simoni, and et al., "An Evaluation of Schemes for Cache Coherence," in *IEEE Computer Architecture Conference - 1988*, pp. 280–289, 1988.

[5] R. Duncan, "A Survey of Parallel Computer Architectures," *IEEE Computer Magazine*, pp. 5–16, February 1990.

[6] M. Tomašević and V. Milutinović, "A Simulation Study of Snoopy Cache Coherence Protocols," in *Hawaii International Conference on System Sciences*, pp. 427–436, 1992.

[7] J. Archibald and J.-L. Baer, "Cache Coherence Protocols: Evaluation Using a Multiprocessor Simulation Model," *ACM Transactions on Computer Systems*, vol. 4, pp. 273–298, November 1986.

A Queuing Model for Finite-Buffered Multistage Interconnection Networks*

Prasant Mohapatra and Chita R. Das
Department of Electrical and Computer Engineering
The Pennsylvania State University
University Park, PA 16802

Abstract

In this paper, we present a queueing model for performance analysis of finite-buffered multistage interconnection networks. The model captures network behavior in an asynchronous communication mode and is based on realistic assumptions. Throughput and delay are computed using the proposed model and the results are validated via simulation. Various design decisions using this model are drawn with respect to delay, throughput, and system power.

1 Introduction

Multistage interconnection networks (MINs) have been proposed as an efficient interconnection medium for multiprocessors. They have been used in various commercial and experimental systems. Behavior of the interconnection network plays an important role in the performance of multiprocessors. For an optimal design, it is necessary to analyze various configurations and constraints of the interconnection network.

Earlier research on MIN performance study have focussed on three types of network models: circuit switched, packet switched with infinite buffer, and packet switched with finite buffer. Study of circuit switched networks has gradually diminished since various packet switching techniques have become more prevalent. Infinite buffer analysis does not necessarily predict realistic behaviors of MINs under various workloads. Recent research effort therefore is directed towards analysis of finite-buffered MINs.

A model for finite buffered MINs should capture the following issues for predicting realistic performance.

• The processors in an MIMD mode operate independent of each other with occasional synchronization. Thus the network model should be based on *asynchronous* message transmission.

*This research was supported in part by the National Science Foundation under grant MIP-9104485.

• The packets are normally of fixed size. Therefore, the time required for transferring a packet from one stage to the next stage is *deterministic*.

• Messages that can not be transmitted from one stage to the next due to the unavailability of buffer space should be *blocked* rather than rejected. Systems like Cedar use blocking of packets.

Prior work on finite-buffered MINs are mainly based on probabilistic models [1-5]. These analyses are valid when all the input/output operations happen at discrete stage cycles. These models do not capture asynchronous behavior especially when the service time of the switching elements (SEs) is more than one clock cycle. A queueing model for finite-buffered asynchronous MINs developed in [6] assumed non-blocking capability and exponential service time for switching elements.

None of the above models has considered all the design issues mentioned earlier. In this paper, we present a queueing model for performance analysis of MINs that considers asynchronous packet switching transmission, finite buffers, deterministic switch service time, and message blocking. The model has been validated via extensive simulation. Average message delay and throughput are used as performance measures to characterize a MIN. Variation of performance with input load and buffer length is discussed.

2 Model Assumptions

The model is based on the following assumptions.

(i) Each processor generates fixed-size messages independently at a rate λ and the intermessage times are exponentially distributed.

(ii) A memory request is uniformly distributed among all the MMs.

(iii) The SEs have deterministic service time (d cycles).

(iv) A packet is blocked at a stage if the destination buffer at the next stage is full. Packets arriving at the first stage of the MIN are discarded if the buffer is full.

Almost all performance studies incorporate assumptions (i) and (ii) to ensure mathematical simplicity. Relaxation of the second assumption to non-uniform traffic is possible and the analysis of a single hot spot traffic model is presented in [7]. Assumption (iii) is based on practical systems like Cedar and BBN Butterfly. Cedar also uses blocking of packets in the MIN and this concept is absorbed in assumption (iv).

3 Queueing Model

The study consists of two parts. First, we present the analysis of an $M/D/1/L$ queue, and then extend the analysis for a network of n queues, where n is the number of stages in the MIN.

3.1 M/D/1/L Queue Analysis

Notations:

L: length of a buffer in the SEs.

p_k: probability that there are k customers in an $M/D/1$ queueing center at steady state.

$p_k^{(L)}$: probability that there are k customers in an $M/D/1/L$ queueing center at steady state.

ρ: traffic intensity at the server $= \lambda \cdot d$.

The analysis of an $M/D/1/L$ queue is described in detail in [7]. Only the results are summarized in this subsection. The probability that there are k customers in an $M/D/1/L$ queueing center is given as

$$p_k^{(L)} = \frac{(1-x)p_k}{\sum_{i=0}^{L} p_i}, \qquad 0 \le k \le L, \qquad (1)$$

where, x denotes the probability that the buffer is full. From [7],

$$x = p_{L+1}^{(L)} = \frac{p_0 - (1-\rho)\sum_{i=0}^{L} p_i}{p_0 + \rho \sum_{i=0}^{L} p_i}. \qquad (2)$$

The values of p_k can be obtained by analyzing the steady state probabilities of an $M/D/1$ queueing center [7].

Using Little's law, the average time, $E[T]$ spent at the center is

$$E[T] = \frac{\sum_{k=1}^{L+1} k p_k^{(L)}}{\lambda(1-x)}. \qquad (3)$$

The denominator captures the effect of blocking by adjusting the arrival rate at a finite-buffered service center.

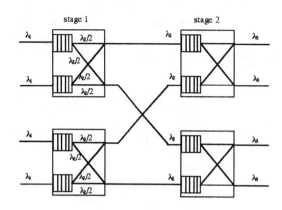

Fig. 1. Model of a (4x4) MIN

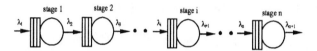

Fig. 2. A Queueing Model of an n-stage MIN

3.2 MIN Analysis

The notations used in Section 3.1 are also used for the MIN analysis with a few modifications as follows.

λ_i: packet arrival rate at stage i, $1 \le i \le n$.

$p_k^{(L)}(i)$: $p_k^{(L)}$ of stage i, $1 \le i \le n$.

ρ_i: traffic intensity at the server $= \lambda_i \cdot d$, $1 \le i \le n$.

x_i: blocking probability at stage i.

The basic model of a (4x4) MIN using (2x2) SEs is shown in Figure 1. The packet arrival and departure rates at each buffer are indicated in the figure. The departure rate is affected by the blocking probability as well as the service time distribution of the server. The uniform memory reference assumption makes all the servers of a particular stage statistically indistinguishable. It is therefore sufficient to analyze one buffer per stage of the MIN. A packet has to travel through a chain of n buffers in an n-stage MIN. A MIN is thus modelled as a chain of n queueing centers as shown in Figure 2.

Characterization of the interdeparture time distribution and hence the departure rate is necessary to analyze the MIN model. We therefore analyze the probability density function (*pdf*) of the interdeparture time of an $M/D/1/L$ queue. Let τ_i be a random variable which represents the time between departures from an $M/D/1/L$ queueing center of ith stage. Let ϕ be the event that the queue is empty after a departure. $f_{\tau_i}(t)$ represents the probability density function of τ_i and $f_{\tau_i|\phi}(t)$ denotes the probability density function of τ_i given that the queue is empty. $f_{\tau_i|\overline{\phi}}(t)$ denotes the probability density of τ_i, given that the queue is non-empty. The state $p_0^{(L)}(i)$ denotes the probability that the queue is empty. Hence, the interdeparture probability density function is given by

$$f_{\tau_i}(t) = f_{\tau_i|\phi}(t)p_0^{(L)}(i) + f_{\tau_i|\overline{\phi}}(t)[1 - p_0^{(L)}]. \quad (4)$$

As the server has a deterministic service time of d cycles, there will be a departure every d cycles when the queue is not empty. Thus

$$f_{\tau_i|\overline{\phi}}(t) = \delta(t - d), \quad (5)$$

where $\delta(t)$ is an impulse function. When the queue is empty, the *pdf* is the density of the service time plus the arrival time. It can be derived as [7],

$$f_{\tau_i|\phi}(t) = \lambda_i e^{-\lambda_i(t-d)}U(t - d), \quad (6)$$

where, $U(t)$ is an unit step function.

Let $E[\tau_i]$ represents the expected value of the interdeparture time of packets from the queueing center. $E[\tau_i]$ can be obtained from equation (4) as

$$E[\tau_i] = \int_0^\infty t \cdot f_{\tau_i}(t)dt = d + \frac{p_0^{(L)}(i)}{\lambda_i}. \quad (7)$$

It is extremely difficult to accurately characterize the nature of interdeparture process. In order to keep the model tractable, we can approximate the interdeparture time distribution from one stage to the next as exponential with an average value of $\lambda_{i+1} = 1/E[\tau_i]$ requests/cycle. It will be shown in Section 4 that this assumption does not induces substantial difference between analytical and simulation results.

Based on our approximation, buffers at each stage of the MIN will have a Poisson arrival process and can thus be modelled as $M/D/1/L$ queueing centers. Using equation (7) and the blocking probability x_i, we get

$$\lambda_i = \begin{cases} \dfrac{\lambda_{i-1}(1 - x_{i-1})}{p_0^{(L)}(i) + \lambda_{i-1}(1 - x_{i-1})d}, & \text{for } 2 \le i \le n; \\ \lambda, & \text{for } i = 1. \end{cases} \quad (8)$$

The above expression is used to compute λ_i starting from $i = 1$ to n. The average time spent at each stage can be computed using equation (3). The average delay for a packet is obtained by summing up the delays of all the stages. The normalized throughput, X, is determined by the output of a buffer in the last stage of the MIN model, and is equal to λ_{n+1}.

4 Performance Evaluation

A simulation model of a MIN was developed in which packets were generated randomly with an exponential distribution of interarrival time by each processor. A uniform random number generator was used to determine the destination memory. Throughput and delay were computed by counting the number of request completions and the average time taken to reach the output port, respectively. Comparisons between the analytical and simulation results for (64x64) and (1024x1024) systems using (2x2) SEs are shown in Figures 3 and 4. The difference between the analysis and the simulation results is within 7%. The curves indicate that the analytical results are fairly accurate.

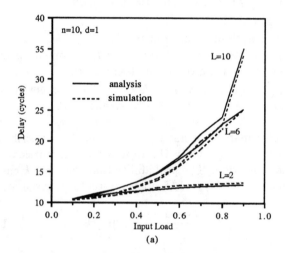

Fig. 3. Delay of a (64x64) MIN

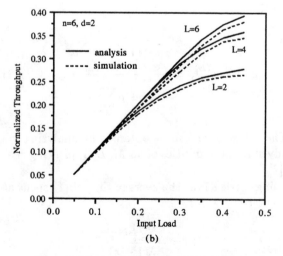

Fig. 4. Throughput of a (1024x1024) MIN

The effect of buffer length on delay of a 256-node MIN is depicted in Figure 5. It is mentioned in [1] that a small buffer length shows performance equivalent to an infinite buffer. It can be inferred from Figure 5 that this is true only when the input load is less. The variation of delay is prominent until the buffer length is considerably high for heavy traffic. The model can be used to determine the minimum buffer length required to get a performance equivalent to the infinite buffer case.

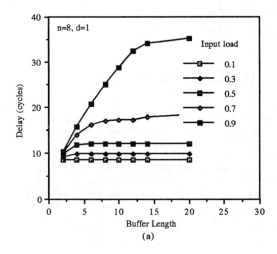

Fig. 5. Effect of Buffer Length on MIN Performance

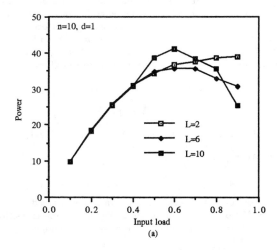

Fig. 6. Variation of System Power

Throughput and delay are not necessarily sufficient measures of system performance. It is observed that higher throughputs result in longer delay. A combined metric called *system power* is sometimes more meaningful. System power is defined as the ratio of throughput to delay. A higher power means either a higher throughput or lower delay. The variation of system power with respect to the input load is shown in Figure 6 for various buffer lengths. It is observed that system power increases with the input load for small buffers. For large buffer size, the throughput

first increases with the input load until it saturates. On the other hand, delay increases monotonically. Thus, after a certain input load, the power reduces. The model can be used for predicting the optimum load to maximize the power of a MIN.

5 Concluding Remarks

A queueing model for evaluating performance of finite-buffered, asynchronous MINs is presented in this paper. The uniqueness of this model compared to previous finite-buffered analyses is that it captures asynchronous operations, deterministic service time of switches, and message blocking. Comparison with simulation results show that the analytical model is highly accurate. Various design alternatives based on performance requirements are discussed. It is difficult to come up with an optimal set of design parameters to satisfy all performance measures. The model can be used to compute suitable values of MIN parameters based on the priorities of performance metrics.

References

[1] H. Jiang, L. N. Bhuyan, and J. K. Muppala, "MVAMIN: Mean Value Analysis Algorithms for Multistage Interconnection Networks," Journal of Parallel and Distributed Computing, pp. 189-201, July 1991.

[2] D. M. Dias and J. R. Jump, "Analysis and Simulation of Buffered Delta Network," IEEE Trans. on Computers, pp. 273-282, Aug. 1981.

[3] D. L. Willick and D. L. Eager, "An Analytical Model of Multistage Interconnection Networks," ACM SIGMETRICS Conf. on Measurement and Modeling of Computer Systems, pp. 192-202, 1990.

[4] T. Lin and L. Kleinrock, "Performance Analysis of Finite-Buffered Multistage Interconnection Networks with a General Traffic Pattern," ACM SIGMETRICS Conf. on Measurement and Modeling of Computer Systems, pp. 68-78, May, 1991.

[5] H. Yoon, K. Y. Lee, and M. T. Liu, "Performance Analysis of Multibuffered Packet-switching networks in Multiprocessor Systems," IEEE Trans. on Comput., vol. C-39, no.3, pp. 319-327, Mar. 1990.

[6] T. N. Mudge and B. A. Makrucki, "An Approximate Queueing Model for Packet Switched Multistage Interconnection Networks," Int. Conf. on Distributed Computing Systems, pp. 556-562, Oct. 1982.

[7] P. Mohapatra, Dependability and Performance Modelling of Parallel Computers, Ph.D. Thesis, The Pennsylvania State University, Aug. 1993.

Composite Performance and Reliability Analysis for Hypercube Systems

Samir M. Koriem   **L. M. Patnaik**

Department of Computer Science and Automation

Indian Institute of Science   Bangalore-560 012   INDIA

Abstract

In this paper we propose a novel technique to model and analyze the performability of parallel and distributed architectures using GSPN-reward models.

1 Introduction

In order to model and analyze the performability of parallel and distributed systems, we propose a technique based on *Generalized Stochastic Petri Nets* (GSPNs). Our performability modeling technique essentially consists of the following.

- A *GSPN performance model* which represents variations in the internal states and environment of the system.

- A *GSPN reliability model* which represents changes in the structure of the system due to the occurrence of faults and their repairs.

- A *GSPN-reward model* for *combining* Markov reliability model with the appropriate performance metric obtained from the performance model.

This paper is organized as follows. In section 2, we use the GSPN technique [6] to describe the behavior of *iPSC/2 hypercube* system [1] and to derive the performance metrics for this system under the workload of concurrent *bitonic sorting algorithm* [8]. In section 3, a reliability model for the hypercube architecture is described using the GSPN technique. We analyze one reliability model based on *disconnected reliability approach* [7]. In section 4, we explain how to use the *GSPN-reward technique* [2] to combine the Markov reliability model with the appropriate performance metric obtained from the performance model. Section 5 concludes this paper.

2 Performance Model

In this section, we use the *GSPN technique* to investigate the computation and communication behavior of the iPSC/2 hypercube [1] under the workload of a concurrent bitonic sort program [8].

Using the Bitonic sort algorithm described in [8], the behavior of the iPSC/2 hypercube system is represented by the GSPN model shown in Figure 1. The following performance measures can be obtained from Figure 1: (i) *mean time to complete the program* (T_{MP}); (ii) *speedup*; and (iii) *effective processor utilization*. In [3], we explain how the GSPN Bitonic model can handle a given problem. Also, the meaning associated with each place and transition is explained in [3].

3 Reliability Model

The objective of this section is to develop a GSPN reliability model based on the *disconnected reliability approach* (DRA) [7] for a binary d-cube of size $N = 2^d$ nodes. This reliability model represents the second part of the performability model, as we mentioned in section 1.

The novelty of our approach is that we develop a simple GSPN reliability model which gives the same results of the DRA [7]. The main advantage of our GSPN reliability model over DRA (Markov chain analysis) is that there is no need to manually enumerate all the possible states since they are automatically generated from the GSPN model. Another advantage of our approach is that we do not require a closed-form solution unlike that in [7].

If the hypercube system has disconnections among the nodes, then the probability of a disconnection becomes an important parameter to calculate the reliability of such a system. Let $P_{rob}(i)$ be the probability of a disconnection occurring at the ith node failure, given that no disconnection had occurred up to the $(i-1)st$ node failure. The *conditional probability* of a single *node disconnection* [7] is computed as follows.

$$\text{For } i < d \qquad P_{rob}(i) = 0 \qquad (1)$$

$$\text{For } i = d \qquad P_{rob}(i) = N / \binom{N}{d} \qquad (2)$$

$$\text{For } i > d \qquad P_{rob}(i) = N * \mathcal{A}_2 * \mathcal{A}_3 * \mathcal{A}_4 / \mathcal{A}_1 \qquad (3)$$

where $\mathcal{A}_1 = \binom{N}{i-1}$, $\mathcal{A}_2 = \binom{d}{d-1}$, $\mathcal{A}_3 = \binom{N-d-1}{i-d}$, $\mathcal{A}_4 = 1/(N-i+1)$.

The disconnected reliability approach is explained through the GSPN model shown in Figure 2. The places and transitions of Figure 2 are described below for a d-cube topology.

- **Places**

p_1: number of hypercube nodes (N);

p_2: minimum number of active nodes $(\mathcal{D} = N/2)$;

p_3: number of node(s) awaiting repair;

p_4: node(s) whose failure causes the whole system to fail.

- **Transitions**

t_1: one node has failed;

t_2: all nodes have failed or a single fault has occurred from which the system cannot recover;

t_3: repair of node(s). In our model, incorporation of repair (through t_3) is easy unlike that in [7]. We consider on-line repair of a node with exponential distribution of repair time given by $1/\mu$.

In the fault-free state, the GSPN model contains N tokens in place p_1 and $\mathcal{D}$ tokens in place p_2. To obtain all the reliability aspects of the DRA, the *rate functions* and *enabling functions* of the transitions and the *variable arc functions* among the places, are calculated as follows. In the following description of Figure 2, $m(p)$ represents the number of tokens in place p.

Arcs	Variable Arc Function
$p_1 \to t_2$	$m(p_1)$
$p_2 \to t_2$	$m(p_2)$
$p_3 \to t_2$	$m(p_3)$
$t_2 \to p_4$	$m(p_1) + m(p_3)$

Transition	Enabling Function
t_2	0 if $(m(p_1) = 0 \wedge m(p_2) = 0 \wedge m(p_3) = 0)$
	1 otherwise

Transition	Rate Function
t_1	$\lambda_i = m(p_1) \lambda C_i; \qquad C_i = C_0 (1.0 - P_{rob}(i))$
	if $([m(p_1) > (N - d)] \wedge [m(p_1) \le N])$
	$\qquad P_{rob}(i)$ as in equation 1
	if $(m(p_1) = [N - d])$
	$\qquad P_{rob}(i)$ as in equation 2
	if $([m(p_1) \le (N - d - 1)] \wedge [m(p_1) > (N - \mathcal{D})])$
	$\qquad P_{rob}(i)$ as in equation 3
t_2	$h_i = m(p_1) \lambda (1.0 - C_i); \qquad C_i = C_0 (1.0 - P_{rob}(i))$
	if $([m(p_1) > (N - d)] \wedge [m(p_1) \le N])$
	$\qquad P_{rob}(i)$ as in equation 1
	if $(m(p_1) = [N - d])$
	$\qquad P_{rob}(i)$ as in equation 2
	if $([m(p_1) \le (N - d - 1)] \wedge [m(p_1) > (N - \mathcal{D})])$
	$\qquad P_{rob}(i)$ as in equation 3
	if $(m(p_1) = [N - \mathcal{D}])$
	$\qquad h_i = m(p_1) \lambda$

where C_0 is a fault coverage factor for the recovery scheme, and λ is a node failure rate per hour. The results derived from the GSPN model of the SRA (shown in Figure 2) can be found in [3].

4 Performability Model

We now provide a detailed discussion on how to combine the following two parts to construct a hypercube performability model.

(i) An appropriate performance metric obtained from the GSPN bitonic sort model that was developed and analyzed in section 2.

(ii) A Markov chain derived from the GSPN disconnected reliability model that was developed and analyzed in section 3.

The performability model of the hypercube system is constructed as follows.

- The GSPN reliability model shown in Figure 2 is able to describe the changes due to the structural state process of the hypercube system by failure-repair events under *disconnected reliability approach.*. This structural state process (referred to as the stochastic process), $\{Z(t), t \ge 0\}$, is a homogeneous finite state continuous-time Markov chain (CTMC). Let $\Omega = \{S_i : i = 0, ..., n\}$ be the finite state space associated with $Z(t)$. At any given time, the hypercube system can be in one of the n states. Regarding the performability analysis, let $\Omega_O \in \Omega$ be the set of states that represent an operational hypercube system and $\Omega_F \in \Omega$ be the remaining set of states that represent a failed system. However, each state (configuration) structure $S_i \in \Omega_O$, $\Omega_O = \{S_i : i = 0, ..., \mathcal{D}\}$, is associated with a set of operational nodes that define the configuration of the hypercube system. The non-operational states are lumped together into a state $S_F \in \Omega_F$.

- To construct a Markov reward model (MRM) for the hypercube system, a *reward rate* (r_i) of the performance

model of the bitonic sort algorithm must be associated with each state $S_i \in \Omega_O$ of the $CTMC$ derived from the GSPN disconnected reliability model. The reward rate r_i indicates how much work the hypercube system completes per unit time when the system is in state $S_i \in \Omega_O$. The reward rate r_i that we are considering is the mean time to complete the bitonic sort program $T_{MP}(i)$. Let us now define a reward rate vector **r** (also called reward structure) over the states of $Z(t)$ such that a reward rate r_i is associated with state $S_i \in \Omega$. In other words, the reward rate r_i represents the performance measure $T_{MP}(i)$ when the system has the structure indicated by $S_i \in \Omega_O$. For instance, the reward rate r_0 for the state S_0 is $T_{MP}(0)$ when N nodes are operational.

Since the time-scale of the performance-related events (i.e. $T_{MP}(i)$) is much faster than the time-scale of the reliability-related events (e.g. node failure, node repair, etc.), the steady-state values from the GSPN bitonic sort model are used to specify the *performance levels* or reward-rates $T_{MP}(i)$, $\{i = 0, ..., \mathcal{D}\}$, for each state $S_i \in \Omega_O$. Therefore, the reward rates are calculated by solving the GSPN performance model with different initial markings $N, N - 1, ..., N/2$. The initial markings represent new operational hypercube configurations obtained after a node failure.

4.1 Performability Analysis

Our *essential problem* now is how to assign the reward rate vector **r** to the failure-repair Markov model. For this purpose, we use an extension of the GSPN technique called *GSPN-reward model* [2]. The basic idea of this technique is that the performance levels (reward rates) are specified at the net level as a function of net primitives such as *the number of tokens in a place* or *the rate of a transition*. For each marking obtained from the net, the reward rate function is evaluated and the result is assigned as the reward rate for that marking. The underlying Markov model is then transformed into a Markov reward model, thus permitting evaluation of performability models and deriving some measures of hypercube behavior.

In our analysis, the GSPN reliability model shown in Figure 2 is used to implement the GSPN-reward model of the hypercube system. Based on the model of Figure 2, the GSPN-reward model is constructed as follows:

- Associate the performance levels of $T_{MP}(i)$, $\{i = 0, ..., \mathcal{D}\}$, with the markings whose place p_1 has J tokens, $\{J = N, ..., \mathcal{D}\}$. These reward rates reflect the performance levels of the system in different configurations.

- Associate a reward rate of zero with other markings.

Once we perform the above two steps, the GSPN model of Figure 2 is changed to GSPN-reward model.

In the GSPN-reward model of Figure 2, we can use the following function that explains the above two steps.

$$r_i = \begin{cases} T_{MP}(i) & \text{If } ([m(p_1) \le N] \wedge [m(p_1) \ge (N - \mathcal{D})]) \\ 0 & \text{Otherwise} \end{cases}$$

For example, $T_{MP}(0)$ is assigned to the net when place p_1 contains N tokens.

4.2 Performability Measures

In this section, based on the MRM obtained from the GSPN-reward model, we define the following measures [9,3] that *combine* the performance and reliability aspects.

The *Expected Reward Rate* $E[X(t)]$ at time t is given by

$$E[X(t)] = \sum_i r_i \, p_i(t) \qquad (4)$$

where $p_i(t) = P_{rob}[Z(t) = i]$ is the probability that the system is in state i at time t.

The *Expected Accumulated Reward* $E[Y(t)]$ in the interval $(0, t)$ is given by

$$E[Y(t)] = \sum_i r_i \int_0^t p_i(u)\,du \qquad (5)$$

The *Expected Time Averaged Accumulated Reward* $E[W(t)]$ can be express as

$$E[W(t)] = \frac{1}{t} \sum_i r_i \int_0^t p_i(u)\,du \qquad (6)$$

The *Expected Interval Availability* $E[A_I(t)]$ is given by

$$E[A_I(t)] = \frac{1}{t} \sum_{i \in \Omega_O} \int_0^t p_i(u)\,du \qquad (7)$$

The *Expected Uptime* $E[U(t)]$ is given by

$$E[A_I(t)] = \frac{1}{t} \sum_{i \in \Omega_O} \int_0^t p_i(u)\,du \qquad (8)$$

4.3 Performability Results

To obtain the hypercube performability measures, we use the performability metrics that were explained in the previous section. The effect of failure and repair on $E[X(t)]$ is explained in Figure 3 and the corresponding effect on $E[W(t)]$ is illustrated in Figure 4. From Figure 3 (the expected performance of the hypercube system at time t) and Figure 4 (the time-averaged performance of the hypercube system over the interval $(0, t)$), we observe the following. Plot I represents the hypercube system without failure and repair. The results of this modeling assumption are that no degradation of performance takes place, so the hypercube performance remains independent of time. In plot VIII, the system has no repair. Plot VII indicates that the system has no repair and $\mathcal{C}_0 = 0.9$. These plots reflect the effect of coverage factor on $E[X(t)]$ and $E[W(t)]$. The expected performance level (or the expected time-averaged accumulated reward) of plot VIII deteriorates more rapidly than that of plot VII.

From plot VII, we observe that the expected performance level (reward rate) gets halved at time $t = 2400$. At time 2400 in Figure 4, the expected time-averaged accumulated reward decreases by only one quarter because $E[W(t)]$ is the time average of $E[X(t)]$ over $(0, t)$. Plot IV ($\mu = 2.0$ and without $\mathcal{C}_0$) and plot VI ($\mu = 0.1$ and without $\mathcal{C}_0$) show the effect of repair strategy on $E[X(t)]$ and $E[W(t)]$. The expected performance level (or the expected time-averaged accumulated reward) of plot III ($\mathcal{C}_0 = 0.9$ and $\mu=2.0$) shows greater improvement compared to plot V ($\mathcal{C}_0 = 0.9$ and $\mu=0.1$). Plot II reflects the effect of increase of the coverage value to 0.999999 (with $\mu=0.1$) on $E[X(t)]$ and $E[W(t)]$. This plot clearly shows that $E[X(t)]$ (or $E[W(t)]$) increases with an increase in the value of $\mathcal{C}_0$. However, it can be observed that both plots I and II are almost identical. Actually, $\mathcal{C}_0$ reflects how well the redundancy schemes and the recovery strategies have been implemented by the system designer.

In Figure 5, the behavior of $E[A_I(t)]$ is described for different $\mathcal{C}_0$ and μ values. This cumulative measure indicates the expected amount of time spent by the system in the operational states during the interval $(0, t)$. Figure 5 shows the following plots: I ($\mathcal{C}_0 = 0.999999$ and $\mu=0.1$), II ($\mu=2.0$ and without $\mathcal{C}_0$), III ($\mu=0.1$ and without $\mathcal{C}_0$), IV ($\mathcal{C}_0 = 0.9$ and without μ),

and V (without μ or $\mathcal{C}_0$). As expected, the results obtained from Figure 5 are similar to the results depicted by the plots of $E[W(t)]$, because $A_I(t)$ can be a special case of $W(t)$ [4].

The performability measures $E[Y(t)]$ and $E[U(t)]$ are compared for different values of the parameters $\mathcal{C}_0$ and μ and are shown in Figure 6. Figure 6 shows the plots I (without μ or $\mathcal{C}_0$), II ($\mathcal{C}_0 = 0.9$ and without μ), III ($\mathcal{C}_0 = 0.9$ and $\mu=0.1$), IV ($\mu=2.0$ and without $\mathcal{C}_0$), and V ($\mathcal{C}_0 = 0.9$ and $\mu=2.0$). These plots clearly demonstrate that $E[Y(t)]$ is always greater than $E[U(t)]$. Actually, $E[Y(t)]$ represents the total computation done by the hypercube system during the interval $(0, t)$.

5 Conclusions

An important contribution of our work is the study of both performance and reliability of concurrent systems in an integrated framework. We have adopted the GSPN technique because of its many attractive features. In our GSPN performance model, important aspects of hypercube such as concurrency, message passing mechanism between the host and the nodes as well as among the nodes, have been modeled in a natural way. These features have been illustrated through the GSPN model of the iPSC/2 hypercube system under the workload of a concurrent bitonic sort algorithm.

To study the reliability aspects of the hypercube system, we have developed a simple GSPN reliability model that gives all the features of the disconnected reliability approach in an easy way. To construct the hypercube performability model, we have used the GSPN-reward model. The performability model is analyzed to obtain various instantaneous and cumulative measures.

References

[1] L.Bomans and D.Roose, "Benchmarking the iPSC/2 Hypercube Multiprocessor," *Concurrency: Practice and Experience*, (Sept. 1989), pp.3-18.

[2] G.Ciardo, J.Muppala and K.Trivedi, "On the Solution of GSPN reward models," *Performance Evaluation*, (July 1991), pp.237-253.

[3] S.M.Koriem "Reliability and Performability Analysis of Distributed Memory Multicomputers," *Ph.D. Dissertation*, I.I.SC., Banagalore, India, (April 1993).

[4] S.M.Koriem, L.M.Patnaik, "Fault-Tolerance Analysis of Hypercube Systems Using Petri Net Theory," To appear in *The Journal of System and Software*.

[5] S.M.Koriem, L.M.Patnaik, "Performability Studies of Hypercube Architectures," *Int. Parallel Processing Symposium*, California, USA, (March 1992).

[6] M.A.Marson, G.Balbo and G.Conte, "A Class of Generalized Stochastic Petri Nets for the Performance Evaluation of Multiprocessor Systems," *ACM Trans. on Computer Systems*, (May 1984), pp.93-122.

[7] W.Najjar and J.Gaudiot, "Reliability and Performance Modeling of Hypercube-Based Multiprocessors," Proc. of the 2nd *Int. MCPR Workshop*, Rome, Italy, (May 1987), pp.305-320.

[8] S.Seidel and L.Ziegler, "Sorting on Hypercubes" Proc. of the *Second Conf. on Hypercube Multiprocessors*, Knoxville, Tennessee, (Sept. 1986), pp.285-291.

[9] R.Smith, K.Trivedi, and A.Ramesh, "Performability Analysis : Measures, an Algorithm, and a Case Study," *IEEE Trans. on Computer*, (April 1988), pp.406-417.

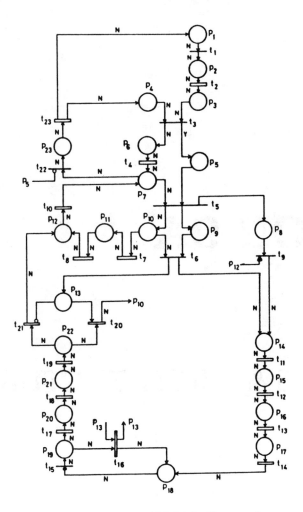

Figure 1. GSPN Model for Concurrent Bitonic Sort Algorithm.

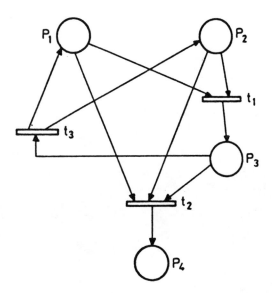

Figure 2. GSPN Model of Disconnected Reliability Approach.

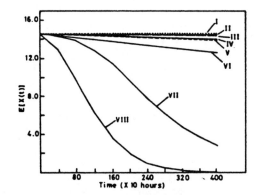

Figure 3. Variation of $E[X(t)]$ with Time for the Hypercube System.

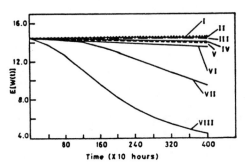

Figure 4. Variation of $E[W(t)]$ with Time for the Hypercube System.

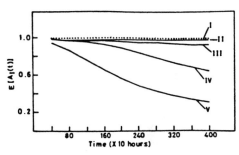

Figure 5. Variation of $E[A_I(t)]$ with Time for the Hypercube System.

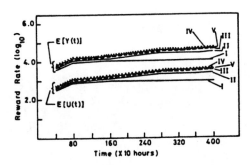

Figure 6. Variation of $E[Y(t)]$ and $E[U(t)]$ with Time for the Hypercube System.

SESSION 9A

DISTRIBUTED SYSTEMS & ARCHITECTURE

Estimation of Execution times on Heterogeneous Supercomputer Architectures

Jaehyung Yang[†], Ishfaq Ahmad[‡], Arif Ghafoor[†]

[†]School of Electrical Engineering
Purdue University
West Lafayette, Indiana 47907

[‡] Computer Science Department
Hong Kong University of
Science and Technology

Abstract

For managing tasks efficiently in a Distributed Heterogeneous Supercomputing System (DHSS) , we require a thorough understanding of applications and their intelligent scheduling within the system. For this purpose, an accurate estimation of the execution time of applications on various architectures is needed. In this paper we present a framework to address this issue. We propose two techniques, called augmented code profiling and augmented analytical benchmarking, to characterize applications and architectures in a DHSS, respectively.

These techniques are based on code profiling and analytical benchmarking, respectively and provide a detailed architectural-dependent characterization of DHSS applications.

1 Introduction

The marriage between high speed networking and supercomputing technology has resulted in the recent development of Distributed Heterogeneous Supercomputing System (DHSS) [7, 12]. The objective is to achieve a superlinear speedup through a computationally powerful environment that is capable of solving many engineering and scientific problems which are intractable on a single supercomputing system. A DHSS is also expected to outperform a homogeneous supercomputing system because irrespective of how powerful a single machine or a set of homogeneous machines might be, such a system cannot satisfy the diverse characteristics of program codes efficiently [9]. Specially, ill-matched codes can degrade the overall performance of these systems. Building a suite of heterogeneous supercomputers with existing machines, having diverse computational characteristics, can provide a more powerful environment for solving complex problems. However, an efficient use of a heterogeneous system requires a thorough understanding of the characteristics of applications, the architectures of machines and their operational features.

A number of DHSS's have been proposed recently, with a few of them already prototyped. The most noticeable are the five gigabit network testbeds, namely; Aurora, Blanca, Casa, Nectar and Vistanet [13]. The functional concept behind the design of the Casa and the Nectar testbeds resembles more closely to a DHSS. However, these testbeds have been built to solve a specific set of applications and they cannot manage a wide variety of applications effectively. The future DHSSs, on the other hand, are expected to serve a large variety of users developing diverse applications which are expected to run concurrently on various machines within a DHSS.

One major challenge for future DHSSs, therefore, is the management of applications that requires finding a suitable match between tasks of these applications and machines. For this purpose, code profiling is used to characterize applications in order to identify those codes which have the identical computational behavior [7]. One approach is to chose a set of "computational structures", which we call *templates*, that can run on individual supercomputers. Very few code profiling methodologies, within the context of DHSS, have been proposed in literature [16, 8]. However, these methodologies have limitations in their applicability. Most of them are based on a rather simplistic and a highly abstract view of parallelism. Such methodologies cannot provide an accurate estimate of the execution time of a task, since they do not take into account the detailed architectural knowledge. Furthermore, scalability and mapping issues cannot be handled properly by these approaches. New code profiling methodologies are needed which can incorporate such issues as well as the detailed architectural characteristics so that they can be used more accurately and intelligently in making scheduling and mapping decisions as they can significantly impact the execution times of applications [5]. However, there is a trade-off between the accuracy of the information generated by a profile and the complexity involved in generating it.

The objective of this paper is to propose a general framework for characterizing application programs and architectures of supercomputers in a DHSS, using a code profiling and an analytical benchmarking technique, respectively. We show how a combination of these approaches provides a systematic methodology for estimating the execution times of applications on various machines. Our approach is based on generating a "Representative Set of Templates"[1] (RST) that allows a user to generate code-profiles which can represent the execution behavior of the task at varying levels of details. Using this set, we propose two architecture-dependent characterizations of DHSS applications, which are called Code Flow Graph (CFG) and Code Interaction Graph (CIG). These graphs possess enough information about applications that makes them useful for managing these applications in a DHSS. In this paper, we do not address the issue of scheduling/mapping. We focus on how to characterize the performance of machines in a DHSS and its applications.

This paper is organized as follows. In the next section, the proposed template based approach is briefly described. Section 3 discusses RST and its use for augmented analytical benchmarking. Section 4 examines augmented code profiling and the estimation process. Section 5 concludes this paper.

2 The Template Based Execution Time Estimation

A parallel program is composed of a number of interacting processes which can run to completion without intervention. These processes have some dependencies and interactions, which can be captured through a computational structure [14], which we call templates. Topological structures of these templates for many known algorithms are generally well-behaved and regular. A number of such templates constitute a task which can run as an independent entity on a supercomputer system. In turn, a number of

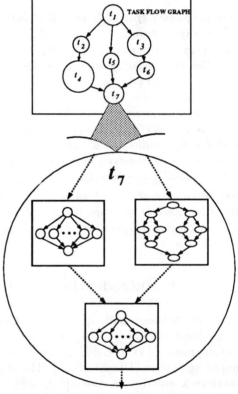

Figure 1: A Task Flow Graph and its template level representation for Code Flow Graph.

---

[1] These templates serve as both benchmarks and code types.

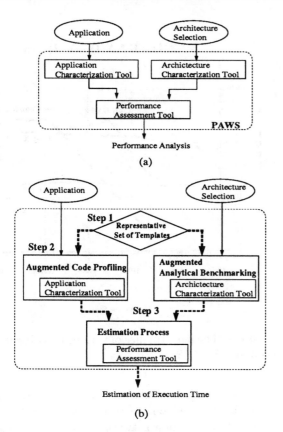

Figure 2: (a) Overall PAWS block diagram. (b) The proposed overall methodology for estimating execution time of a DHSS application.

such tasks constitute the overall distributed application of DHSSs. Such a application is represented as a Task Flow Graph (TFG) or a Task Interaction Graph (TIG) [3]. The overall structural configuration of a TFG is illustrated in Figure 1 Using a code profiling technique and analytical benchmarking, we can generate a CFG (CIG) from TFG (TIG), respectively.

The generation of a CFG (CIG) requires code profiling and analytical benchmarking information. One possible methodology to generate a CFG (CIG) is to use our earlier proposed approach, where a tool, called Performance Assessment Window System (PAWS), can be used to assess the performance of an application on a supercomputer system [10]. In the PAWS, the architecture of a machine is characterized in a hierarchical form and detailed low level timing information is used to assess its performance. However, the process is quite exhaustive and may not be viable for large applications which does not permit high profiling overhead. Furthermore, this tool does not consider concurrent execution of tasks across heterogeneous systems. The methodology we are proposing here is a two level approach and provides a more

practical solution where an application is represented in forms of templates as shown in Figure 1. These templates, augmented with certain parameters, provide a mechanism to estimate the overall execution time of the application. Also, the proposed approach is hybrid, in the sense that it combines both fine grain and coarse grain benchmarks. The parametric characterizations of an architecture at an operation level as proposed in PAWS, is a fine grain characterization. By combining the concept of templates with such a benchmarking, and representing an application in terms of templates, we can have a more powerful mechanism to estimate the execution time of an application. Such an approach can consist of three steps. First, we need to identify some "representative templates", so that every task can be expressed structurally in terms of some of these templates. In the second step, we need to develop a code profiling mechanism to identify templates in a task. Finally, in the third step the estimate of the execution time need to be obtained by using the benchmark results for these templates. These benchmarks can be generated by either analyzing templates and using their performance results or using the approach described in the PAWS [10]. Figure 2 (b) shows the overall process involved in this approach.

In this paper, we emphasize the process of generating RST (i.e. code types), and discuss the techniques used for Augmented Code Profiling, Augmented Analytical Benchmarking and Estimation processes, as shown in Figure 1.

3 A Representative Set of Templates

Most of the existing parallel algorithms can be categorized into a number classes based on their structures [14]. These structures can be represented by graphs. Such a representation is useful for capturing the behavioral information of parallel programs, including parallelism, precedence relationship among processes, communication interaction, etc. For this purpose, we introduce the concept of "Representative Set of Templates" (RST). Basically, through RST, we can identify all the possible computational structures that exist for known parallel algorithms. The structures (templates) need to be augmented with some parameters that can help us to assess the performance of the templates on a given machine. Some parameters are also needed to specify scalability and mapping information such as the type of problems (Fast Fourier Transform, Linear System Equation Problem, Ordinary Differential Equation Problem, etc.), the size of

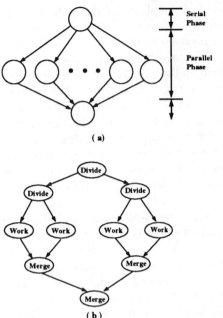

Figure 3: (a) Partitioning Template. (b) Multiphase Template

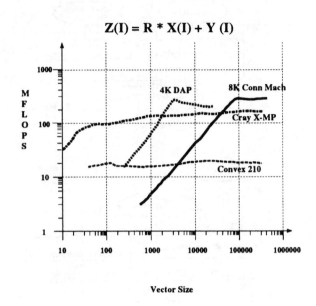

Figure 4: The performance of various machine models on 32-bit SAXPY

input data set, etc. Most of the templates can have some regular structures [14]. Examples of such templates are shown in Figure 3. The template in Figure 3(a) is called *partitioning template* and the one in Figure 3(b) is called *multiphase template*. As can be seen from this figure, a template must be augmented with additional topological parameters, such as total number of processes in the template, the depth of the template, etc. Formally, we need the following sets to define augmented template.

RST
$\mathcal{S}$ = set of problems
$\mathcal{P}$ = set of topological information
$\mathcal{N}$ = set of acceptable input data

Then a augmented template is described by a tuple (t, s, p, n), where $t \in RST$, $s \in \mathcal{S}$, $p \in \mathcal{P}$, and $n \in \mathcal{N}$.

3.1 Augmented Analytical Benchmarking

Most of the existing benchmark programs are architecture-independent and cannot provide realistic and meaningful profiles about machines. This is due to the fact that such programs may not be mapped properly on the machine, and, instead of yielding benchmark profiles, they can even result in a speedup of less than 1, that is, a performance worse than a uniprocessor. For example, analyses have shown that

if the standard molecular motion computation algorithm is executed on a supercomputer with multistage interconnection network, such as the Butterfly System, or on a shared bus interconnection system, the speed up approaches zero as we increase the number of processors beyond a certain value [5]. It is, therefore, highly desirable that benchmarks should be able to capture the effect of architectural features of machines. For such purpose, we propose a two level benchmark scheme for characterizing machine architecture based on the proposed RST. The first level is the template level analytical benchmarking and the next level is the operational level benchmarking. This method, we call augmented analytical benchmarking, provides a systematic way of generating performance profile of machines.

3.1.1 Template Level Analytical Benchmarking

There exist a number of methodologies to benchmark parallel machines, such as *Kernel, Partial (Trace) Benchmarks, Synthetic Benchmarks*, etc [4]. A number of codes for benchmarking the performance of parallel machines have been proposed. These include Livermore Loops, Linpack, etc [6]. In a DHSS environment, an application is decomposed into multiple tasks that run separately on different machines. It is, therefore, important that analytical benchmarking for a DHSS should be able to estimate the performance of a machine on each part of the application. Also, as we have already mentioned, a benchmark program must take into account the architectural characteristics of machines. However, the existing benchmarking programs are not specially designed to measure the architecture specific performance, rather their objective is to measure the overall performance of each machine under a simulated application environment. Some examples of such benchmarks can be found in Perfect Club [11], although some benchmarks in this case also are still being developed.

Template level analytical benchmarking is a coarse grain benchmarking methodology, that is used for measuring the overall performance of machines for each template type. Since the performance of a machine depend on the parameter (t, s, p, n), the benchmarking should be carried out for various values of these parameters. For example, the benchmark results for various values of the input vector sizes for the SAXPY problem is shown in Figure 4 [7]. Since detailed benchmarking for all possible values of each parameter is impractical, we can use a regression model based on data collected for selected parameter sizes. This model can provide a generalized profile so that it can be used to estimate the execution time for a template, with parameters for which benchmark results are not available. The regression function can be expressed as follows:

$$\mathcal{B}: \quad \text{RST} \times \mathcal{M} \rightarrow t, t \in \mathcal{R}^+$$

$\mathcal{R}^+$ contains positive real numbers representing regression values, and $\mathcal{M}$ represents the set of machine models in a DHSS.

3.1.2 Operational Level Benchmarking

Since templates can only provide the overall description of a task, detailed effects of various aspects of machine architecture may not be observed at this level. Also, for the algorithms which have irregular and unpredictable structures, we may have to carry out detailed analyses to estimate their execution time. Also, the error may be introduced in the process of regression analysis. In order to overcome these problems, we can use a fine grain benchmarking, similar to the one used in the PAWS.

For characterizing architectures of machines in a DHSS, we propose an analytical benchmarking technique based on the sets of operations, similar to ones describe in the PAWS, where the timing data for the

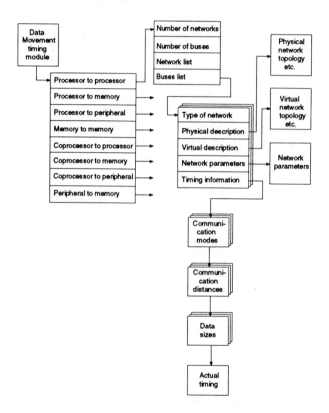

Figure 5: A Hierarchical Structure of PAWS for Generating Timing information of Data Movement.

lowest level of the hierarchy needs to be collected. In this approach, a machine architecture is functionally partitioned into four categories, namely;

1. Computation

2. Data Movement and Communication

3. Input/Output

4. Control.

Each category is continuously partitioned into subsystems until a subsystem is fine enough to be characterized in terms of raw timing information. This information can be organized further in a hierarchal form with the raw timing information contained in the leaf nodes of this extended hierarchy. PAWS uses an integrated approach based upon low level benchmarking that completely determines the operation and behavior of the machine for each subsystem, and analytical models which are application dependent. For more accurate analysis of the machine architecture, a graphical language, called IF1, is used as a description language for the applications. IF1 is an acyclic graphical language and can capture the flow of operations and operands. As an example of this fine grain analysis, the hierarchy for the Data Movement function is shown in Figure 5.

4 Augmented Code Profiling for Execution Time Estimation

In most of the cases, code profiling is an on-line process, that incurrs a run-time overhead. A "detailed" operational profile may take into account all the important architectural characteristics of a machine, such as the type of parallelism, the interconnection topology, the memory organization scheme, etc. The generation of such a profile requires a detailed analysis of the task with respect to the architectural characteristics defined at the selected level. Although such a profile provides very useful information for efficiently scheduling/mapping a task via accurately matching it to a machine, it can only be generated at the cost of an increased overhead associated with the analysis of the task. A "coarse" template level profile, on the other hand, can be generated with a relatively low overhead. However, such a profile may not be accurate enough for scheduling/mapping tasks effectively. This "accuracy vs. complexity" trade-off depends on the level selected. This selection can be a part of the user-specified processing requirements.

In this section, we propose a methodology for augmented code profiling based on the RST. The existing code profiling methods classify a code to a specific type [8, 16]. However, in our approach we identify composite templates of various code types within a task as shown in Figure 1. This detailed characterization of tasks is quite desirable since a real world task may indeed consist of several different codes (templates), in addition to a major template type. Though minor template types constitute a small portion in a task, they should not be ignored because even such a small portion of the unmatched code may become the bottleneck in executing a task as described by the *Amdahl's Law*. Therefore, the code profiling should identify each composite template instead of classifying the overall task as a single template. The identified templates can then be augmented with their parameters. If a finer analysis is desired, then the operational level benchmarking, similar to the one used in PAWS can be chosen.

The overall process of code profiling should consider the effect of the compiler as each compiler differs in its ability to optimize the code and exploit the possible parallelism. For example, [17] shows two different performances of the same Cray X-MP for Livermore kernels running under two different compilers, CFT '82 and W/X.14. In a way, the proposed augmented code profiling is a compiler-dependent code profiling process.

We now formally elaborate on the phases involved in the code profiling process which is a two-phase process. During the first phase, the profile called, Augmented Template Graph for a task t is generated. For this purpose, template types in a task are identified, which are then augmented with parameters, thus generating a augmented template graph. If a finer analysis is desired, then the next phase is carried out where each augmented template is further analyzed in terms of its operational behavior. This analysis results in an IF1 graph as mentioned earlier.

Estimating the execution time of a task on all the machines in a DHSS is a bottom-up multi-phase process, which depends on the selected level for code profiling. If a template-level profiling is chosen, the estimation of the execution time requires the identification of all the templates in the task and their interconnectivity. Once, all the templates have been evaluated in terms of their execution times on every machine as discussed in previous sections, a methodology is needed to estimate the execution time of the overall task on all the machines using some type of aggregation approach. One possible approach is to carry out critical path analysis of a task structure that consists of templates.

If the operational level code profiling is selected, then the results are first aggregated at the template level and then to the task level. Aggregation at the operational level will require analyzing IF1 graphs with timing values [10], obtained for every machine. The tasks in a TFG (TIG) can then be labeled with a vector, giving estimation of execution time of the tasks on various machines. This graph is know as a Code Flow Graph (Code Interaction Graph) [8], which can then be used for scheduling and mapping of the applications in a DHSS [8].

5   Conclusion

We have presented a methodology for characterizing heterogeneous supercomputer architectures and applications in a DHSS. Specifically, we have proposed a augmented code profiling and augmented analytical benchmarking techniques using a "representative set of templates". The proposed technique enables us to estimate the execution time of applications on heterogeneous architectures in a DHSS.

6   Acknowledgement

This work was partially supported by an NSF grant #CDA-9121771 and by a grant from the Purdue Research Foundation.

7   References

[1] M. Berry, D. Chen, et. al. "The Perfect Club Benchmarks: Effective Performance Evaluation of Supercomputers," *International Journal of Supercomputer Applications*, Vol. 3, No. 3, Fall 1989, pp. 5-40.

[2] D. H. Bailey, E. Barszcz, et. al. "The NAS Parallel Benchmarks-Summary and Preliminary Results" *Proc. 1991 Supercomputing*, Nov. 1991, pp.158-165.

[3] N. Bowen, C. N. Nicolau, and A. Ghafoor, "On the Assignment Problem of Arbitrary Process System to Heterogeneous Distributed Computer Systems," *IEEE Transactions on Computers*, Vol. 41, No. 4, March 1992, pp. 257-273.

[4] T. M. Conte and W. W. Hwu, "Benchmark characterization," *IEEE Computer*, Vol. 24, No. 1 January 1991, pp. 48-56.

[5] Z. Cvetanovic, "The Effects of problem partitioning, allocation, and granularity on the performance of multiple-processor systems," *IEEE Transactions on Computers*, Vol. 36, No. 4, April 1987, pp. 421-432.

[6] J. Dongarra, J. L. Martin, and J. Worlton, "Computer benchmarking: paths and pitfalls" *IEEE Spectrum*, Vol. 24, No. 7, July 1987, pp. 38-43.

[7] R. F. Freund, and D. S. Conwel, "Superconcurrency: a form of distributed heterogeneous supercomputing," *Supercomputing Review*, Vol. 3, No. 10, October. 1990, pp. 47-50.

[8] A. Ghafoor, and J. Yang,"Distributed Heterogeneous Supercomputing Management System", To be appear in the *IEEE Computer*, Vol 26. No. 6, June 1993.

[9] D. Menascé, "Scheduling tasks in heterogeneous systems," *Technical Report*, Dept. of Computer Sc., University of Maryland, College Park, 1990.

[10] D. Pease, A. Ghafoor, at el., "PAWS: A performance evaluation tool for parallel computing systems," *IEEE Computer*, Vol. 24, No. 1, January 1991, pp. 18-29.

[11] K. A. Robbins, and S. Robbins, "Dynamic behavior of memory reference streams for the Perfect Club benchmarks," *Proc. International Conf. on Parallel Processing*, Chicago, August 1992, pp. I48-I52.

[12] L. Stapleton, "Meet the Metacomputer: where the network is the computer," *Supercomputing Review*, Vol. 4, No. 3, March 1991, pp. 42-44.

[13] "Gigabit network testbeds" *IEEE Computer*, Vol. 23, No. 9, September 1990, pp. 77-80.

[14] S. Madala, and B. Sinclair, "Performance of Synchronous Parallel Algorithms with Regular Structures," *IEEE Transactions on Parallel and Distributed Systems*, Vol. 2, No. 1, January 1991, pp. 105-116.

[15] J. E. Smith, "Characterizing computer performance with a single number" *Communications of the ACM*, Vol. 31, No. 3, October 1988, pp. 1202-1206.

[16] M. Wang, S. Kim, M. A. Nichols, R. F. Freund, H. J. Siegel, and W. G. Nation, "Augmenting the optimal selection theory for superconcurrency," *Proc. of the Workshop on Heterogeneous Processing*, March 1992, pp. 13-22.

[17] J. E. Worlton, "Understanding supercomputing benchmarks." *Datamation*, Vol. 30, No. 4, September 1 1984, pp. 121-130

[18] M. Wu, and D. Gajski, "Hypertool: A Programming Aid for Message-Passing Systems," *IEEE Transactions on Parallel and Distributed Systems*, Vol. 1, No. 3, July 1990, pp. 330-343.

Adaptive Deadlock-Free Routing in Multicomputers Using Only One Extra Virtual Channel

Chien-Chun Su and Kang G. Shin
Real-Time computing Laboratory
Department of Electrical Engineering and Computer Science
The University of Michigan
Ann Arbor, MI 48109–2122
{ccsu,kgshin}@eecs.umich.edu

Abstract: *We present three protocols defining the relationship between messages and the channel resources requested: request-then-hold, request-then-wait, and request-then-relinquish. Based on the three protocols, we develop an adaptive deadlock-free routing algorithm called the $3P$ routing. The $3P$ routing uses shortest paths and is fully-adaptive, so messages can be routed via any of the shortest paths from the source to the destination. Since it is a minimal or shortest routing, the $3P$ routing guarantees the freedom of livelocks.*

The $3P$ routing is not limited to a specific network topology. The main requirement for an applicable network topology is that there exists a deterministic, minimal, deadlock-free routing algorithm. Most existing network topologies are equipped with such an algorithm. In this paper, we present an adaptive deadlock-free routing agorithm for n-dimensional meshes by using the $3P$ routing. The hardware required by the $3P$ routing uses only one extra virtual channel as compared to the deterministic routing.

1 Introduction

Distributed-memory, MIMD (multiple-instruction-multiple-data) multicomputers usually consist of a large number of nodes, each with its own processor and local memory. These nodes use an interconnection network to exchange data and synchronize with one another. Thus, the performance of a multicomputer depends strongly on network latency and throughput.

There are two types of message routing: (1) *deterministic routing* that uses only a single path from source to destination, and (2) *adaptive routing* that allows more freedom in selecting message paths. Most commercial multicomputers use deterministic routing because of its deadlock freedom and ease of

implementation. However, adaptive routing can reduce network latency, increase network throughput, and tolerate component failures. But the flexibility of adaptive routing may introduce deadlocks and/or livelocks. A deadlock occurs when a message waits for an event that will never happen. In contrast, a livelock keeps a message moving without progressing toward its destination.

A routing algorithm is said to be *minimal* if the number of hops of the routing path between two nodes is minimum and every hop brings the message closer to its destination. A minimal, fully-adaptive routing algorithm allows the message to be routed via any of the shortest paths between its source and destination. In this paper, we present a new fully-adaptive, deadlock-free, and minimal routing algorithm by using only one extra virtual channel as compared to the deterministic routing. This algorithm is based on the wormhole routing, which has higher transmission efficiency and requires less buffers than other switching methods. For the proposed routing algorithm, we assume that a deadlock-free, minimal routing function exists even if this extra virtual channel is not used. This assumption is reasonable because many popular topologies are equipped with such a routing function, e.g., *e-cube* routing for the hypercube, xy routing for the mesh, and the virtual-channel routing [2] for the torus.

Several adaptive, deadlock-free routing algorithms have been proposed in recent years. In [2], Dally and Seitz proposed the concept of virtual channel to develop deadlock-free routing algorithms. Virtual channels are logical abstractions that share the same physical link. They are time-multiplexed over a single physical link, so one separate queue must be maintained in a node for each virtual channel. Virtual channels are used to remove the cycles in a channel-dependency graph, thus providing deadlock freedom in message transmissions. However, the algorithms in [2] are deterministic.

The work described in this paper was supported in part by the National Science Foundation under Grant MIP-9203895. The opinions, findings, and recommendations expressed in this publication are those of the authors, and do not necessarily reflect the views of the NSF

In [6], Linder and Harden extended the concept of virtual channel to multiple, virtual interconnection networks that provide adaptability, deadlock-freedom and fault-tolerance. Each link is shared by many virtual channels, and the number of virtual channels used depends on how many virtual networks are needed. These virtual channels can be divided into several groups or virtual networks. Message passing inside a virtual network is deadlock-free and messages are constrained to travel through virtual networks in a defined order. When the message is blocked in a virtual network, it can keep going forward via another virtual network, thereby increasing routing adaptability. The chief disadvantage of this method is that many virtual channels (in general, an exponential function of the network dimension) are required, e.g., a k-ary n-cube needs $2^{n-1}(n+1)$ virtual channels.

In [4], Glass and Ni proposed partially-adaptive routing algorithms for 2D and 3D meshes without adding physical or virtual channels. They first investigated the possible deadlock cycles on 2D and 3D-meshes, then proposed some prohibited turns of these cycles to prevent deadlocks. However, if the minimal routing is required, then there exists only a single routing path for at least a half of source-destination pairs. Because this algorithm cannot route messages along any of shortest paths in the network, it is called partially-adaptive routing. Also, the higher the network dimension, the more source-destination pairs, each with only a single routing path, will result. The authors of [5] extended the same concept to n-dimensional meshes, k-ary n-cubes and hypercubes. In k-ary n-cubes, if $k > 4$, nonminimal routing should be used, thus taking more hops than needed and possibly introducing livelocks.

In [1], Chien and Kim proposed a planar-adaptive routing algorithm which limits the routing freedom to two dimensions at a time. The reduced freedom makes it possible to prevent deadlocks with only a fixed number of virtual channels, which is independent of network dimension. The hardware overhead is much less than that of the algorithm in [6].

In this paper, we propose an adaptive, deadlock-free, and livelock-free routing algorithm with less hardware overhead (except for partially-adaptive algorithms) and better adaptability than others. Section 2 presents three protocols that will be used to propose an adaptive deadlock-free routing algorithm. Section 3 describes the proposed routing algorithms for n-dimensional meshes. The proposed algorithm is compared with several other adaptive algorithms. The paper concludes with Section 4.

2 New Protocols for Adaptive, Deadlock-free Routing

In message-passing multicomputers, a message may be broken into one or more packets for transmission. According to the property of wormhole routing, a packet contains one or more flow-control digits (flits). The first flit of a packet (head flit) has to build the transmission path between the source and destination. Each flit of a packet following the head flit advances as soon as the preceding flit moves along (i.e., flits pipelining) and gets blocked when the required channel resources are unavailable.

A simple examination of the protocol for requesting channel resources in multicomputers leads to two cases:

1. Request-then-wait : While the requested channels are held by the other packets, the requesting packet will block and wait for the channels to be available.

2. Request-then-hold : While the requested channels are available, the packet will hold the channel and route the flits forward.

Now, we propose to add a third protocol:

3. Request-then-relinquish : Once a packet fails to get its requested channels, it terminates the request and does *not* wait for these channels either. In other words, no packet waits for a channel using this protocol.

Based on the above three protocols, we propose a new adaptive deadlock-free routing algorithm. First, we assume that:

1. The interconnection network can be divided into two virtual interconnection networks, VIN_1 and VIN_2, where VIN_1 supports deadlock-free minimal routing and VIN_2 uses one virtual channel (called *extra virtual channel*) to share the bandwidth of a physical link with the channels in VIN_1. The channels of VIN_2 are used to enhance routing adaptability.

2. Each virtual channel in VIN_1 is assigned a channel number. Also, packets requesting for the virtual channels in VIN_1 should obey strict increasing or decreasing order of channel numbers.

3. Only request-then-hold and request-then-relinquish protocols are used on the extra virtual channel (VIN_2), i.e., a requesting packet

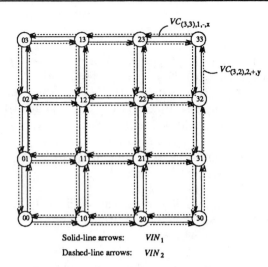

Figure 1: Two virtual interconnection networks in a 2-D mesh.

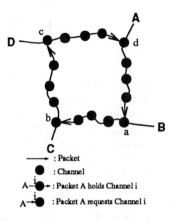

Figure 2: A deadlock of four packets.

never waits for the extra virtual channel. Also, only request-then-wait and request-then-hold protocols are used on the virtual channels in VIN_1.

Fig. 1 shows VIN_1 (solid-line arrows) and VIN_2 (dashed-line arrows). In this case, each directed physical link is shared by two virtual channels. In case VIN_1 needs to use two virtual channels over each directed link (e.g., k-ary n-cubes), a total of three virtual channels are required on each physical link. The proposed adaptive routing algorithm, which can be applied to various topologies, is described as follows.

1. A requesting packet first checks if any extra virtual channel is available. Also, the minimum distance between the current location of the head flit and the packet's destination must be reduced after taking this hop (i.e., minimal routing). If these two conditions hold, then use the request-then-hold protocol else use the request-then-relinquish protocol, and search for the extra virtual channel of a different dimension. Repeat this search until the packet finds a channel of VIN_2 or no suitable channel can be found.

2. If no suitable channel of VIN_2 can be found in Step 1, then the packet waits for the virtual channel of VIN_1 according to an increasing or decreasing order of channel numbers.

For convenience, this proposed algorithm is called the $3P$ (three protocols) routing.

Theorem 1: The $3P$ routing is deadlock-free.

Proof: We prove by contradiction. Assume there exists a deadlock for the proposed routing algorithm, then we can get a channel-dependency graph with a circular wait among the channel resources as shown in Fig. 2, where the arrow lines represent the packets and the black circles indicate the channels requested or held. This is a deadlock of four packets; one can extend this proof to those cases with more packets involved.

According to the proposed routing algorithm, the extra virtual channels never uses the request-then-wait protocol, so the channels a,b,c,and d in the graph are virtual channels of VIN_1. Since the deadlock-free routing in the virtual channels of VIN_1 should obey a strict order (say increasing) of requested channel numbers, the relationships among channel a, b, c, and d are summarized as:

$$a < b \text{ for packet B}$$
$$b < c \text{ for packet C}$$
$$c < d \text{ for packet D}$$
$$d < a \text{ for packet A.}$$

Thus, a<b<c<d<a, a contradiction. Therefore, no deadlock exists. □

In fact, the channels between a and b, b and c, c and d, d and a can be the channels of VIN_1 or VIN_2. If we remove or add a channel between two blocked channels, this change does not affect the deadlock freedom. In the next section, we apply this routing algorithm to n-dimensional meshes.

3 Application and Comparison

Before discussing the details of adaptive routing algorithms for different topologies, it is convenient to label each output virtual channel (VC) of a node with $VC_{node,VIN,direction,dimension}$, where $node$ specifies which processing node to be used; $VIN = 1(2)$ indicates VIN_1 (VIN_2); $direction = +(-)$ represents positive (negative) direction; and $dimension$ specifies which dimension to be used. For example, in Fig. 1, $VC_{(3,3),1,-,x}$ represents the output channel of node (3,3) of VIN_1 in the negative direction of dimension x.

Due to the limited space, we only present an adaptive, deadlock-free routing algorithm for n-dimensional meshes by using the $3P$ routing. For other topologies, such as k-ary n-cubes, C-wrapped hexagonal meshes [7] and so on, a similar method can be used, though the virtual channels required in VIN_1 may be different. An n-dimensional mesh consists of $k_0 \times k_1 \times \cdots \times k_{n-2} \times k_{n-1}$ nodes, where $k_i \geq 2$ is the number of nodes along dimension i. Each node X is represented by n coordinates, $(x_0, x_1, \ldots, x_{n-2}, x_{n-1})$, where $0 \leq x_i \leq k_i - 1$, $0 \leq i \leq n - 1$. Two nodes X and Y are neighbors if and only if $x_i = y_i$ for all i, $0 \leq i \leq n - 1$, except one, say j, such that $y_j = x_j \pm 1$. Thus, each node has from n to $2n$ neighbors, depending on its location in the mesh [4]. If X and Y are neighbors, then the channel of dimension i at node X is in positive direction when $x_i = y_i - 1$, or in negative direction when $x_i = y_i + 1$.

The xy routing [4] in meshes can be extended to n-dimensional meshes (called *extended xy* routing). That is, all packets should follow the dimension order, $0 \to 1 \to \ldots \to n - 2 \to n - 1$. The *extended xy* routing is deadlock-free, minimal and deterministic. Therefore, based on the *extended xy* routing, the proposed $3P$ routing can be applied to n-dimensional meshes. Based on the assumption of $3P$ routing, we need two virtual interconnection networks ($VINs$): VIN_1 supports *extended xy* routing and VIN_2 is used to enhance adaptability. Therefore, the bandwidth of each link needs to be shared by two virtual channels.

Adaptive routing algorithm for n-dimensional meshes:

Input: Source Node S $= (s_0, s_1, \ldots, s_{n-1})$;
Destination Node D $= (d_0, d_1, \ldots, d_{n-1})$;
Current Intermediate Node C
$= (c_0, c_1, \ldots, c_{n-1})$;
Initial: Routing Tag R $= (r_0, r_1, \ldots, r_{n-1})$

$$= (d_0 - s_0, d_1 - s_1, \ldots, d_{n-1} - s_{n-1});$$

Step 1. Update R; /*$r_i := r_i + 1$ or $r_i := r_i - 1$*/
Step 2. If (R==0), flits arrive at destination;
Step 3. If (((any $r_i > 0$) && ($VC_{C,2,+,i}$ available))
|| ((any $r_i < 0$) && ($VC_{C,2,-,i}$ available))),
    if $r_i > 0$ then send packet via $VC_{C,2,+,i}$
        else send packet via $VC_{C,2,-,i}$;
/*i can be chosen randomly or by any
    selection function */
Step 4. $i = min\{j : r_j \neq 0\}$;
    /*Find the lowest dimension $i \exists r_i \neq 0$. */
Step 5. If $r_i > 0$ then request $VC_{C,1,+,i}$
        else request $VC_{C,1,-,i}$;

In Step 1, the routing function used to update R is that (1) if the flit comes from $VC_{neighbor,*,+,i}$ then $r_i := r_i - 1$ and (2) if the flit comes from $VC_{neighbor,*,-,i}$, then $r_i := r_i + 1$, where * means "don't care." Since the routing tags can be used to specify the routing paths, it is not necessary to carry the source and destination addresses in the packets. In Step 3, request-then-hold and request-then-relinquish protocols are used to request the virtual channels of VIN_2. In Step 5, request-then-hold and request-then-wait protocols are used to request the channel of VIN_1.

Theorem 2: The $3P$ routing on n-dimensional meshes is minimal and deadlock-free. □

In n-dimensional meshes, the number of shortest paths for a source–destination pair is $(|r_0| + |r_1| + \cdots + |r_{n-1}|)!/|r_0|!|r_1|! \cdots |r_{n-1}|!$. Because the $3P$ routing on n-dimensional meshes can route packets via *any* available virtual channel of VIN_2 along shortest paths, it is fully-adaptive. The proposed algorithm only needs two virtual channels over each physical link. We can apply the methods proposed in [6] (abbreviated as the L-H algorithm in Fig. 3) to n-dimensional meshes, but it requires 2^{n-1} virtual channels. The planar-adaptive routing algorithm [1] (C-H algorithm in Fig. 3), which is not fully-adaptive, needs three virtual channels. The partially-adaptive routing algorithms in [4, 5] (G-N algorithm in Fig. 3) do not add any virtual channel, but they are not fully-adaptive.

Fig. 3 shows a table that compares the virtual-channel requirements of the $3P$ routing and other adaptive deadlock-free routing algorithms for several topologies. The proposed $3P$ routing requires only one additional virtual channel as compared to the deterministic routing. Furthermore, it is minimal and fully-adaptive. It can also be used to construct an adaptive deadlock-free routing algorithm for C-wrapped H-meshes [7]. The number of virtual channels required by the $3P$ routing on H-meshes is the same as that of k-ary n-cubes.

The $3P$ routing turns out to have a flavor somewhat similar to the one proposed in [3], although both have been developed independently. However, the method proposed in [3] dealt only with adaptive routing. The fault-tolerant $3P$ routing can be derived by changing the protocols of links around the faulty nodes or links. We will report the details of fault-tolerant routing in a forthcoming paper.

Topologies or properties \ Algorithms	Determ.	L-H	C-K	G-N	3P
2D mesh	1	2	2	1	2
Torus	2	6	4	#	3
n-D mesh	1	2^{n-1}	3	1	2
k-ary n-cube	1	$2^{n-1}(n+1)$	6	#	3
Hypercube	1	2^{n-1}	3	1	2
H-mesh	2	*	*	*	3
adaptability	no	fully	planar	partial	fully
minimal	yes	yes	yes	yes##	yes

*: Not mentioned in the published paper.
#: If k<5, then use 1 virtual channel, otherwise use nonminimal routing.
##: The class of k-ary n-cubes uses minimal routing only if k<5.

Figure 3: Comparison between virtual-channel requirements and routing properties for adaptive deadlock-free routing algorithms.

4 Conclusion

Communication efficiency is one of the most important factors to consider when designing a multicomputer system. The interprocessor communication speed strongly depends on the routing strategy, processor and data-link speed, and network topology. In this paper, we proposed three protocols of wormhole routing defining the relationship between messages/packets and channel resources: request-then-hold, request-then-wait, and request-then-relinquish. Based on these three protocols, an adaptive deadlock-free routing algorithm (the $3P$ routing) is proposed, which is then applied to various network topologies. Specifically, we presented adaptive routing algorithms by applying the $3P$ routing for n-dimensional meshes. We compare the $3P$ routing with the existing adaptive routing algorithms. For n-dimensional meshes and hypercubes, the $3P$ routing requires two virtual channels on each physical link. For k-ary n-cubes and C-wrapped hexagonal meshes, it requires three virtual channels. As compared to the deterministic routing, such as the xy routing, *extended xy* routing, e cube routing, and virtual channel routing [2], the $3P$

routing requires only one additional virtual channel regardless of network size/dimension. In addition to its deadlock freedom, the $3P$ routing is also minimal and fully-adaptive. The minimal routing ensures livelock-freedom. The fully-adaptiveness enables messages/packets to be transmitted via any of the shortest paths. Either the existing adaptive routing algorithms are just partially-adaptive, or require an excessive amount of hardware. By contrast, the proposed $3P$ routing uses little additional hardware despite its advantage of being fully-adaptive and minimal.

There are two directions of future work for the $3P$ routing. First, we need to implement the $3P$ routing with a compact hardware design. Second, it is necessary to evaluate the performance improvement by using the $3P$ routing.

References

[1] A. A. Chien and J. H. Kim, "Plannar-adaptive routing: Low-cost adaptive networks for multiprocessors," *in 19th Annual International Symposium on Computer Architecture*, pp. 268–277, 1992.

[2] W. J. Dally and C. L. Seitz, "Deadlock-free message routing in multiprocessor interconnection networks," *IEEE Trans. on Computers*, vol. 36, no. 5, pp. 547–553, May 1987.

[3] J. Duato, "Improving the efficiency of virtual channels with time-dependent selection functions," *in Proc. Parallel Architectures and Languages Europe*, pp. 635–650, June 1992.

[4] C. J. Glass and L. M. Ni, "Adaptive routing in mesh-connected networks," *in Proceedings of the 1992 International Conference on Distributed Computing Systems*, pp. 12–19, 1992.

[5] C. J. Glass and L. M. Ni, "The turn model for adaptive routing," *in Proceedings of the 1992 International Symposium on Computer Architecture*, pp. 278–287, 1992.

[6] D. H. Linder and J. C. Harden, "An adaptive and fault tolerant wormhole routing strategy for k-ary n-cubes," *IEEE Trans. on Computers*, vol. 40, no. 1, pp. 2–12, Jan. 1991.

[7] K. G. Shin, "HARTS: A distributed real–time architecture," *IEEE Computer*, vol. 24, no. 5, pp. 25–34, May 1991.

A Hybrid Shared Memory/Message Passing Parallel Machine[†]

Matthew I. Frank   and   Mary K. Vernon

Computer Sciences Department
University of Wisconsin–Madison
Madison, WI 53706
{mfrank, vernon}@cs.wisc.edu

Abstract

Current and emerging high-performance parallel computer architectures generally implement one of two types of communication mechanisms: shared memory (SM) or message passing (MP). In this paper we propose a hybrid SM/MP architecture together with a hybrid SM/MP programming model, that we believe effectively combines the advantages of each system. The SM/MP architecture contains both a high-performance coherence protocol for shared memory, and message-passing primitives that coexist with the coherence protocol but have no coherence overhead. The SM/MP programming model provides a framework for safely and effectively using the SM/MP communication primitives. We illustrate the use of the model and primitives to reduce communication overhead in SM systems.

1. Introduction

Two data communication mechanisms currently dominate in existing and emerging large-scale parallel computers. In the Message Passing (or Distributed Memory) model, each processor has its own private main memory and communicates with other processors only by sending and receiving messages. The Shared Memory model comes in two main flavors. in the Static Shared Memory (or Non-Uniform Memory Access, NUMA) model, each processor is physically adjacent to a portion of global memory and communicates with other processors by reading and writing locations in it's own or other portions of the shared memory. In the Dynamic Shared Memory (i.e., the Cache-Coherent or Distributed Shared Memory) model, shared data resides in a physically distributed memory and is (automatically) replicated, on demand, to provide processors with fast access to local copies.

Dynamic Shared Memory systems are easier to program in that shared read-only structures may be efficiently accessed by all processors without re-naming, and shared writable objects may migrate from processor to processor as necessary to balance the computational load. However, the data replication comes at the cost of 1) overhead for keeping the copies of shared data coherent, 2) fixed size data transfers (corresponding to a cache block or a page in virtual memory), and 3) a more complex programming model that includes some understanding of how the fixed-size copies of shared data are kept coherent. The ultimate viability of dynamic SM systems will depend on several factors, including the efficiency of the programming model, coherence mechanisms, and communication primitives, and the simplicity of the required hardware and/or compiler support. We address some of these issues in this paper.

Previous work on dynamic SM systems has focused on 1) mechanisms that guarantee that accessible local copies are coherent within some memory consistency model, and 2) the use of pre-fetching, update coherence primitives, and/or multi-threading to hide memory latencies. In this paper we take a different approach. We propose a simple message-passing facility for overlapping communication and computation in dynamic SM systems. A key idea is that a copy of a shared memory block that is not guaranteed to stay coherent can be viewed as a *message* that can be sent from one cache or memory to another. Destination processors can *read* such messages using a special read operation, so as not to accidentally read a message when reading a coherent value was intended. Like standard MP primitives, our proposed primitives and programming model expose the communication overhead of the machine so that programmers and compiler writers can easily evaluate communication costs. Unlike standard MP primitives, our proposed primitives are tightly integrated in the shared memory architecture (with only minor additional hardware support), such that shared objects need not be renamed, messages need not interrupt destination processors, and messages can be used in some cases simply to reduce coherence overhead.

In section 2 we provide some background and more detailed motivation for our proposed SM/MP primitives. In section 3 we define the new primitives and the SM/MP programming model, and in section 4 we discuss the implementation of the primitives in an SM/MP architecture. Section 5 compares our proposed primitives with various ideas in the previous literature.

[†]This work was supported in part by the National Science Foundation (CDA-9024618, CDA-9024144).

2. Background: Dynamic Shared Memory

Previous studies have observed that the dynamic shared memory programming model must include the notion of *non-uniform* memory access times, since programs developed using uniform memory access semantics have notoriously poor performance [Ande91, Cher91a, Hill90, Lin90, Mart89]. What is required is an approximate model of the coherence operations that provides a framework for assessing trade-offs between logically partitioning the data among the processes and balancing the computational load among the processes. We provide an informal model which provides a basis for our SM/MP programming model, and then comment on some disadvantages of the SM communication primitives.

Due to the benefits of read sharing as well as the overhead of updating all copies on all write operations, the following rules provide a minimal model for most high-performance coherence protocols:[1]

1. An *ordinary* read operation retrieves a *shared* copy of the page or block from main memory unless the processor already has a *shared* or *exclusive* copy.

   Side effect: If another processor has an exclusive copy of the page or block, the exclusive copy will be changed to a shared copy.

2. An *ordinary* write operation retrieves an *exclusive* copy of the page or block from main memory, unless the processor already has an exclusive copy.

   Side effect: All other *shared* or *exclusive* copies of the page or block are destroyed.

Retrieving, modifying and/or destroying copies at remote memories involves communication. As in the MP model, non-local operations are viewed as *expensive*, the exact cost being machine-specific.

Using the above model, efficient SM programs will often have similar partitioning of the computational load as their MP counterparts, due to similar models of when communication occurs. However, there are several disadvantages of the shared memory communication primitives. Ordinary read and write operations allow communication and computation to be overlapped only if multithreading is used. Prefetching may be used to address this problem, but requires the consumer(s) to request the data (at an appropriate time) and requires communication for the producer to regain write access to

---

[1]Note that the model can be augmented or adapted to systems with additional communication primitives such as read-exclusive and write-update operations, as well as to systems that use compiler-assisted coherence protocols in which incomplete data-dependence information may cause extra, unnecessary invalidation messages.

the data. Alternatively, an update protocol might be used to update rather than invalidate the consumers' copies. However, in hardware coherence protocols this forwarding is done each time a processor writes a word in the block; and in any case, all copies are updated whenever forwarding is done. In iterative algorithms where not all consumers need to receive the data in every iteration, the update operation has particularly high overhead. The above coherence overheads include unnecessary traffic over the interconnect as well as unnecessary memory access latencies.

3. The SM/MP Programming Model

In this section we propose extensions to the dynamic SM programming model that 1) allow communication to take place with no coherence overhead and 2) facilitate maximal overlap of communication and computation. An efficient implementation of the model that can be integrated with any coherence protocol is outlined in section 4.

3.1. MP Primitives for Dynamic Shared Memory

We propose a new type of copy of a shared memory block or page, called *possibly-stale*. These copies serve as messages that can be sent from one processor to another using the MP-send and MP-read operations defined below. In some cases, the communication implemented by MP-send and MP-read will only work properly if a program can insure that sequential MP-sends to some particular destination will execute in the order they are generated. We define an MP-sync operation that can be used to guarantee this ordering. For cases where the process that computes a value does not know the identity of the consumer, and/or cases where the consumer can more effectively schedule the message transmission, we define an MP-prefetch operation that allows the consumer to fetch a message. Thus, the following primitives are defined to manipulate possibly-stale blocks:

- **MP-send**: The MP-send operation creates a new copy of a specified block or page and sends the copy to a specified processor. The new copy is of type possibly-stale. The destination processor keeps the new copy unless it already has a shared or exclusive copy of the block.

- **MP-read**: The MP-read operation returns a value from a shared, exclusive, or possibly-stale copy of the block. If the processor does not have a copy of the block, the MP-read operation retrieves a new copy of type possibly-stale.

- **MP-prefetch**: The MP-prefetch operation retrieves a copy of the block, of type possibly-stale, unless the processor already has a copy.

- **MP-sync**: The MP-sync operation completes when all previous MP-operations issued by the processor have completed. MP operations that are issued after the MP-sync operation will be delayed until the MP-sync completes.

Typically, the processor issuing an MP-send operation has recently modified the location and thus has an *exclusive* copy of the block. As under the MP model, the processor issuing the MP-send may re-write the block without incurring any coherence overhead since the processor retains the block in state *exclusive*. The new *possibly-stale* copy is allowed to become incoherent, but can only be read using the MP-read primitive. (An ordinary read or write operation by a processor will replace a *possibly-stale* copy with a *shared* or *exclusive* copy, respectively.) From this perspective the *possibly-stale* copy is a message, and the MP-send and MP-read primitives can be used together to implement a restricted but useful form of message passing communication.

MP-send operations are treated as write operations in the memory consistency model; thus the processor may not be required to wait for these operations to complete except at specified synchronization points. Additionally, the programmer must keep in mind that the MP-read operation can return the value of a possibly stale copy, or the value of a coherent copy of the data. In this sense, the primitives must be used in a way that, for a correct SM/MP program, deleting the MP-send operations and replacing MP-read operations with ordinary reads will yield a program that is functionally the same.

3.2. Examples

We present two examples of the use of the SM/MP model. In both examples, explicit MP operations appear in the program, but might alternatively be generated by a compiler. Also, explicit MP-send operations operate on data objects; we assume that the compiler will optimize these operations by appropriately deleting duplicate operations for the same block or page. The first example is an iterative chaotic SOR algorithm where communication is reduced by calculating a schedule, either deterministically or probabilistically, which determines how often a process sends updated values to each other process (e.g. [Fuen92]). Due to the chaotic nature of the algorithm, processes are allowed to use stale values if they have not received the update. Note that MP-prefetch operations can be used instead of the MP-send operations to implement a similar communication schedule. In this example, the MP primitives are highly efficient, whereas standard prefetching or update-write operations would have significantly higher overhead.

The second example is one in which two processes communicate via a circular buffer, as is often done in pipelined and other course-grain dataflow algorithms (e.g., [Schn89]). In contrast to the previous example, the possibly-stale copies are known to be up-to-date at the time of the MP-read operations. Thus, the MP primitives are used to implement the shared buffer synchronization and to reduce coherence overheads as compared with write-update protocols and standard prefetching.

```
initialize x, A;
Spawn P threads;
task k:
  Apply-MP-read x;
  repeat until x converges
    for each i in thread k's partition of x
    x_i := f(A, x);
    compute subset-of-P to receive x_i
    MP-send (x_i , subset-of-P);
```

Figure 1. Chaotic SOR: SM/MP Algorithm (One Process)

```
Buffer[B] of items;
/* producer's buffer index */
i = 1, copy_of_i = 1;
/* consumer's buffer index */
j = 1, copy_of_j = 0;

Producer
  Apply-MP-read copy_of_j;
  repeat
    while not(i == copy_of_j)
      i := (i + 1) mod B;
      create value in Buffer[i];
      MP-send(Buffer[i], Consumer);
    every Nth iteration
      MP-sync;
      copy_of_i := i;
      MP-send(copy_of_i, Consumer);

Consumer
  Apply-MP-read copy_of_i, Buffer;
  repeat
    while not(j == copy_of_i)
      j := (j + 1) mod B;
      use item in Buffer[j];
    every Nth iteration
      MP-sync;
      copy_of_j := j;
      MP-send(copy_of_j, Producer);
```

Figure 2. Communication Using Circular Buffers: SM/MP Implementation

4. The SM/MP Architecture

The communication primitives defined in section 3 can be integrated with dynamic shared memory coherence protocols with very little increase in hardware complexity. Below we illustrate this for a generic directory-based cache coherence protocol with cache block states *exclusive*, *shared*, and *invalid*.

Table 1 specifies the actions taken by the cache in response to MP requests issued by the processor as well as MP operations that come in from the network. A new cache block state, *possibly stale* is used to implement the MP operations. No other new storage is required in the memory system. In particular, the state of the blocks in the directory is never modified by the MP operations.

MP-send requests are translated into remote MP-put requests if the processor has a shared or exclusive copy of the block. Otherwise, the cache forwards the MP-send request to the directory.

For a network MP-put request, if the block containing the word isn't present in the cache, new space is allocated for the block and the new data is stored in state *possibly stale*. If the block is in state *possibly-stale*, the new data overwrites the existing block. If the block in state *shared* or *exclusive* (or some other coherent state), the copy in the destination cache is either coherent with the copy from which the MP-put was generated, or is more recent than that copy, or is going to be invalidated by the write operation that generated the value in the MP-put. In all of these cases, it is safe to ignore the incoming MP-put.

For processor MP-read requests, the cache responds with the data if the block is in any valid state. Otherwise, the cache issues a remote MP-get request, which returns a copy of the data that will not be kept coherent by the hardware. A processor MP-prefetch request leads to a remote MP-get request unless the processor has a coherent copy of the block, in which case the MP-prefetch is ignored.

Finally, an ordinary processor read or write request purges the block if it is in state Possibly Stale, and fetches a coherent block. When a block in state *possibly stale* is deleted from the cache, no write-back occurs.

The proposed MP primitives integrate simply and easily with hardware cache coherence protocols. Implementation can similarly be integrated with distributed shared memory protocols (a.k.a. virtual shared memory) and/or with compiler-assisted coherence protocols.

5. Comparison with Previous Approaches

Prefetching techniques have been the principal approach advocated in the literature for reducing read latencies (i.e., for overlapping communication and computation) in large-scale SM systems. Synchronized prefetching techniques have been proposed [Good89] for

Table 1: State Transitions for MP Requests in the SM/MP Coherence Protocol

Request	Cache Block State	Next State	Action
Processor MP-send	Shared or Exclusive	Unchanged	forward MP-put request to remote processor
	Possibly Stale or Invalid	Unchanged	forward MP-send request to directory
Network MP-Put	Shared or Exclusive	Unchanged	forward Ack to sender
	Possibly Stale or Invalid	Possibly Stale	insert block & forward Ack to sender
Processor MP-read	Not Invalid	Unchanged	return value
	Invalid	Possibly Stale	forward MP-get request to directory
Processor MP-prefetch	Shared or Exclusive	Unchanged	no action
	Possibly Stale or Invalid	Possibly Stale	forward MP-get request to directory
Processor Read	Possibly Stale	Shared	forward get-S request to directory
Processor Write	Possibly Stale	Exclusive	forward get-X request to directory
Network Invalidate	Possibly Stale	Possibly Stale	no action

simulating producer-initiated data transfers (in certain cases) in dynamic shared memory systems. These operations allow a prefetch to be issued early and to remain pending in the memory system until a new value is released by another processor. Recent examples include QOLB [Jame90], Notify [Cher91b], and cooperative prefetch [Hill92]. The advantages of the SM/MP primitives include: 1) simpler hardware support, 2) more general and efficient support for multiple consumers, 3) lower overhead and a simpler programming model in many cases (such as those illustrated in section 3.2). Whether synchronized prefetching or synchronized MP-prefetching would be desirable in an SM/MP system is an open question.

Write-update operations, which update rather than invalidate all copies of a block whenever a particular processor writes the block, are available in some systems [Leno92] and have recently been advocated as a technique for overlapping communication and computation in SM systems [Cart91, Rost93]. Our SM/MP primitives have some features in common with write-update primitives. Key differences include: 1) the explicit use of possibly stale copies allows the MP-send operations to be optimized and used more selectively than write-updates, 2) the MP-prefetch primitive, and 3) the SM/MP programming model that aids in identifying cases where the MP operations might be advantageous.

Lee and Ramachandran [Lee91] have proposed a selective WRITE-GLOBAL primitive, as well as primitives that read and write only the local cache, to support implementing a weak-consistency coherence model in software. They also propose that the software use the local-only operations for private variables, to avoid false-sharing conflicts in some cases where private and shared data are assigned to the same cache block.

Kranz et. al. have recently proposed integrating message-passing primitives into dynamic shared memory systems [Kran93]. Their principle motivations were to bundle data into large messages, and to combine synchronization with data transfer. Key differences in the SM/MP approach include: 1) the MP primitives are embedded in the shared memory hardware, and 2) SM/MP messages do not interrupt the destination processor. It remains an open question whether the SM/MP model should be augmented to include an MP-send-and-interrupt primitive.

6. Summary and Conclusions

In this paper we have proposed a set of message passing primitives for dynamic shared memory systems. We have developed the hybrid SM/MP architecture by carefully extending SM systems to incorporate key performance features of MP systems. This involved introducing a new state for cache blocks or memory pages, called possibly-stale, that can only be read by a special MP-read operation, as well as the view that blocks in this state are *messages*. The two most significant advantages of the proposed primitives are the simplicity of their implementation, and the simplicity of the model they support for overlapping communication and computation. We have also shown that the primitives allow the elimination of unnecessary coherence overhead in certain important cases. We thus believe that these primitives may be useful in improving the viability of Dynamic Shared Memory systems for parallel computing.

Key questions we are currently investigating include the performance benefits of the proposed SM/MP primitives for various applications, whether there exist examples where a variant of synchronized prefetch can improve the performance of an SM/MP architecture (thus justifying the memory system storage for pending requests), whether an interrupting MP-send operation is desirable, and whether applying existing compiler technology for distributed memory machines to SM/MP systems can yield improved performance as compared with existing compilers for dynamic shared memory systems.

References

[Ande91] Anderson, R. J. and L. Snyder, "A Comparison of Shared and Nonshared Memory Models of Parallel Computation", *Proc. of the IEEE*, Vol. 79, No. 4 , April 1991, pp. 480-487.

[Cart91] Carter, J. B., J. K. Bennett and W. Zwaenepoel, "Implementation and Performance of Munin", *Proc. 13th ACM Symp. on Operating System Principles*, Pacific Grove, CA , pp. 152-164, October 1991.

[Cher91a] Cheriton, D. R., H. A. Goosen and P. Machanick, "Restructuring a Parallel Simulation to Improve Cache Behavior in a Shared Memory Multiprocessor: A First Experience", *Int'l. Symp. on Shared Memory Multiprocessing*, Tokyo , pp. 109-118, April 1991.

[Cher91b] Cheriton, D. R., H. A. Goosen and P. D. Boyle, "Paradigm: A Highly Scalable Shared-Memory Multicomputer Architecture", *Computer,* Vol. 24, No. 2 , February 1991, pp. 33-46.

[Fuen92] Fuentes, Y. O. and S. Kim, "Parallel Computational Microhydrodynamics: Communication Scheduling Strategies", *AIChE Journal,* Vol. 38, No. 7 , July 1992, pp. 1059-1078.

[Good89] Goodman, J. R., M. K. Vernon and P. J. Woest, "A Set of Efficient Synchronization Primitives for a Large-Scale Shared-Memory Multiprocessor", *Proc. 3rd Int'l. Conf. on Architectural Support for Programming Languages and Operating Systems*, Boston , pp. 64-75, April 1989.

[Hill90] Hill, M. D. and J. R. Larus, "Cache Considerations for Multiprocessor Programmers", *Communications of the ACM,* Vol. 33, No. 8 , August 1990, pp. 97-102.

[Hill92] Hill, M. D., J. R. Larus, S. K. Reinhardt and D. A. Wood, "Cooperative Shared Memory: Software and Hardware for Scalable Multiprocessors", *5th Int'l. Conf. on Architectural Support for Programming Languages and Systems*, Boston , pp. 262-273, October 1992.

[Jame90] James, D. V., A. T. Laundrie, S. Gjessing and G. S. Sohi, "Distributed-Directory Scheme: Scalable Coherent Interface", *IEEE Computer,* Vol. 23, No. 6 , June 1990, pp. 74-77.

[Kran93] Kranz, D., K. Johnson, A. Agarwal, J. Kubiatowicz and B. Lim, "Integrating Message-Passing and Shared-Memory: Early Experience", to appear *Symp. on Principles and Practice of Parallel Programming (PPoPP)*, May 1993.

[Lee91] Lee, J. and U. Ramachandran, "Architectural Primitives for a Scalable Shared Memory Multiprocessor", *3rd ACM Symp. on Parallel Algorithms and Architectures*, pp. 103-114, July 1991.

[Leno92] Lenoski, D., J. Laudon, K. Gharachorloo, W. Weber, A. Gupta, J. Hennessy, M. Horowitz and M. S. Lam, "The Stanford Dash Multiprocessor", *Computer,* Vol. 25, No. 3 , March 1992, pp. 63-79.

[Lin90] Lin, C. and L. Snyder, "A Comparison of Programming Models for Shared Memory Multiprocessors", *Int'l. Conf. on Parallel Processing,* Vol. II , August 1990, pp. 163-170.

[Mart89] Martonosi, M. and A. Gupta, "Tradeoffs in Message-Passing and Shared-Memory Implementations of a Standard Cell Router", *Proc. Int'l. Conf. on Parallel Processing,* Vol. III , August 1989, pp. 88-96.

[Rost93] Rosti, E., E. Smirni, T. D. Wagner, A. W. Apon and L. W. Dowdy, "The KSR1: Experimentation and Modeling of Poststore", to appear *Proc. of the 1993 ACM Sigmetrics Conference*, May 1993.

[Schn89] Schneider, D. A. and D. J. DeWitt, "A Performance Evaluation of Four Parallel Join Algorithms in a Shared-Nothing Multiprocessor Environment", *Proc. 1989 SIGMOD Conf.*, Portland, Oregon , June 1989.

Scalability Study of the KSR-1 *

Umakishore Ramachandran   Gautam Shah   S. Ravikumar

Jeyakumar Muthukumarasamy

College of Computing,

Georgia Institute of Technology, Atlanta, GA 30332

e-mail: rama@cc.gatech.edu

Abstract – *There has been concern in the architectural community regarding the scalability of shared memory parallel architectures owing to the potential for large latencies for remote memory accesses. KSR-1 is a recently introduced commercial shared memory parallel architecture, and the scalability of KSR-1 is the focus of this research. Our key conclusions are as follows: The communication network of KSR-1 is fairly resilient in supporting simultaneous remote memory accesses from several processors. The multiple communication paths realized through this pipelining help in the efficient implementation of tournament-style barrier synchronization algorithms. The architectural features of KSR-1 such as the poststore and prefetch are useful for boosting the performance of parallel applications. The network does saturate when there are simultaneous remote memory accesses from a fully populated (32 node) ring.*

1 Introduction

Scalability is a term that is used to signify whether a given parallel architecture shows improved performance for a parallel algorithm with increased number of processors. Performance improvement with added processors is thus used as a measure of scalability. The notion of *speedup*, defined as the ratio of execution time of a given program on a single processor to that on a parallel processor, is perhaps the earliest metric that has been widely used to quantify scalability. A drawback with this definition is that an architecture would be considered non-scalable if an algorithm running on it has a large sequential part. There have been several recent attempts at refining this notion and defining new scalability metrics (see for instance [7, 5]).

With the evolution of parallel architectures along two dimensions, namely shared memory and message passing, there has been a considerable debate regarding their scalability given that there is a potential for large latencies for remote operations in both cases as the size of the machine grows. The communication latency in message passing machines is explicit to the programmer and not implicit depending on the memory access pattern of an algorithm as in the case of shared memory machines. Thus there is a tendency to believe that message passing architectures may be

more scalable than its architectural rival as is evident from the number of commercially available message passing machines [2]. Yet, there is considerable interest in the architectural community toward realizing scalable shared memory multiprocessors. Indeed the natural progression from sequential programming to shared memory style parallel programming is one of the main reasons for such a trend.

KSR-1 is a shared memory multiprocessor recently introduced by Kendall Square Research (Section 2). The focus of this paper is to study the scalability of KSR-1. We first obtain low level measurements of read/write latencies for the different levels of the memory hierarchy (Section 3.1); we then implement and measure (Section 3.2) the costs of shared memory style synchronization primitives (locks and barriers); and finally we implement two parallel algorithms believed to be the computational kernels of several scientific and numerical applications and report on their observed performance (Section 3.3). Concluding remarks are presented in Section 4.

2 Architecture of KSR-1

KSR-1 is a 64-bit cache-only memory architecture (COMA) based on an interconnection of a hierarchy of rings [9]. The ring at the lowest level in the hierarchy can contain up to 32 processors. There are 32 MBytes of second-level cache (called *local-cache*) and 0.5 MBytes of first-level cache (called *sub-cache*) present at each node. The processing node issues two instructions per cycle. The CPU clock speed is 20MHz and the machine has a peak performance of 40 MFLOPS per processing node.

The architecture provides a sequentially consistent shared memory model. The COMA model is implemented using the ALLCACHE^{TM} memory in KSR-1. A distinguishing characteristic of the ALLCACHE^{TM} memory is that there is no fixed location for any System Virtual Address. An invalidation-based cache coherence protocol is used to maintain sequential consistency. The *get_sub_page* instruction allows synchronized exclusive access to a sub-page, and the *release_sub_page* instruction releases the exclusive lock on the sub-page. KSR-1 also provides *prefetch* and *poststore* instructions. The poststore instruction sends the updated value on the ring and all place holders for this sub-page in the local-caches receive the new value.

The interconnection network uses a unidirectional

---

*This work is supported in part by an NSF PYI Award MIP-9058430 and an NSF Grant MIP-9200005

slotted pipelined ring with 24 slots in the lowest level ring (two address interleaved sub-rings of 12 slots each), and hence there could be multiple packets on the ring at any given instance. The ring protocol ensures round-robin fairness and forward progress. The lowest level ring has a capacity of 1 GBytes/sec. Each node in KSR-1 has a hardware *performance monitor* that gives useful information such as the number of sub-cache and local-cache misses and the time spent in ring accesses. Our experiments are run on a one level 32-node KSR-1.

3 Experimental Study

We have designed experiments aimed at evaluating the low-level architectural features of KSR-1. These experiments include measuring the effectiveness of the hardware cache coherence scheme in servicing read write misses at different levels of the cache hierarchy; the utility of get_sub_page instruction for implementing read (shared) and write (exclusive) locks; and the usefulness of poststore instruction for implementing barrier synchronization algorithms.

The notion of scalability is meaningful only in the context of an algorithm-architecture pair. An approach to determining the scalability of any given architecture would be to implement algorithms with known memory reference and synchronization patterns and measure their performance. The measurements would allow us to determine how the features of the underlying architecture interacts with the access patterns of the algorithms, and thus give us a handle on the scalability of the architecture with respect to these algorithms. For this purpose we have implemented two computational kernels that are important in several numerical and scientific applications. We analyze their performance with respect to the architectural features of KSR-1 in Section 3.3.

3.1 Latency Measurements

We have conducted simple experiments to verify the latencies to access various levels of the memory hierarchy as detailed in [4]. Our measurements mostly agree with the published values [9]. The writes to sub-cache and local cache are slightly more expensive than reads. The plausible explanation for this result is that writes incur replacement cost in the sub-cache. As expected, varying the number of processors does not increase the latency for reads and writes to the sub-cache and local cache.

We may expect that for higher number of processors, the contention to use the ring might cause higher latencies. One of the outcomes of this experiment confirming the expectation is that the ring latency increases (about 8% for 32 processors) when multiple processors simultaneously access remote data. We also observe a cost associated with the allocation of blocks (depending on the access pattern); when the stride of access exhibits more spatial locality this allocation cost gets amortized over the multiple sub-blocks that are accessed within the same block.

3.2 Synchronization Measurements

The second set of measurements involves implementing synchronization mechanisms. Locks and barriers

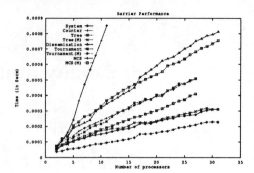

Figure 1: Performance of the barriers on the KSR

are typical synchronization mechanisms used in multiprocessors. KSR-1 hardware does not guarantee FCFS to resolve lock contention but does guarantee forward progress due to the unidirectionality of the ring. The exclusive lock is not efficient in applications that exhibit read sharing. We have implemented a version of read-write lock using the KSR-1 exclusive lock primitive. We have experimented with a synthetic workload of read and write lock requests and as expected, our locks perform better that the hardware exclusive lock in the presence of read sharing. An interesting outcome of the experiment was that even when the workload has writers only, our implementation performs better than the naive hardware exclusive lock. This result is surprising considering the software overhead (maintaining the queue of requesters) in our queue-based lock algorithm. The result can be partly explained due to the interaction of the operating system.

We have implemented five barrier synchronization algorithms [8] on the KSR-1. In our implementation, we have aligned (whenever possible) mutually exclusive parts of shared data structures on separate cache lines so that there is no false sharing. The results are shown in Figure 1. KSR-1 owing to its pipelined ring interconnect has multiple communication paths. The latency experiments clearly show that simultaneous distinct accesses on the ring do not increase the latency for each access. Thus we expect algorithms which can exploit simultaneous communication to perform well. A detailed analysis of the observed performance can be found in [4], and we highlight one key observation: Although it is stated that the MCS algorithm may be the best for large-scale cache coherent multiprocessors in Reference [8], our results on KSR-1 indicate that the tournament(M) algorithm is better than the MCS(M) algorithm. This is due to the fact that there are multiple communication paths available in the architecture and due to the false sharing inherent in the MCS algorithm.

3.3 NAS Kernels

The Numerical Aerodynamic Simulation (NAS) parallel benchmark [1] consists of five kernels and three applications which are considered to be representative of several scientific and numerical applications. We have implemented three of the five kernels on KSR-1 as part of our scalability study. The first one is the Embarrassingly Parallel (EP) kernel, which evaluates integrals by means of pseudo-random trials and is used in

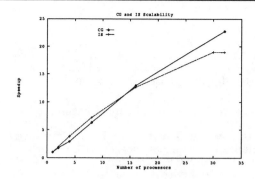

Figure 2: Speedup for CG and IS

many Montecarlo simulations. As the name suggests, it is highly suited for parallel machines, since there is virtually no communication among the parallel tasks. Our implementation [3] showed linear speedup, and given the limited communication requirements of this kernel, this result was not surprising.

For the rest of this section we focus on the other two kernels, namely, the conjugate gradient method (CG) and the integer sort (IS) since both these kernels revealed interesting insight into the scalability of the KSR-1 architecture. In both the kernels, there is considerable amount of true data sharing among the processors; and further in both kernels one data structure is accessed in sequence but another is accessed based on the value in the first data structure. Thus a considerable part of the data access patterns in both the kernels is data dependent.

3.3.1 Conjugate Gradient

The CG kernel computes an approximation to the smallest eigenvalue of a sparse symmetric positive definite matrix. On profiling the sequential code we observed that over 90% of the time is spent in a sparse matrix multiplication routine. Since most of the time is spent only in this multiplication routine, we parallelized only this routine for this study. The sequential code uses a sparse matrix representation based on a column start, row index format. In this format the result vector is computed in a piece-meal manner owing to the indirection in accessing it. Therefore, there is a potential for increased cache misses in carrying out this loop. We can expect better locality for cache access if an element of the result vector is computed in its entirety before the next element is computed. Thus we modified the sparse matrix representation to a row start column index format. This new format also helps in parallelizing this loop, without need for synchronization among the processors for updating the result (with the original format, the obvious way of parallelizing required synchronization on updates to the result).

On a single processor, the CG algorithm yields 1 MFlop for a problem size of n = 14K, with 2M non-zero elements. The relatively poor absolute performance of the algorithm can be attributed to the data access pattern that may not be utilizing the sub-cache and local-cache efficiently, and to the limited size of the two caches. This reason is also confirmed by measuring the number of accesses (local-cache and remote caches) using the hardware performance monitor.

Table 1 gives the speedup and the measured serial

Procs	Time (secs)	Speedup	Serial Fraction
1	1638.859	1.000	-
2	930.477	1.761	0.13551
4	565.221	2.899	0.12651
8	259.552	6.314	0.03814
16	126.519	12.953	0.01568
32	72.008	22.759	0.01309

Table 1: Conjugate Gradient, datasize=$n = 14000, nonzeros = 2030000$

fraction[1] [7] for the CG algorithm on the KSR-1. Figure 2 shows the corresponding speedup curve for the algorithm. Up to about 4 processors, the insufficient sizes of the sub-cache and local-cache inhibits achieving very good speedups. However, notice that relative to the 4 processor performance the 8 and 16 processor executions exhibit superunitary [6] speedup. This superunitary speedup can be explained by the fact that the amount of data that each processor has to deal with fits in the respective local-caches, and as we observed earlier the algorithm design ensures that there is very limited synchronization among the processors. We notice that there is a drop in speedup when we go from 16 to 32 processors. We attribute this drop to the increase in the number of remote memory references in the serial section of the algorithm. With 32 processors, since each processor now works on a smaller portion of the data, the processor that executes the serial code has more data to fetch from all the processors thus increasing the number of remote references. To prove that our hypothesis is indeed the case, we modified the implementation using poststore to propagate the values as and when they are computed in the parallel part. Since poststore allows overlapping computation with communication, the serial code would not have to incur the latency for fetching all these values at the end of the serial part. Using poststore improves the performance (3% for 16 processors), but the improvement is higher for lower number of processors. For higher number of processors, the ring is close to saturation due to the multiple (potentially simultaneous) poststores being issued by all the processors thus mitigating the benefit.

Given this result, we can conclude that the CG algorithm scales extremely well on this architecture. KSR-1 has an architectural mechanism for selectively turning off sub-caching of data. Given that the CG algorithm manipulates three huge vectors, it is conceivable that this mechanism may have been useful to reduce the overall data access latency and thus obtain better performance. However, there is no language level support for this mechanism which prevented us from exploring this hypothesis. The "prefetch" primitive of KSR-1 allows remote data to be brought into the local-cache. There is no architectural mechanism for "prefetching" data from the local-cache into the sub-cache. Given that there is an order of magnitude dif-

---
[1] Serial fraction represents the ratio of the time to execute the serial portion of the program to the total execution time of the program. The measured serial fraction is a ratio derived using the execution times and incorporates the above algorithmic as well as any other architectural bottlenecks.

ference in the access times of the two, such a feature would have been very useful.

3.3.2 Integer Sort

Integer Sort (IS) is used in "particle-in-cell method" applications. The kernel is implemented using a bucket sort algorithm. In this algorithm, each key is read and count of the bucket to which it belongs is incremented. A prefix sum operation is performed on the bucket counts. Lastly, the keys are read in again and assigned ranks using the prefix sums. This sequential algorithm proves to be quite challenging for parallelization considering trade-offs with respect to the computation-communication ratio. An efficient parallel algorithm is given in [4] and used in this study.

In our implementation, the amount of simultaneous network communication increases with the number of processors. As we observed in Section 3.1, the network is not a bottleneck when there are simultaneous network transfers until about 16 processors. Barrier synchronization (using the system barrier) is used at the conclusion of the phases and we know from the synchronization measurements that barrier times increase with the number of processors. However, from the measurements we know that the time for synchronization in this algorithm is negligible compared to the rest of the computation. Therefore, we contend that the slow-down from 16 to about 30 processors is more due to the algorithm than any inherent architectural limitation. However, we notice a performance drop from 30 to 32 processors, which we attribute to the ring saturation effect with simultaneous network accesses that we mentioned earlier (see Section 3.1). Thus the only architectural bottleneck that affects the performance of this algorithm at higher number of processors is the bandwidth of the ring. The reason for the extremely good speedups observed for up to 8 processors can be explained by the efficient utilization of the caches by the algorithm. These caching effects dominate initially as we increase the number of processors aiding the performance. However, the inherent communication bottlenecks in the algorithm result in a loss of efficiency at higher number of processors. This reasoning is confirmed by the data from the hardware performance monitor which shows that the latencies for remote accesses start increasing as the number of processors increase.

4 Concluding Remarks

We have designed and conducted a variety of experiments on KSR-1, from low level latency and synchronization measurements to the performance of two numerical computational kernels. Thus we believe that this study has been a fairly good assessment of the scalability of the architecture. There are several useful lessons to be reported as a result of this study.

The sizes of the sub-cache and local-cache is a limitation for algorithms that need to manipulate large data structures as is evident from our results. The prefetch instruction of KSR-1 is very helpful and we used it quite extensively in implementing both CG and IS. However, the prefetch only brings the data into the local-cache. It would be beneficial to have some prefetching mechanism from the local-cache to the sub-cache, given that there is roughly an order of magnitude difference between their access times, Further, (as we observed in the CG algorithm) the ability to selectively turn off sub-caching would help in using it better depending on the access pattern of a given algorithm.

The ring network, due to its pipelined nature, shows remarkable resilience in being able to sustain simultaneous remote accesses. This feature, in fact, improves the performance of certain barrier algorithms such as the tournament and MCS. Pipelining of the network can also be exploited with a suitable choice of data structures for shared variables as we demonstrated in the CG and IS algorithms. The architecture aids computation-communication overlap with the prefetch and poststore instructions both of which resulted in improving the performance in our experiments. The ring does become a bottleneck for large numbers of processors since the number of slots available on a single level 0 ring is 24.

The results of our experiments show that the architectural features of KSR-1 for the most part aids the scalability of parallel programs. In fact, superunitary speedup is observed for IS over a certain range of processors. Further, developing parallel applications using the shared memory paradigm is a definite advantage judging from our experience on the KSR-1. However, one has to be careful in the use of certain primitives. For example, poststore seems especially useful (for overlapping computation and communication) in situations where prefetching may not be a big help. However, our results show that indiscriminate use of this primitive can be detrimental to performance due to the increased communication saturating the network for large number of processors. The results reported in this study are on a level 0 ring KSR-1. Future work include experiments with multi-level rings.

References

[1] D. H. Bailey, E. Barszcz, and J. T. Barton et al. The NAS parallel benchmarks - summary and preliminary results. In *Supercomputing*, pages 158–65, 1991.

[2] Thinking Machines Corporation. *The Connection Machine CM-5 Technical Summary*, October 1991.

[3] Steve Breit et al. Implementation of EP, SP and BT on the KSR-1. Summer work at KSR, 1992.

[4] U. Ramachandran et al. Scalability study of the KSR-1. Technical Report GIT-CC 93/03, Ga Tech, 1993.

[5] John L. Gustafson. Reevaluating Amdahl's law. *Communications of the ACM*, 31(5):532–533, May 1988.

[6] D. P. Helmbold and C. E. McDowell. Modelling speedup (n) greater than n. *IEEE Transactions on Parallel and Distributed Systems*, 1(2):250–6, 1990.

[7] A. H. Karp and H. P. Flatt. Measuring parallel processor performance. *Communications of the ACM*, 33(5):539–543, May 1990.

[8] J. M. Mellor-Crummey and M. L. Scott. Algorithms for scalable synchronization on shared-memory multiprocessors. *ACM Transactions on Computer Systems*, 9(1):21–65, February 1991.

[9] Kendall Square Research. Technical summary, 1992.

Personalized Communication Avoiding Node Contention on Distributed Memory Systems

Sanjay Ranka* and Jhy-Chun Wang*

School of Computer and Information Science
Syracuse University
Syracuse, NY 13244
ranka@top.cis.syr.edu

Manoj Kumar

IBM T.J. Watson Research Center

Yorktown Heights, NY 10598

Abstract

In this paper, we present several algorithms for performing all-to-many personalized communication on distributed memory parallel machines. Each processor sends a different message (of potentially different size) to a subset of all the processors involved in the collective communication. The algorithms are based on decomposing the communication matrix into a set of partial permutations. We study the effectiveness of our algorithms both from the view of static scheduling as well as runtime scheduling.

1 Introduction

For distributed memory parallel computers, load balancing and reduction of communication are two important issues for achieving a good performance. It is important to map the program such that the total execution time is minimized. There is a large class of scientific problems, which are irregular in nature, achieving a good mapping is considerably more difficult. Further, the nature of this irregularity may not be known at the time of compilation, and can be derived only at runtime. The handling of irregular problems requires the use of runtime information to optimize communication and load balancing [2]. These packages derive the necessary communication information based on the nonlocal data required for performing the local computations.

In this paper we develop and analyze several simple methods of scheduling all-to-many personalized com-

*Work was supported in part by NSF under CCR-9110812 and in part by DARPA under contract # DABT63-91-C-0028. The content of the information does not necessarily reflect the position or the policy of the Government and no official endorsement should be inferred.

munication. The cost of the scheduling algorithm can be amortized over several iterations as the same schedule can be used several times. Assuming a system with n processors, our algorithms take as input a $n \times n$ communication matrix COM. $COM(i,j)$ is equal to a positive integer m if processor P_i needs to send a message (of m unit) to P_j, otherwise $COM(i,j) = 0$, $0 \le i, j < n$. Thus, row i of COM represents the *sending vector* of processor P_i, which contains information about destination and size of different messages. Our algorithms decompose the communication matrix COM into a set of partial permutations, $pm^1, pm^2, \cdots, pm^l$, l is a positive integer, such that if $COM(i,j) = m$ then there exists a k, $1 \le k \le l$, that $pm_i^k = j$.

Assuming that the processors perform their operation in a synchronous fashion, the time taken to complete a permutation depends on the largest message in the permutation. Since the message sizes in one permutation may vary widely, we develop several schemes to reduce the variance of message size within one permutation. This is done by splitting large messages into smaller pieces, each of which is sent in different phases.

With the advent of new routing methods [4], the distance to which a message is sent is becoming relatively less and less important. Thus assuming no link contention, permutation is an efficient collective communication primitive. Clearly, this is not going to be the case for all architectures. The paths of two messages may have a common link. This may sequentialize the transfer of the two messages (specially for machines that use circuit switching routing). In this paper, we have not addressed link contention. The algorithms developed in this paper can be extended to the architectures where link contention is an important issue by decomposing into partial permutation which avoid link contention. The cost of these algorithms would depend on the topology as well as the routing method.

We show that our algorithms are inexpensive enough to be suitable for static as well as runtime scheduling. If the number of times the same communication schedule is used is large (which happens for a large class of problems), the fractional cost of the scheduling algorithm is quite small. Further, compared to the naive algorithms, our algorithm can result in significant reduction in the total amount of communication.

The rest of the paper is organized as follows. Section 2 gives the definitions and assumptions made for developing our message scheduling algorithms, and an overview of CM-5. Section 3 presents a simple asynchronous communication algorithm. Section 4 develops algorithms that will avoid node contention and discusses their time complexity. Section 5 presents experimental results for a 32 node CM-5. Finally, conclusions are given in Section 6.

2 Preliminaries

A $n \times n$ communication matrix COM can be decomposed into a set of communication phases, cp^k, $1 \leq k \leq l$, l is a positive integer, such that

$$COM(i,j) = m, \ m > 0 \ \Rightarrow \ \exists!k, \ 1 \leq k \leq l, \ cp_i^k = j \ .$$

We define the kth communication phase as:

$$cp_i^k = j, \ i = 0, 1, \ldots, n-1, \ and \ 0 \leq j < n$$

if processor P_i need to send message to processor P_j at the kth phase, otherwise $cp_i^k = -1$.

Thus, *node contention* can be formally defined as:

$$\exists k, \ 1 \leq k \leq l, \ cp_{i_1}^k = j_1 \ and \ cp_{i_2}^k = j_2$$

$$\Rightarrow i_1 \neq i_2 \ and \ j_1 = j_2 \neq -1 \ ,$$

where $i_1, i_2 = 0, 1, \ldots, n-1$ and $0 \leq j_1, j_2 < n$.

A *partial permutation* pm^k is a communication phase that,

$$pm_{i_1}^k = j_1 \ and \ pm_{i_2}^k = j_2 \ ,$$

$$i_1, i_2 = 0, 1, \ldots, n-1 \ and \ 0 \leq j_1, j_2 < n \ ,$$

$$i_1 = i_2 \ \Leftrightarrow \ j_1 = j_2 \ .$$

$pm_i^k = -1$ if P_i does not send a message.

Thus, by definition, permutation avoids node contention. We will use permutation as the underlying communication primitive.

Asynchronous_Send_Receive()
For all processor P_i, $0 \leq i \leq n-1$, *in parallel do*
    allocate buffers for incoming messages;
    send out all outgoing messages to other processors;
    confirm incoming messages from other processors.

Figure 1: Asynchronous Communication Algorithm

We make the following assumptions for developing our algorithms and their complexity analysis. (1) each permutation can complete in $(\tau + M\varphi)$ time, where τ is the communication startup latency, M is the maximum size of any message sent in one permutation, and φ represents the inverse of the data transmission rate; (2) each node sends and receives approximately equal number of messages, which is denoted as density d, where $d \leq n$; (3) each processor can only send/receive one message at a time.

3 Asynchronous Communication (AC)

The most straightforward approach is using asynchronous communication. The algorithm is given in Figure 1. This naive approach is expected to perform well when the density d is small. The worst case time complexity of this algorithm is difficult to analyze as it will depend on the congestion and contention on the nodes and the network.

4 Methods Avoiding Node Contention

Our scheduling algorithms assume the availability of a global communication matrix COM. A concatenation operation can be performed on the sending vector of each processor to derive this matrix at runtime. Concatenate operation has efficient implementation on architectures like fat-tree, hypercubes and meshes [1, 5].

4.1 Linear Permutation (LP)

In this algorithm (Figure 2), in the kth iteration, each processor P_i sends a message to processor $P_{(i \oplus k)}$ and receives a message from $P_{(i \oplus k)}$, where $0 < k < n$. This algorithm assumes that the number of processors, n, is a power of 2. One can easily extend this algorithm when n is not a power of 2.

Linear_Permutation()

For all processor P_i, $0 \le i \le n - 1$, *in parallel do*

  for k = 1 to n-1 do

    $j = i \oplus k$;

    if $COM(i,j) > 0$ *then* P_i sends a msg to P_j;

    if $COM(j,i) > 0$ *then* P_i receives a msg from P_j;

  endfor

Figure 2: Linear Permutation Algorithm

RS_NH()

1. Use matrix COM to create a $n \times d$ matrix $CCOM$;

2. In each row k, $0 \le k < n$, build a heap $heap_k$ based on the entries $CCOM(k,j)$'s corresponding message size, where $0 \le j < d$;

3. *Generate_Permutations().*

Figure 3: Random Scheduling using Heaps (RS_NH) Algorithm

4.2 Random Scheduling using Heaps (RS_NH)

The RS_NH algorithm is depicted in Figure 3. The detailed description of this algorithm is given in [6].

Assuming the maximum message size allowed to be sent in one iteration is M_{thresh} (each iteration may have different value of M_{thresh}, which is decided by the function *Decide_Size()*). Suppose only k messages are smaller than M_{thresh}, then $(n-k)$ partial messages (with remaining message sizes) are put back in the heap, we set $\lambda = \frac{k}{n}$. We have designed several methods to choose the value of λ. For details the reader is referred to [6].

5 Experimental Results

We have implemented our algorithms on a 32 node CM-5. Besides the algorithms we describe in this paper, one algorithms is also involved in the test: **RS_N**: This is essentially the same as RS_NH algorithm, but no message splitting is done (and hence no heap operation is required).

All the algorithms are executed in a loosely synchronous fashion. We did not explicitly use global synchronization to enforce synchronization between communication stages in any of the algorithms proposed above.

Generate_Permutations()

For all processor P_i, $0 \le i \le n - 1$, *in parallel do*

Repeat

  1. Set vectors *send = receive = −1*;

  2. $x = random(1..n)$;

    for y = 0 to n-1 do

      $i = (x + y) \bmod n$; $j = 0$; $S = \phi$;

      while $(send(i) = -1$ AND $j \le prt(i))$ *do*

        $k = CCOM(i,l)$, where

          $l = Heap\_Extract\_Max(heap_i)$;

        if $(receive(k) = -1)$ *then*;

          $send(i) = k$; $receive(k) = i$;

        endif

        $S = S \cup CCOM(i,l)$; $Heap\_Remove(heap_i,l)$;

        $j = j + 1$;

      endwhile

      /* M_{ik} is $CCOM(i,k)$'s corresponding message size */

      For all entries, $CCOM(i,k)$, in S (except the last one),

        $Heap\_Insert(heap_i, M_{ik})$;

    endfor

  3. $M_{thresh} = Decide\_Size()$;

  4. *if* $(send(i) \ne -1)$ *then* P_i sends a message, no bigger than M_{thresh}, to $P_{send(i)}$;

    if $(receive(i) \ne -1)$ *then* P_i receives a message from $P_{receive(i)}$;

  5. For each row k which sent a complete message at this iteration, decreases $prt(k)$ by 1; For each row l which only sent partial message, add the remainder of the message back to its proper location in $heap_l$;

Until all messages are sent.

Figure 4: Procedure Generate_Permutations()

The data sets for our experiments can be classified into two categories: (1) random generated data sets, the procedure used to create these test sets is described in [6]; (2) the second test set contains communication matrices generated by graph partitioning algorithms [3]; the samples represent fluid dynamics simulations of a part of a airplane with different granularities. We only present the result of 53961-point sample. In this set, The number of messages sent/received by each node may be different.

Table 1 shows the results of $d = 8$. Results show that RS_NH provide a considerable improvement over other approaches. This clearly shows the usefulness of heap structures and thresholding to reduce the variance of messages in one permutation. Figure 2 shows the result for a 53961-point sample. The result has similar behavior as the first test set, which reveal that even if the number of messages in each row is non-uniform, our algorithms maintain their characteristics and performance. The RS_NHs are superior when the *msg_unit* becomes large, which in turn means that it is worth the extra effort (of using heap and message

Msg_unit	AC	LP	RS_N	RS_NH
comm*				
16	3.820	7.943	3.380	3.839
64	8.124	11.463	5.455	5.879
256	24.873	26.771	15.101	15.176
1024	89.027	83.063	57.825	53.744
4096	301.681	282.814	222.201	207.420
8196	830.939	967.832	592.921	467.793
comp†	0	0.091	3.211	16.872
# iters‡	0	31.0	10.1	11.2

*: the total communication cost in milliseconds.

†: the scheduling cost in milliseconds.

‡: number of iterations (permutations) to complete the scheduling.

Table 1: Experimental results for density $d = 8$, the minimum message size in each level is Msg_unit bytes, and the maximum size is $32 \times Msg\_unit$ bytes.

breaking) to reduce the variance of message sizes in each permutation.

6 Conclusion

In this paper, we have developed several algorithms for scheduling all-to-many personalized communication with non-uniform message sizes. The performance of asynchronous communication algorithm (AC) depends on the network congestion. This algorithm is only suitable for small message sizes. The linear permutation algorithm (LP) is very straightforward, it introduces little computation overhead, but it needs to go through same number of communication phases $(n - 1)$ even if the density d is small.

The RS_NH algorithms are found to be very useful in handling non-uniform messages. The use of a heap structure so that the bigger messages will be scheduled earlier, and the decomposition of large messages into smaller messages give a significant reduction of the total time required for communication.

One of the issues which we have not addressed in this paper is link contention. On the CM-5, link contention does not have a major effect on the communication cost of the schedules generated by our algorithms. We are currently developing algorithms for architectures on which link contention is an important issue.

Msg_unit	AC	LP	RS_N	RS_NH
comm*				
16	16.103	17.941	12.907	12.895
32	26.826	27.349	20.965	19.512
64	48.367	46.552	37.662	33.479
128	87.700	80.769	69.874	61.605
256	163.598	149.746	135.387	116.832
512	300.644	280.240	256.659	224.406
comp	0	0.097	6.059	30.358
# iters	0	31.0	18.05	20.05

Table 2: Experimental results for test set 2: 53961-point, the minimum message size in each level is Msg_unit bytes, and the maximum size is $276 \times Msg\_unit$ bytes.

References

[1] Zeki Bozkus, Sanjay Ranka, and Geoffrey C. Fox. Benchmarking the CM-5 multicomputer. In *Proceedings of the Frontiers of Massively Parallel Computation*, 1992. to appear.

[2] R. Das, R. Ponnusamy, J. Saltz, and D. Mavriplis. Distributed memory compiler methods for irregular problems - data copy reuse and runtime partitioning. In J. Saltz and P. Mehrotra, editors, *Compilers and Runtime Software for Scalable Multiprocessors*. Elsevier, Amsterdam, The Netherlands, 1991. to appear.

[3] Nashat Mansour. *Parallel Genetic Algorithms with Application to Load Balancing for Parallel Computing*. PhD thesis, Syracuse University, Syracuse, NY 13244, 1992.

[4] Lionel M. Ni and Philip K. McKinley. A survey of wormhole routing techniques in direct networks. *IEEE Computer*, 26(2):pp.62–76, February 1993.

[5] Sanjay Ranka and Sartaj Sahni. *Hypercube Algorithms with Applications to Image Processing and Pattern Recognition*. Springer-Verlag, 1990.

[6] Sanjay Ranka, Jhy-Chun Wang, and Manoj Kumar. All-to-many personalized communication on distributed memory machines. Technical Report SU-CIS-92, Syracuse University, September 1992.

SESSION 10A

MEMORY AND DISKS

DESIGN OF ALGORITHM-BASED FAULT TOLERANT SYSTEMS WITH IN-SYSTEM CHECKS*

Shalini Yajnik and Niraj K. Jha
Department of Electrical Engineering
Princeton University
Princeton, NJ 08544

Abstract – *To improve the reliability of compute-intensive applications run on multiprocessor architectures, fault tolerance is introduced into the system with on-line detection and location of faults. This can be achieved by a low-cost scheme, called Algorithm-based fault tolerance (ABFT), which encodes data at the system level and modifies the algorithm to operate on the encoded data. The resultant encoded output data is checked for correctness by some checks. In this paper we present an extended model for representing and designing ABFT systems. The model takes into consideration the processors evaluating the checks. We propose a design method which considers the processors computing the checks to be a part of the ABFT system and guarantees concurrent error detection even in the presence of faults in these processors, unlike most methods presented earlier.*

1. INTRODUCTION

An important consideration in high-performance multiprocessor systems is to ensure the correctness of the results of the computations. **Algorithm-Based Fault Tolerance** [1]-[12] is a technique which imparts concurrent error detection/location capability to a multiprocessor system. This scheme encodes the input data at the system level and modifies the algorithm to run on the encoded input data. The output data is also in encoded form. The scheme uses checks to evaluate the correctness of the encoded output data.

Banerjee and Abraham [3] presented a graph-theoretic model to represent ABFT systems. Nair and Abraham [4] proposed a matrix-based model which was derived from the graph-theoretic model. Most earlier work in the design of general ABFT systems [5, 6, 7, 8], with the exception of [9], assumes that the checking operations are performed by processors outside the system which are either fault-free or have some means of exposing their own faults, such as the self-checking property. This assumption makes the design process easier. However, the checking operations are usually an integral part of an ABFT system and, in many applications, are likely to be performed on the system processors. We call such checks *in-system checks*. When the system processors on which they are computed fail, these checks become unreliable. There-

*Acknowledgments: This work was supported by Office of Naval Research under Contract no. N00014-91-J-1199.

fore, the accuracy of the computations is dependent on the reliability of the processors performing the checking operations as well. Banerjee and Abraham [10] introduced check evaluating nodes in their graph model and showed how to analyze such a system for fault tolerance. But they did not show the mapping of checks onto the system processors which compute them, such that the fault tolerance properties of the system could be preserved.

This paper develops an extended model for characterizing ABFT systems. This model considers the processors computing the checks to be part of the total system. We propose a method for the design of such extended ABFT systems with the desired fault-detecting capability. The approach taken in [9] to attack this problem is different in the sense that a randomized algorithm is used to obtain an ABFT system which can inherently tolerate faults in processors computing the checks, by adding more checks than would be required if the checks were always assumed to be fault-free. In this paper we take a direct and deterministic approach. Our method can actually be used in conjunction with the method in [9] to reduce the number of processors computing the checks.

The rest of the paper is organized as follows. In Section 2, we present a brief overview of the graph-theoretic and the matrix-based models. We also introduce our extended graph-theoretic model and give the necessary definitions and concepts. In Section 3, we give our design method for ABFT systems. Finally, in Section 4 we give the conclusions.

2. PRELIMINARIES

This section defines the terms used to describe ABFT systems. We also present a brief overview of the graph-theoretic and matrix-based models used to represent these systems.

The traditional system-level fault model is used here [1]. A fault in a processor is assumed to manifest itself as an error in at least one of the data elements computed by the processor. Suppose the system consists of n processors $p_1, p_2,..., p_n$, producing a set of k data elements $d_1, d_2,..., d_k$. An ABFT system contains, in addition to the processors and the data elements, a set of checks $c_1, c_2,..., c_q$. The set of data elements that a processor produces or a check checks is said to be their respective data sets. The result of a computation is checked by one or more other processors in the system. These processors are called *check_computing* processors. We assume that all the processors in the system are capable of performing either useful computation, or a checking operation or both. A set of faulty processors in the system is said to be a *fault pattern*. The set of erroneous data elements produced by a fault pattern is said to be an *error pattern*.

The inputs of the system are encoded using a system-level code, such as checksums. The algorithm is modified to operate on this encoded input and produce encoded output data. We refer to the redundant part of the output data as *check_data*. A *check* on a set of output data elements is a function computed on those elements and compared with a check_data element. We assume the checks to be bounded [3], i.e. the number of data elements a check is assumed to be able to handle is bounded. This assumption is different from the one made in [5, 9] where the checks are assumed to operate on any number of data elements. A (g, h) check is defined on g data elements and the result of the checking operation is as follows [10].

- The check outputs a 0 if the check_computing processor is non-faulty and all the data elements in its data set are error-free.
- The check outputs a 1 if the check_computing processor is non-faulty and at least one data element in its data set is in error, but the number of data elements which are erroneous in its data set does not exceed h.
- The check is unpredictable if
  1. The check_computing processor is faulty or
  2. The number of erroneous elements in its data set is greater than h.

Definition 1 *A system is said to be t-fault (t-error) detecting if for every fault (error) pattern of size t or less there is at least one check which evaluates to a '1'.*

2.1 The Graph-theoretic Model

An ABFT system can be represented by a tripartite graph called the **PDC** graph [3]. This graph has three types of nodes, the processor nodes P, the data nodes D and the check nodes C. There is an edge between a processor node $p_i \in$ P and a data node $d_j \in$ D if processor p_i is involved in the computation of data d_j. There is an edge between a data node $d_j \in$ D and a check node $c_k \in$ C if d_j is checked by check c_k. When used for the purpose of designing an ABFT system, this model does not take into account the check_computing processors, i.e. the processors which compute the checks are not included in P. As pointed out before, most previous design methods using this approach assume that check_computing processors are fault-free or capable of exposing their faults by self-checking or some other such means. We propose an extended model to take care of the check_computing processors, which could be possibly faulty, too. According to our model, an ABFT system is represented by two graphs, the **PDC** graph, which is a tripartite graph and the check evaluation (**CE**) graph, which is a bipartite graph. The graphs have six types of nodes:

- **Info_processor nodes** which perform useful computations.
- **Real_data nodes** which are the result of the useful computations.

- **Check nodes** which represent the checks.
- **Check_data nodes** which represent the redundant part of the encoded output data.
- **Code_processor nodes** which perform computations on the encoded input to produce the check_data nodes.
- **Check_computing processor nodes** which perform the checking operations. In general, these can be of the info_processor node type or of the code_processor node type.

The PDC graph consists of the info_processor nodes, the code_processor nodes, the real_data nodes, the check_data nodes and the check nodes. The CE graph consists of check_computing processor nodes and the check nodes. Each code_processor node in the PDC graph is connected to one or more check_data nodes since it performs computation on the encoded input and produces part of the redundant section of the encoded output. In the CE graph there exists an edge between a check_computing processor node p_i and a check node c_j if c_j is evaluated on p_i. Thus this graph represents the mapping of checks onto the check_computing processors. First, we design the PDC graph, assuming that the checks in the system are fault-free and then we design the CE graph in such a way as to maintain the fault detectability of the system even in the presence of failures in check_computing processors.

Each check in the system has at least one check_data element in its data set. One such check_data element is referred to as the *check_compare* element of that check. A check computes some function of the data elements in its data set and compares its check_compare element with the result of the computation. A check_data element is assumed to be computed on only one code_processor.

2.2 The Matrix-based Model

Nair and Abraham [4] proposed a matrix-based model for ABFT systems. They broke up the tripartite PDC graph of the graph-theoretic model into two bipartite graphs, the PD graph and the DC graph. The two bipartite graphs were represented as PD and DC matrices. The PD matrix has processors as rows and data elements as columns, and the DC matrix has data elements as rows and checks as columns. The product of the PD and DC matrices is another matrix with processors as rows and checks as columns. This is called the PC matrix. The ij^{th} entry in the PC matrix corresponds to the number of paths from processor p_i to check c_j in the original tripartite graph. We use the matrix model in the construction of the CE graph later on in the paper. Our model is the same except that besides the original set of processors and data elements, we also include the code_processors and the check_data elements in the three matrices. We next give a few definitions from [4, 12].

Definition 2 ^rPD *is defined as the matrix whose rows are formed by adding r different rows of matrix PD, for all possible combinations of r rows, and setting all non-zero entries in the resulting matrix to 1.*

Definition 3 ^rPC *is the matrix obtained by the product of the matrices* ^rPD *and DC.*

3. FAULT DETECTION

A procedure for designing t-fault detecting systems has to consider the fact that a single fault pattern can give rise to several error patterns. Nair and Abraham [4] proposed a two-stage design procedure. They introduced the concept of a *unit system*. A unit system is a system in which each processor has only one data element in its data set. The PD matrix is an identity matrix and the DC matrix is designed for the desired fault detectability. The unit systems are extended to form the desired final *composite* ABFT system with the same detectability. The procedure given in [4] cannot be used if the processors share data elements. Vinnakota and Jha [8] gave a procedure which allowed sharing of data elements among processors. We propose a three-stage design procedure which allows sharing of data elements among processors as well as allows the check_computing processors in the system to be faulty. A t-fault detecting unit system is first designed under our extended ABFT model, assuming checks are fault-free. Its copies are created and the final composite system is obtained by superposition of the original and the copies. However, this is not enough if the check_computing processors are also allowed to be faulty. We also have to design the CE graph which shows the mapping of checks to check_computing processors in order to ensure that there is at least one check implemented on a non-faulty processor for every possible error pattern, which can be handled by the system. The following section gives the methods which can be used to form the unit system when the checks are assumed to be bounded.

3.1 Constructing the unit system

The real_data elements in the original PD graph are distributed among the template unit system and its copies. The method of distributing the real_data among the unit systems is shown later in Section 3.2. In this section, we assume that the distribution has already been done. We are given a template unit PD graph consisting of m processors and m real_data elements. Each info_processor in the unit system is connected to exactly one real_data element. A fault in a processor of the unit system means an error in the corresponding data element. Therefore the cardinality of the fault patterns will be the same as the cardinality of the resulting error patterns. In this case designing t-fault detecting systems will be the same as designing t-error detecting systems. We can take advantage of the methods proposed by Rosenkrantz and Ravi in [7] for this purpose. The resulting system has to satisfy the constraint that each check should have at least one check_data element in its data set. The methods given in [7] have to be modified to introduce check_data elements in the DC graph. Let n be the total number of data elements in the unit system. In such systems, the number of real_data elements is equal to the number of info_processor nodes m and the number of check_data elements is equal to the number of checks q. Only one check_data element can be mapped to one code_processor in a unit system. So the number of code_processors introduced is equal to the number of checks. Therefore, $n = m + q$.

Theorem 1 *For $t = 1$, the number of checks sufficient for constructing the unit system is $\left\lceil \frac{m}{g-1} \right\rceil$.*

Method of unit system construction: Construct data sets of each check by taking $g - 1$ elements from the real_data set and an unused element from the check_data set. Each data element is present in exactly one check's data set and therefore, any single fault will be caught by the checks. □

The following lemma from [7] is stated here without proof.

Lemma 1 Suppose a collection of checks T satisfies the following three conditions:

1. Each data element is checked by at least one check in T.
2. For each pair of checks in T, at most one data element is checked by both the checks.
3. Each check in T checks only one data element not checked by any other check.

Then T is a valid scheme for detecting 2 faults. □

Theorem 2 *For $t = 2$, $g > 2$, $m \geq (g-1)^2$, the number of checks sufficient for constructing the unit system is $\left\lceil \frac{2m}{g-1} \right\rceil$.*

Method of unit system construction: Let $A =$ set of real_data elements in the unit system, where $|A| = m \geq (g-1)^2$. Let $B =$ set of check_data elements in the unit system, where $|B| = q = \left\lceil \frac{2m}{g-1} \right\rceil$. Arrange the set A of real_data elements in a grid of $(g-1)$ columns. The total number of rows is at least $(g-1)$. Construct **row checks** by taking $(g-1)$ real_data elements of a row in A and a single unused element from the check_data set B. If the last row has less than $(g-1)$ elements in it, take elements starting from the top of the first column to complete the row check. Construct **column checks** by taking an unused element from B and $(g-1)$ elements of a column from A starting from where the last row/column check ended, going to the top of the next column if there are no elements left in the present column. There will be g elements in each check except the last one which may have less than g data elements. This method constructs a set of checks which satisfies the conditions of Lemma 1, therefore the above scheme is valid for 2-fault detection.

Example 1 Let $m = 17$, $t = 2$, $g = 5$. Then $|B| = \left\lceil \frac{2m}{g-1} \right\rceil = 9$ and $|A| = 17$.

Let $A = [a_1, a_2, ..., a_{17}]$ and $B = [b_1, b_2, ..., b_9]$.
A arranged in a grid with 4 columns:

$$
\begin{array}{cccc}
a_1 & a_2 & a_3 & a_4 \\
a_5 & a_6 & a_7 & a_8 \\
a_9 & a_{10} & a_{11} & a_{12} \\
a_{13} & a_{14} & a_{15} & a_{16} \\
a_{17} & & &
\end{array}
$$

Checks:

Row checks	Column checks
$r_1 = [a_1, a_2, a_3, a_4, b_1]$	$c_1 = [a_{13}, a_{17}, a_2, a_6, b_6]$
$r_2 = [a_5, a_6, a_7, a_8, b_2]$	$c_2 = [a_{10}, a_{14}, a_3, a_7, b_7]$
$r_3 = [a_9, a_{10}, a_{11}, a_{12}, b_3]$	$c_3 = [a_{11}, a_{15}, a_4, a_8, b_8]$
$r_4 = [a_{13}, a_{14}, a_{15}, a_{16}, b_4]$	$c_4 = [a_{12}, a_{16}, b_9]$
$r_5 = [a_{17}, a_1, a_5, a_9, b_5]$	□

Theorem 3 *For $t = 2$, $g > 2$, $m < (g-1)^2$, the number of checks sufficient for constructing the unit system is $\lceil 2\sqrt{m} \rceil$.*

Method of unit system construction: Let $A = $ set of real_data elements in the unit system, where $|A| = m < (g-1)^2$. Let $B = $ set of check_data elements, where $|B| = \lceil 2\sqrt{m} \rceil$. If $m \leq \lceil \sqrt{m} \rceil \times \lfloor \sqrt{m} \rfloor$ then construct a grid of elements of A in a $\lceil \sqrt{m} \rceil \times \lfloor \sqrt{m} \rfloor$ fashion, else construct a grid in a $\lceil \sqrt{m} \rceil \times \lceil \sqrt{m} \rceil$ fashion. Construct **row checks** by taking all the elements of a row in A and an unused element from B. There are $\lceil \sqrt{m} \rceil$ such checks. Each of these checks will have less than or equal to g data elements. Construct **column checks** with all the elements of a column and an element from B. Each of these checks will also have less than or equal to g data elements in them. This scheme also satisfies the conditions of Lemma 1 and, therefore, is a valid 2-fault detection scheme. $\square$

Theorem 4 *For $t = 3$, $g > 2$, $m \geq (g-1)^2$, the number of checks sufficient for the construction of the unit system is $\left\lceil \frac{m}{g-1} \right\rceil + \left\lceil \frac{m + \lceil \frac{m}{g-1} \rceil}{g-1} \right\rceil$.*

Proof and method of unit system construction:
Let B be the set of $\left\lceil \frac{m + \lceil \frac{m}{g-1} \rceil}{g-1} \right\rceil$ check_data elements. Let A be the set of the real_data elements in the unit system and the rest of the $\left\lceil \frac{m}{g-1} \right\rceil$ check_data elements.

$|A| = m + \left\lceil \frac{m}{g-1} \right\rceil \geq m + \frac{m}{g-1} \geq g(g-1)$. Construct a grid of data elements of A with $(g-1)$ columns. The grid will have at least g rows. Number the elements columnwise, i.e. go down the first column, then the second column and so on. Place the check_data elements present in A at positions g, $2g$, $3g$,... in the grid and if the number of elements in A is not a multiple of g then place a check_data element at the bottom of the last column. Construct **row checks** with a row of A and an unused element of B. The last row check might have less than g data elements. Construct **column checks** with g elements by going down a column and going to the next column if the present column is used up. The last column check also might have less than g elements. The total number of checks is thus $\left\lceil \frac{m}{g-1} \right\rceil + \left\lceil \frac{m + \lceil \frac{m}{g-1} \rceil}{g-1} \right\rceil$. The resulting set of checks satisfies the conditions of Lemma 1 and hence will detect 2 faults. Now assume that 3 faults are present. If all the 3 faults are in B, then they will be caught by different row checks. If A has a single fault and B has 2 faults, then the fault in A will be caught by a column check which contains no element of B. If A has 2 faults in different rows and B has a single fault, then at least one of the faults will be caught by a row check which contains an error-free data element from B. If A has both the faults in the same row, the column checks will detect the faults. If A has all the 3 faults then there will always be at least one row or column check that contains exactly one erroneous data element. It can be seen that no two data elements in a row of A will appear together in a column check because the grid has at least g rows. $\square$

Theorem 5 *For $t = 3$, $m < (g-1)^2$, the number of checks sufficient for the unit system construction is $(\lceil 2\sqrt{m} \rceil + 1)$.*

Method of unit system construction: Let B be the set of $(\lceil \sqrt{m} \rceil + 1)$ check_data elements. Let A be the set of m real_data elements and the rest of the check_data elements. If $m \leq \lceil \sqrt{m} \rceil \times \lfloor \sqrt{m} \rfloor$ then the number of check_data elements in A is $\lfloor \sqrt{m} \rfloor$. In that case, construct a grid of elements of A in a $(\lceil \sqrt{m} \rceil + 1) \times \lfloor \sqrt{m} \rfloor$ fashion. If $m > \lceil \sqrt{m} \rceil \times \lfloor \sqrt{m} \rfloor$ then the number of check_data elements in A is $\lceil \sqrt{m} \rceil$. In this case, construct a grid of elements of A in a $(\lceil \sqrt{m} \rceil + 1) \times \lceil \sqrt{m} \rceil$ fashion. Construct **row checks** by taking a row of data elements from A and an unused element from B. There will be $(\lceil \sqrt{m} \rceil + 1)$ such checks. Construct **column checks** with all the elements of a column of A. Each of the row and the column checks will have less than or equal to g data elements. $\square$

For $t \geq 4$, we will use a hierarchical construction method. An analysis of the hierarchical approach has been given before in [11]. In all previous methods we have assumed that each check has a distinct check_compare element, so that the number of check_data elements was equal to the number of checks. But in the next procedure, since we are using hierarchical construction methods, the checks can share the same check_compare element. So the number of check_data elements will no longer be equal to the number of checks.

Theorem 6 *For $t \geq 4$, $r = \lceil \log_2(\frac{t}{3}) \rceil$, $m = (g-1)^{r+2}$, the number of checks sufficient for the unit system construction is $(r+2)g^{r+1} - g^r$.*

Proof: Let A be the set of all the $(g-1)^{r+2}$ real_data elements in the unit system and B be the set of all the check_data elements. The number of check_data elements required, as will be shown later, is $g^{r+2} - (g-1)^{r+2}$. Therefore, the total number of data elements is g^{r+2}. Then by [7], the number of checks sufficient for the system is $(r+2)g^{r+1} - g^r$. $\square$

Method of unit system construction: We will base our construction procedure on the method given in [7].

1. Partition the set A into $(g-1)^r$ sets A_1, A_2,..., $A_{(g-1)^r}$, each containing $(g-1)^2$ data elements. For each of the sets we need to detect only 3 faults.

2. By Theorem 4, a set A_i needs $(2g-1)$ checks to detect 3 faults. Add $(2g-1)$ distinct check_data elements from B to each set A_i, $1 \leq i \leq (g-1)^r$. Let us term these augmented sets as A_{1i}, $1 \leq i \leq (g-1)^r$. Each set A_{1i} has g^2 data elements.

3. Construct row and column checks for each A_{1i}. There will be $(2g-1)(g-1)^r$ such checks.

4. Construct B', a subset of B, by removing all the $(2g-1)(g-1)^r$ check_data elements used in Step 2. So $|B'| = g^{r+2} - g^2(g-1)^r$.

5. Partition the set B' into $[g^r - (g-1)^r]$ sets B_{11}, B_{12},..., $B_{1[g^r-(g-1)^r]}$, of g^2 elements each. For each of these sets too, we need to detect 3 faults. But these sets already have the check_data elements needed for construction of the checks.

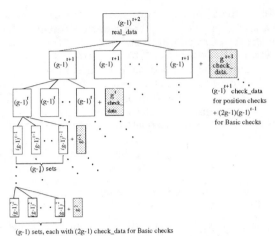

(g-1) sets, each with (2g-1) check_data for Basic checks

Figure 1: Hierarchical construction

6. Construct the row and column checks for each set B_{1j}, $1 \leq j \leq [g^r - (g-1)^r]$. There are $(2g-1)(g^r-(g-1)^r)$ such checks. We define the checks constructed in Step 3 and Step 6 as *Basic checks*.

7. For $k = 1$ to r, do the following:

   (a) Group together $(g-1)$ of the A_{ki} sets and one B_{kj} set. Let us define these groups as the *A-groups*. There will be $(g-1)^{r-k}$ such groups.

   (b) The B_{kj} sets, not used in the earlier step, are grouped together, with each group having g sets. We term these groups as the *B-groups*.

   (c) Number the data elements in each of the A_{ki} and B_{kj} sets from 1 to g^{k+1}.

   (d) Construct *position checks* T_{ij}, for group j, by taking the ith element from each set in the group. Since each set has g^{k+1} data elements, there will be g^{k+1} such checks for each group.

   (e) Combine all the g sets in an A-group j, $1 \leq j \leq (g-1)^{r-k}$, to form one set $A_{(k+1)j}$. Each of these sets will have g^{k+2} data elements.

   (f) Combine all the g sets in a B-group j, to form one set $B_{(k+1)j}$. Each of these sets, too, will have g^{k+2} data elements. □

Claim: The number of check_data elements sufficient for the unit system is $g^{r+2} - (g-1)^{r+2}$.

Proof of Claim: If we look at the construction procedure in the top-down manner as shown in Figure 1, we see that at every stage we partition a given set into $(g-1)$ sets of the same size and add a set of check_data elements to this group of $(g-1)$ sets. In the last stage each real_data set has $(g-1)^2$ data elements. There are $(g-1)^r$ such sets. To each of these sets we add $(2g-1)$ check_data elements to construct the Basic checks.

The number of check_data elements required by the real_data sets for the Basic checks $= (2g-1)(g-1)^r$. Now each real_data set, after being augmented by the $(2g-1)$ check_data elements, has g^2 data elements. Therefore, the number of check_data elements required for position checks for each group of $(g-1)$ sets $= g^2$. If we do this for every stage, we get the following, where N is the total number of check_data elements.

$$
\begin{aligned}
N &= (2g-1)(g-1)^r + (g-1)^{r-1}g^2 + \\
&\quad (g-1)^{r-2}g^3 + \ldots + (g-1)g^r + g^{r+1} \\
&= (2g-1)(g-1)^r + g^{r+1}\left[\frac{1 - \left(\frac{g-1}{g}\right)^r}{1 - \frac{g-1}{g}} \right] \\
&= (2g-1)(g-1)^r + g^2[g^r - (g-1)^r] \\
&= g^{r+2} - (g-1)^{r+2} \quad \square
\end{aligned}
$$

Example 2 Let $m = 10^4$, $g = 11$, $t = 12$, $r = 2$. Then $|A| = (11-1)^4 = 10,000$ and $|B| = 11^4 - 10^4 = 4641$.

Partition A into 100 sets $A_1, A_2, \ldots, A_{100}$ of size 100 each. We require that each of these sets should have 3-fault detectability. Add 21 check_data elements from B to each A_i, $1 \leq i \leq 100$, to form A_{1i}. Then $|B'| = 4641 - 2100 = 2541$. Partition B' into 21 sets, B_{11}, $B_{12}, \ldots, B_{1(21)}$ of size 121 each. For each of these sets, construct the Basic checks. There will be $121 \times 21 = 2541$ such checks.

Group the A_{1i} and B_{1j} sets into 11 groups in the following manner:

$$
\begin{aligned}
A'_1 &= (A_{11}, A_{12}, \ldots, A_{1(10)}, B_{11}) \\
A'_2 &= (A_{1(11)}, A_{1(12)}, \ldots, A_{1(20)}, B_{12}) \\
\ldots & \\
A'_{10} &= (A_{1(91)}, A_{1(92)}, \ldots, A_{1(100)}, B_{1(10)}) \\
B'' &= (B_{1(11)}, B_{1(12)}, \ldots, B_{1(20)}, B_{1(21)})
\end{aligned}
$$

Construct position checks for each of the groups. Each group will have 121 position checks. So the number of position checks $= 121 \times 11 = 1331$.

Now consider each group as a set of 1331 elements. Group all the 11 groups together and construct position checks. There are 1331 such checks. Therefore, the total number of checks is $2541 + 1331 + 1331 = 5203$ which is also equal to $(r+2)g^{r+1} - g^r = 4 \times 11^3 - 11^2$.

Theorem 7 *For $t \geq 4$, $r = \lceil \log_2(\frac{t}{3}) \rceil$, $m < (g-1)^{r+2}$, the number of checks sufficient for the unit system construction is $r(x + \lceil 2\sqrt{x} \rceil + 1)g^{r-1} + (\lceil 2\sqrt{x} \rceil + 1)g^r$, where $x = \frac{m}{(g-1)^r}$.*

The method of construction is the same as in the case of $m = (g-1)^{r+2}$, except for the construction of Basic checks for those sets which have less than $(g-1)^2$ data elements at the rth stage. For these sets we construct the Basic checks using the construction method based on Theorem 5.

For the case of $t \geq 4$, $g > 2$, $r = \lceil \log_2(\frac{t}{3}) \rceil$, $m > (g-1)^{r+2}$, partition the set of real_data elements into sets containing $(g-1)^{r+2}$ data elements (the last one may contain less than $(g-1)^{r+2}$ data elements). For each set containing $(g-1)^{r+2}$ data elements we use the construction procedure based on Theorem 6. For the set containing less than $(g-1)^{r+2}$ data elements, we use the construction procedure based on Theorem 7.

3.2 Composite System Construction

It is assumed here, as in [8], that the processors in the original non-fault-tolerant system can share data elements, but each processor produces at least one data element which no other processor in the system affects. We will refer to such data elements as the *distinguishing data* elements of the processor. The code_processors compute only the check_data elements and we assume that there is no sharing of check_data elements among the code_processors. The checks used in our method are $(g, 1)$ checks. Banerjee and Abraham [3] have shown that even if h of the $(g, 1)$ checks in an ABFT system are combined into a single (gh, h) check, the system has the same fault detection capabilities. So all designs can be confined to systems which have $(g, 1)$ checks. For designing the complete ABFT system, the unit template system has to be constructed first. From the unit template system, copies are created and then these systems are combined to form the composite ABFT system using some rules. The method of construction is based on the design process given in [8]. Before going into the design process, we state a definition from [8].

Definition 4 *A data element is defined as filled if it has been used in or added to the tripartite graph of the unit system, else it is called unfilled.*

Given the original non-fault tolerant PD graph:

1. Construct the PD part of the template unit system, where each processor is connected to exactly one of its distinguishing data elements. If more than one distinguishing data element exist, then one can be chosen at random.
2. Construct copies of the template unit system such that all the unfilled data elements in the system are filled. If there are no unfilled data elements left for a processor, then reuse an already-filled element. For details of this part the readers are referred to [8].
3. Using the methods given in Section 3.1, design the DC part of the template unit system such that it has the desired fault detectability. Add check_data nodes, code_processor nodes and check nodes to the template unit PD graph.
4. In the copies of the template, add the code_processor nodes, check_data nodes and check nodes with these edges connected in the same way as in the unit template system. The check nodes and the check_data nodes in the copies are numbered differently though. □

Given the template system and its copies, the composite system is formed by superimposing the unit systems, as follows.

1. The data sets of the info_processor nodes are the same as in the original non-fault tolerant PD graph. The data set of a code_processor is the union of the data sets of that processor in all the unit systems.
2. While merging the DC part of the graphs, the set of checks to which a data element is connected is formed by taking the union of the set of checks to which it is connected in each of the unit systems.

After this, the checks whose data sets (excluding the check_compare element) are found to be identical are merged into a single check, and if the check_compare element of the removed check is not connected to any other check, it is removed from the composite system. □

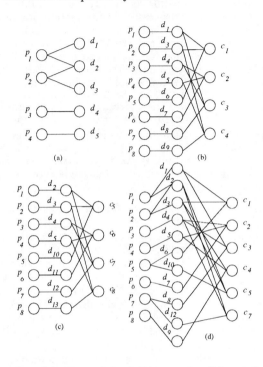

Figure 2: (a) The original PD graph, (b) a 2-fault detecting template unit system, (c) a copy of the template system, (d) the composite system.

Example 3 Consider the implementation of some algorithm on a parallel system with the PD graph as shown in Figure 2(a). Suppose we wish to design a 2-fault detecting version of this system. A template system for the PD graph, which is 2-fault detecting (with $g = 3$), is shown in Figure 2(b). The real_data set of the template system is $\{d_1, d_3, d_4, d_5\}$. Each member of this set is a distinguishing element. A single copy needs to be formed to fill all the data elements. This is shown in Figure 2(c). By Theorem 2 we require four checks to detect two faults in the template system. Therefore, we need to add four check_data nodes $\{d_6, d_7, d_8, d_9\}$, and consequently four code_processor nodes p_5, p_6, p_7 and p_8 to the graph. The real_data set of the copy is $\{d_2, d_3, d_4, d_5\}$. The template system and the copy are now composed to form the final system which is shown in Figure 2(d). In the final system, check c_6 is equivalent to check c_2 and check c_8 is equivalent to check c_4. So checks c_6 and c_8 are removed. □

The following theorem from [8] is stated here without proof.

Theorem 8 *If the template system is t-fault detecting, the composition of the template system and the copies is also t-fault detecting.*

3.3 The Check Evaluation Graph

We are given a composite system which is t-fault detecting, assuming that the checks are fault-free. We now need to map checks to the processors in such a way as to maintain t-fault detectability of the system even when the checks may themselves fail. If the check-computing processors in the system are allowed to be faulty, then a fault pattern is *detectable* if and only if for every error pattern that the fault generates, there is at least one non-faulty processor evaluating a check which has a single erroneous data element in its data set. We will use both the matrix model and the graph model for system description and design. We assume that all the processors in the system have the capability to perform the checking operations. We give some definitions and theorems from [4].

Definition 5 *A row of ^rPC is said to be completely detectable if and only if the fault pattern represented by the row is detectable for all possible error patterns produced by that fault.*

Theorem 9 *A system is t-fault detecting if and only if the matrices ^iPC, for $i = 1, 2, ..., t$, are completely detectable, i.e. all the rows of each of the matrices are completely detectable.*

For a row of ^rPC to be detectable, there should be at least one entry in the row which is less than or equal to the error-detectability (h) of the check used and the corresponding check should be evaluated on a processor which is not in the set of faulty processors defined by the row. This should be true for all the rows of each of the matrices ^iPC, for $i = 1, 2, ..., t$. Since we want to take care of all the possible error patterns that a fault pattern can generate, we consider the PC matrices of each of the unit systems, instead of operating on the PC matrix of the final composite system. First we form the PD matrix for each of the unit systems, with processors as rows and data elements as columns. There is a '1' in position (i, j) of the PD matrix if processor p_i has data element d_j in its data set. All the other positions have '0's. Then we form the DC matrix of the unit system with data elements as rows and checks as columns. There is a '1' in position (i, j) in the matrix if data element d_i is checked by check c_j. All the other positions have '0's. The PC matrix is obtained by multiplying the PD and DC matrices.

We define the *processor set* of a check to be the set of processors, any one of which can be used to evaluate the check. We need to find the processor sets of each of the checks such that the mapping of checks to processors maintains the fault detectability of the system. Here, we propose an algorithm to perform this mapping.

Algorithm for mapping of checks:

1. For each of the unit systems, construct a PC matrix. Tag each row of the matrix with the corresponding processor. Let d be the number of unit systems. Term the PC matrix of the ith unit system as PC_i, $1 \le i \le d$. A check in the final system has a column in each of the PC matrices. Throw out the columns corresponding to the checks which are redundant and have been removed from the composite system.

2. For each of the PC_i matrices, construct ^rPC_i, $1 \le r \le t$. Each row in ^rPC_i is a combination of r rows of PC_i. Tag each row in ^rPC_i with a set of r processors corresponding to the r rows whose combination it is.

3. Initialize the processor sets of each check to all the processors in the system.

4. Columnwise concatenate all the matrices ^rPC_i, $1 \le r \le t$, $1 \le i \le d$, to form one matrix PF.

5. Arrange the rows in the final matrix PF in an increasing order of the number of '1's, and number the rows from 0 onwards. Number the columns from 0 to $q-1$, where q is the total number of checks in the system.

6. In each row of PF, mark the '1' which appears first in or after the column i *modulo* q, where i is the row number.

7. For all rows of PF starting from the first do
    $i = 0$; throw $= 0$;
    While (throw $= 0$ & $i \le q - 1$){
       if (the ith column entry is '1') and
       (the set of processors in the row tag $\bigcap$
       the ith column check's processor set is ϕ)
       then {
          throw the row out of the matrix;
          throw $= 1;$}
       $i = i + 1;$}
    if (throw $= 0$) then do the following
    (a) Let present_check $=$ Marked '1' column.
    (b) If (processor set of present_check) $-$ (set of processors in the row tag) $= \phi$, label the marked '1' as *bad* and mark a '1' in that row which is not labeled bad. Go to Step 7(a). Else, remove those processors from the present_check's processor set which are also in the row tag of the present row.
    (c) Throw the present row out of the matrix.

8. Map each check to any of the processors in its processor set. Here, we can consider the communication overhead and/or the processor load while assigning the checks to the processors. □

In the above method, we go through each row in the matrix and make sure that the check corresponding to the marked '1' is being evaluated on a fault-free processor. We continue modifying the processor sets of each of the checks as we go down the rows. At any given stage, the rows which have at least one valid '1' entry, i.e. one '1' entry whose corresponding check does not have any of the processors in the row tag in its processor set, are thrown out of the matrix. Thus the solution which is generated at the end of the construction makes sure that each row has at least one '1' whose corresponding check is being evaluated on a fault-free processor. The complexity of the algorithm is $O(dqn^t)$.

Example 4 Consider the PD graph of the system given in Figure 2(a). Suppose we want to make the system 1-fault detecting. The unit systems consist of

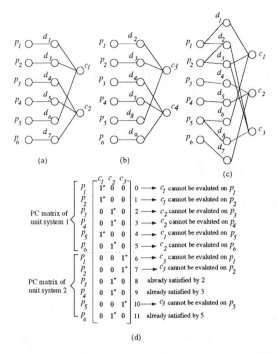

Figure 3: (a) Unit system 1, (b) unit system 2, (c) the composite system, (d) the PF matrix.

system such that the fault detectability of the composite system is not affected. This mapping method would be applicable to any other method as well which could derive a composite system with the desired fault detectability properties.

References

[1] K.H. Huang and J.A. Abraham, "Algorithm-based fault tolerance for matrix operations," IEEE Trans. Comput., vol. C-33, pp. 518-528, June 1984.

[2] J.Y. Jou and J.A. Abraham, "Fault tolerant matrix arithmetic and signal-processing on highly concurrent computing structures," Proc. IEEE, vol. 74, no. 5, pp. 732-741, May 1986.

[3] P. Banerjee and J.A. Abraham, "Bounds on algorithm-based fault tolerance in multiple processor systems," IEEE Trans. Comput., vol. C-35, no. 4, pp. 296-306, Apr. 1986.

[4] V.S.S. Nair and J.A. Abraham, "A model for the analysis of fault tolerant signal processing architectures," in Proc. Int. Tech. Symp. SPIE, San Diego, pp. 246-257, Aug. 1988.

[5] D. Gu, D.J. Rosenkrantz and S.S. Ravi, "Design and analysis of test schemes for algorithm-based fault tolerance," in Proc. Int. Symp. Fault-Tolerant Comput., Newcastle-upon-Tyne, U.K., pp. 106-113, June 1990.

[6] V.S.S. Nair and J.A. Abraham, "A model for the analysis, design and comparison of fault-tolerant WSI architectures," in Proc. Workshop Wafer Scale Integration, Como, Italy, June 1989.

[7] D.J. Rosenkrantz and S.S. Ravi, "Improved bounds on algorithm-based fault tolerance," in Proc. Annual Allerton Conf. Comm., Cont. and Comput., Allerton, IL, pp. 388-397, Sept. 1988.

[8] B. Vinnakota and N.K. Jha, "Design of multiprocessor systems for concurrent error detection and fault diagnosis," in Proc. Int. Symp. Fault-Tolerant Comput., Montreal, pp. 504-511, June 1991.

[9] R. Sitaraman and N.K. Jha, "Optimal design of checks for error detection and location in fault tolerant multiprocessor systems," accepted for publication in IEEE Trans. Comput..

[10] P. Banerjee and J.A. Abraham, "Concurrent fault diagnosis in multiple processor systems," in Proc. Int. Symp. Fault-Tolerant Comput., Vienna, pp. 298-303, June 1986.

[11] V.S.S. Nair and J.A. Abraham, "Hierarchical design and analysis of fault-tolerant multiprocessor systems using concurrent error detection," in Proc. Int. Symp. Fault-Tolerant Comput., Newcastle-upon-Tyne, U.K., pp. 130-137, June 1990.

[12] B. Vinnakota and N.K. Jha, "Diagnosability and diagnosis of algorithm-based fault tolerant systems," accepted for publication in IEEE Trans. Comput..

4 real_data elements. In order to detect a single fault in the system we add two checks with $g = 3$ (from Theorem 1). The unit systems and the final system are shown in Figure 3. Construct the PC matrices of the two unit systems, PC_1 and PC_2, and concatenate them as shown in Figure 3(d) to form matrix PF. Number the rows from 0 to 11. In row 0, there is a '1' entry for check c_1. In order to make this entry valid, check c_1 should not be evaluated on p_1. Similarly, the other rows are handled. In row 2, there is a '1' entry corresponding to check c_2. If this entry is to be valid, check c_2 cannot be evaluated on processor p_3. This also validates the '1' entry in row 8. If we go through the whole matrix PF, we get the following processor sets for the checks.

Check	Processor Set
c_1	$\{p_3, p_4, p_6\}$
c_2	$\{p_1, p_2, p_5\}$
c_3	$\{p_3, p_4, p_6\}$

We can then map a check to any one of the processors in its processor set. One possibility is to map c_1 to p_4, c_2 to p_5, and c_3 to p_6. $\Box$

4. CONCLUSIONS

In this paper we have given methods for designing ABFT systems in which the checks are not assumed to be present outside the system, but are a part of the total system and any fault in the check_computing processors is also detected by the system. We first design unit systems which also include processors which compute the redundant part of the system-level code. We combine these unit systems into a composite system. We then map the checks on to the processors in the

A Cache Coherence Protocol for MIN-Based Multiprocessors With Limited Inclusion*

Mazin S. Yousif, Chita R. Das and Matthew J. Thazhuthaveetil

Department of Electrical and Computer Engineering
The Pennsylvania State University
University Park, PA 16802

Abstract

In this paper, we look into a feasible approach to incorporating caches into selected switching elements of a multistage interconnection network (MIN)-based multiprocessor. Along with the processor private caches, these switch caches form a two-level cache hierarchy. Selected switch caches within a particular stage of the MIN are connected by a coherence control bus, through which a write-invalidate cache coherence protocol is maintained. Considering scalability and practicality issues, only limited inclusion between the two cache levels is enforced. A simulation-based performance study is conducted to analyze the impact of the protocol on system performance. Comparison between limited and strict inclusion shows that system performance declines with limited inclusion.

1 Introduction

Hiding memory latency with private caches is viable approach to designing shared-memory multiprocessors. Unfortunately, private caches also introduce the *cache coherence* problem [1]. A number of cache coherence protocols for general network-based multiprocessors have been proposed in the literature [1]. Schemes for MIN-based multiprocessors complement the MIN architecture with either a coherence control bus connecting all private caches [2]; or with caches and directories in all Switching Elements (SEs) of the MIN [3]. The coherent bus scheme is not scalable, and the inclusion of caches in all SEs is not practical. In this paper, we propose a hierarchical two-level cache design for MIN-based multiprocessors.

The approach in [3] adds caches and global directory to all the SEs in the MIN, which involves substantial hardware overhead and redesign of the entire MIN. The coherence problem is avoided by allowing only one copy of each modifiable shared data block outside of main memory. To deal with the above mentioned problem, we looked into the concept of incorporating caches into the SEs of *a selected stage* of a MIN [4]. These caches are referred to as *switch caches*. Switch caches back up the processor private caches forming a two-level cache hierarchy. Our design differs from the approach of [3] in that, (*i*) switch caches are added to selected switching elements in a MIN only, (*ii*) there could be multiple copies of a shared block outside of main memory, thus improving access times to shared data, and (*iii*) switch caches are assumed to be finite.

The two-level cache coherence scheme proposed in [4] was based on strict inclusion. Switch cache set associativity must increase with MIN size to satisfy inclusion. One approach to get around this restrain is to limit the set associativity to a practical value. (A set associativity of 16 has been used in the Facom M-382 [5].) This is the main theme of this work. We assume *limited inclusion* by bounding switch cache set associativity to 16, as discussed in Section 2. This is, however, at the expense of extra coherence control signals, as explained in Section 3. The impact of the cache coherence protocol on system performance is analyzed through a discrete-event simulation study in Section 4. The last section concludes this work.

2 Architectural Description

The proposed architecture, a shared-memory multiprocessor, is shown in Figure 1 with 16 nodes and a 16×16 baseline MIN. Each node contains a processor with a private cache and a portion of the system's globally accessible main memory, as in the BBN Butterfly [6]. Moreover, as the nodes are on one side of the network, the outputs are wrapped around and connected to the corresponding inputs, as illustrated by the barrel shape in Figure 1.

Consider a unique path MIN of size $N \times N$, made up of $n(= \log_2 N)$ stages of $\frac{N}{2}$ (2×2) switches, such as

*This research was supported in part by the National Science Foundation under grant MIP-9104485.

the MIN shown in Figure 1. For any selected stage i of the MIN, $0 \leq i < n$, we can divide the $\frac{N}{2}$ switches into *switch groups* with the following property: *Each switch within a given switch group provides a path to a set of 2^{n-i} MIN outputs; any path to a member of the set passes through a switch of the group, and terminates in a member of the output set.* For example, in stage 1 of Figure 1, switches 0, 1, 2 and 3 form a switch group providing access to the MIN output set $\{0-7\}$; and switches 4, 5, 6 and 7 form another switch group providing access to the MIN output set $\{8-15\}$.

Caches are added to the switching elements of a selected stage of the MIN with split-transaction *coherence buses* connecting the switch caches within a switch group, as shown in Figure 1. The switch caches in a group when combined form a unified second level shared cache. A coherence bus is used for coherence control signals transfer and blocks transfer among switch caches of a group. Thus, switch cache controllers snoop at all activity on their coherence bus. The number of switch caches in each group is given by 2^{n-i-1}, where i, $0 \leq i < n$, represents the stage where switch caches are added. The number of switch groups in stage i and the number of coherence buses are given by 2^i. A switch cache holds shared data blocks from exactly 2^{n-i} memory modules.

A hierarchical two-level cache system, like ours, requires the *inclusion property* between local caches and switch caches [7]. In our scheme, we will <u>not</u> strictly enforce inclusion. We will bound the set associativity to a maximum of sixteen. Table I lists some values for the switch cache size and set associativity. Note that in Table I, local and switch caches have equal block sizes, and the size of the MIN is not mentioned because it is irrelevant. Note also that inclusion is enforced in the shaded area of Table I. For switch cache placements that require higher set associativity, a sixteen-way set associativity will be maintained with a block replacement policy from the relevant set to make room for an incoming block. Various approaches to alleviate this problem are proposed in [7, 8].

The selection of a stage to incorporate switch caches into is a critical design parameter. If stage 0 is chosen, $\frac{N}{2}$ switch caches need to be connected on the sole coherence bus. This would not be feasible for large values of N. At the other extreme, if stage $(n - 1)$ is chosen, the benefit of switch caches would be limited due to the longer transit time through the MIN. Due to space limitation, we report system performance results when switch caches are incorporated into stage $\lceil \frac{\log_2 N}{2} \rceil$. Performance results for other stages such as $\lfloor \frac{\log_2 N}{2} \rfloor$, $(\lfloor \frac{\log_2 N}{2} \rfloor - 1)$ are reported in [4].

3 The Cache Coherence Protocol

This is a write-invalidate protocol. Coherence related information is maintained in status bits contained in each local cache directory entry as well as in each switch cache directory entry. The local status bits encode three states: *Local-Valid* (valid, potentially shared in local caches); *New-Master* (modified, only copy, requires write back on replacement); and *Local-Invalid*. The protocol guarantees that a New-Master copy is the only updated copy of the block in the system. Other copies of the block in local caches, if any, are Local-Invalid.

Each switch cache tag directory entry contains a valid bit, a status bit and an $(i + 1)$-bit address field, where i is the stage number of the MIN where switch caches are placed. A valid bit is required to show the presence of a block in a switch cache. The status bit encodes two states per block: *CoMaster-Valid* or *CoMaster-Invalid*. With this protocol, no shared block ever has more than one switch cache copy, and every currently active shared block may have exactly one switch cache copy. A switch cache copy is CoMaster-Valid if it is consistent with copies of the block in local caches, which will necessarily be Local-Valid. The CoMaster-Invalid state is used to identify a block that has been modified in local caches, which will be in state New-Master; the switch cache connected directly to that node, referred to as the *home switch cache* contains the CoMaster-Invalid copy. The local node containing this dirty copy is identified by the address field of the home switch cache tag directory entry of the CoMaster-Invalid copy of the block. Each shared block in a memory module directory is tagged with a *presence* bit to indicate whether a switch cache copy of a block exists or not. Description of the protocol along with all relevant transactions are in [4].

4 Performance Study

The impact of the coherence protocol on system performance is evaluated through extensive discrete-event simulations. These simulations are driven by stochastically-generated traces. Each simulation represents an exact flow of processor requests in the adopted architecture with cache coherence enforced. In the simulations, we assumed a workload model of two referencing streams – the first stream is made up of references to private blocks and the second stream consists of references to shared blocks. We use a *Least Recently Used Stack Model* (LRUSM) for both streams, but with different locality properties.

It is assumed that the service time of a switching element is 2 cycles; memory access time is 12 cycles;

local cache access time is 1 cycle; switch cache access time is 3 cycles coherence bus service time is 2 cycles; average processor thinking time is 2 cycles. We also assume that the probability of a processor read is 0.75; probability of a processor write is 0.25; probability that a private block is modified is 0.3; degree of sharing is varied from 0.05 to 0.2. Number of shared blocks is 2048 for a system with 16 nodes; this number doubles when system size doubles. It is assumed that a local cache is direct mapped with 512 blocks; local and switch caches have equal block size of 16 bytes; the size and set associativity of a switch cache for the selected stage number, ($\lceil \frac{\log_2 N}{2} \rceil$), for different system sizes, are similar to those in Table I.

Processor and bus utilizations for different system sizes and degrees of sharing were obtained, as shown in Figures 2 and 3. The closer the switch caches are to the processor end, the higher the processor and bus utilizations. Processor utilization is high due to the smaller delay incurred due to a miss. Coherence bus utilization increases as a result of more activity on a bus due to the smaller number of buses and the larger number of switch caches connected on a bus. For small size systems, it is preferable to add caches to a stage of switching elements closer to the processor side since the number of switch caches connected to a bus is small. As system size increases, the stage where caches provide improved performance moves away from the processor end.

Figures 2 and 3 show that system performance is dependent on the stage number where switch caches are added. Bus utilization will be the limiting factor as to where switch caches can be placed. Figure 4 gives the closest stage of the MIN where switch caches can be incorporated, and which will result in acceptable bus utilization. The two-tuple sets in Figure 4 give the stage number and the number of switch caches in a switch group. It is shown that for systems with up to 128 nodes, it is possible to place switch caches in the first stage of the MIN. This stage will move away from the processor side as system size increases, as shown in Figure 4 [4].

A performance comparison between limited and strict inclusion was also conducted [4]. It is observed that with limited inclusion, a 1024 node system suffers from around 14% increase in coherence bus utilization along with around 18% drop in processor utilization, when switch cache are incorporated in stage $\lceil \frac{\log_2 N}{2} \rceil$, and a degree of sharing of 10%. This is attributed to (i) extra coherence control signals, and (ii) more replacements of switch cache blocks because of the smaller switch cache size.

5 Conclusions

In this paper, we studied the effectiveness of incorporating caches in the switching elements of a selected stage in a MIN. When used in a MIN-based multiprocessor with a private cache per processor, it results in a two-level cache hierarchy. We implemented limited inclusion since set associativity of the higher level cache is restricted to 16. This assumption was at the expense of extra coherence control signals such as increased block transfers and invalidations. Simulations were conducted to quantify the negative impact of this limited inclusion on system performance. However, unlike strict inclusion, this scheme is scalable and practical.

References

[1] S. J. Eggers, *Simulation Analysis of Data Sharing in Shared Memory Multiprocessors*, Ph.D. Dissertation and Technical Report, UCB/CSD 89/501 University of California, Berkeley, April 1989.

[2] M. S. Algudady, C. R. Das and M. J. Thazhuthaveetil, "A Write-Update Cache Coherence Protocol for MIN-Based Multiprocessors with Accessibility-Based Split Caches," *Proc. of Supercomputing'90*, November 1990, pp. 544-553.

[3] H. E. Mizrahi, J. L. Baer, E. D. Lazowska and J. Zahorjan, "Extending the Memory Hierarchy into Multiprocessor Interconnection Networks: A Performance Analysis," *Proc. of 1989 Int. Conf. on Par. Proc.*, vol. I, August 1989, pp. 41-50.

[4] Mazin S. Yousif, Chita R. Das and Matthew J. Thazhuthaveetil, *Analysis of a Hierarchical Cache Coherence Protocol for MIN-Based Multiprocessors With Limited Inclusion*, Technical Report, TR-93-123, Dept. of ECE, The Pennsylvania State University, 1993.

[5] A. Hattori, M. Koshino, and S. Kamamito, "Three-Level Hierarchical Storage System for the FACOM M-380/382," *IFIP Proc. of Information Processing*, 1983, pp. 693-697.

[6] BBN Advanced Computers, Inc., Butterfly GP1000 Switch Tutorial.

[7] J. L. Baer and W. H. Wang, "On the Inclusion Properties for Multi-Level Cache Hierarchies," *Proc. of 15th Ann. Int. Symp. on Comp. Arch.*, May 1988, pp. 73-80.

[8] A. W. Wilson Jr., "Hierarchical Cache/Bus Architecture for Shared Memory Multiprocessors," *Proc. of 14th Ann. Int. Symp. on Comp. Arch.*, May 1987, pp. 244-252.

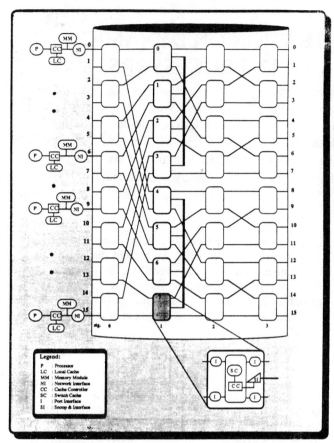

Figure 1: The Proposed Multiprocessor Architecture with a 16 × 16 Baseline MIN

Table I: List of Switch Cache Size and Associativity

Local Cache Size = 512 Blocks		
MIN Stage #	S.C.* Set Associativity	S.C.* Size (Blocks)
0	2	1024
1	4	2048
2	8	4096
3	16	8192
4	16	8192
5	16	8192
5	16	8192

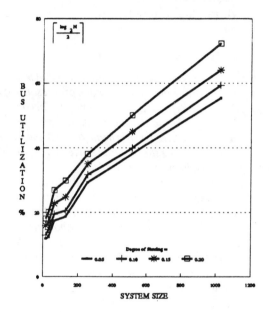

Figure 2: Coherence Bus Utilization (%) for Different Switch Cache Placements

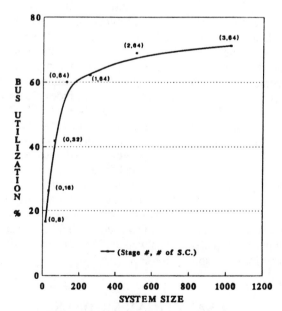

Figure 4: Bus Utilization (%) for Possible Closest Switch Cache Locations

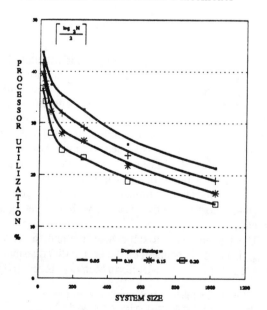

Figure 3: Processor Utilization (%) for Different Switch Cache Placements

IMPACT OF MEMORY CONTENTION ON DYNAMIC SCHEDULING ON NUMA MULTIPROCESSORS

M. D. Durand*
Bellcore
445 South Street
Morristown, NJ, 07960
durand@bellcore.com

T. Montaut   L. Kervella
IRISA
Campus de Beaulieu
35042 Rennes Cdx, France
montaut, kervella@irisa.fr

W. Jalby
MASI, Université de Versailles
45, av des Etats Unis
78035 Versailles Cdx, France
jalby@seti.inria.fr

Abstract – *Self-scheduling is a method for task scheduling in parallel programs, in which each processor acquires a new block of tasks for execution whenever it becomes idle. To get the best performance, the block size must be chosen to balance the scheduling overhead against the load imbalance. To determine the best block size, better analytical models of self-scheduling are needed.*

We present an experimental study of self-scheduling on a BBN TC2000, a NUMA machine. Previously published models to predict running time and optimal block size were tested using our experimental results. Although the models gave good predictions for small block sizes, for large block sizes the models fail, underestimating the running time by almost a factor of two. We present an upper bound on the running time and use it to explain this failure.

Keywords: *Dynamic scheduling, load balancing, memory performance, NUMA multiprocessors, self-scheduling.*

INTRODUCTION

Self-scheduling is an effective dynamic scheduling technique for balancing the load when task duration is unpredictable. A program decomposed into N parallel subtasks is executed by p processors. The subtask durations are a set of independent, identically distributed (*iid*) random variables with density f, distribution F, mean μ and standard deviation, σ. As processors become idle, they repeatedly acquire blocks of K tasks for execution[a]. The cost, h, of acquiring a block of tasks is fixed. The basic problem is to choose an optimal block size, K_{opt}, sufficiently large to amortize the overhead, h, without creating a load imbalance.

The performance of self-scheduling and the choice of K_{opt} is highly dependent on the probability distribution of task duration. On Non-Uniform Memory Access (NUMA)

machines, memory latency is a major component of these distributions. In this paper, we present the results of an experimental study of self-scheduling and discuss the effectiveness of self-scheduling algorithms and performance models in light of this new data. Our results reveal both the strengths and an unexpected weakness of these models.

RELATED WORK

Self-scheduling has been implemented in runtime systems on a number of multiprocessors [9, 1, 3], indicating that it is a method of practical as well as theoretical interest. Analytical models to predict running time and optimal block size of self-scheduling for fixed block size have been presented [5, 6]. These will be described in the next section.

Finally, a number of authors have proposed algorithms for self-scheduling with varying block size, in which the block size is reduced as the algorithm progresses so that all processors will finish more or less at the same time. In guided self-scheduling (GSS) [8], each time a processor becomes idle, it acquires $1/p$th of the remaining tasks. In trapezoidal self-scheduling (TSS) [10], the block size is decreased linearly rather than exponentially as in GSS. Both GSS and TSS are designed primarily to counter load imbalance due to uneven processor starting times and algorithmic variations in task duration. In a new method, Factoring[4], a probabilistic analysis was used to determine the rate at which K should be reduced. Since this analysis depends only on p, μ, σ and the number of remaining tasks, it is applicable to a wide variety of task duration distributions. Experimental studies [4, 10] showed that TSS and factoring generally performed as well as GSS and substantially better for some applications.

ANALYTICAL MODELS

The earliest and most comprehensive model of fixed-size block scheduling was presented by Kruskal and Weiss[5]. They estimate the running time of a block scheduled pro-

---

*Durand's contribution to this work was performed in part at IRISA and Ecoles des Mines de Paris and supported by the North Atlantic Treaty Organization under a Grant awarded in 1990.

[a]The last processor to request work may receive an odd-sized block if N is not evenly divisible by K.

gram to be

$$E[T] \approx \frac{N}{p}\mu + \frac{Nh}{pK} + \sigma\sqrt{2K\ln(p)}, \qquad (1)$$

when p and K are large and $K \gg \log p$ and

$$E[T] \approx \frac{N}{p}\mu + \frac{Nh}{pK} + \sigma\sqrt{2K\ln(\frac{p\sigma}{\sqrt{K}\mu})}, \qquad (2)$$

when $K \ll N/p$ and $\sqrt{K}/p$ is small. In these equations, the first term describes the time required for p processors to execute N tasks in a perfectly efficient system. The second term describes the scheduling overhead. The third term corresponds to the load imbalance which occurs due to the variation in task duration. Equation 1 shows the tradeoff between load imbalance and overhead: as K increases, the overhead decreases by a factor of $1/K$, whereas the load imbalance increases as $\sqrt{K}$.

Kruskal and Weiss estimate optimal block size by finding the value of K which minimizes Equation 1. Taking the derivative and setting it equal to zero yields

$$K_{opt} = \left(\frac{\sqrt{2}Nh}{\sigma p\sqrt{\ln(p)}}\right)^{2/3}. \qquad (3)$$

Unfortunately, the original equation is only valid when K is large. Hence, the value of K_{opt} derived from Equation 3 may not be valid for all values of K.

A similar approach to estimating the running time was presented by Madala and Sinclair [6], in which they used a simpler but less accurate estimate of load imbalance. Translated into the notation used in the current paper, their expression yields

$$E[T] \leq \frac{N}{p}\mu + \frac{Nh}{pK} + \sigma\sqrt{K}\frac{p-2}{\sqrt{2p-3}} + K\mu + h. \qquad (4)$$

The first two terms are identical to those given by Kruskal and Weiss in Equation 1. The third term represents the load imbalance. The final two terms represent the running time required to allocate and execute a single block of tasks. These two terms appear because, in their analysis, Madala and Sinclair present an upper bound. Madala and Sinclair also discussed the problem of optimal block size. However, because they did not present an analytically soluble expression for block size, we were not able to use their method in this study.

Kruskal and Weiss' estimate is quite accurate but is only valid for large K. Madala and Sinclair use a much cruder estimate which is, however, simpler and valid everywhere. Using order statistics[2] to compute the load imbalance more precisely, we present a refined version of Kruskal and Weiss' estimate:

$$E[T] \simeq E[T_i] + (\frac{N-Kp}{p})E[d_i]$$
$$+ \left\lceil\frac{N}{K}\right\rceil\frac{h}{p} + E_p[F^{*K}], \qquad (5)$$

as well as an upper bound

$$E[T] \leq E[T_i] + \frac{\lfloor(N-1)/K\rfloor K}{p}E[d_i]$$
$$+ \frac{\lfloor(N-1)/K\rfloor + p}{p}h + E_p[F^{*K}]. \qquad (6)$$

Here

$$E_p[G = F^{*K}] = p\int_{-\infty}^{+\infty} x[G(x)]^{p-1}g(x)dx, \qquad (7)$$

where G, the K-fold convolution of F, is the distribution function of the running time of a block of tasks. A complete development of these equations can be found in [7].

Equations 5 and 6 have the advantage of being valid everywhere and yet much more accurate than equation 4. The first term, the expectation of processor starting times, reflects the fact that not all processors start simultaneously. When an analytical expression for the last term, $E_p[F^{*K}]$, does not exist, it can be computed numerically. Although more difficult to compute than the previous models, Equation 5 allows us to gauge the sensitivity of running time estimates to inaccuracies in the computation of the load imbalance because the expression used in the last term gives a more precise estimate.

EXPERIMENTATION

In this section we present an experimental study of the effects of memory contention on task duration on a BBN TC2000. A number of factors can cause variable task duration: memory contention, if-statements, dynamic data structures and algorithmic constructs. We chose to study memory contention for two reasons. First, on NUMA machines, the effects of memory contention overwhelm the other factors. Second, the impact of memory contention on task duration is extremely difficult to predict analytically.

The **BBN TC2000** is a MIMD multiprocessor with a hybrid shared/distributed memory structure. Individual processor boards are connected by a multi-stage interconnection network. The global memory is distributed across the boards, but has a single address space. Each board includes 16 Mbytes of memory, a Motorola 88100 processor and switch interface circuitry and control circuitry, connected by a T-bus. Thus the CPU can access data stored on its local memory via the T-bus. Data references to distant memory locations are decoded by the switch interface circuitry and sent across the switch. Logically, the programmer can ignore the difference between global and local memory, but global accesses are slower than local ones. Self-scheduling and memory interleaving are supported by the Uniform System (US), a runtime system provided with the TC2000.

Experiments were run on a dedicated machine to measure the impact of the memory storage strategy, the degree of global activity and the number of processors on task duration. A synthetic task was designed to test these parameters:

```
Task(x)
int x ;
 {
  ELOG_LOG (START_TASK,x) ;
  /* Global memory accesses */
  for(i=0 ;i<gmax ;i++)
     G[i][x]=B[i][x] ;
  ELOG_LOG (END_GLOBAL,x) ;
  /* Local calculations */
  for (i=0 ;i<lmax ;i++)
     { a=a*a+1 ;
       a=a/a+1 ;
     }
  ELOG_LOG (END_TASK,x) ;
 }
```

The global loop is executed $gmax$ times and accesses the arrays G and B, which are stored in global memory. In the local loop, executed $lmax$ times, each processor accesses its own private variable, a, stored in its local memory. The degree of network contention is a function of the ratio of global to local memory accesses, which is controlled by the relative values of $gmax$ and $lmax$. This ratio determines what percentage of that time is spent performing memory operations across the network.

For each experiment, $N = 50p$ identical tasks were executed. The duration of each task and the associated overhead were measured using the event logging instructions at the beginning and end of the task. Because of the non-deterministic nature of parallel machines, each experiment was run five times.

The parameters of the 30 experiments in the test suite are shown in Table 1.

Memory Storage Strategy	hotspot (**H**), interleaved (**I**)
Network contention	Global/local accesses
low (**l**)	(gmax, lmax) = (80,650)
medium (**m**)	(gmax, lmax) = (160,325)
high (**h**)	(gmax, lmax) = (320,162)
Number of processors	p = 10, 20, 30, 40, 50

Table 1: Experimental Parameters

Two types of memory storage strategy were used, *interleaved* and *hotspot*. In the interleaved strategy, the rows of the matrices G and B are evenly distributed across the memories on the processing units working on the current problem. In the hotspot strategy, G and B are stored in a single global memory. Three different degrees of contention and five values of p were considered. Because of the lack of hardware support for cache coherency, cache behavior was not studied.

The results showed that for interleaving experiments, task duration is largely insensitive to network activity except when the contention is very high (large values of p and $gmax$). Hotspot experiments, however, are sensitive to network contention and exhibit large standard deviations. Histograms of task duration are shown in Figures 1 and 2. Figure 1 shows the task duration for an interleaving experiment with low contention. The secondary and

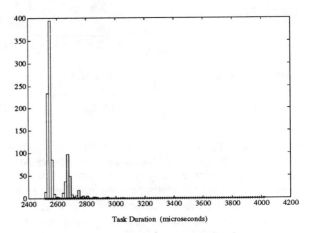

Figure 1: Task duration histogram: Experiment Il20

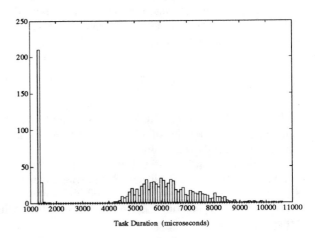

Figure 2: Task duration histogram: Experiment Hm20

tertiary peaks to the right are due to hard clock interrupts ($\sim$100 μsecs every $\sim$10 msecs) and soft clock interrupts ($\sim$200 μsecs every $\sim$100 msecs), respectively. Although all interleaving histograms exhibited this general shape, as $gmax$ and p increase the peaks grow wider and move to the right.

The results of a hotspot experiment are shown in Figure 2. The two distinct peaks are due to a lack of symmetry in the roles of the processors. The narrow peak on the left corresponds to tasks that were executed on $P0$, the processor where the global data is stored. $P0$ has fast access to the data via the T-bus resulting in much shorter and constant task duration. The other processors access data through the switch resulting in longer, more variable memory latencies as seen in the wide, asymmetric bell curve on the right. All hotspot experiments have this general shape. As the contention increases, the second peak becomes wider and flatter and the gap between the two distributions increases.

In addition to task duration, **total running time** was measured for each experiment. These running times were compared with the running times predicted by the models in the previous section, as shown in Figures 3 and 4[b].

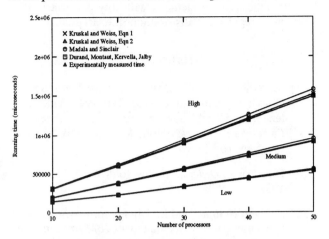

Figure 3: Predicted and actual times for hotspot data.

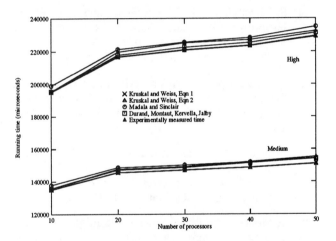

Figure 4: Predicted and actual times for interleaved data.

In the hotspot experiments, the first two estimates of Kruskal and Weiss (Equations 1 and 2) and our estimate gave the closest values, with predicted running time within 1% of true running time and often better. Madala and Sinclair's estimate was not quite as good, with predicted running times differing from true running times by as much as 6% in the worst cases. Because it is an upper bound, Equation 4 has a tendency to overestimate the running time. Madala and Sinclair's worst performance occurs when the contention is high. This is because the bound they used to estimate load imbalance gives the best approximation when the underlying distribution is a high, narrow exponential [2]. As the contention increases, the distribution becomes lower and flatter, resembling a high, narrow exponential less and less. In the interleaving experiments, the estimate of Madala and Sinclair is much more successful, with their

estimate and our estimate both generating predictions on the order of 1% of the true running time. Kruskal and Weiss' Equations 1 and 2 give results on the order of 2% of true running time.

Finally, we tested **block scheduling** by determining K_{opt} from Equation 3 using experimentally measured values of h and σ. We then re-executed the same suite of experiments, allocating work in blocks of $K = K_{opt}$ tasks. The running times are reported in Table 2. In

	Hotspot				Interleaving		
	K=1	K = K_{opt}			K=1	K = K_{opt}	
	T	K	T		T	K	T
Hl10	1.423	14	1.460	Il10	1.405	34	1.721
Hl20	2.259	14	1.830	Il20	1.481	32	1.679
Hl30	3.320	13	2.508	Il30	1.513	28	1.489
Hl40	4.356	13	3.326	Il40	1.505	26	1.423
Hl50	5.429	13	4.082	Il50	1.518	25	1.369
Hm10	1.924	14	1.599	Im10	1.358	54	1.427
Hm20	3.705	12	2.918	Im20	1.476	51	1.165
Hm30	5.541	11	4.754	Im30	1.492	46	2.077
Hm40	7.347	10	5.675	Im40	1.522	39	1.803
Hm50	9.231	10	6.998	Im50	1.553	34	1.598
Hh10	3.051	11	2.541	Ih10	1.953	84	2.273
Hh20	6.000	9	4.734	Ih20	2.197	75	2.260
Hh30	8.970	9	7.017	Ih30	2.252	60	1.904
Hh40	11.91	9	9.386	Ih40	2.270	62	1.978
Hh50	14.87	8	11.56	Ih50	2.324	20	1.965

Table 2: Block scheduling times (10^5 microsecs).

hotspot experiments, with the exception of experiment Hl10, block scheduling led to substantial performance improvement, running between 20% to 25% faster for most experiments. For these experiments, Equation 3 gives good results. However, the results for the interleaving experiments are not so successful. Approximately half of these experiments ran *slower* with the "optimal" block size than with self-scheduling. To understand this surprising phenomenon, look at Figure 5, which shows running time as a function of block size for experiment Im30. The lower two curves are the expected running times given by Equations 1 and 5. The upper curve is the upper bound on the running time given in Equation 6. The jagged line in the middle shows experimentally measured running times for values of K between 1 and 50.

As the figure shows, when K is large, the running time is highly sensitive to the relative values of N, K and p. If N/K is not an exact multiple of p, a few processors will have to execute an extra block, resulting in a large load imbalance. This is why the running time for $K = 45$, which does not divide N evenly, approaches the upper bound, whereas the running time for $K = 50$ is close to the estimates given by Equations 5 and 1. Kruskal and Weiss' model does not take this type of load imbalance into account. It is designed to address load imbalance problems associated with task duration variance.

[b]The low contention data was not presented in this figure because it is very similar to the medium contention data and renders the figure unreadable

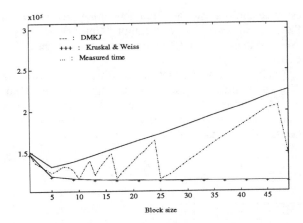

Figure 5: Total execution time (microsecs)

In Equation 3 we see that K_{opt} is proportional to $(h/\sigma)^{2/3}$. Hence, when h/σ is small, the model works well because it operates in the range where K is small and the upper and lower bounds are close together. Kruskal and Weiss' model performs well for the hotspot experiments where large values of σ lead to smaller values of K_{opt}. In the case of the interleaving experiments, the large values of the ratio h/σ push the model into the region where a small change in K can lead to a major change in the running time. Here a small error in selecting K_{opt} can be disastrous. For example, if we choose $K = 47$ instead of $K = 50$, the program runs 78% slower!

CONCLUSION

In this paper, we presented an experimental study of the impact of memory latency on task duration on a BBN TC2000. We discussed the relationship between the shape of the task duration distributions generated and the features of the TC2000 architectural design. We applied analytical models of self-scheduling to our data and discussed their strengths and weaknesses in predicting running time. We also used the models to predict the optimal block size, K_{opt}, and studied the behavior of block scheduling experimentally.

Our experimental results show that on NUMA machines task duration distributions have large variance and are irregular and unpredictable. This indicates that dynamic scheduling is imperative for these machines. Running time predictions from the fixed block scheduling analytical models of Kruskal and Weiss and others were compared with true running times and were found to perform well when the block size, K, is one. However, Kruskal and Weiss' model does not predict the running time accurately for large K because it does not reflect the fact that the running time is very sensitive to how work is spread across the processors. One solution to this problem is to modify Kruskal and Weiss' formula to consider whether Kp divides N evenly when K is large. A second approach is to use a variable block size algorithm which reduces the block size to smooth out irregular finishing times. Our experimental results indicate these methods will only perform well if they take irregular task distributions into account.

References

[1] Ralph Butler, James Boyle, Terrence Disz, Barnett Glickfeld, Ewing Lusk, Ross Overbeek, James Patterson, and Rick Stevens. *Portable Programs for Parallel Processors*. Holt, Rinehart and Winston, Inc, New York, 1987.

[2] H.A. David. *Order Statistics*. John Wiley, New York, 1981.

[3] S. Flynn Hummel and E. Schonberg. Low-Overhead Scheduling of Nested Parallelism. *IBM Journal of Research and Development*, pages 743–65, November 1991.

[4] S. Flynn Hummel, E. Schonberg, and L. E. Flynn. Factoring: A Practical and Robust Method for Scheduling Parallel Loops. In *1991 Supercomputing Conference*, pages 610–619, November 1991.

[5] C. Kruskal and A. Weiss. Allocating Independent Subtasks on Parallel Processors. *IEEE Transactions on Software Engineering*, 11(10):1001–16, October 1985.

[6] Sridhar Madala and James B. Sinclair. Performance of Synchronous Parallel Algorithms with Regular Structures. *IEEE Transactions on Parallel and Distributed Systems*, 2(1):105–116, January 1991.

[7] T. Montaut. Ordonnancement dynamique d'un programme décomposé en sous-tâches indépendantes. analyse des performances. Master's thesis, Institut de Recherche en Informatique et Systèmes Aléatoires, September 1991.

[8] C. Polychronopoulos and D. Kuck. Guided Self-Scheduling: A Practical Scheduling Scheme for Parallel Supercomputers. *IEEE Transactions on Computers*, 36(12):1425–39, 1987. Special Issue on Supercomputing.

[9] Robert H. Thomas and Will Crowther. The Uniform System: An approach to runtime support for large scale shared memory parallel processors. In *Proceedings of the 1988 International Conference on Parallel Processing*, pages 245–254, August 1988.

[10] T. H. Tzen and L. M. Ni. Dynamic Loop Scheduling for Shared-Memory Multiprocessors. In *Proceedings of the 1991 International Conference on Parallel Processing*, pages II–247–II–250, August 1991.

Reliability Evaluation of Disk Array Architectures *

John A. Chandy and Prithviraj Banerjee
Center for Reliable and High-Performance Computing
University of Illinois at Urbana-Champaign
Urbana, IL 61801

Abstract - Numerous redundant disk organizations have been proposed and used to provide increased performance and reliability from the I/O subsystems architecture, and once a disk fails in such a system, different forms of sparing and reconstruction have also been proposed. In this paper, we offer a comprehensive evaluation of the relationship between various disk array architectures, various disk reconstruction strategies, and their reliability under various failure rate and repair rate distributions. Specifically, we perform the evaluations of the reliabilities (through a measure of system mean time to failure) of different redundant disk organizations with variations in each of the following orthogonal directions. We also consider the effect of the reconstruction strategy on the reliability. A third and important dimension to our study involves the use of realistic failure rates and repair rates of disks that are dependent on the workload and the organization of the disks. Finally, we perform our study of scalability of the results to varying system sizes, by specifically addressing disk organizations with 16 to 1024 disks. The paper's contribution is in presenting a uniform framework for evaluation of these multidimensional studies, as well as offering an improved model for predicting disk array reliability taking into account system load as well as disk organization.

1 Introduction

The design of Redundant Arrays of Inexpensive Disk (RAID) systems arose out of a need to address two issues facing I/O storage systems — namely, performance and reliability. Performance was gained through the use of multidisk arrays and reliability was obtained from the use of redundancy. The latter has proven to be very important in the acceptance of RAID systems in commercial environments. Reliability of the mass storage system is more than just desirable – it is in fact necessary and required for critical I/O intensive applications such as transaction processing and supercomputing applications. Even in less crucial environments such as simple file server usage, the reliability of the storage subsystem is important.

Over the last few years, numerous redundant disk organizations, specifically RAID levels 0 through 5, have been proposed and used in many commercial parallel disk systems [1, 2, 3, 4, 5]. Once a disk fails in such a system, different forms of sparing and reconstruction have been proposed. With all of these choices, it is very difficult to answer questions regarding which scheme of redundant disk organization should be used with which reconstruction scheme, how the performance of these systems scale with different system sizes, and exactly how much reliability improvement one actually gets from these redundancy schemes.

In this paper, we will offer a comprehensive evaluation of the relationship between various disk array architectures, various disk reconstruction strategies, and their reliability under various failure rate and repair rate distributions. Specifically, we perform the evaluations of the reliabilities (through mean time to failure of the system measure) of different redundant disk organizations with variations in each of the following orthogonal directions. The four different disk array organizations we consider are basic RAID level 0 (with no parity), RAID level 1 (using mirroring), RAID level 5 (using parity) [4], and disk arrays based on block designs [6]. For each type of disk organization, an extremely key feature is the scheme of reconstruction and sparing used. In our study we investigate the use of manual sparing, hot sparing, parity sparing and distributed sparing [6, 7]. A third dimension to our study involves the use of realistic failure rates and repair rates of disks. While previous researchers have used relatively simple failure rate and repair rate distributions, we propose the use of more realistic failure rates and repair rates that depend on the system workload. Finally, we perform our study of scalability of the results to varying system sizes, by specifically addressing disk organizations with 16, 32, 64, 128, 256, 512 and 1024 disks. The paper's contribution is in presenting a uniform framework for the evaluation of these multidimensional studies, as well an improved model for predicting disk array reliability taking into account system load as well as disk organization.

2 Reliability Models Under Consideration

In this section, we will provide an analytical model for the reliability evaluation of redundant disk systems. We first review the reliability models proposed by other researchers, and then propose our more realistic workload- and organization-dependent reliability model.

2.1 A simple reliability model

For our analysis, we are concerned only with the failure rates of the actual disks themselves. Since we are ignoring the failure of other I/O subsystem components, the reliability analysis presented here should be regarded as optimistic with regards to the entire system failure rate. The failure rate of a nonredundant disk array can contribute a great deal to the entire subsystem failure rate. If, however, through the use of redundancy the disk array failure

*Acknowledgement: This research was supported in part by the Office of Naval Research under contract N00014-91-J-1096 and in part by the Joint Services Electronics Program under contract N00014-84-C-1049.

rate is reduced to sufficiently low levels, it can be rendered insignificant as compared to subsystems more prone to failures such as power supplies and fans [8]. The systems to be evaluated include RAID 0, RAID 1, RAID 5, and block designs. Most of our discussions will be in terms of mean time to failure (MTTF), the standard measure of system reliability.

In [4], Patterson et al., presented a general approximate framework for the evaluation of mean time to failure of disk arrays. In their approach it is assumed that failures are independent and follow an exponential distribution. Using the assumption that $MTTF_{disk}$ is probably much larger than the time to repair the disk, they arrived at the following relation:

$$MTTF_{sys} = \frac{MTTF_{disk}^2}{G(G-1)MTTR} \qquad (1)$$

where $MTTF_{sys}$ is the mean time to failure of a single parity group with G disks, and $MTTR$ is the time to repair a failed disk in such a system. The MTTF of an n-group system can be approximated by dividing $MTTF_{sys}$ by n. Gibson and Patterson extended this model with detailed Markov analysis to include dependent failures as well as different sparing strategies [9]. Malhotra and Trivedi offered a simplified Markov-base model of reliability [10], and their model attempts to take into account uncovered faults and false alarms and does not make the linearization approximation. Their $MTTF_{sys}$ expression is as follows:

$$MTTF_{sys} = \frac{\frac{1}{MTTR} + \frac{G(1+p(1-a))-1}{MTTF_{disk}}}{\frac{G}{MTTF_{disk}}\left(\frac{1-p}{MTTR} + \frac{(G-1)(1-a)}{MTTF_{disk}}\right)} \qquad (2)$$

The probability that a disk fault is covered is represented by p, and a is the probability that a fault can be predicted and thus allow data to be copied back before failure. Notice that as $p \to 1$, $a \to 0$, and $MTTF_{disk} \gg MTTR$, the Malhotra-Trivedi model is essentially the same as the Patterson model. These assumptions are not unreasonable as missed errors are rare, and a equals zero on most systems since prediction of faults is rarely implemented. Also, $MTTF_{disk}$ is generally around 40,000 hours as compared to $MTTRs$ of up to 24 hours. Since these assumptions are generally valid, we will refer only to the Patterson model from now on.

2.2 Workload-dependent failure rate models

The analysis described up to now does not take into account possible varying MTTFs and MTTRs. Particularly, the system load can affect the failure rate as well as the time to repair. Because of the mechanical nature of current magnetic disks, it is likely that increased activity will induce a failure of the disk. Iyer et al. were able to establish a concrete relationship between the amount of system load and system failure rates for computer systems [11]. They suggested possible environmental effects that are caused by increased utilization, namely, *electrical noise, temperature rise,* and *mechanical stress.* All of these aspects can contribute to disk failure.

The above study offered two possible models to correlate failure rates with utilization rates. The first was a linear model in which the failure rate $= \frac{1}{MTTF} = A_l + B_l\lambda$.

It was found that better correlations were obtained with a nonlinear exponential regression model: failure rate $= A_e e^{B_e \lambda}$. We were not able to duplicate their experiments with respect to disk failure, but their study showed that paging which is I/O intensive was a major contributor to hardware failures. We therefore chose A's and B's that were similar in range to those presented in [11] (Table 1). These coefficients are chosen such that at zero load, the $MTTF$ is simply the nominal $MTTF_{disk}$, Note, also, that the values are by no means exact, but rather intend to show the possible effect of load-based failure rates. The MTTF is expressed in hours, and λ is in terms of requests per second.

Table 1: Coefficients for failure model

$A_l = \frac{1}{MTTF_{disk}}$	$B_l = \frac{.001}{MTTF_{disk}}$
$A_e = \frac{1}{MTTF_{disk}}$	$B_e = .001$

2.3 Workload-dependent repair rate models

The second aspect of system MTTF is the mean time to repair. This parameter is also affected by the system load rate. Remember that the mean time to repair is the reconstruction time of the disk, and with high system load, the reconstruction time will take longer. More user requests will delay any reconstruction requests, and in most reconstruction implementations, all reconstruction requests are submitted to a lowest priority queue, causing even more of a delay on high-load systems. The effect of system load on mean time to repair has been established in several studies [12, 13, 14].

In their analysis, Muntz and Lui arrived at an estimate of reconstruction time that had the following form [12]:

$$T = \frac{A}{\lambda}ln\left[1 - \frac{B\lambda}{\mu + C\lambda}\right] \qquad (3)$$

where λ is the normal load request rate and μ is the disk service rate.

As discussed in [13], this model is somewhat inaccurate due to its treatment of the disk service rate and parity generation polices. Even after incorporating a better realization of the parity generation process, it still fails to be an accurate model of the reconstruction process. Because of the complexity of the reconstruction procedure, with multiple priority queues for servicing disks, it is nearly impossible to accurately model the behavior, and simulations become a much better tool. What we have done is to use the simulation results from previous studies to drive our reliability simulations [6, 14].

We have used these detailed simulations to obtain estimates of reconstruction time behavior for different types of reconstruction strategies: hot sparing, parity sparing and distributed sparing, and various disk organizations. An exponential model was constructed to represent reconstruction time and it was fairly accurate over the range of load rates that we will be investigating. The basic form is exponential in that $T = \frac{A}{\mu}e^{B\lambda}$. Note that as the loads become heavier, this model breaks down, since it does not

Table 2: Reconstruction times

Sparing Strategy	Predictive model	
	RAID5	Block Designs
Hot	$\frac{I}{\mu}e^{\frac{\lambda}{70+(10n-4.75)D}}$	$\frac{I}{\mu}e^{\frac{\lambda}{13(D+2)}}$
Parity	$\frac{I(nG+1)}{\mu nG}e^{\frac{\lambda}{88+(7n-.8)D}}$	$\frac{I(D+5)}{2\mu(D+2)}e^{\frac{\lambda}{8(D+2)}}$
Distributed	$\frac{InG}{\mu(nG+1)}e^{\frac{\lambda}{73+5.25(2n-1)D}}$	

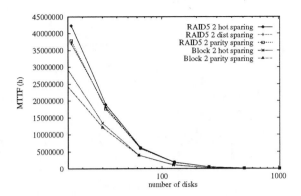

Figure 1: MTTF for arrays
varying sparing strategies.
(constant failure rate, request rate = 200 reqs/s)

approach a vertical asymptote as the disk system saturates. The Muntz and Lui model does have this behavior. However, such high loads are often unrealistic especially on large disk arrays, and we will not be investigating that range.

The resulting simulation models are summarized in Table 2. In our study, I was the number of disk tracks needed to be reconstructed — in our case our simulated disk system had 17612 tracks, μ was the disk service rate in tracks/s. Using split access operation [15], the service time for one track was taken to be the full rotational latency namely 16.6 ms, thus the service rate μ was 60 tracks/s. Note that in a load-free system reconstruction can proceed uninhibited and take the minimum time of 294 s on a hot sparing system. The other two sparing systems have correction factors to account for configuration differences. The parity sparing factor is there because of the additional overhead of generating parity for the newly merged group, and distributed sparing systems take less time because there is less data to reconstruct, since the spare blocks do not have to be copied. The block design systems we are using have interdisk interaction factors of 1/2. The estimated equations are known to be accurate (through simulations) on configurations up to 64 disks for RAID5, and up to 62 disks on block-designed organizations. Larger systems are assumed to follow extrapolations of the proposed models.

3 Simulation Results

Having identified clearly the various dimensions of study, specifically, the choice of disk organizations, the choice of reconstruction strategy, the choice of the failure rate and repair rate models, we are now ready to proceed with our study of disk systems of varying sizes.

We constructed a reliability simulation procedure using facilities available in CSIM [16]. Using these tools we evaluated several systems by measuring 5000 failure points for various disk array configurations and sparing strategies. For the manual sparing systems, we represented reconstruction of the disk as a random time from 4 to 24 h during which the repairman will arrive, making an mean time to repair of 14 h. For automatic sparing systems, we use the reconstruction times arrived at in our previous study for systems smaller than 64 disks. For larger arrays, the reconstruction time models presented earlier are employed.

3.1 Impact of disk organization and sparing strategy

Several studies have investigated impact of disk organization on reliability [9, 10]. In this section, we address the

use of automatic sparing and reconstruction techniques, and look at the various strategies under consideration (hot, parity, and distributed). Figure 1 shows these sparing strategies for 2 parity RAID5 and block-designed configurations. RAID5 systems naturally have better reliability than block designed systems because of their ability to handle double faults as long as the faults are in separate parity groups. Because block designed systems use interleaved parity groups, the double parity disks can not handle dual faults. The same is true of any other interleaved parity group system such as parity declustering [13].

It is clear that hot sparing is the best scheme for RAID5 systems and though not evident on the graph, the same behavior is true at larger sizes. Hot sparing is also desirable for block designed systems at small disk array sizes. Remember, however, that the hot sparing systems use one extra disk that helps in reliability but does not help in system performance. The parity and distributed sparing systems do not have this extra cost because they use interleaved sparing strategies. The proper choice of sparing strategy can dramatically alter the system reliability.

3.2 Impact of workload-dependent failure rates

In this section, we investigate the effect of the load-based failure models presented earlier. In Figure 2, a RAID5 2-parity manual sparing system is shown for the different failure rates. With the parity system, there is a noticeable difference between the constant failure rate systems and the more realistic models. The Patterson model is superimposed on this graph to show its validity for the constant failure rate model, but it fails to correctly predict behavior for the more realistic non-constant failure rate models. Figure 3 shows the effect as system load changes for 64 disk 2-parity arrays with automatic sparing. It is clear that in high-load systems, the higher failure rate due to increased activity must be taken into account when determining overall system reliability. Remember that our coefficient choices for the linear and exponential model were not based on any known failure rates. Figure 4 shows the effect as the base coefficient (B_l and B_e) is changed using 0.01, 0.001, and 0.0001. It is clear that the choice of coefficients can significantly affect the reliability estimates,

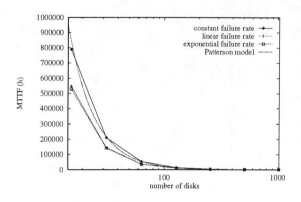

Figure 2: MTTF for RAID5 2-parity arrays,
different failure models
(manual sparing, request rate = 200 reqs/s)

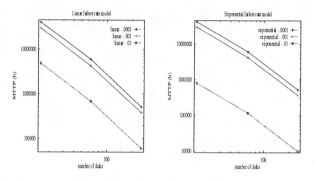

Figure 4: MTTF for 2-parity arrays,
varying model coefficients
linear and exponential failure models
(hot sparing, request rate = 200 reqs/s)

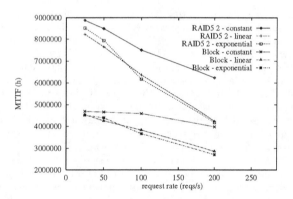

Figure 3: MTTF for different failure models.
(64 disks, 2 parity, hot sparing)

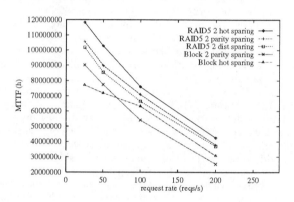

Figure 5: MTTF varying load-dependent repair rate.
(16 disks, 2 parity, constant failure rate)

so more work must be done to arrive at good estimates of these coefficients. At any rate, the usefulness of a load-based failure rate model is evident.

3.3 Impact of workload on repair rates

As shown in the previous section, high disk access rates can increase the failure rates of the disks. In addition, increased disk activity can also increase the mean time to repair for automatic reconstruction systems. Figures 5 and 6 show the effect of system load for 2-parity RAID5 and block-designed configurations. As the system load increases from 25 rps to 200 rps, the MTTF decreases to up to a third of expected MTTF as predicted by a simple model. If either of the non-constant failure rate models are added in, the falloff is even more dramatic as in Figure 3. This dropoff is simply due to the longer reconstruction time as a result of the higher I/O load. High system load can significantly affect the overall system reliability, and, in environments where the system activity will be great, it may be necessary to improve reliability through the use of extra parity or better sparing strategies. These plots show the importance of considering system load when determining overall system reliability.

Notice also that as the disk array size grows larger, the gap between block designed systems and RAID5 systems grows. As the size grows larger, the ability to tolerate dual faults becomes more crucial, and thus the disparity between the two organizations. Also, in both figures, it is clear that hot sparing becomes a more attractive strategy as the system load increases. This behavior occurs because the parity sparing reconstruction method is able to improve the performance of normal requests and this effect shows at lower request rates. However, as the request rate increases, parity sparing loses its benefit because the spare writeback will tend to interfere with normal request patterns, while hot sparing will not have this problem.

4 Conclusions

In this paper, we have presented reliability evaluations of several disk array configurations. It is clear that disk arrays must have some form of redundancy, and just as important, as the disk arrays grow larger, automatic sparing is not only desirable but necessary to retain high system availability. Block-designed systems and other parity declustered systems will suffer poorer reliability then equivalent RAID5 systems. We have also presented models that attempt to simulate failure rates and repair rates

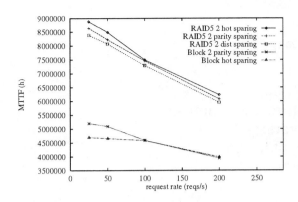

Figure 6: MTTF varying load-dependent repair rate. (64 disks, 2 parity, constant failure rate)

based on system load and disk organization, and demonstrated that under these realistic models, the mean time to failure of disks will be less than that of previously proposed models based on constant failure and repair rates.

References

[1] M. Y. Kim, "Synchronized disk interleaving," *IEEE Transactions on Computers*, vol. C-35, pp. 978–988, November 1986.

[2] K. Salem and H. Garcia-Molina, "Disk striping," in *International Conference on Data Engineering*, pp. 336–342, 1986.

[3] A. L. N. Reddy and P. Banerjee, "An evaluation of multiple-disk I/O systems," *IEEE Transactions on Computers*, vol. 38, pp. 1680–1690, December 1989.

[4] D. A. Patterson, G. A. Gibson, and R. H. Katz, "A case for redundant arrays of inexpensive disks (RAID)," in *Proceedings of 1988 ACM SIGMOD International Conference on Management of Data*, pp. 109–116, June 1988.

[5] J. Gray, B. Host, and M. Walker, "Parity striping of disk arrays: Low-cost reliable storage with acceptable throughput," in *Proceedings of 16th International Conference on Very Large Data Bases (VLDB)*, pp. 148–161, August 1990.

[6] A. L. N. Reddy, J. A. Chandy, and P. Banerjee, "Design and evaluation of gracefully degradable disk arrays," *Journal for Parallel and Distributed Computing*, vol. 17, pp. 28–40, January 1993.

[7] J. Menon and R. L. Mattson, "Comparison of sparing alternatives for disk arrays," in *Proceedings of International Symposium on Computer Architecture*, pp. 318–331, May 1992.

[8] M. Schulze, G. Gibson, R. Katz, and D. Patterson, "How reliable is a RAID?," in *Proceedings of the Spring 1989 COMPCON*, pp. 118–123, 1989.

[9] G. A. Gibson and D. A. Patterson, "Designing disk arrays for high data reliability," *Journal for Parallel and Distributed Computing*, vol. 17, pp. 4–27, January 1993.

[10] M. Malhotra and K. S. Trivedi, "Reliability analysis of redundant arrays of inexpensive disks," *Journal for Parallel and Distributed Computing*, vol. 17, pp. 146–151, January 1993.

[11] R. K. Iyer, S. E. Butner, and E. J. McCluskey, "A statistical failure/load relationship: Results of a multicomputer study," *IEEE Transactions on Computers*, vol. C-31, pp. 467–476, July 1982.

[12] R. R. Muntz and J. C. S. Lui, "Performance analysis of disk arrays under failure," in *Proceedings of 16th International Conference on Very Large Data Bases (VLDB)*, pp. 162–173, August 1990.

[13] M. Holland and G. A. Gibson, "Parity declustering for continuous operation in redundant disk arrays," in *Proceedings of Fifth International Conference on Architectural Support for Programming Languages and Operating Systems*, pp. 23–35, October 1992.

[14] J. A. Chandy and A. L. N. Reddy, "Failure evaluation of disk array organizations." to appear in *Proceedings 13th International Conference on Distributed Computing Systems*, May 1993.

[15] A. L. N. Reddy, "A study of I/O system organizations," in *Proceedings of International Symposium on Computer Architecture*, pp. 308–317, May 1992.

[16] H. D. Schwetman, "CSIM reference manual (Revision 14)," March 1990.

Prime-way Interleaved Memory

De-Lei Lee*

Department of Computer Science
York University
North York, Ontario
Canada M3J 1P3

Abstract – In this paper, we study prime-way interleaved memory in the context of vector multiprocessor systems. Practical, efficient techniques for implementation of prime-way interleaved memory are presented. A method for minimizing access contention among multiple vector streams is described.

1. Introduction

The general technique to produce high memory bandwidth for main memory of most vector multiprocessors, called P-way interleaved memory, is to pipeline the memory by partitioning memory into P independent memory modules that can be active in some overlapped fashion [15]. A P-way interleaved memory has a memory space of NP memory locations to store NP words, and each memory location is uniquely identified by a memory address A in the range $0 \leq A \leq NP - 1$, where N is the number of memory locations in a single memory module. The two equations below are typically employed to map a memory address A to memory location L of memory module M:

$$L = \lfloor \frac{A}{P} \rfloor \qquad (1)$$

$$M = A \bmod P. \qquad (2)$$

In vector processor systems, a vector of words is usually specified by three parameters (SA, S, VL) describing the set of addresses $\{SA + i \times S \mid 0 \leq i \leq VL - 1\}$ for words of the vector, where SA, S, and VL are the start address, stride, and length of the vector, respectively. A sequence of one-word requests to P-way interleaved memory for words at addresses $SA, SA+S, \cdots, SA+(VL-1)S$ is directed to E distinct memory modules in the sequence $(SA \bmod P, (SA + S) \bmod P, \cdots, (SA + (E-1)S) \bmod P)$ *periodically*, where $E = P/\gcd(P, S)$ and gcd stands for the greatest common divisor function [6]. Thus, whenever S is not relative prime to P, E becomes only a fraction of P; memory contention occurs within the vector access stream if E is less than memory module access cycle time. As vector stride S is application dependent, the choice of P becomes important in the design of P-way interleaved memory.

The most common style of P-way interleaved memory is the use of $P = 2^k$ memory modules [14]. Such a choice of P makes the computation of equations (1) and (2) trivial. Since P is a power of 2, when a vector has an even stride, access contention may occur and memory bandwidth is degraded accordingly [15]. Moreover, when the memory is accessed via multiple, concurrently operating memory ports, each requesting access to a memory location at every clock cycle, access contention may occur among multiple vector access streams as well [10]. Such access contention has now been recognized as the primary source of memory bandwidth degradations in vector multiprocessors [1,13]. While designing a main memory that can totally avoid access contention is impossible, little is known on how to design one that can minimize it. In practice, main memory of most vector multiprocessors is neither capable of minimizing access contention within single access stream nor among multiple access streams; when access contention occurs, memory bandwidth decreases accordingly and performance penalty is accepted as a fact of life [6,15].

The notion of prime-way interleaved memory is due to Budnik and Kuck [2]. The choice of prime P maximizes the number of vector strides that are relative prime to P. A definite advantage of using prime-way interleaved memory in the context of vector multiprocessors is the minimized access contention within a single vector access stream. The real problem arises when P is prime, since computing equation (1) involves division by the prime number. Such an operation requires not only extensive hardware but also multiple clock cycles to compute, making the use of prime-way interleaved memory counterproductive in practice. An early attempt to implement prime-way interleaved memory includes a work by Lawrie and Vora [7] in the context of array processors, where all P memory modules operate in lock-step to access a vector of P words at a time in parallel. They recognized that while equation (2) affects the behavior of prime-way interleaved memory, the role of equation (1) is to ensure 100% utilization of memory locations. Lawrie and Vora used the equation $L = \lfloor \frac{A}{2^{\lceil \log P \rceil}} \rfloor$ in place of equation (1). As the equation is not bijective, their approach avoided division at the expense of up to 50% underutilization of memory space.

*This research was supported in part by grants from the Natural Science and Engineering Research Council of Canada and the Ontario Center for Excellence.

The purpose of this paper is two-fold. First, we give a method for implementing prime-way interleaved memory. Our method ensures full utilization of memory space. Furthermore, we show how to make prime-way interleaved memory capable of minimizing access contention among multiple vector access streams by way of memory access ordering. The next section describes how to implement prime-way interleaved memory. Section 3 introduces the notion of memory access ordering, and shows how to implement it. Concluding remarks are drawn in Section 4.

2. Prime-way Interleaved Memories

Two specific types of prime-way interleaved memories are described in this section, and denoted as Mersenne type (M-type) and Fermat type (F-type) memories, respectively. Techniques for the M-type and F-type memories can be generalized to general type (G-type) prime-way interleaved memories comprising any prime number of memory modules. We shall assume mapping of memory addresses to memory modules is determined by equation (2) throughput this section.

2.1. The M-type Memories

An M-type memory has a memory space of NP memory locations, where $P = 2^k - 1$ is the number of memory modules and $N = 2^m$ is the number of memory locations in a single memory module. Each memory location within a memory module is uniquely identified by a location index L of m bits in the range $0 \leq L \leq N - 1$. Each memory location in the memory space is uniquely identified by a memory address A of n bits in the range $0 \leq A \leq NP - 1$, where $n = m + k$.

The following equations are used to map memory address A to memory location L of memory module M:

$$L = \lfloor \frac{A}{2^k} \rfloor, \; if \; A \bmod 2^k \neq 0; \tag{3}$$

$$L = 2^{m-k}P + (\frac{A}{2^k} \bmod 2^{m-k}), \; otherwise. \tag{4}$$

The use of equation (2) will map N memory addresses to each memory module, and memory addresses mapped to memory module M for all $0 \leq M \leq P - 1$ are of the form $A = i \times P + M$ for all $0 \leq i \leq N - 1$. In what follows, we show that the N memory addresses mapped to memory module M are further mapped into the N memory locations of the memory module in a bijective manner by the use of equations (3) and (4), thereby ensuring full utilization of memory locations. The lemmas used in this section are presented without proofs. Their proofs are contained in [8]. The following lemma regards the role of equation (3).

Lemma 1 *Let $A \equiv M \pmod{P}$. Then,*

1. $\lfloor \frac{A}{2^k} \rfloor = \lfloor \frac{A+P}{2^k} \rfloor - 1$, *if $A \bmod 2^k \neq 0$;*

2. $\lfloor \frac{A}{2^k} \rfloor = \lfloor \frac{A+P}{2^k} \rfloor$, *if $A \bmod 2^k = 0$.*

Note that $A + P$ is the next successive memory address of A mapped to memory module M. The first part of the lemma reveals that the use of equation (3) maps two successive memory addresses into two consecutive memory locations of memory module M when neither one is evenly divisible by 2^k. Furthermore, the second part of the lemma indicates that when A is evenly divisible by 2^k, the use of equation (3) maps the successive memory address $A + P$, which is not evenly divisible by 2^k, into memory location $\lfloor \frac{A}{2^k} \rfloor$. Consequently, the use of equation (3) maps all those memory addresses which are not evenly divisible by 2^k into contiguous locations of memory module M starting from location zero without conflict. The next lemma gives the exact number of such memory addresses.

Lemma 2 *Let $A \equiv M \pmod{P}$. Then, there are exactly $2^{m-k}P$ such A's that $A \bmod 2^k \neq 0$.*

So, the use of equation (3) will map $2^{m-k}P$ memory addresses into the contiguous $2^{m-k}P$ memory locations of memory module M starting from location zero in a bijective manner. Next, we show that the use of equation (4) will map the remaining 2^{m-k} memory addresses not covered by equation (3) into the remaining locations of memory module M in a bijective manner. We will use $a \mid b$ to denote that a evenly divides b.

Lemma 3 *Let $A = i \times P + M$ where $0 \leq i \leq 2^m - 1$; and $2^k \mid A$. Then, $\frac{A}{2^k} \bmod 2^{m-k}$ is a bijective mapping.*

Notice that the value of L obtained from equation (4) is greater than or equal to $2^{m-k}P$. Accordingly, the use of equation (4) maps the 2^{m-k} memory addresses not covered by equation (3) into the remaining 2^{m-k} contiguous locations of memory module M starting from location $2^{m-k}P$ in a bijective manner. Consequently, we have established the following result.

Theorem 1 *Use of equations (2), (3) and (4) will map all the NP memory addresses into distinct memory locations of the M-type memory.*

Thus, we have successfully avoided the use of equation (1) in the construction of prime-way interleaved memory. We shall now turn our attention to the computation of equations (3) and (4). Computing the location index L from equation (3) is trivial and takes $O(1)$ gate delay as the divider is a power of 2. Inspection of equation (4) reveals that the first term is a binary number of m bits in which the most significant k bits are all 1's and the remaining $(m - k)$ bits are all 0's. The second term of equation (4) is less than 2^{m-k} and is a binary number of m bits in which the most significant k bits are all 0's and the remaining $(m - k)$ bits are the least significant $(m - k)$ bits of the result of A divided by 2^k. So, the computation of the location index L from equation (4) is equivalent to forming a binary number of m bits in which the most significant k bits are all 1's, and the remaining $(m - k)$ bits are obtained from the memory address A by removing the most significant k bits and the least significant k bits. This can be done in $O(1)$ gate delay as well. Deciding whether A is evenly divisible by 2^k, i.e., if the least significant k bits of A are all 0's, can be done in at most $O(\log \log P)$ gate delay using $O(\log P)$ gates. Thus, we have the following theorem.

Theorem 2 *Equations (3) and (4) can be computed in $O(\log \log P)$ gate delay using $O(\log P)$ gates.*

2.2. The F-type Memories

A F-type memory has a memory space of NP memory locations, where $P = 2^k + 1$ is the number of memory modules and $N = 2^m$ is the number of memory locations in a single memory module. Each memory location within a memory module is uniquely identified by a location index L of m bits in the range $0 \leq L \leq N-1$. Each memory location in the memory space is uniquely identified by a memory address A of n bits in the range $0 \leq A \leq NP - 1$, where $n = m + k + 1$.

The following equations are used to map memory address A to memory location L of memory module M:

$$L = \lfloor \frac{A}{2^{k+1}} \rfloor, \ if \ A \bmod 2^{k+1} \geq P - 2; \qquad (5)$$

$$L = 2^{m-k-1}(2^{k+1} - 1 - (A \bmod 2^{k+1}))$$

$$+ (\lfloor \frac{A}{2^{k+1}} \rfloor \bmod 2^{m-k-1}), \ otherwise. \qquad (6)$$

The next three lemmas can be used to prove the validity of equations (5) and (6).

Lemma 4 *Let $A \equiv M \pmod{P}$. Then,*

$$1. \lfloor \frac{A}{2^{k+1}} \rfloor = \lfloor \frac{A + P}{2^{k+1}} \rfloor - 1, if A \bmod 2^{k+1} \geq P - 2;$$

$$2. \lfloor \frac{A}{2^{k+1}} \rfloor = \lfloor \frac{A + P}{2^{k+1}} \rfloor, if A \bmod 2^{k+1} < P - 2.$$

Lemma 5 *Let $A \equiv M \pmod{P}$. Then, there are exactly $2^{m-k-1}P$ such A's that $A \bmod 2^{k+1} \geq P - 2$.*

Lemma 6 *Let $A = i \times P + M$ where $0 \leq i \leq 2^m - 1$; and $A \equiv C \pmod{2^{k+1}}$. Then, $\lfloor \frac{A}{2^{k+1}} \rfloor \bmod 2^{m-k-1}$ is a bijective mapping.*

Using the three lemmas above, it is not difficult to prove the following theorem.

Theorem 3 *Use of equations (2), (5), and (6) will map all the NP memory addresses into distinct memory locations of the F-type memory.*

The computation of the location index L from equations (5) and (6) can be carried out in a way similar to that for equations (3) and (4). The following theorem runs parallel to Theorem 2.

Theorem 4 *Equations (5) and (6) can be computed in $O(\log \log P)$ gate delay using $O(\log P)$ gates.*

2.3. The General Case

Techniques described for constructing the M-type and F-type memories can be generalized to the G-type memories in which the number of memory modules is prime of the form $P = 2^k \pm \Delta$, wherein Δ is any integer not restricted to 1. Two equations similar to equations (3) and (4) can be used to determine L when P is of the form $2^k - \Delta$, and the equations can be computed with the same time and hardware complexity as that of equations (3) and (4).

Likewise, two equations similar to equations (5) and (6) can be used to determine L when $P = 2^k + \Delta$, and the equations can be computed with the same time and hardware complexity as that of equations (5) and (6). Space limit prevents detailed description of the G-type memories, interested readers are referred to [8].

Before closing this section, it should be pointed out that there is a family of functions that are equivalent to the specific functions described in this section for the bijective mapping of memory addresses to locations of a memory module. Any function in the family can be obtained through algebraic manipulation of the specific functions and can be used to determine L from A with the same time and hardware complexity [8].

3. Memory Access Ordering

As mentioned earlier, when main memory of vector multiprocessors is accessed via multiple memory ports, access contention can occur both within single vector access stream and among multiple vector access streams. The order in which memory modules are referenced by a vector stream determines how the access stream interacts with others, and the interaction among the multiple access streams in turn determines the degree of access contention among them [10]. The interaction among multiple vector access streams is inherently complex, since each vector access stream may reference memory modules in a different order from others as the vector can have practically arbitrary start address and stride.

Memory access ordering is a way to minimize access contention among multiple vector access streams by *making all active memory ports reference all the memory modules in the same order while serving multiple vector access streams* [8]. This notion is realizable in prime-way interleaved memory, since there is such a guarantee that *all* the P memory modules will be touched *periodically* by a vector with any stride that is not a multiple of the prime number. Memory access ordering is generally not applicable to interleaved memory having power-of-two memory modules. This is because either various fractions of the memory modules will be referenced by each vector access stream if mapping of addresses to memory modules is determined by equation (2), or a vector access stream will reference the memory modules in some pseudo-random fashion rather than periodically if mapping of addresses to memory modules is determined by a different equation that minimizes access contention within a single vector access stream [4,9,11,12]. In the remainder of this section, we show how to realize the notion of memory access ordering in prime-way interleaved memory. We will assume that a vector (SA, S, VL) must be in a vector register before its use by a vector processor or its store to memory via a memory port.

One way for a memory port to serve a vector access stream is to access the words of the vector in the increasing order of their addresses, i.e., access the word at address SA first, then the word at address $SA+S$, and so on. In this case, the P memory modules will be referenced in the sequence $\langle SA \bmod P, (SA + S) \bmod P, \cdots, (SA+(VL-1)S) \bmod P \rangle$. When $VL = P$ and S is not a multiple of P, the memory module sequence referenced is a permutation of the sequence $\langle 0, 1, \cdots, P - 1 \rangle$. While the start address SA determines the first memory module in the memory module sequence, the order in which memory modules are referenced is determined by vector stride S. Accessing words of a vector in the increasing order of their addresses at each memory port makes it impossible for all memory ports to reference the memory modules in the same order, since stride of the vector requested by each access stream can be arbitrarily different from one another.

We say a sequence of P addresses for the words of a vector (SA, S, P) is an optimized address sequence if the sequence of addresses references the memory modules in the sequence $\langle SA \bmod P, (SA + 1) \bmod P, \cdots, (SA+P-1) \bmod P \rangle$. An optimized address sequence references the P memory modules in a sequence that is a cyclically shifted version of the sequence $\langle 0, 1, \cdots, P - 1 \rangle$. While the start address SA determines the first memory module in the memory module sequence, the order in which the memory modules are referenced by an optimized address sequence is fixed, and independent of vector stride S. Thus, making every memory port access words of a vector in optimized address sequence will allow all memory ports to request the memory modules in the same order, thereby fulfilling the notion of memory access ordering. While the general idea is quite simple, successfully implementing it requires that each memory port must be able to generate addresses for a vector in optimized address sequence, one address every clock period after some initial latency, in order to keep up with the maximum access rate. We shall first present an algorithm for generating addresses in optimized address sequence, and then show how to implement the algorithm efficiently.

For any integer s in the range from 0 to $P - 1$, define I_s to be the multiplicative inverse of s $\pmod P$, i.e., I_s is an integer in the range from 1 to $P - 1$ such that $I_s \times s \equiv 1 \pmod P$. In the case of $s = 0$, I_s is defined to be 1 for purpose explained shortly. Without loss of generality, assuming vector stride S is positive, the following theorem points to a way of generating addresses for the words of a vector (SA, S, P) in optimized address sequence.

Theorem 5 Let $s = S \bmod P \neq 0$, $I = I_s$, $A \in Q = \{SA + i \times S \mid 0 \leq i \leq P\}$, and $A \equiv F \pmod P$. If $A+I\times S \notin Q$, then $A-(P-I)S \in Q$ and $A-(P-I)S \equiv F + 1 \pmod P$.

Proof. $A \in Q$ implies that $A = SA + i \times S$ for some i, $0 \leq i \leq P$. $A + I \times S \notin Q$ implies that $A + I \times S > SA + P \times S$. It follows that $SA + 2P \times S > SA + (i + I)S > SA + P \times S$ due to $I < P$, hence $1 \leq i + I - P \leq P - 1$. So, $SA+(i+I-P)S = A+(I-P)S \in Q$ and $A+(I-P)S \equiv F + 1 \pmod P$. □

It follows directly from Theorem 5 that the algorithm below generates addresses for the words of the vector (SA, S, P) in optimized address sequence $\langle A_0, A_1, \cdots, A_{P-1} \rangle$ correctly.

Algorithm 1.

1. $A_0 := SA$;
2. $R1 := I \times S$;
3. $R0 := (P - I) \times S$;
   for $i := 0$ to $P - 2$ {
4.   if $A_i + I \times S \in Q$
5.   then $A_{i+1} := A_i + R1$;
6.   else $A_{i+1} := A_i - R0$;
   }

We now improve the algorithm for greater computational efficiency. First, we introduce a set of P control vectors defined as $CV = \{CV_s \mid 0 \leq s \leq P-1\}$. Each control vector in CV is a precomputed, binary number of P bits. The purpose of using control vector CV_s is to eliminate the membership testing performed at step 4 of Algorithm 1 as follows. The ith bit of control vector CV_s, denoted by $CV_s[i]$, is used to indicate whether $A_i + I \times S$ is a member of Q or not at the ith iteration of the algorithm, depending upon $CV_s[i]$ equal to 1 or 0. CV_0 is needed for vectors whose strides are multiple of P. Thus, in light of Theorem 5, each control vector CV_s, where $0 \leq s \leq P - 1$, can be constructed as follows. Below, $Q = \{\bar{i} \times s \mid 0 \leq i \leq P\}$ and $I = I_s$.

1. $A := 0$;
   for $i := 0$ to $P - 1$ {
2.   if $A + I \times s \in Q$
3.   then $\{CV_s[i] := 1;\ A := A + I \times s;\}$
4.   else $\{CV_s[i] := 0;\ A := A - (P - I) \times s;\}$
   }

The set of P control vectors CV can be implemented with a combinational circuit using $O(P)$ gates and each CV_s can be generated as a function of s in $O(1)$ gate delay [8]. Note Algorithm 1 makes use of I_s, i.e., the multiplicative inverse of s $\pmod P$, and $P - I_s$, where $s = S \bmod P$. The entire set of multiplicative inverses, $\{I_s \mid 0 \leq s \leq P - 1\}$ contains P integers each of $\lceil \log P \rceil$ bits. And the set of integers $\{I'_s = P - I_s \mid 0 \leq s \leq P - 1\}$ has the same collection of elements as the multiplicative inverse set. I_0 is needed for vectors whose strides are multiple of P. The set of P integers can be precomputed, and the two sets can be implemented with a combinational circuit using $O(\log P)$ gates, so that I_s and I'_s can be generated as a function of s in $O(1)$ gate delay [8].

Having thus introduced the set of control vectors, and the set of precomputed constants, Algorithm 2, which is a modified version of Algorithm 1, can now be used to generate addresses for words of the vector in optimized address sequence $\langle A_0, A_1, \cdots, A_{P-1} \rangle$ more efficiently.

Algorithm 2.

1. $A_0 := SA$;
2. $R1 := I_s \times S$;
3. $R0 := I'_s \times S$;
   for $i := 0$ to $P - 2$ {
4.   if $CV_s[i] = 1$
5.   then $A_{i+1} := A_i + R1$;
6.   else $A_{i+1} := A_i - R0$;
   }

While the operations for generating addresses in optimized address sequence are presented in a sequential manner in Algorithm 2 for illustrative purposes, actual implementation requires overlapping of the operations in order to generate an address every clock period after some initial latency. Note the multiplication of S and I_s can be done in approximately the same amount of time as the addition operation using a Wallace tree [3,16], as I_s is only of $\lceil \log P \rceil$ bits, e.g., I_s is of 5 bits when P is 31. A pipelined hardware implementation of the algorithm is contained in [8]. Using Algorithm 2 as a building block, an algorithm can be constructed to generate addresses in optimized address sequence for the more general case where vector length VL is an arbitrary integer not necessarily equal to P [8].

4. Concluding Remarks

The original motivation for prime-way interleaved memory was to minimize access contention within a single access stream. An additional advantage explored in this paper is the minimized access contention among multiple access streams in the context of vector multiprocessors.

The techniques described in Section 2 for realizing prime-way interleaved memory have implicitly assumed that each memory module consists of a single memory bank. The more general case is to have 2^b memory banks per memory module, where b is any integer not restricted to 0. This is a rather useful generalization since the maximum number of vector access streams a prime-way interleaved memory can support is now determined by two design parameters: the prime number as well as the number of banks per memory module. Thus, a prime-way interleaved memory can be designed to support a different number of vector access streams by varying the number of memory banks within a memory module, while the prime number is fixed.

Such a generalization can be readily obtained as follows. Index the banks in each memory module from 0 to $2^b - 1$, and let each bank have 2^{m-b} memory locations. Such a P-way interleaved memory has a memory space of NP locations with each location uniquely identified by an address A in the range $0 \leq A \leq NP - 1$, where $N = 2^m$ is the total number of locations in a single memory module. This memory system maps a memory address A to location $\lfloor \frac{L}{2^b} \rfloor$ of memory bank $L \bmod 2^b$ of memory module $A \bmod P$, where L is a binary number of m bits generated from A using equations (3) and (4) if P is of the form $2^k - \Delta$, or equations (5) and (6) if P is of the form $2^k + \Delta$, where $\Delta = 1$. The $\Delta > 1$ case is treated in [8]. Computing equations $\lfloor \frac{L}{2^b} \rfloor$ and $L \bmod 2^b$ is trivial as they involve division by power-of-two. Using the results from Section 2, one can show the above set of equations maps the NP addresses into memory locations of the prime-way interleaved memory in a bijective manner. The results presented in Section 3 are directly applicable to this generalization without any modification.

The following lists the first four Mersenne primes, $2^2 - 1$, $2^3 - 1$, $2^5 - 1$, $2^7 - 1$, and the first four Fermat numbers, $2 + 1$, $2^2 + 1$, $2^4 + 1$, $2^8 + 1$ [5]. Thus, $3, 5, 7, 17, 31, 127, 257$ may be considered good prime numbers for prime-way interleaved memory, using the techniques described in this paper. The list of prime numbers appears to provide a very good coverage of possible numbers of memory modules one could choose in practice, especially when the above generalization is taken into consideration.

References

[1] D.H. Bailey. Vector computer memory bank contention. *IEEE Trans. Comput.*, C-36:293–298, March 1987.

[2] P. Budnik and D.J. Kuck. The organization and use of parallel memories. *IEEE Trans. Comput.*, C-20:1566–1569, Dec. 1971.

[3] K. Hwang. *Computer arithmetic: principles, architecture and design*. Wiley, New York, 1979.

[4] D.T. Harper III and D. A. Linebarger. Conflict-free vector access using a dynamic storage scheme. *IEEE Trans. Comput.*, C-40:276–283, 1991.

[5] D.E. Knuth. *Seminumerical algorithms*. Addison-Wesley, 1981.

[6] P.M. Kogge. *The architecture of pipelined computers*. Hemisphere Publishing Corporation, 1981.

[7] D.H. Lawrie and C.R. Vora. The prime memory system for array access. *IEEE Trans. Comput.*, C-31:435–442, Oct. 1982.

[8] D.-L. Lee. Method for implementation of prime-way interleaved memory comprising any prime number of memory modules. *U.S. Patent* (pending).

[9] A. Norton and E. Melton. A class of linear transformations for conflict-free power-of-two stride access. *Proc. Int'l Conf. Parallel Processing*, pages 247–254, Aug. 1987.

[10] W. Oed and O. Lange. On the effective bandwidth of interleaved memories in vector processor systems. *IEEE Trans. Comput.*, C-34:949–957, Oct. 1985.

[11] R. Raghavan and J.P. Hayes. On randomly interleaved memories. *Proc. Supercomputing '90*, pages 49–58, Nov. 1990.

[12] B.R. Rau. Pseudo-randomly interleaved memory. *Proc. 18th International Symposium on Computer Architecture*, pages 74–83, May 1991.

[13] K.A. Robbins and S. Robbins. Bus conflicts for logical memory banks on a Cray Y-MP type processor system. *Proc. Int'l Conf. Parallel Processing*, pages 21–24, Aug. 1991.

[14] R.M. Russell. The Cray-1 computer system. *CACM*, 21:63–72, 1978.

[15] H.S. Stone. *High-performance computer architecture*. Addison Wesley, 1990.

[16] C.C. Wallace. A suggestion for a fast multiplier. *IEEE Trans. Electronic Compututers*, EC-13:14–17, 1964.

SESSION 11A

ROUTING ALGORITHMS AND RING, BUS STRUCTURES

Reducing the Effect of Hot Spots by Using a Multipath Network

Mu-Cheng Wang, Howard Jay Siegel, Mark A. Nichols†, and Seth Abraham

Parallel Processing Laboratory, School of Electrical Engineering
Purdue University, West Lafayette, IN 47907-1285 USA

Abstract -- *One type of interconnection network for a medium to large-scale parallel processing system (i.e., a system with 2^6 to 2^{16} processors) is a buffered packet-switched multistage interconnection network (MIN). It has been shown that the performance of these networks is satisfactory for uniform network traffic. More recently, several studies have indicated that the performance of MINs is degraded significantly when there is hot spot traffic, that is, a large fraction of the messages are routed to one particular destination. A multipath MIN is a MIN with two or more paths between all source and destination pairs. This research investigates how the Extra Stage Cube multipath MIN can reduce the detrimental effects of tree saturation caused by hot spots. Simulation is used to evaluate the performance of the proposed approach. The objective of this evaluation is to show that, under certain conditions, the performance of the network with the usual routing scheme is severely degraded by the presence of hot spots. With the proposed approach, although the delay time of hot spot traffic may be increased, the performance of the background traffic, which constitutes the majority of the network traffic, can be significantly improved.*

1. INTRODUCTION

A variety of approaches to the design of an interconnection network that supports communication among the processors and memories of a medium to large-scale parallel processing system (i.e., a system with 2^6 to 2^{16} processors working together to solve a problem) have been proposed and/or implemented. These approaches are often based on a multistage interconnection network (MIN) topology consisting of multiple stages of switches. A multistage cube network has been used or proposed for use in systems capable of MIMD operations such as the BBN Butterfly Plus [6], IBM RP3 [16], PASM [21, 22], and NYU Ultracomputer [8]. The class of multistage cube network topologies includes: baseline, butterfly, flip, generalized cube, indirect binary n-cube, multistage shuffle-exchange, omega, and SW-banyan (S=F=2, L=n) networks [19, 20]. Here, the class of multistage cube networks will be represented by the generalized cube topology.

Let N be the number of processors in a multiprocessor system and n be the size (number of input and output ports) of the network switches used. The advantages of the multistage cube network approach include: the number of network components is proportional to $N\log_n N$, efficient distributed control schemes, network partitionability, variations on the generalized cube topology make available multiple simultaneous source/destination paths [2], and the ability to employ a variety of different implementation techniques [19]. Previous studies (e.g., [7, 12, 28]) have shown that packet-switched MINs can provide reasonable performance with small-sized switch buffers

This research was supported by the Naval Ocean Systems Center under the High Performance Computing Block, ONT.

†The author is currently with NCR Corporation, San Diego, CA.

when a moderate traffic load of uniform network traffic is assumed. These advantages, coupled with good overall network performance, make an MIN topology appealing [20].

A physically distributed memory system composed of processors, memory, and a buffered packet-switched network is assumed in this study (see Fig. 1). One processor and its associated memory module form a processing element (PE). It is assumed that memory on all PEs is accessed as a single shared address space. Similar system configurations can be found in the IBM RP3 [16] and BBN Butterfly Plus [6]. PEs satisfy memory references through the network if the referenced address is not located on the attached memory module. From the viewpoint of a program, the only difference between a reference to memory on the local PE and memory on another PE is that the remote reference takes more time to complete.

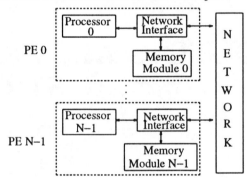

Fig. 1: A physically distributed memory.

A common network traffic model employed within MIN performance studies is one where the stream of memory requests from each PE to all of the memory modules is independent and identically distributed. However, this uniform traffic model does not capture the effect of traffic requests to a single shared variable, for instance, when the shared variable is used for interprocess synchronization, e.g., a semaphore. When a significant fraction of traffic is being directed at a single destination PE, causing a "hot spot" [17], the network performance is degraded substantially due to a number of the network buffers becoming saturated during a short period of time.

This research investigates how a multipath MIN (i.e., a MIN with two or more paths between PE source and destination pairs) can reduce the effects of network buffer saturation caused by hot spots. The multiprocessor systems assumed herein support limited multitasking, so, when a job encounters a hot spot, the system can perform a context-switch that preempts the current job for another one that is not processing a hot spot. The main idea behind the approach proposed here is to use the extra paths to balance the network traffic when a hot spot occurs. The target network considered in this research is the extra stage cube (ESC), which has exactly two disjoint paths between any source and destination [2, 3, 19]. The performance of the proposed approach (i.e., the mean delay time for both the uniformly distributed traffic and the hot spot traffic) was evaluated by simulation. The objective of this evaluation is to show that, under certain conditions, the performance of the network with the usual routing scheme is severely degraded by the presence of hot spots. With the proposed approach, although the delay time of hot spot traffic may be increased, the performance of the background traffic, which constitutes the majority of the network traffic, can be significantly improved.

Researchers have considered the use of extra stage networks for fault tolerance (e.g., [27]) and performance improvement under a uniform distribution assumption [5]. The goal of the study presented here is different: to develop techniques that can, under the operating environment assumptions made, use the extra stage to improve performance when a hot spot occurs.

The following section will introduce the topology and operating policy of the ESC network. Section 3 will describe the assumed operating environment and will discuss how system performance is degraded by hot spots. Then, the proposed approach to reduce the adverse effects of hot spots is described and evaluated in Section 4. Section 5 summarizes the results.

2. NETWORK MODEL

The ESC [2, 3, 19] is a multipath MIN and is formed from the multistage cube network [20] by adding an extra stage to the input side of the network, along with multiplexers and demultiplexers at the input (extra) stage and output stage, respectively. In addition, dual I/O links to and from the devices using the network are required.

In the ESC network, PE j sends data into the network through network input port link j and receives data from network output port link j. For each $n \times n$ switch, the m-th switch output link, $0 \le m < n$, has the same link number as the m-th switch input link. There are $\log_n N + 1$ stages, numbered from $\log_n N$ to 0, each containing N/n switches. For $0 \le i < \log_n N$, each switch at stage i of the network has n input links that differ only in the i-th base n digit of their link numbers. Each switch in stage $\log_n N$, also referred to as the extra stage or input stage, is connected like the output stage (stage 0) and has n input links which differ only in the 0-th base n digit of their link labels. As a result, when traversing the ESC, only stages $\log_n N$ and 0 can move a message from a switch input link to a switch output link that differs in the 0-th base n digit of the link label.

Fig. 2 shows an ESC network with N=8 inputs and outputs, 2×2 switches (i.e., n=2), and four stages. The input stage and output stage can each be enabled or disabled (bypassed). A stage is enabled when its switches are being used to provide interconnection; it is disabled when its switches are being bypassed. The ESC with 2×2 switches is designed to tolerate any single fault and is robust in the presence of multiple faults [4].

Normally, the network will be set so that the input stage is disabled and the output stage is enabled. The resulting structure matches that of the multistage cube network. Enabling both the input stage and output stage provides two disjoint paths between any source and destination. The ESC can be controlled in a distributed fashion by a simple extension of the routing tags used for the multistage cube network.

In this study, a packet-switched ESC network is assumed. Each switch is assumed to have a finite buffer associated with each output port. Each output buffer in a switch has a capacity of S fixed length packets. During each network cycle, one packet is transferred from a nonempty output buffer of a switch to an output buffer in the next stage. Each PE memory module can accept a packet from the network each network cycle.

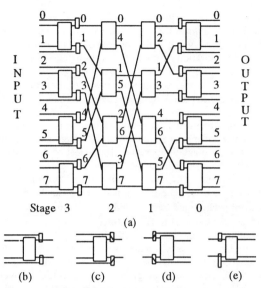

Fig. 2: (a) ESC network for N = 8; (b) input stage enabled; (c) input stage disabled; (d) output stage disabled; (e) output stage enabled.

Each output buffer can accept up to n requests from input ports simultaneously if there is enough space available for those requests. Because each output buffer is finite, there is a conflict when the number of incoming packets destined for the same output port exceeds the current packet space available for that buffer by $R \ge 1$ packets. To solve the conflict, R randomly chosen packets are blocked. A blocked packet will remain in the previous stage output buffer and will be resubmitted during the next cycle. To insure message preservation throughout the system, there is an infinite buffer associated with each PE. This buffer can hold blocked messages that cannot enter the network due to insufficient buffer space. Consequently, no PE-generated messages will be discarded.

Several variations on the basic switch buffer design are possible [9, 24]. Only output buffers with the first in first out (FIFO) policy are considered in this research. The degradation produced by hot spots is inherent to the blocking operation of the network, which is, in general, present for any buffering policy [14]. Moreover, the proposed approach is applicable to other organizations, such as those in [9, 24].

3. HOT SPOTS

It is usually assumed that memory requests are randomly directed at the network outputs (e.g., [7, 12, 28]), which is actually a reasonable assumption. Although the requests generated by PEs cooperating on a single problem are not independent, the presence of a large number of PEs and a number of different problems (e.g., a multi-tasking environment) will tend to randomize the data requests. Also, the hardware or software can hash (map) logical to physical memory locations, thereby guaranteeing that requests appear random and are distributed across all memory modules [8]. There is, however, one exception: multiple memory requests generated from different PEs can be directed at the same memory location during a short period of time. It is possible for such requests to be queued up at the associated memory module, which then becomes a bottleneck. Recall that n is the size of the switches used. When the output queue in a stage i switch is full, it can cause the queues in the n stage i+1 switches that feed it to fill. In turn, those n stage i+1 switches can cause the n^2 switches feeding them to fill. Eventually, a tree of saturated switches results, with the stage 0 switch at the congested memory module as the root. Tree saturation degrades the performance of all PEs in a system, including those not accessing the congested memory module. Such a memory module has been termed a hot spot [17] and has been studied by various researchers (e.g., [13, 16, 26]).

Hot spots can occur for many reasons. It is assumed in this study that a hot spot in a multiprocessor system is the result of a global synchronization operation. In the rest of this discussion, the global synchronization requests are denoted as synchronization traffic and the remaining uniform traffic as background traffic. The background traffic can be subdivided into hot background traffic and nonhot background traffic. These are background traffic destined and not destined for the hot spot memory module, respectively.

Naturally, all PEs will not request global synchronization at the same time (otherwise, synchronization would not be necessary). The distribution of global synchronization requests is modeled by a normal distribution, characterized by mean μ and standard deviation σ (as in [1]). Values for these parameters depend on the details of the multiprocessor implementation and the specific application under consideration. When σ is close to zero, the synchronization requests will be generated in a single cycle; as it increases, the hot spot traffic distribution will eventually resemble the background traffic distribution.

To perform a global synchronization, all PEs in the system will independently send a synchronization message to a specific PE called the coordinator. When $N - 1$ of these requests have been received by the coordinator, all PEs in the system are informed via a broadcast that the synchronization operation is complete. To increase both system performance and job throughput and to reduce PE idle time, it is assumed that each PE will continue other work while the synchronization message is in progress. One way to do this is to perform a job context switch. More precisely, the job issuing the global synchronization request will be put into the waiting queue (i.e., the job will be deactivated) and another job in the ready queue will become activated during the next cycle. This permits the PEs to start computation on another job while the previous job waits for its global synchronization request to be fully processed. Thus, by performing a job context switch, it is assumed that each PE will continue to generate background traffic at a constant rate irrespective of whether any global synchronization operation is in progress (e.g., [17, 26]). It is assumed that the architecture is designed for fast context switches (e.g., HEP [11]), so that context switch time is much smaller than the time to fully process a global synchronization request.

It is assumed also, for simplicity, that there is only *one* synchronization operation in existence at a time (as in [1]). Each must be completely finished before another one can begin. Because a multitasking environment is assumed, there is no reason to prohibit other active jobs from issuing synchronization requests. It is done here to simplify the simulations and analysis. However, because of the structure of the proposed approach to reduce degradation due to a single hot spot, the approach will also be beneficial in the case of multiple hot spots.

Many different approaches, such as the combining networks proposed for the IBM RP3 [16] and NYU Ultracomputer [8], or the repetition filter memory proposed for the Columbia CHoPP [23], have been suggested to eliminate the hot spot problem.

The basic idea of these schemes is to incorporate some hardware in the interconnection network to trap and combine data accesses if they are directed at the same memory location and they meet at a specific switch. The delay time experienced by a request traversing a buffered multistage network enhanced with combining under hot spot traffic has been studied in [10, 15]. Combining requests reduces communication traffic and thus decreases the average amount of queue space used, which leads to a lower average network delay time. However, the hardware required for such schemes is extremely expensive [17]. An inexpensive approach proposed in [26] uses a software combining tree to decrease memory contention and prevent tree saturation by initially distributing a single synchronization operation over many memory modules, and then combining collections of synchronization signals. In this approach, a hot spot request must traverse the interconnection network more than once, whereas in a hardware combining network the request will traverse the network only once.

The use of feedback schemes in a MIN with distributed routing control to reduce degradation to memory requests not destined for the hot spot was proposed in [18]. In this approach, each processor avoids sending requests destined for a hot module into the network if tree saturation is detected. If these problem-causing requests can be held outside the network, the severity of the tree saturation problem can be reduced. Another approach to alleviate the impact of tree saturation, proposed in [25], employs multiple queues at each switch input port (one separate queue for each switch output port). Every switch has a fixed amount of storage. Each queue in a switch is permanently allocated a slot and also shares available storage with other queues in a switch dynamically; no single queue may exhaust the entire storage pool. Thus, it can sustain the nonhot-spot traffic flow even in the presence of saturation trees and thereby improve the overall network performance.

The approach proposed herein differs from the above schemes by using a multipath MIN to reduce the interference between synchronization and background traffic. The goal of this paper is to demonstrate the potential of this multipath MIN approach. It is very difficult, and beyond the scope of this paper, to directly compare this scheme with all of the various previously proposed methods due to the differences in hardware costs and traffic models.

4. PROPOSED APPROACH

4.1 Definitions of Parameters

Because there is only one unique path associated with each source/destination pair in the multistage cube network [20], a message will suffer a long delay if it must go through the saturated area in the network caused by a hot spot. There are multiple paths available for a message if an ESC network is used. However, the existence of multiple paths through the network may not alleviate this problem if each message chooses its routing path independently. Even with dynamic rerouting, as congestion begins, the messages to the hot spot will themselves be rerouted and in the steady state will saturate all the alternative paths [17]. If all synchronization messages are forced to choose some predetermined path to the destination and if background messages can use alternate paths that do not involve any switches in the saturated area, as proposed here, the delay time of background messages can be improved significantly. The saturation tree problem cannot be completely eliminated by separating the background and synchronization traffic when they traverse through the network because the combined arrival rates of the synchronization and background traffic still exceed the hot spot processing rate. Thus, the hot spot will remain the bottleneck of the network. Most synchronization messages will be enqueued inside the network before being processed. However, the research herein shows that the interference between the synchronization and background traffic can be reduced under this condition. The goal of this research is to improve the delay time of the background traffic when a hot spot occurs.

The performance measures addressed are the average delay time of synchronization messages (μ_{SYN}), the average delay time of all background messages ($\mu_{BG(Tot)}$), and the average delay time of hot background messages ($\mu_{BG(HS)}$). Let γ be the background traffic generation rate and assume $0 \le \gamma < 1$, i.e., on average, each PE will generate a message during any cycle with probability γ. Background message generation at a PE is non-deterministic and independent of other PEs.

Let $DT_t(i)$ denote the delay time for message i of type t, where t refers to either synchronization (i.e., t=SYN) or background traffic (i.e., t=BG). Due to the FIFO policy assumed, a message will wait in a PE or switch output buffer until all previous messages have been serviced. Thus, the delay time of a message is the total time a message spends waiting in the output buffers of a PE and the ESC network switches (exclusive of traversal time). That is, $DT_t(i)$ = {the time message i of type t arrives at the destination} − {the time it is generated} − {switch traversing time, i.e., $\log_n N + 1$ cycles }. Because N−1 synchronization messages must traverse the network,

$$\mu_{SYN} = \sum_{i=1}^{N-1} DT_{SYN}(i)/(N - 1) .$$

To compute the average delay time of the background traffic in the presence of a hot spot, not all background messages generated within the simulation period are considered. Define the active session to be the period from the time the first synchronization message is generated to the time the last synchronization message arrives at the destination. Only the background messages generated within the active session are used to compute the average delay time of the background traffic in the presence of a hot spot. Recall that the distribution of synchronization messages is modeled by a normal distribution, characterized by mean μ and standard deviation σ. Given μ=3000, σ=10, and γ=0.5, the typical active session may last for approximately 325 cycles if the extra stage is bypassed in a 256×256 ESC with 4×4 switches and output buffer size equal to 12. Define β to be the set of all background messages generated within the active session. $|\beta|$ is the cardinality of β, i.e., the total number of background messages generated within the active session. Similarly, define β' to be the set of all background messages generated within the active session where the destination happens to be the same as the hot spot destination (i.e., the hot background messages). Then,

$$\mu_{BG(Tot)} = \sum_{i \in \beta} DT_{BG}(i)/|\beta| , \quad \mu_{BG(HS)} = \sum_{j \in \beta'} DT_{BG}(j)/|\beta'| .$$

The approach proposed here was evaluated by simulation. A unidirectional finite-buffered packet-switched ESC network connected from PE to PE (i.e., a PE-to-PE network configuration [19]) is considered, with both input and output stages enabled. Specifically, an ESC network with 256 I/O ports and 4×4 switches was simulated. Given the distribution of synchronization messages and the background message generation rate, the performance measures (i.e., μ_{SYN}, $\mu_{BG(Tot)}$, $\mu_{BG(HS)}$) were obtained from simulating the network environment for 125 active sessions. μ_{SYN}, $\mu_{BG(Tot)}$, and $\mu_{BG(HS)}$ were

calculated based on the delay times of the corresponding type of messages generated within all 125 active sessions.

Given a source and destination pair, there are n paths that are distinct from stages $\log_n N - 1$ to 1 in terms of the switches (and associated links) used [3, 19]. The input stage allows a message to select one of these paths from source to destination. Formally, it can be described as follows. Recall that for $1 \le i \le \log_n N - 1$, each switch at stage i of the network has n input links that differ only in the i-th base n digit of their link numbers. Each switch in the input and output stages has n input links that differ only in the 0-th base n digit of their link numbers. Consequently, only the input and output stages can move a message from a switch input link to a switch output link that differs in the 0-th base n digit of the link label. Thus, once the input stage output link has been selected for a message, this output link label and all link labels of every switch traversed by the message from stages $\log_2 N - 1$ to 1 have the same 0-th base n digit. Because there are n output links available for a message at the input stage, there are n paths (including both switches and links) that are distinct from stages $\log_2 N - 1$ to 1. Thus, to determine the routing path for a message, only the setting of the input stage switches is considered in the following discussion.

4.2 The Hot Section Approach

A synchronization message is pending if a PE has generated a synchronization message and is still waiting for the response. It is assumed that a local hot spot flag is associated with each PE, and is set to indicate there is a pending synchronization request in that PE. The global synchronization response can be broadcast from the coordinating PE only if all N−1 synchronization messages have been received. Once the synchronization message's response arrives, the local hot spot flag will be reset.

Let the network outputs be partitioned into K sections of equal size, where K is a power of two, and let the i-th section, for $0 \le i < K$, contain the N/K network outputs i(N/K)+j, for $0 \le j < N/K$. Define the section including the hot spot to be the hot section. Also, define the messages destined for the hot spot to be hot spot messages. Obviously, hot spot messages will include both synchronization and hot background traffic. Based on the idea described above, the hot section approach for a message entering the input stage of the network is shown in Fig. 3. This procedure is executed by each PE independently.

```
if ( the generated message is a synchronization message )
   { set the routing tag to use the upper output link of the switch
     at the input stage and set the local hot spot flag to true }
else /* this is a background message */
   if ( the local hot spot flag is true )
     if ( destination address ∈ hot section )
       if ( destination address == hot spot address )
         { set the routing tag to use the upper output
           link of the switch at the input stage }
       else
         { set the routing tag to use an output link that is not
           the upper output of the switch at the input stage }
     else /* destination address ∉ hot section */
       { set the routing tag to set the input stage switch to
         straight }
   else /* there is no pending hot message */
     { set the routing tag to set the input stage switch to straight }
```

Fig. 3: The hot section approach for setting the input stage.

For PE A, consider a background message destined for the hot section but not the hot spot address. If this message arrives at the upper input link of an input stage switch, it will be randomly sent to one of the output links other than the uppermost link. If this message enters any other switch input link, it will be sent straight through the switch.

Recall from Section 2 that only stages $\log_n N$ and 0 can move a message from a switch input link to a switch output link that differs in the 0-th base n digit of the link label. Therefore, all input/output links of a switch from stage $\log_n N - 1$ to 1 must have the same 0-th base n digit. By forcing hot spot messages to take the upper output link of the input stage switch, only switches whose input/output links' 0-th base n digit = 0 from stages $\log_n N - 1$ to 1 will be utilized by hot spot messages. Because stages $\log_n N - 1$ to 0 form a multistage cube network topology, all network outputs are reachable from any stage $\log_n N - 1$ switch.

If K = 1, there is only one hot section and all network outputs are resident in the hot section. By preventing the nonhot background messages from taking the upper output link of the input stage switch, no switches whose input/output links' 0-th base n digit = 0 from stages $\log_n N - 1$ to 1 will be used by these background messages. Thus, 1/n of the ESC network is reserved exclusively for hot spot messages.

Although 1/n of the switches from stage $\log_n N - 1$ to stage 1 in an ESC network are reserved for hot spot messages when K=1, not all such reserved switches are actually utilized. Due to the saturation tree structure and the topology of the ESC network, the number of switches used by hot spot messages at stage i is only 1/n of the number of switches used at stage i+1 where $1 \le i \le \log_n N - 1$. Fig. 4 provides an example of this phenomenon. Assuming the hot spot is located at the network output port 0 in Fig. 4, the number of switches used by hot spot messages at stage 1 is one, which is only one half of the switches used at stage 2, which, in turn, is only one half of those used at stage 3. This occurs even though the number of switches reserved for these messages is the same at all stages.

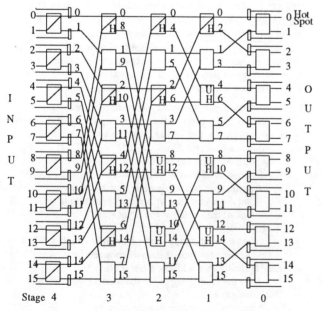

Fig. 4: The switches labeled ''H'' are reserved for hot spot messages. Those switches that are also labeled ''U'' are reserved but not used.

Increasing the K value will force some nonhot background messages to pass through the saturation tree. Thus, it will reduce the number of reserved but unused switches. For example, in Fig. 4, if K is chosen to be 2, all network outputs are partitioned into two sections and the hot section consists of the output ports 0 to 7. By using the proposed approach, a background message generated by an even-numbered PE and not destined for the hot section will go through a reserved switch (a switch labeled ''H'') at stage 3, then enter a reserved but unused switch (i.e., a switch labeled ''U'' and ''H'') at stages 2 and 1. The switches with ''H'' only in stages 2 and 1 will not be used by such a message (recall the destination for this message is one of ports 8 to 15). Naturally, for these switches reserved and used by hot spot messages, the closer they are to the input stage, the less congested they will be. Thus, when K=2, the delay time of the nonhot background messages that traverse through the reserved switches at stage 3 will not increase significantly. However, this can improve the utilization of the unused switches reserved for hot spot messages and balance the network traffic more evenly.

If K equals N, the hot section includes only the hot spot, and all nonhot background messages simply go straight across the input stage (i.e., no switches are reserved exclusively for hot spot messages). In addition to some nonhot background messages, the upper output link of a switch at the input stage is also used to deliver hot spot messages, causing the upper output link to be overloaded.

Depending on the background traffic generation rate (γ), the distribution of synchronization messages (σ), the output buffer size (S), and the switch size used (n), an optimal K value between 1 and N should exist. When the network environment is known, the optimal K value can be determined via simulation.

4.3. Simulation Results

The effect of different values of K is investigated first. The average delay times for background and synchronization traffic is examined by simulating a 256×256 network with 4×4 switches and an output buffer size of 12 packets. Results for two background message generation rates ($\gamma=0.7$ and $\gamma=0.8$) are shown in Fig. 5. The distribution of synchronization messages was modeled by a normal distribution with $\mu=3000$ and $\sigma=10$. Fig. 5 shows the delay time of synchronization messages (μ_{SYN}), background messages ($\mu_{BG(Tot)}$), and hot background messages ($\mu_{BG(HS)}$). Recall that that $\mu_{BG(Tot)}$ is the average delay for the background traffic, including both hot and nonhot background messages. It will be shown later that $\mu_{BG(Tot)}$ represents the delay time of the majority of network traffic.

It can be seen that the minimum $\mu_{BG(Tot)}$ is observed when K=4, while μ_{SYN} does not show significant change for different K values in the range shown. This phenomenon occurs because of the different distribution assumptions for synchronization and background messages. Consistent with the definition of the active session, the simulations show that most synchronization messages arrive at the network input ports earlier than background messages generated within the active session. An example of this is provided in Fig. 6. Thus, the conflicts among the synchronization messages are the main reason that synchronization messages experience a long delay. Different K values have only limited effect on the delay of synchronization messages.

Increasing the K value will reduce the number of reserved but unused switches and provides more switches for nonhot background messages to use. Thus, the average delay of back-

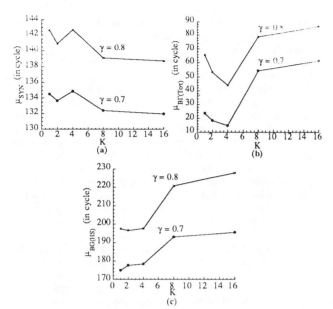

Fig. 5. For $N = 256$, $n = 4$, $S = 12$, and $\sigma = 10$, the average delay time for different K values, when $\gamma = 0.7$ and 0.8, of a (a) synchronization message; (b) background message; and (c) hot background message. The values on the vertical axes vary among the graphs.

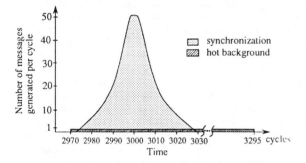

Fig. 6: Number of synchronization and hot background messages generated per cycle during active session.

ground traffic can be improved. However, increasing the K value also forces more nonhot background messages to pass through the saturation tree caused by synchronization messages. Consequently, more nonhot background messages will be blocked by synchronization messages and the delay time for background traffic will increase. As shown in Fig. 5(b), when $K < 4$, the benefit obtained from reducing the number of reserved but unused switches is larger than the adverse effects caused by forcing more nonhot background traffic to traverse the saturation tree. Thus, when $K < 4$, $\mu_{BG(Tot)}$ decreases as K increases. When $K \geq 4$, the benefit obtained from further reducing the unused switches is completely offset by the adverse effects. Consequently, $\mu_{BG(Tot)}$ increases as K increases.

Because more nonhot background messages will use the switches in the saturation tree as K becomes larger, the conflicts between the hot background messages and other background messages will increase. Thus, as shown in Fig. 5(c), $\mu_{BG(HS)}$ becomes larger when increasing the K value. Because the distribution of background traffic is assumed to be uniform, K has more impact on the hot background messages than synchronization messages.

As stated previously, the optimal K value depends on the task being executed and system environment. It can be tuned based on knowledge of the system and expected characteristics of the tasks.

To further evaluate the performance of the hot section approach with K=4, a 256×256 ESC network with the same configuration as described above was simulated. Different distributions of synchronization messages (σ) and background traffic generation rates (γ) were simulated using two different routing schemes: bypassing the input stage and the hot section approach with K=4. The former approach incurs no time delay for the input stage in the simulations and is equivalent to using a multistage cube network (i.e., constructed without the extra stage). It also corresponds to how the ESC could be used normally when there are no faults. The measures being analyzed were μ_{SYN}, $\mu_{BG(Tot)}$, and $\mu_{BG(HS)}$. These simulation results are shown in Fig. 7.

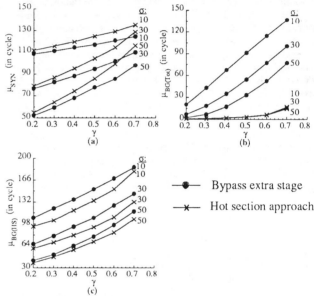

Fig. 7: For $N = 256$, $n = 4$, $S = 12$, and $K = 4$, the average delay time as a function of γ for different values of σ for a (a) synchronization message; (b) background message; and (c) hot background message. The values on the vertical axes vary among the graphs.

As can be seen from Fig. 7(b), the average delay time of background messages ($\mu_{BG(Tot)}$) is reduced significantly when compared to the bypass scheme. This is because the overlap between the routing paths chosen by synchronization and nonhot background messages is reduced substantially.

Consider the cases when $\sigma=10$ and $\sigma=30$. Notice that as γ increases, the hot background messages (Fig. 7(c)) experienced more delay than synchronization messages (Fig. 7(a)) with either the bypass or hot section approach. For example, for the bypass scheme with $\sigma=10$ and $\gamma=0.5$, the delay for the hot background messages ($\mu_{BG(HS)}$) was 151, which is 30% more than the delay for synchronization messages ($\mu_{SYN}=116$). This phenomenon occurs because of the different distribution assumptions for synchronization and background messages. Most of the hot background traffic is introduced into the network after all of the synchronization messages have already caused the congestion, as shown in Fig. 6. Because both classes of messages use the same routing paths, both will experience significant delays relative to the nonhot background messages.

From Fig. 7(a), it is interesting to note that with the hot section approach, μ_{SYN} is larger than the corresponding average delay in the bypass scheme. Also, with the hot section approach, $\mu_{BG(HS)}$ is reduced when compared to the bypass scheme as shown in Fig. 7(c). This phenomenon is a result of the following.

Define the height of the saturation tree ($\underline{H_t}$) to be the length (number of arcs) of the longest path from the root to a leaf node. The saturation tree is completely balanced if it has n^i nodes at depth i, where $0 \leq i \leq H_t$. Recall that within the active session, most of the hot background messages are generated after all of the synchronization messages. Ideally, most of the hot background messages will not be received at the destined node until all synchronization messages have been accepted. However, if the saturation tree is not completely balanced, a hot background message may arrive at a nonleaf node X while there are still some synchronization messages enqueued in other subtrees of node X. Both classes of messages will use the same path from node X to the hot spot. Thus, those synchronization messages will be delayed by one network cycle by the hot background message. For example, consider a hot background message M_1 that arrives at leaf node B at depth i, where $i < H_t$ as shown in Fig. 8. M_1 needs to wait at an output buffer of node B until all messages in front of it have been served. When M_1 is accepted by the parent node A, the synchronization messages enqueued in other subtrees (e.g., in nodes C and F) will be delayed by one cycle. For the bypass case, if M_1 is a nonhot background message, it will have less impact on the delay of synchronization messages compared to the case when it is a hot background message. This is because nonhot background messages will leave the saturation tree eventually (i.e., they are not going to the hot spot).

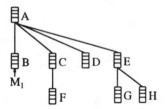

Fig. 8: Saturation tree in ESC network with 4×4 switches.

To delay any synchronization messages, a hot background message must enter a nonleaf node earlier enough such that there are still some synchronization messages enqueued in other subtrees of that node. Although the nonhot background messages have less impact on the delay of synchronization messages, they can block the hot background messages in the bypass case.

With the hot section approach, in addition to hot spot messages, only the nonhot background messages that have the minimum overlap with the saturation tree are allowed to use the reserved subnetwork. For example, for K=4, N=256, and n=4, no nonhot background messages will use the saturation tree switches in stages 0, 1, and 2. However, with the bypass scheme, the network is shared by all messages. The number of nonhot background messages that must traverse the saturation tree will increase (compared to the hot section case), which causes the hot background traffic to be blocked. Consequently, the possibility that a hot background message can interfere with the synchronization messages is reduced. Thus, the hot section approach provides more opportunity for the hot background

messages to delay the synchronization messages. It is expected that μ_{SYN} will be larger in the hot section approach than in the bypass scheme. For hot background messages, if the chance to delay the synchronization messages is increased, more messages can traverse the network without suffering a long delay. Consequently, $\mu_{BG(HS)}$ is reduced.

Whether it is reasonable to sacrifice synchronization message delay time to improve the delay time of background messages is dependent on which one is more important to users. The number of synchronization messages is N−1 and it takes at least N−1 cycles to accept all synchronization messages at the destination. (The actual time to accept all the synchronization messages is dependent on the interference between the background and synchronization traffic and the way synchronization messages are generated, i.e., σ.) Thus, by definition, the active session is at least N−1 cycles long. The total number of background messages generated within the active session equals $N \times \gamma \times$ |time of active session| $\geq N \times \gamma \times (N-1) \simeq N^2 \times \gamma$, if N is large. For $N \times \gamma \gg 1$, it is obvious that the number of synchronization messages is relatively small compared to the number of background messages. Thus, this approach increases the average delay of the N−1 synchronization messages, but it reduces the delay time of at least $N^2 \times \gamma$ background messages by approximately the same amount. Because the value of this trade-off depends on the tasks being executed, it can be applied whenever the user deems it appropriate.

As σ increases, the synchronization messages are generated within a wider range of network cycles and eventually they will resemble the background traffic distribution. Thus, as σ becomes larger, the saturation tree problem caused by the synchronization messages becomes less serious, and consequently, μ_{SYN} and $\mu_{BG(HS)}$ decrease substantially when employing either the bypass or hot section approach (see Fig. 7(a) and (c)). With the bypass scheme, when the saturation tree problem becomes less serious, $\mu_{BG(Tot)}$ becomes smaller because more background messages can pass through the network without being blocked by the saturation tree. However, with the hot section approach, there is only a limited overlap between the routing paths chosen by the hot spot messages and nonhot background messages. Thus, the average delay time of a background message is not affected significantly when increasing the value of σ.

The impact of buffer on the message average delay time when hot spots occur was considered and evaluated in [29]. It is quite interesting to note that given the same traffic generation rate (γ), the hot section approach using a buffer size of eight can outperform the bypass scheme using a buffer size of 20. This implies that the hardware cost of an extra stage in an ESC network may be offset by reducing the size of the output buffers. However, one important benefit obtained from adding an extra stage to the multistage cube network is to provide a fault tolerant capability when switches or links become faulty. Thus, in an environment when hot spots occur, the additional hardware cost of an extra stage can be justified when the benefits gained are considered.

Also in [29] a modification of the hot section approach is examined. In this approach, K=1 and hot background messages are routed in the same way as nonhot background messages with the hot section approach. Simulation results showed that the modification degraded the background traffic delay as compared to the hot section approach with K=1.

5. CONCLUSIONS

This study examined a technique for using the ESC network to improve network performance when hot spot situations occur. Simulation was used to evaluate the proposed technique under certain assumptions about the execution environment and the network structure. The results of these simulations were reported and discussed.

In a multipath ESC network, if all synchronization messages are forced to choose some predetermined path to the destination and if the nonhot background messages can use the paths that do not involve any switches in the saturated area or have only limited overlap, the network performance can be improved significantly. Although the proposed approach may increase the delay time for the synchronization messages, it substantially reduces the delay time of the background messages that constitute the majority of the network traffic. Simulation results have shown that, for the operating environment assumptions made, the proposed scheme can be used to: (1) control tree saturation among network switch output buffers, (2) reduce the delay of memory requests that are not to the hot memory module, and (3) increase overall system bandwidth.

All analyses presented were based on the framework of a 256×256 ESC network with 4×4 switches. Based on the reasoning used in the analysis of the simulation results, the proposed approach should behave similarly for larger size ESC networks. It is expected that the same idea can be applied to other multipath networks with minor modifications. Furthermore, although a single shared address space was assumed here, the technique is also applicable to message passing systems. Future research on this topic includes evaluating the proposed approach under different operating environment assumptions.

ACKNOWLEDGMENTS: The authors acknowledge useful discussions with Prof. J. K. Antonio, Prof. J. A. B. Fortes, Dr. W. G. Nation, Mr. G. Saghi, and Prof. V. Rego.

REFERENCES

[1] S. Abraham and K. Padmanabhan, "Performance of the direct binary n-cube network for multiprocessors," *IEEE Trans. Comput.,* v. 38, Jul. 1989, pp. 1000-1011.

[2] G. B. Adams III, D. P. Agrawal, and H. J. Siegel, "A survey and comparison of fault-tolerant multistage interconnection networks," *Computer,* v. 20, Jun. 1987, pp. 14-27.

[3] G. B. Adams III and H. J. Siegel, "The extra stage cube: a fault-tolerant interconnection network for supersystems," *IEEE Trans. Comput.,* v. C-31, May 1982, pp. 443-454.

[4] G. B. Adams III and H. J. Siegel, "Modifications to improve the fault tolerance of the extra stage cube interconnection network," *1984 Int'l Conf. Parallel Processing,* Aug. 1984, pp. 169-173.

[5] C. Y. Chin and K. Hwang, "Packet switching networks for multiprocessors and data flow computers," *IEEE Trans. Comput.,* v. C-33, Nov. 1984, pp. 991-1003.

[6] W. Crowther, J. Goodhue, R. Thomas, W. Milliken, and T. Blackadar, "Performance measurements on a 128-node butterfly parallel processor," *1985 Int'l Conf. Parallel Processing,* Aug. 1985, pp. 531-540.

[7] D. M. Dias and J. R. Jump, "Analysis and simulation of buffered delta networks," *IEEE Trans. Comput.,* v. C-30, April 1981, pp. 273-282.

[8] A. Gottlieb, R. Grishman, C. P. Kruskal, K. P. McAuliffe, L. Rudolph, and M. Snir, "The NYU Ultracomputer — designing an MIMD shared-memory parallel computer," *IEEE Trans. Comput.,* v. C-32, Feb. 1983, pp. 175-189.

[9] M. J. Karol, M. G. Hluchyj, and S. P. Morgan, "Input vs. output queueing on a space-division packet switch," *IEEE Global Telecomm. Conf.,* Dec. 1986, pp. 659-665.

[10] B.-C. Kang, G. Lee, and R. Kain, "Performance of multistage combining networks," *1991 Int'l Conf. Parallel Processing,* Aug. 1991, pp. 550-557.

[11] J. S. Kowalik, *Parallel MIMD Computation: The HEP Supercomputer and Its Applications,* The MIT Press, Cambridge, MA, 1985.

[12] C. P. Kruskal and M. Snir, "The performance of multistage interconnection networks for multiprocessors," *IEEE Trans. Comput.,* v. C-32, Dec. 1983, pp. 1091-1098.

[13] M. Kumar and G. F. Pfister, "The onset of hot spot contention," *1986 Int'l Conf. Parallel Processing,* Aug. 1986, pp. 28-34.

[14] T. Lang and L. Kurisaki, "Nonuniform traffic spots (NUTS) in multistage interconnection networks," *J. Parallel Distrib. Comput.,* v. 10, Sep. 1990, pp. 55-67.

[15] G. Lee, "A performance bound of multistage combining networks," *IEEE Trans. Comput.,* v. 38, Oct. 1989, pp. 1387-1395.

[16] G. F. Pfister, W. C. Brantley, D. A. George, S. L. Harvey, W. J. Kleinfelder, K. P. McAuliffe, E. A. Melton, V. A. Norton, and J. Weiss, "The IBM Research Parallel Processor Prototype (RP3): introduction and architecture," *1985 Int'l Conf. Parallel Processing,* Aug. 1985, pp. 764-771.

[17] G. F. Pfister and V. A. Norton, "'Hot spot' contention and combining in multistage interconnection networks," *IEEE Trans. Comput.,* v. C-34, Oct. 1985, pp. 933-938.

[18] S. L. Scott and G. S. Sohi, "Using feedback to control tree saturation in multistage interconnection networks" *16th Ann. Symp. Comput. Arch.,* May 1989, pp. 167-176.

[19] H. J. Siegel, *Interconnection Networks for Large-Scale Parallel Processing: Theory and Case Studies, Second Edition,* McGraw-Hill, New York, NY, 1990.

[20] H. J. Siegel, W. G. Nation, C. P. Kruskal, and L. M. Napolitano, Jr., "Using the multistage cube network topology in parallel supercomputers," *Proc. of the IEEE,* v. 77, Dec. 1989, pp. 1932-1953.

[21] H. J. Siegel, W. G. Nation, and M. D. Allemang, "The organization of the PASM reconfigurable parallel processing system," *1990 Parallel Computing Workshop,* sponsored by the Computer and Information Science Department at the Ohio State University, 1990, pp. 1-12.

[22] H. J. Siegel, T. Schwederski, J. T. Kuehn, and N. J. Davis IV, "An overview of the PASM parallel processing system," in *Computer Architecture,* D. D. Gajski, V. M. Milutinovic, H. J. Siegel, and B. P. Furht, eds., IEEE Computer Society Press, Washington, DC, 1987, pp. 387-407.

[23] H. Sullivan, T. R. Bashkow, and K. Klappholz, "A large-scale homogeneous, fully distributed parallel machine," *Fourth Ann. Symp. Comput. Arch.,* Mar. 1977, pp. 105-124.

[24] Y. Tamir and G. L. Frazier, "High-performance multi-queue buffers for VLSI communication switches," *15th Ann. Symp. Comput. Arch.,* Jun. 1988, pp. 343-354.

[25] N.-F. Tzeng, "Alleviating the impact of tree saturation on multistage interconnection network performance," *J. Parallel Distrib. Comput.,* v. 12, Jun. 1991, pp. 107-117.

[26] P. C. Yew, N. F. Tzeng, and D. H. Lawrie, "Distributing hot-spot addressing in large-scale multiprocessors," *IEEE Trans. Comput.,* v. C-36, Mar. 1987, pp. 388-395.

[27] K. Yoon and W. Hegazy, "The extra stage gamma network," *13th Ann. Symp. Comput. Arch.,* Jun. 1986, pp. 175-182.

[28] H. Yoon, K. Y. Lee, and M. T. Liu, "Performance analysis of multibuffered packet-switching networks in multiprocessor systems," *IEEE Trans. Comput.,* v. C-39, Mar. 1990, pp. 319-327.

[29] M. C. Wang, H. J. Siegel, M. A. Nichols, and S. Abraham, *Using the Extra Stage Cube Multipath Network to Reduce the Impact of Hot Spots,* Tech. Rep. TR-EE 92-25, EE School, Purdue, Jul. 1992.

HARDWARE SUPPORT FOR FAST RECONFIGURABILITY
IN PROCESSOR ARRAYS *

M. Maresca*, H. Li* and P. Baglietto*

Dist - University of Genoa
via Opera Pia 11a - 16145 Genova Italy
Tel. +39-10-353.2739 Fax +39-10-353.2948
E_mail mm@dist.dist.unige.it

HaL Computer Systems, USA

Abstract - *Massively parallel computers are implemented by means of modules at different packaging levels. This paper discusses a hierarchical node clustering scheme (HNC) for packaging a class of reconfigurable processor arrays called Polymorphic Processor Array (PPA) which uses circuit-switching-based routers at each node to deliver a different topology at every instruction. The PPA family suffers from an unknown signal delay between arbitrary two nodes connected by the circuited-switched paths. This either forces the hardware clock to compromise to the worst signal delay or makes the software dependent on the system size. The use of the HNC scheme allows to obtain communication speed-up and automatic control, at the compiler level, over signal propagation delay.*

1 INTRODUCTION

A Polymorphic Processor Array (PPA) [1] is a two-dimensional mesh connected computer, in which each node is equipped with a switch able to interconnect its four NEWS ports. PPA changes the switch setting, as part of the instruction, to speed up the data exchange between nodes; it shortens the distance between the nodes that have to communicate by short-circuiting all the intermediate nodes between the source and destination nodes.

PPA belongs to the class of reconfigurable processor arrays, such as the Mesh with Reconfigurable Bus (MRB) [2], the Processor Array with Reconfigurable Bus Systems (PARBS) [3], the Gated Connection Network (GCN) [4] and the Polymorphic-Torus [5]. The characteristic that distinguishes PPA from MRB, PARBS and GCN is that of providing only a limited set of reconfiguration capabilities: PPA allows the short-circuiting to occur only along rows and columns, whereas

it does not support diagonal or arbitrary shape paths. At the first glance such a restriction seems to limit the power of PPA, however, as to be explained in the paper, it leads to a more programmable architecture, thanks to the possibility of implementing effective hardware solutions, namely Hierarchical Node Clustering, supporting the data transmission along the short-circuited paths.

In PPA, like in MRB, PARBS and GCN, it is assumed that communication between any pair of nodes take an amount of time independent of the distance between the two nodes, which is equivalent to hypothesize that both the links and the switches are ideal; this is usually referred to as a "computation model". In reality, it is evident that in a physical system the propagation delay between two nodes is a function of the distance between them and besides, considering that reconfigurable processor arrays are proposed for modeling massively parallel computers, the worst case distance between two nodes may be very high. The gap between the idealized model and the implementation makes it impossible to write programs featuring the same complexity as the corresponding algorithms, unless proper design techniques are used to implement the basic operations of the idealized model.

In this paper we propose a design technique called Hierarchical Node Clustering (HNC) that reduces the maximum propagation delay in PPA from linear to logarithmic. The HNC scheme is based on the observation that the implementation of a massively parallel computer consists of a hierarchy of packages, each of which is characterized by a different propagation delay. Such a hierarchy can be exploited in the implementation of massively parallel computers based on hierarchical topologies such as tree, pyramid and hypercube. For example, considering that the $\log n$ steps of a scan operation [6] in an n-processor tree or

---

* The work described in this paper has been supported in part by research grant from MURST (Ministero della ricerca Scentifica e Tecnologica) and from CNR (Consiglio Nazionale delle ricerche - National Program on Information Systems and Parallel Computing and Strategic Project on Neural Networks).

hypercube exhibit a different degree of locality, it is possible to take advantage of such a locality and allow each step to be executed within the time actually needed instead of compromising to the longest delay required by the final reduction. On the contrary non hierarchical topologies such as ring and mesh always incur worst signal delay because any communication involves all the wires at all packaging levels. This paper shows how a non-hierarchical topology, such as PPA, can be benefited from the HNC scheme.

The paper is organized as follows. In the next section we introduce the PPA computation model, and in section 3 we describe how such a model can be implemented by taking advantage of the HNC technique. In section 4 we analyze the propagation delay in PPA both without HNC and with HNC, and in section 5 we discuss the cost and the benefit of HNC. In section 6 we provide some concluding remarks.

2 POLYMORPHIC PROCESSOR ARRAY COMPUTATION MODEL

A Polymorphic Processor Array (PPA), whose logical architecture is shown in Fig. 1, consists of a stack of three planes: the processor plane P, the memory plane M and the switch plane S. The P-plane is a two-dimensional array of processing elements (PE_{ij}) able to perform arithmetic and logical computations, the M-plane is a two-dimensional array of random access memories (M_{ij}), and the S-plane is an array of switch-boxes (S_{ij}) interconnected by a two-dimensional torus, where $i, j = 0, 1, ..., n - 1$. Each set (PE_{ij}, M_{ij}, S_{ij}) is referred to as a PPA node and is called N_{ij}. PPA is a SIMD architecture. Similar to other SIMD architectures, a central program controller broadcasts instructions and data to the P-plane and the memory addresses to the M-plane. However,

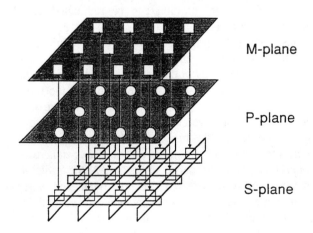

Fig. 1 - Polymorphic Processor Array Architecture.

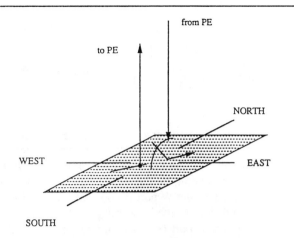

Fig. 2 - PPA switch-box

unlike other SIMD architectures, the PPA program controller also broadcasts a switch instruction to the S-plane for the control of data communication.

A switch-box S_{ij}, shown in Fig. 2, consists of an ideal switch (no delay) which can be oriented toward any of the four mesh directions (we shall refer to the orientation of the switch as the *switch-box orientation*) and which can be opened or shorted under program control (we shall refer to the OPEN/SHORT configuration of the switch as the *switch-box configuration*). The PPA computation model can be completely specified by the orientation and configuration of the switch-box.

The switch-box orientation, which can be N, E, W or S, is programmed by the central program controller through the switch instruction, therefore is identical for all switch-boxes. On the contrary, the switch-box configuration, which can be OPEN or SHORT, depends on local data and can be different in each switch-box. (For example the statement "*where (local_data == 3) configuration = OPEN elsewhere configuration = SHORT* " sets the switch-box configuration to OPEN in the nodes in which the element of *local_data* is 3 and sets the switch-box configuration to SHORT in the other nodes).

The operation of each PPA node depends on both the orientation and the configuration of the corresponding switch-box. At each instruction the PEs corresponding to switch-boxes in the OPEN configuration output a datum toward the current orientation, while the PEs corresponding to switch-boxes in the SHORT configuration let the data pass through them along the current orientation. Independent of the configuration, all the PEs load the datum incoming from the direction opposite to the current orientation (E and W are opposite directions and N and S are opposite directions).

Referring to the example in Fig. 3, in which the switch-box orientation is E, each node, independent of

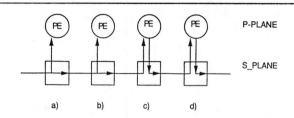

a)　　　b)　　　c)　　　d)

Fig. 3 - PPA data movement toward the E direction.

the switch-box configuration, receives a message from the W port. Depending on the switch-box configuration, in each node either the W port is connected to the E port (nodes *a* and *b*), or the processing element itself is connected to the E port (nodes *c* and *d*); nodes *c* and *d* also send a datum toward the E direction.

As described, the PPA distinguishes itself from the other computation models currently proposed for reconfigurable processor arrays in the following aspects:

- *Directionality* : the PPA switch-box implements directional connections from one specific port to another, as opposed to MRB [2] and PARBS [3], that support bidirectional communication.

- *One-dimensional interconnection* : the PPA switch-box allows the propagation of data either along the vertical direction (N→S or S→N) or

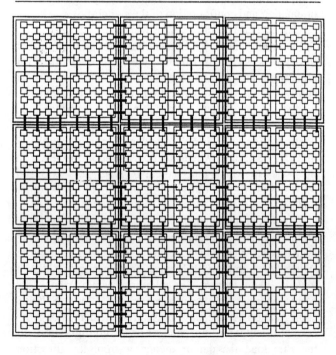

$p-1 = 2, n_0 = 4, n_1 = 2, n_2 = 3, n = 24$

Fig. 4 - Mapping PPA onto a multiple package implementation.

along the horizontal direction (E→W or W→E), but not simultaneously. This is in contrast with GCN [4], MRB [2] and PARBS [3], where the interconnection between non-opposite ports (N→E, N→W, S→E, S→W, E→N, E→S, W→N and W→S) is also supported.

These two important properties are the necessary conditions for the implementation of HNC, to be described next.

3 HIERACHICAL NODE CLUSTERING IN PPA IMPLEMENTATION

A PPA system of n^2 nodes can be implemented by p different packaging levels (for example the chip level, the multi-chip level and the board level), each of which is characterized by a propagation delay. We recursively define a module at packaging level $i+1$ as a two-dimensional array of $n_i \times n_i$ modules at packaging level i. In general n_i^2 modules at packaging level i are contained in each module at packaging level $i+1$ ($i = 0, 1, ... , p-2$)

and $n^2 = \prod_{i=0}^{p-1} n_i^2$ is the total number of nodes of the PPA

system. A module at packaging level 0 corresponds to a PPA node. Fig. 4 shows an example in which a module at packaging level 1 (e.g. chip) contains 4×4 ($n_0 = 4$) nodes, a level 2 module (e.g. multi-chip carrier) contains 2×2 chips, and a level 3 module (e.g. board) contains 3×3 carriers.

The HNC scheme consists of adding a switch-box to each PPA module at each packaging level; under proper control, such a switch-box is able to bypass the module where the switch-box resides. As a result, the signal that, without HNC, would have to ripple through the nodes with SHORT configuration can be redirected via the module switch-box directly as shown in Fig. 5. Consequently, the propagation delay in a module at level i is reduced from the sum of the delays of the modules at level i - 1 to the delay of the module switch-box at level i.

In the following we show the design of the module switch-box; we first give the switch-box design of the lowest packaging level, then generalize the design for the switch-boxes at the higher levels. Due to the PPA characteristics of *directionality* and *one-dimensional interconnection* introduced in section 2, the switch-box design can be considered for horizontal and vertical communication separately. From now on we will focus on a single PPA row, without loss of generality. Following the rule, the n nodes of a PPA row are hierarchically packaged in such a way that n_0 nodes are contained in each module at level 1, $n_0 n_1$ nodes are

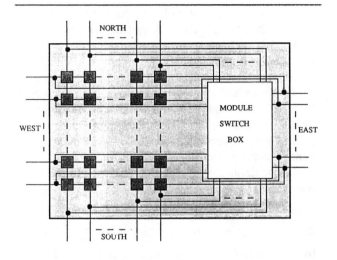

■ Submodule at level i-1

All paths are $\prod_{k=0}^{i-1} n_k - 1$ bits wide

Fig. 5 - Structure of a module at generic packaging level.

contained in each module at level 2, and in general $\prod_{j=0}^{i-1} n_j$ nodes are contained in each module at level i, $i = 1, 2, ...$

, p -1. The total number of nodes in the row is $n = \prod_{i=0}^{p-1} n_i$.

Fig. 6 illustrates the switch-box design for the lowest level module. The figure is self-explanatory on the decoding of the orientation and the configuration. It also contains the bypassing circuit consisting of eight tri-state gates. The characteristic of *directionality* of the PPA switch-boxes allows for the use of electrically active elements, i.e. tri-state gates, instead of passive elements, such as pass-transistors, transmission-gates, and pre-charge/pull-down circuits like in [4] to implement the PPA switch-boxes. The benefits of tri-state implementation include the known fixed delay of each gate and the automatic regeneration of the signal, both of which support scalability.

A switch-box at generic packaging level i, $i = 1, 2, ...$, p -1, shown in Fig. 7, has

- $4\prod_{k=0}^{i-1} n_{k-1}$ nearest-neighbor ports Nj , Ej , Wj , Sj

$j = 0, 1, ... , \prod_{k=0}^{i-1} n_{k-1} - 1$;

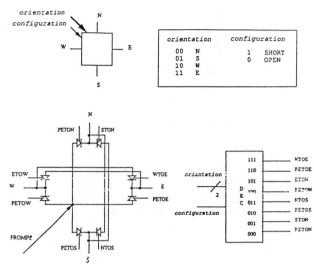

Fig. 6 - Node switch-box.

- $\prod_{k=0}^{i-2} n_k n_{i-1}^2$ input control signals C_{lm}^{i-1}, l, m = 0, 1,

... , $\prod_{k=0}^{i-2} n_k n_{i-1}^2 - 1$ corresponding to the configuration of the modules at packaging level i -1 included in it;

- $\prod_{k=0}^{i-1} n_k$ output status signals C_j^i , j = 0, 1, ... ,

$\prod_{k=0}^{i-1} n_k - 1$, corresponding to its configuration.

Considering that a switch-box associated with a module at packaging level i bypasses a row or a column of modules at packaging level i -1 contained in it only when all the switch-boxes of such modules are in the SHORT configuration, it is possible to derive the following rules that give the C^i's as functions of the C^{i-1}'s [1].

- case 1: orientation is E or W

$$C_h^1 = \underset{k=0}{\overset{n_0-1}{AND}} C_{hk} \qquad\qquad h = 0, 1, ..., n_0\text{-}1$$

---

[1] The convention followed here is that "1" denotes a SHORT configuration and "0" denotes an OPEN configuration.

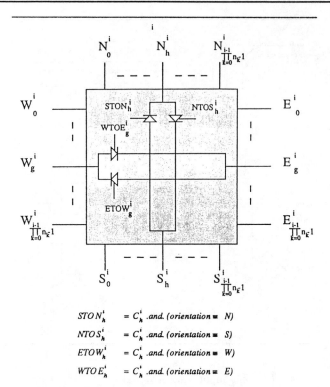

$$STON_h^i = C_h^i \,.and. \,(orientation \equiv N)$$

$$NTOS_h^i = C_h^i \,.and. \,(orientation \equiv S)$$

$$ETOW_h^i = C_h^i \,.and. \,(orientation \equiv W)$$

$$WTOE_h^i = C_h^i \,.and. \,(orientation \equiv E)$$

Fig. 7 - Switch-box at generic packaging level i.

$$C_h^i = \mathop{AND}_{k=0}^{n_{i\text{-}1}\text{-}1} \left[C_{h \bmod \prod_{m=0}^{i-2} n_m}^{i-1} \right]_{\left\lfloor \frac{h}{\prod_{m=0}^{i-2} n_m} \right\rfloor, k}$$

$$h = 0, 1, ... , \prod_{m=0}^{i-1} n_m - 1$$

- case 2: orientation is N or S

$$C_h^1 = \mathop{AND}_{k=0}^{n_0\text{-}1} C_{kh} \qquad h = 0, 1, ... , n_0\text{-}1$$

$$C_h^i = \mathop{AND}_{k=0}^{n_{i\text{-}1}\text{-}1} \left[C_{h \bmod \prod_{m=0}^{i-2} n_m}^{i-1} \right]_{k, \left\lfloor \frac{h}{\prod_{m=0}^{i-2} n_m} \right\rfloor}$$

$$h = 0, 1, ... , \prod_{m=0}^{i-1} n_m - 1$$

4 PROPAGATION DELAY ANALISYS

In this section we introduce a propagation delay model for PPA implementation, that takes into account the use of modules at different packaging levels. We perform the analyses for PPA systems with and without HNC. Such analyses are important because the delay information can be used by the compiler to control the scheduling of the instructions. More specifically, in a real system a single clock period may be not long enough to perform the required communication, therefore it becomes important to know exactly how many clock periods are required. Knowing the exact delay avoids blindly adopting the worst case delay.

In our definition, the PPA propagation delay model corresponds to a symmetric matrix T of size $n \times n$, each element of which, T_{sd}, corresponds to the propagation delay between node N_{is} and and node N_{id}, as well as between node N_{si} and node N_{di}, $i = 0, 1, ... , n\text{-}1$. The following two analyses adopt this notation.

4.1 Propagation Delay without HNC

Let t_i^{link}, $i = 0, 1, ... , p\text{-}1$, be the propagation delay along the interconnection link between two neighbor modules at packaging level i and let t_i^{switch}, $i = 0, 1, ... , p\text{-}1$, be the propagation delay of a module switch-box at packaging level i. The propagation delay between two generic nodes s and d $(1 \le s \le n\text{-}2, \, 2 \le d \le n\text{-}1, \, d > s)$ is given by

$$T_{sd} = (d\text{-}s)t_0^{link} + (d\text{-}s\text{-}2)t_0^{switch} +$$

$$+ \sum_{i=0}^{p-1} \alpha_i(s,d)(t_{i+1}^{link} - t_i^{link}) \qquad (1)$$

where

$$\alpha_i(s,d) = pos \left[\left| \frac{d}{\prod_{j=0}^{i} n_j} \right| - \left\lceil \frac{s}{\prod_{j=0}^{i} n_j} \right\rceil + 1 \right]$$

with $\quad pos(x) \equiv x \quad$ if $x \ge 0$
$\qquad \quad pos(x) \equiv 0 \quad$ if $x < 0$

The first two terms of equation (1) refer to the delay between the source and the destination assuming that only one packaging level (i.e. level 0) is implemented. When a new packaging level $i+1$ is introduced, the propagation delays t_i^{link} along the links that interconnect modules at packaging level $i+1$ must be replaced by t_{i+1}^{link}. The term $\alpha_i(s,d)$ represents the number of boundaries of modules at packaging level i located between nodes s and d. An example of this calculation is shown in Fig. 8a.

4.2 Propagation Delay with HNC

The propagation delay between two generic nodes s and d is given by

$$T_{sd}^{HNC} = T_{sd} +$$

$$\sum_{i=0}^{p-1} \beta_i(s,d)(t_{i+1}^{switch} - n_i t_i^{switch} - (n_i-1)t_i^{link}) \quad (2)$$

where

$$\beta_i(s,d) = pos \left| \left\lfloor \frac{d}{\prod_{j=0}^{i} n_j} \right\rfloor - \left\lceil \frac{s}{\prod_{j=0}^{i} n_j} \right\rceil \right|$$

and $\alpha_i(s,d)$ and $pos\,(x)$ are defined as above.

The delay T_{sd}^{HNC} is derived by taking T_{sd} as a starting point. When a module at level $i+1$ is bypassed the following corrections to T_{sd} are required:

1) the delay of the module switch-box at level $i+1$ (i.e. t_{i+1}^{switch}) is added;

2) the sum of the delays of the switch-boxes at level i (i.e. t_i^{switch}) is subtracted;

3) the sum of the delays of the links at level i (i.e. t_i^{link}) is subtracted.

The term $\beta_i(s,d)$ represents the number of modules at packaging level i entirely contained between nodes s and d. An example of this calculation is shown in Fig. 8b.

5 SIMULATION RESULTS AND DISCUSSION

5.1 Communication Speed-Up

The most evident benefit of the HNC technique is the reduction of the propagation delay between any pairs of PPA nodes. While in a PPA system without HNC, the propagation delay between two nodes is a linear function of their distance, in a PPA system adopting HNC the propagation delay between two nodes can be reduced up to a logarithmic function of their distance. Such a limit, shown in Fig. 9, is reached when each cluster at level i, $0 < i < p$ contains two clusters at level i-1 and the propagation delay is the same at every packaging level.

The logarithmic reduction occurs only in an idealized situation. In reality, a larger size of the clusters is supported by the packaging technology. Fig. 10

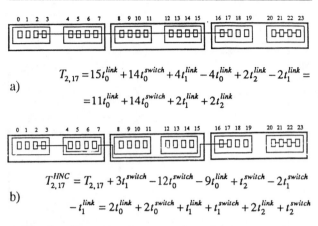

a)
$$T_{2,17} = 15t_0^{link} + 14t_0^{switch} + 4t_1^{link} - 4t_0^{link} + 2t_2^{link} - 2t_1^{link} =$$
$$= 11t_0^{link} + 14t_0^{switch} + 2t_1^{link} + 2t_2^{link}$$

b)
$$T_{2,17}^{HNC} = T_{2,17} + 3t_1^{switch} - 12t_0^{switch} - 9t_0^{link} + t_2^{switch} - 2t_1^{switch}$$
$$- t_1^{link} = 2t_0^{link} + 2t_0^{switch} + t_1^{link} + t_1^{switch} + 2t_2^{link} + t_2^{switch}$$

Fig. 8 - Example of computation of the propagation delay: a) without HNC, b) with HNC.

illustrates the propagation delay in a realistic case. In such a case we have assumed we have only two packaging levels, namely the integrated-circuit level (IC) and the printed-circuit-board (PCB) level, both characterized by switch-box propagation delay and interconnection link propagation delay of 4 nsec. In each IC there reside 16×16 nodes and in each PCB we package 16×16 IC's; this constructs a processor array of $256 \times 256 = 64K$ nodes. In Fig. 10 we compare the curves of the propagation delays without HNC (curve (1)) and with HNC (curve (2)). Due to the fact that HNC can only bypass IC's, the effect of HNC on the delay curve is only a reduction of the slope, instead of a trasformation from a linear function to a logarithmic function. Further improvement can be made toward logarithmic reduction by placing additional switch-boxes, at the board level, that bypass groups of short-circuited chips. Curve (3) shows how the propagation delay is improved by placing bypass switch-boxes every two, four and eight chips. Fig. 10 shows how HNC helps to reduce the propagation delay of a PPA system in a controllable and quantizable manner. Depending on the maximum acceptable propagation delay, a proper number of clustering levels can be chosen to guarentee the worst delay and the delay for certain desired communication topologies. The upper bound of the propagation delay is limited by the logarithmic behaviour shown in Fig. 9 where every pair of clusters at any level is grouped into a cluster at the upper level.

5.2 Scalability

Scalability is always an important issue for the design of a parallel system. For reconfigurable processor array, the scalability hampers both the hardware and the software. Without the insight and the control on the delay time, the system clock depends on the system size.

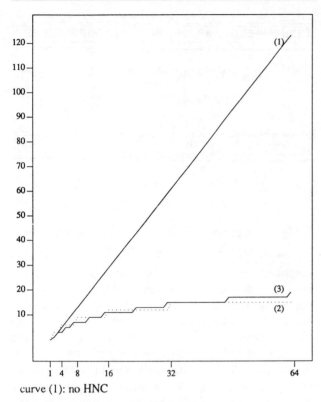

curve (1): no HNC

curve (2): $p = 8$, $n_i = 2$, $t_i^{link} = t_i^{switch} = 1$ logical time unit,
          $i = 0, 1, ..., p\text{-}1$

curve (3): reference logarithmic curve $(3\log(d\text{-}s))$

Fig. 9 - Propagation delay (in logical time units) versus
distance: ideal model.

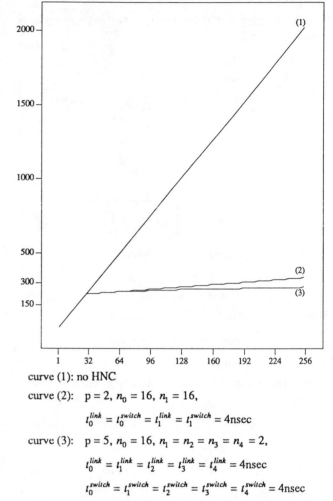

curve (1): no HNC

curve (2): $p = 2$, $n_0 = 16$, $n_1 = 16$,

$$t_0^{link} = t_0^{switch} = t_1^{link} = t_1^{switch} = 4\text{nsec}$$

curve (3): $p = 5$, $n_0 = 16$, $n_1 = n_2 = n_3 = n_4 = 2$,

$$t_0^{link} = t_1^{link} = t_2^{link} = t_3^{link} = t_4^{link} = 4\text{nsec}$$

$$t_0^{switch} = t_1^{switch} = t_2^{switch} = t_3^{switch} = t_4^{switch} = 4\text{nsec}$$

Fig. 10 - Propagation delay (in nanoseconds) versus distance:
realistic case.

Similarly, the program depends on the system size and needs to be re-written when the size changes. Such a system design is not scalable and is a poor model for programmers.

The difficulty to maintain scalability is not unique to PPA. For a bus-based multiprocessor system, a protocol and complex hardware are needed to support scalability. Even so, the bus-based system usually fails quickly and exhibits poor performance. For a parallel computer based on high-degree network such as a hypercube, the difficulty of scalability lies on the linear growing of the wires. Even for a sequential system, when a large amount of memory is built for the system, the memory access time becomes "unscalable", consequently complex cache system is popularly adopted to maintain the scalability.

An interesting observation here is that when "size" grows, the curse due to "size" damages the scalability from different aspects: memory access time for a sequential processor, wiring complexity for a high-degree network, or for reconfigurable processor array the undeterministic delay time. The hierarchical node clustering scheme is a solution to remedy the "unscalable" aspect of PPA.

5.3 Impact on programming

With the experience in developing massively parallel processing hardware in the past decade, experts in the field have gained enough confidence that one can build a large massively parallel hardware with controlled and careful engineering. However, it is a challenge to develop system software and programs for a massively parallel computer. As a result, hardware architect has less hesitance to either include or exclude hardware features only if it helps to ease the software development.

As discussed in the previous sections, the HNC scheme provides a controlled insight into the signal delay in a very large PPA. A very immediate benefit of HNC is that it shortens the signal delay hence the system clock if the architect adopts the worst-case delay for choosing the clock. What is more important is that, with the timing information provided by HNC, one can choose a clock cycle shorter than the worst-case delay.

Specifically, for position-dependent reconfiguration, the delay is known at compile time. It is therefore possible for the compiler to allocate enough, but just enough, time to accommodate the delay. This removes the burden of the worst case for a reconfigurable system in general. Similarly, for regular configurations such as tree and pyramid (which uses 2's power distances for reconfiguration), the delay is usually pre-calculated and stored in compiler tables. Only for the data-dependent reconfiguration, the delay is not known at the compile time and therefore the maximum delay needs to be used.

The allowance of code optimization is another benefit offered by HNC. Since in PPA computation and communication can be conducted independently, if the communication delay is known at the compile time, it is possible to pack other computations that are not dependent on the on-going communication. This is similar to the code optimization for Superscalar in which multiple instructions are issued and allowance for code compaction exists when there is an independence between consecutive instructions.

We reminded the reader in the previous sections that there are many models for reconfigurable processor array and the model discussed in this paper is not the most powerful one considering that it only supports communication along the row or column direction. However, it is this restriction that supports the HNC scheme, which possesses nice properties of known delay at compile time. Thus, the programmer is released from the timing burden of a reconfigurable system with the help of a compiler. Furthermore, there exist chances for code optimization. From the software viewpoint, a simpler and less powerful model becomes a better design solution for a massively parallel system.

6 CONCLUDING REMARKS

In this paper we have analyzed the propagation delay in reconfigurable processor arrays. While in reconfigurable processor array computation models communication between arbitrary nodes is assumed to be instantaneous, it is not in implementations. The gap between computation models and implementations can be filled by the Hierarchical Node Clustering technique introduced in this paper, which allows the bypassing of clustered nodes of various sizes in long distance communication, and the management of the signal delay in a controlled manner.

The key benefit of narrowing the gap between a model and its implementation is the ease of programming. Using the HNC scheme makes PPA

scalable; the signal delay time can be controlled and be more independent of the size of PPA hence PPA programs can be written subject to "the model" without concerning with "the implementation". With the significant reduction of the signal delay, the programming subject to the model is efficient and does not have to compromise with the worst case delay.

The application of HNC in PPA allows to significantly shorten the propagation delay between any pair of nodes. Coupling HNC with the possibility of overlapping computation and communication allows the compiler to implement compaction techniques, in such a way to optimize the code generated and to eliminate (or at least to reduce) the number of idle states to be inserted after instructions requiring long distance communication.

Not any reconfigurable processor array can take advantage of the proposed technique. HNC requires that communication be carried out only along one of the two dimensions at a time, namely either horizontally or vertically; the automatic spreading of data along arbitrary multi-shaped paths is not allowed. This reveals an interesting relationship between the "power of the model" and "programming". The "row/column only" PPA is a less powerful model that exhibits a limited number of reconfiguration capabilities. Nevertheless, with the HNC scheme, it becomes a scalable architecture and enjoys the programming efficiency.

The HNC technique contributes to the hardware aspect by reducing the signal delay; its most important value, however, lies in the profound impact on programming massively parallel computers.

REFERENCES

[1] M. Maresca, Polymorphic Processor Array, Technical Report - International Computer Science Institute, Berkeley, April 1991.

[2] R. Miller , V. K. Prasanna-Kumar, D. Reissis and Q. F. Stout, Meshes with Reconfigurable Buses, Proc. MIT Conference on Advanced Research in VLSI pp. 163-178, (1988).

[3] B. F. Wang and G. C. Chen, Constant Time Algorithms for the Transistive Closure and Some Related Graph Problem on Processor Arrays with Reconfigurable Bus System, IEEE Trans. on Parallel and Distribuited Systems Vol. 1, Num 4, pp. 500-507, (Oct. 1990).

[4] D. B. Shu and J. G. Nash, The Gated Interconnection Network for Dynamic Programming, Concurrent Computations , S. K. Tewsburg, B. W. Dickinson and S. C. Schwartz (editors), Plenum Publishing Company, pp. 645-658.

[5] H. Li and M. Maresca, Polymorphic-Torus Network, IEEE Trans. on Computers Vol. 38, Num. 9, pp. 1345-1351, (Sept. 1989).

[6] G. E. Blelloch, Scans as Primitive Parallel Operations, IEEE Trans. on Computers Vol. C-38, Num. 11, (1989).

Closed Form Solutions for Bus and Tree Networks of Processors Load Sharing A Divisible Job

S. Bataineh*, T. Hsiung and T. G. Robertazzi

Departments of Electrical Engineering
Jordan University of Science and Technology*
Irbid P.O. Box 3030, Jordan
State University of New York at Stony Brook
Stony Brook NY 11794

ABSTRACT

Optimal load allocation for load sharing a divisible job over processors interconnected in either a bus or a tree network is considered. The processors are either equipped with front-end processors or not so equipped. Closed form solutions for the minimum finish time and the optimal data allocation for each processor are obtained. The performance of large symmetric tree networks is examined by aggregating the component links and processors into a single equivalent processor. This allows an easy examination of large tree networks. In addition it becomes possible to find a closed form solution for the optimal amount of data that is to be assigned to each processor in the tree network in order to achieve the minimum finish time.

1 INTRODUCTION

A divisible job is a job that can be arbitrarily split in a linear fashion among a number of processors. Applications include the processing of very large data files such as occurs in signal and image processing, Kalman filtering and experimental data processing. Most work to date on load sharing has involved indivisible jobs, that is jobs that can only be assigned to a single processor [6, 7, 8, 9, 10, 11, 12, 13]. Only a small amount of recent work has examined jobs that can be assigned to multiple processors [17, 18, 19].

Load sharing of a divisible job among a number of processors which are connected together by an interconnection network such as a tree network and a bus network was examined in detail in [2, 3, 4, 15, 16, 20, 21], respectively. A set of recursive equations were developed to calculate the optimal fractions of the load that have to be assigned to each processor in the network in order to achieve the minimum finish time. The processors were assumed to have different speeds.

In this paper, the processors are all assumed to have the same speed. This enables one to find a closed form equation by which one can calculate the optimal fractions of the load that has to be assigned to each processor in the network in order to achieve the minimum finish time. Moreover, compact simple expressions for the minimum finish time for different networks are

also obtained.

This paper is organized as follows. In the second section tree networks where the processors have front-end processors are examined. Due to space limitations the case without front-ends and the case of bus architectures appears in [22].

2 TREE NETWORK

2.1. Introduction

Consider a tree network of communicating processors. In the tree we have three types of nodes (processors): root, intermediate and terminal nodes. Each tree has one root node that originates the load. An intermediate node can be viewed as a parent of lower level nodes with which it has a direct connection. Also it is a child of an upper level node with which it has a direct connection. The terminal nodes can only be children nodes. The kind and the number of levels in a particular tree determine its' size, that is the total number of nodes in that tree. The kind of a tree is determined by the number of nodes that a parent node has. A parent in a "binary" tree would have two children. The root is assumed to be level 0 and its children would be in level 1 and so on. The lowest level is $N-1$. Every processor can only communicate with it's children processors and parent processor.

In this work, we discuss two types of trees. One is where processors are equipped with front-end processors. Therefore, communication and computation can take place in each processor at the same time. In the second type of tree, processors do not have front-end processors. That is, processors can either communicate or compute but not do both at the same time.

In [2] a finite tree, where processors have different speeds, for the above two cases was discussed. However closed form solution for the minimum finish time were not presented.

In this paper, each of the processors in the tree are assumed to have the same computational speed, $\frac{1}{w}$. The communication speed between a parent processor and each of its children is also assumed to have the same value, $\frac{1}{Z}$. This assumption enables us to collapse the tree into one equivalent node that preserves the same characteristics as the original tree. This allows an easy examination of large tree networks. In addition, it becomes possible to find a closed form solution for the optimal amount of data that is to be assigned to each processor in order to achieve the minimum finish time and also to find a numerical solution to the minimum finish time.

The following definitions are used in this paper. The fraction of the total measurement data that is assigned to the ith processor by the originating processor is α_i. The time that it takes the ith processor to process the entire load when $w = 1$ is T_{cp}. The time for arbitrary w is wT_{cp}. Similarly, the time that it takes a processor to transmit the entire load over a link when $Z = 1$ is T_{cm}. The time for arbitrary Z is ZT_{cm}.

2.2. The Tree Network With Front-End Processors

In this section all the processors in the tree possess front-end processors. That is, each processor can communicate and compute at the same time. This fact will help to reduce the finish time. We will collapse the whole tree into one equivalent node. We start from the terminal nodes(the last level in the tree, level $N - 1$) and move up to the root processor(the first level in the tree, level 0). Similarly we will encounter two cases in our aggregation process: the first case occurs at the last two levels where all processors have the same speed; the second case occurs for the children at level k and their parents at level $k - 1$, $k = 1, 2, \ldots, N - 2$, where all processors, except the parent, have the same speed. In the following, we will discuss analytically the two cases.

The timing diagram of case one is the same as the bus network timing diagram discussed in subsection 2.3 of [22]. The results there can be used to obtain an expression for w_{eqt} which is stated below. Here w_{eqt} is a constant that is inversely proportional to the speed of an equivalent processor that replaces all the processors in the subtree where all the children are terminal nodes and preserves the characteristics of the original system.

$$w_{eqt} = w \frac{r_t^{n-1}(r_t - 1)}{r_t^n - 1} \qquad (2.1)$$

where

$$r_t = \frac{wT_{cp} + ZT_{cm}}{wT_{cp}}$$

This equation is obtained by equating (2.51) of [22] and $w_{eqt}T_{cp}$.

The timing diagram of the second case shows that this is the same as the bus network discussed in subsection (2.3) of [22] where all processors except the the first have the same speed. The time that takes each processor to process its share is computed by the following set of equations:

$$T_1 = \alpha_1 w T_{cp} \qquad (2.2)$$
$$T_2 = \alpha_2 Z T_{cm} + \alpha_2 w_{eq} T_{cp} \qquad (2.3)$$
$$T_3 = (\alpha_2 + \alpha_3) Z T_{cm} + \alpha_3 w_{eq} T_{cp} \qquad (2.4)$$
$$T_4 = (\alpha_2 + \alpha_3 + \alpha_4) Z T_{cm} + \alpha_4 w_{eq} T_{cp} \qquad (2.5)$$

$$\vdots$$

$$T_n = (1 - \alpha_1) Z T_{cm} + \alpha_n w_{eq} T_{cp} \qquad (2.6)$$

The fractions of the total measurement load should sum to one

$$\alpha_1 + \alpha_2 + \cdots + \alpha_n = 1 \qquad (2.7)$$

The optimal values of $\alpha's$ that has to be assigned to each processor in order to achieve the minimum finish time is given by the following set of equations:

$$\alpha_{n-1} = \alpha_n r_i \qquad (2.8)$$
$$\alpha_{n-2} = \alpha_{n-1} r_i \qquad (2.9)$$

$$\vdots$$

$$\alpha_3 = \alpha_4 r_i \qquad (2.10)$$
$$\alpha_2 = \alpha_3 r_i \qquad (2.11)$$
$$\alpha_1 = \alpha_2 c \qquad (2.12)$$

where $r_i = \frac{w_{eq}^i T_{cp} + Z T_{cm}}{w_{eq}^i T_{cp}}$

and $c^i = \frac{w_{eq}^i T_{cp} + Z T_{cm}}{w T_{cp}}$

Here i indicates the level of children nodes being considered. It should be noted that, to achieve the minimum finish time, α_i is solved for by equating T_i to T_{i+1} [3, 4]. The equations can be written in terms of of α_n, r_i, and c^i as follows:

$$\alpha_j = \begin{cases} \alpha_n r^{n-j}, & \text{if } j = 2, 3, \ldots, n-1 \\ \alpha_n r_i^{n-2} c^i, & \text{if } j = 1. \end{cases} \qquad (2.13)$$

Using (2.7) and (2.13) α_n can be found as a function of r_i and c.

$$\alpha_n = \frac{r_i - 1}{(c^i + 1)r_i^{n-1} - c^i r_i^{n-2} - 1} \qquad (2.14)$$

Now all other optimal values of $\alpha's$ can be computed using (2.13) Since $\alpha_1 = \alpha_n r_i^{n-2} c$, α_1 can be expressed in terms of r_i and c as follows:

$$\begin{aligned} \alpha_1 &= \frac{r_i - 1}{c^i(r_i^{n-1} - r_i^{n-2}) + r_i^{n-1} - 1}(r_i^{n-2} c^i) \\ &= \frac{r_i^{n-1} - r_i^{n-2}}{r_i^{n-1} - r_i^{n-2} + \frac{1}{c^i}(r_i^{n-1} - 1)} \qquad (2.15) \end{aligned}$$

In order to find w_{eqi}, we equate (2.2) to $w_{eqi}T_{cp}$. Here w_{eqi} is a constant that is inversely proportional to the speed of an "equivalent" processor that will replace all processors in the subtree where the children are parents in the original tree and preserves the characteristics as the original system.

$$w_{eqi} = w\alpha_1 \qquad (2.16)$$

Substituting the value obtained for α_1 in the above equation, we find that:

$$w_{eqi} = w\left(\frac{r_i^{n-1} - r_i^{n-2}}{r_i^{n-1} - r_i^{n-2} + \frac{1}{c^i}(r_i^{n-1} - 1)}\right) \qquad (2.17)$$

Starting at level $N-1$, one can use equation (2.1) to reduce the original tree by one level and then move up to level $N-2$. Starting from the subtrees where children are at level $N-2$ and up to the root processor one uses equation (2.17) to find $w_{eq_{total}}$. Here $w_{eq_{total}}$ is a constant that is inversely proportional to the speed of an "equivalent" processor that will replace the whole tree while preserving the same characteristics as the original system. Computing $w_{eq_{total}}$, the minimum finish time T_{ftnf} can be written as follows:

$$T_{ftnf} = T_{cp} w_{eq_{total}} \qquad (2.18)$$

and the maximum throughput is

$$\gamma = \frac{1}{T_{ftnf}} \qquad (2.19)$$

3 CONCLUSION

In this work closed form solutions for minimum finish time are obtained for several types of bus architectures and tree network architectures. The performance of these architectures are examined and the effect of the link speed is studied. Processing time for tree networks is only slightly improved as the number of children per node increases, especially if the link speed is slow. Moreover there is a point of diminishing returns for performance (finish time) as the size of a tree is increased.

4 ACKNOWLEDGEMENT

The research in this paper was supported by the SDIO/IST under the Office of Naval Research grant N00014-91-J4063

REFERENCES

[1] Cheng, Y.C. and Robertazzi, T.G., "Distributed Computation with Communication Delays", *IEEE Transactions on Aerospace and Electronic Systems*, Vol. 24, No. 6, Nov. 1988, pp. 700-712.

[2] Cheng, Y.C. and Robertazzi, T.G., "Distributed Computation for Tree Network with Communication Delays", *IEEE Transactions on Aerospace and Systems*, Vol. 26, No. 3, May 1990, pp. 511-516.

[3] Bataineh, S. and Robertazzi, T.G., "Distributed Computation for a Bus Networks with Communication Delays", *Proceedings of the 1991 Conference on Information Sciences and Systems*, The Johns Hopkins University, Baltimore MD, March 1991, pp. 709-714.

[4] Bataineh, S. and Robertazzi, T.G., "Bus Oriented Load Sharing for a Network of Sensor Driven Processors", *IEEE Transactions on Systems, Man and Cybernetics*, Sept. 1991, Vol.21, No. 5, pp. 1202-1205

[5] Hsiung, T. and Robertazzi, T.G., "Performance Evaluation for Distributed Communication Systems for Load Balancing", SUNY at Stony Brook, College of Engineering and Applied Science Technical Report No. 612, Dec. 17, 1991, Available from T. Robertazzi.

[6] Baumgartner, K.M. and Wah, B.W., "GAMMON: A Load Balancing Strategy for Local Computer Systems with Multiaccess Networks", *IEEE Transactions on Computers*, Vol. 38, No. 8, August 1989, pp. 1098-1109.

[7] Bokhari, S.H, "Assignment Problems in Parallel and Distributed Computing", Kluwer Academic Publishers, Boston, 1987.

[8] Lo, V.M., "Heuristic Algorithms for Task Assignment in Distributed Systems", *IEEE Transactions on Computers*, Vol. 37, No.11, Nov. 1988, pp. 1384-1397.

[9] Ramamritham K., Stankovic, J.A. and Zhao, W., "Distributed Scheduling of Tasks with Deadlines and Resources Requirements", *IEEE Transactions on Computers*, Vol. 38, No. 8, August 1989, pp. 1110-1122.

[10] Shin, K.G. and Chang, Y-C., " Load Sharing in Distributed Real-Time Systems with State Change Broadcasts", *IEEE Transaction on Computers*, Vol. 38, No. 8, August 1989, pp. 1124-1142.

[11] Stone, H.S., "Multiprocessor Scheduling with the Aid of Network Flow Algorithms", *IEEE Transaction on Software Engineering*, Vol . SE-3, No. 1, Jan. 1977, pp. 85-93.

[12] Mirchandaney, R. Towsley, D. and Stankovic, J.A., "Analysis of the Effects of Delays on the Load Sharing", *IEEE Transactions on Computers*, Vol. 38, No. 11, Nov. 1989, pp. 1513-1525.

[13] Ni, L.M. and Hwang, K., "Optimal Load Balancing in a Multiple Processor System with Many Job Classes", *IEEE Transaction on Software Engineering*, Vol. SE-11, No. 5, May 1985, pp. 491-496.

[14] Sohn, J. and Robertazzi, T.G., "Optimal Load Sharing for a Divisible Job on a Bus Network", *Proceedings of the 1993 Conference on Information Sciences and Systems*, The Johns Hopkins University, Baltimore MD, March 1993.

[15] Bharadwaj, V., Ghose, D. and Mani, V., "Design and Analysis of Load Distribution Strategies for Infinitely Divisible Loads in Distributed

Processing Networks with Communication Delays", *Dept. of Aerospace Engineering, Indian Institute of Science, Bangalore India Technical Report 422-GC-01-92*, Oct. 1992.

[16] Bharadwaj, V., Ghose, D. and Mani, V., "A Study of Optimality Conditions for Load Distribution in Tree Networks with Communication Delays", *Dept. of Aerospace Engineering, Indian Institute of Science, Bangalore India Technical Report 423-GI-02-92*, Dec. 1992.

[17] Du, J. and Leung, J.Y.-T., "Complexity of Scheduling Parallel Task Systems", *SIAM Journal on Discrete Mathematics*, Vol. 2, Nov. 1989, pp. 473-487.

[18] Blazewicz, J., Drabowski, M. and Weglarz, J., "Scheduling Multiprocessor Tasks To Minimize Schedule Length", *IEEE Transactions on Computers*, Vol. C-35, May 1986, pp. 389-393.

[19] Zhao, W. Ramamritham, K. and Stankovic, J.A., "Preemptive Scheduling Under Time and Resource Constraints", *IEEE Transactions on Computers*, Vol. C-36, Aug. 1987, pp. 949-960.

[20] Bataineh, S. and Robertazzi, T., "Ultimate Performance Limits for Networks of Load Sharing Processors", *Proceedings of the 1992 Conference on Information Sciences and Systems*, Princeton University, Princeton N.J., March 1992, pp. 794-799.

[21] Kim, H.J., Jee, G.-I. and Lee, J.G., "Optimal Load Distribution for Tree Network Processors", *submitted for publication*.

[22] Bataineh, S. and Robertazzi, T.G., "Closed Form Solutions for Bus and Tree Networks of Processors Load Sharing a Divisible Job", *SUNY at Stony Brook CEAS Tech. Rep. 627*, May 27, 1992. Available from T. Robertazzi.

The Message Flow Model for Routing in Wormhole-Routed Networks *

Xiaola Lin, Philip K. McKinley, and Lionel M. Ni

Department of Computer Science
Michigan State University
East Lansing, Michigan 48824-1027

Abstract

In this paper, we introduce a new approach to deadlock-free routing in wormhole-routed networks called the *message flow model*. We first establish the necessary and sufficient condition for deadlock-free routing based on the analysis of the message flow on each channel. We then show how to use the model to prove that a given adaptive routing algorithm is deadlock-free. Finally, we use the method to develop new, efficient adaptive routing algorithms for 2D meshes and hypercubes.

1 Introduction

In massively parallel computers (MPCs), regardless of how well the computation is distributed among the processors, communication overhead can severely limit speedup. In order to reduce network latency and minimize buffer requirements, the *wormhole routing* switching strategy [1] is being used in most current MPCs. For a survey of wormhole routing in direct networks, please refer to [2].

A routing technique is adaptive if, for a given source and destination, which path is taken by a particular packet depends on dynamic network conditions, such as the presence of faulty or congested channels. A *deadlock* occurs when two or more messages are delayed forever due to an acquired-waiting cyclic dependency among their requested resources. In a wormhole-routed network, the small buffers associated with each channel are resources.

The *channel dependence graph* (CDG) model was proposed [1] to assist in the development deadlock-free deterministic routing algorithms for wormhole-routed networks. The CDG for a particular network and routing algorithm is a directed graph $G(V, E)$,

where the vertex set $V(G)$ corresponds to all the unidirectional channels in the network, and the edge set $E(G)$ includes all the pairs of connected channels defined by the routing algorithm. A deterministic routing algorithm is deadlock-free if and only if its CDG is acyclic. As will be shown, an adaptive routing algorithm with cycles in its CDG can still be deadlock-free. The conservative nature of the CDG model with respect to adaptive routing may lead to inefficiencies in the deadlock-free routing algorithms that are developed with it.

In this paper, we introduce a new approach to deadlock-free routing in wormhole-routed networks called the *message flow model*. We use the model to prove that a given adaptive routing algorithm is deadlock-free. We then use the method to develop new, efficient adaptive routing algorithms for 2D meshes and hypercubes. Due to the lack of space, we only present the main results of this research, please refer to [3] for details.

2 The Message Flow Model

A key component of a routing algorithm is the *routing function* [1], which is defined as follows.

Definition 1 *A routing function R_A for a routing algorithm A is a $C \times V \times C$ relation. The triple $(c_i, v_d, c_j) \in R_A$ implies that if the destination node of the packet is v_d and the header flit of the packet has arrived at a router from channel c_i, then it may be forwarded along channel c_j when using algorithm A.*

Given a routing function R, a channel c_i, and destination node v_d, the *forwarding set* $F_R(c_i, v_d)$ is defined as the subset of channels, $\{c_j | (c_i, v_d, c_j) \in R\}$.

Unlike the CDG model, which represents the *potential* routing of messages from one channel to another, the message flow model represents the *actual*

---

*This work was supported in part by the NSF grants CDA-9121641 and MIP-9204066 and by an Ameritech Faculty Fellowship.

flow of messages among channels by accounting for the manner in which a blocked message waits for a channel to become available.

In this work, our assumptions are a superset of those in [1]. We also assume that the channel allocated algorithm used is free from starvation.

Definition 2 *Given a routing function R, a channel c is* deadlock-immune *for a message M if and only if, once M occupies the channel c, M will eventually leave channel c and release it. For convenience, if a message M can never use channel c, c is also defined to be* deadlock-immune *for message M. A channel is said to be* deadlock-immune *if and only if it is deadlock-immune for all messages when using the routing function R.*

Since a message may have arbitrary length by the assumption in [1], if a channel c is deadlock-immune for a message M, then once M has occupied c, M will not only eventually release c, but M is also guaranteed to eventually arrive at its destination.

Theorem 1 [3] *A routing function R is deadlock-free if and only if every channel in $C(I)$ is deadlock-immune when using routing function R.*

Corollary 1 *If the CDG for routing function R is acyclic, then each channel in $C(I)$ is deadlock-immune when using R.*

However, the converse is not necessary true, as will be shown later. The following theorem indicates how to identify deadlock-immune channels for a given routing function.

Theorem 2 [3] *Given a routing function R, a channel c_i is deadlock-immune if, for any node $v_d \in V$, $F_R(c_i, v_d) = \emptyset$.*

Corollary 2 *Given a routing function R, a channel $c_i = (v_j, v_d)$ is deadlock-immune for any message with destination v_d.*

If all of the channels in $F_R(c_i, v_d)$ are busy when the header flit has arrived from channel c_i, the message may wait until any of the channels in $F_R(c_i, v_d)$ becomes available, or the message may wait for a particular channel in $F_R(c_i, v_d)$ to become available. We will focus on the latter waiting strategy because, as will become clear later, if a given routing function is deadlock-free when using the latter waiting strategy, it is also deadlock-free when using the former waiting strategy.

Definition 3 *Let M be a message arriving on channel c_i and destined for node v_d, and let R be*

the routing function. The waiting channel, denoted $f_R(c_i, v_d)$, where $f_R(c_i, v_d) \in F_R(c_i, v_d)$ and $F_R(c_i, v_d) \neq \emptyset$, is defined as the channel on which M will wait if all of the channels in $F_R(c_i, v_d)$ are busy.

Let $P_R(c_i, v_d)$ denote the set of all of the paths from channel c_i to node v_d permitted under R. Formally, $P_R(c_i, v_d) = \{p | p = c_i c_{i+1} c_{i+2} \ldots c_{i+q}$, where $c_{i+q} = (v_{i+q}, v_d)$ and $c_{j+1} \in F_R(c_j, v_d)$ for $i \leq j < i + q\}$. The following theorem shows how to identify a deadlock-immune channel given other known deadlock-immune channels.

Theorem 3 [3] *Given a routing function R, a channel c_i is deadlock-immune if, for every $v_d \in V$ with $F_R(c_i, v_d) \neq \emptyset$, and every channel $c_j = (v_j, v_{j+1}), v_{j+1} \neq v_d$, and c_j in path $p, p \in P_R(c_i, v_d)$, channel $f_R(c_j, v_d)$ is deadlock-immune.*

Corollary 3 *For a channel c_i, a message M with v_d as its destination, $F_R(c_i, v_d) \neq \emptyset$, c_i is deadlock-immune for M if, for any channel $c_j = (v_j, v_{j+1}), v_{j+1} \neq v_d$, and c_j in path $p, p \in P_R(c_i, v_d)$, channel $f_R(c_j, v_d)$ is deadlock-immune.*

Corollary 4 *If, for all $c_i \in C, v_d \in V$, $f_R(c_i, v_d)$ is deadlock-immune when $F_R(c_i, v_d) \neq \emptyset$, then every channel in C is deadlock-immune, that is, the routing function R is deadlock-free.*

3 Adaptive Unicast Routing for 2D Meshes

In this section, we consider an enhanced version of the *double Y-channel algorithm* for adaptive routing in 2D mesh networks. The original algorithm, which was presented in [4], requires that an additional pair of channels be added to the Y dimension, as shown in Figure 1(a). This network can be partitioned into two subnetworks, namely, the *east* subnetwork and the *west* subnetwork; each pair of nodes neighboring in the Y dimension are connected by a pair of channels, while each pair of nodes neighboring in the X dimension are connected by a single channel. The east subnetwork is shown in Figure 1(b). A message with source $v_s = (x_s, y_s)$ and destination $v_d = (x_d, y_d)$ is called an *eastbound* message if $x_d > x_s$ and is called a *westbound* message if $x_d < x_s$. If $x_d = x_s$, then the message may be classified as either an eastbound or westbound message. Any westbound message will be routed through west subnetwork, and any eastbound message through the east subnetwork. The double Y-channel algorithm is a minimal fully adaptive routing algorithm. If the channels are labeled appropriately, for example as shown in Figure 1(b), then a message may follow any shortest path between the source and destination and be guaranteed to traverse channels in monotonic order. Under these conditions, the CDG of the algorithm is acyclic and, therefore, deadlock cannot occur.

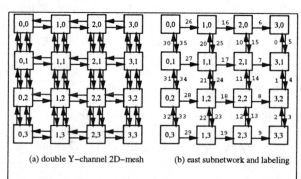

Figure 1. An example of the double Y-channel routing algorithm in a 4 × 4 mesh.

The message flow model may be used to increase the adaptivity of the double Y-channel routing algorithm without adding more virtual channels. Specifically, a westbound message may also safely use Y-channels in the east subnetwork, if such a channel is available when the message arrives at the corresponding router. However, when a westbound message is blocked, it must always wait for a channel in west subnetwork.

The CDG of this *enhanced double Y-channel routing algorithm* may contain a cycle [3]. However, no acquired-waiting cycle situation will ever happen [3].

Theorem 4 [3] *The enhanced double Y-channel routing algorithm is deadlock-free.*

4 Multicast Adaptive Routing in 2D Meshes

In this section, we use the message flow model to study a fully adaptive multicast (FAM) algorithm for the 2D mesh. This algorithm also uses double Y-channels in the mesh.

Deterministic multicast routing algorithms have been proposed in [5], where it was shown that that a *path-based* model of multicast is a practical method for wormhole-routed networks [5].

The proposed multicast routing algorithm (FAM) operates as follows. The destination set is divided into two subsets, D_W and D_E. D_W contains the destination nodes on the west side of the source node s, that is, if $s = (x_0, y_0)$, $D_W = \{d | d = (x, y), x \leq x_0\}$.

The subset D_E contains the rest of the destination nodes, $D_E = D - D_W$. The multicast is performed using two multicast paths, one for D_W and one for D_E, that originate at the source node. Without loss of generality, we will study the west subnetwork, which is defined in the previous section.

At the source node, the destination nodes are ordered in such a way that the source message is routed adaptively to the west towards the destination nodes, that is, the destination nodes are visited in descending order according to their x-coordinates. If more than one destination node has the same x-coordinate, then the message will be sent first north (in increasing y-coordinate order) then south (in decreasing y-coordinate order). Figure 2 shows an example of FAM routing in a 5×5 mesh. At the source node $(0, 2)$, the destinations are ordered as follows: $((3,3), (2,4), (2, 2), (0,0))$. Two destinations, $(2, 2)$ and $(2, 4,)$, have the same x-coordinate, and $(2, 4)$ is visited before $(2, 2)$. Figure 2 shows a multicast path that may be selected by the FAM routing algorithm. The formal description of the FAM routing algorithm is given in [3].

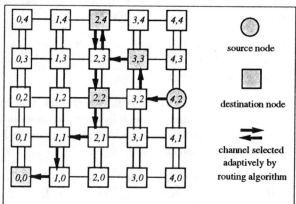

Figure 2. An example of routing path selected by the FAM algorithm.

When a FAM-routed multicast message is blocked, it always waits for a horizontal channel if such a permissible channel exists. The CDG for the FAM algorithm contains cycles [3]. However, such a cycle does not cause an acquired-waiting cycle [3].

Theorem 5 [3] *The FAM multicast routing algorithm is deadlock-free.*

The message flow model can be used to show that any routing algorithm whose CDG is acyclic is deadlock-free [3]. Thus, we claim that the message flow model is more powerful than the CDG model because it can also be used to show the deadlock-free property for a large class of routing algorithms whose CDGs contain cycles.

5 Adaptive Unicast Routing in Hypercubes

Assume that there are two virtual channels for each physical channel in a hypercube. One channel of each pair is identified as the *free* channel and the other as the *fixed* channel. When a message is blocked, it always waits for a fixed channel. A fixed channel will be used in such a way that it is easily shown to be deadlock-immune; a free channel can be used without restriction for fully adaptive routing. By Corollary 4, if each fixed channel is deadlock-immune, then so is each free channel.

Given an n-cube, the set of all fixed channels can be divided into n subsets, $C_0, C_1, \ldots, C_{n-1}$, where $C_k = \{(v_i, v_j) | v_i \text{ and } v_j \text{ differ only in the } k\text{-th bit}$ (from right to left)$\}$. A channel in C_k is referred to as a k-dimension channel.

Algorithm: The FF Algorithm for Message Routing in an n-cube
Input: the local node v_i, destination of the message v_d.
Output: an outgoing channel of v_i to forward the message.
Procedure:

1. If $v_i = v_d$, send the message to the local processor of v_i. Stop.

2. If any free channel $c_j, c_j = (v_i, v_j)$, such that v_j lies along a shortest path from v_i to v_d, is available, then forward the message along channel c_j to node v_j. Stop.

3. Suppose that k is the position of the most significant bit in which v_i and v_d differ. Select the k-dimension fixed channel $c_k, c_k = (v_i, v_k)$ (E-cube routing). If c_k is available, forward the message along c_k to node v_k and stop; otherwise, wait for c_k until it becomes available.

Figure 3. The FF algorithm for message routing in an n-cube.

The distributed routing algorithm, the FF (free-

fixed) algorithm, is given in Figure 3.

A message routed by the FF algorithm may alternate between the two sets of channels (free channel set and fixed channel set). Clearly, the FF algorithm provides much better adaptivity than the previous routing algorithm.

Because the free channels are used with no restriction for minimal fully adaptive routing, the CDG of FF routing algorithm will contain cycles. However, the following theorem holds.

Theorem 6 [3] *The FF routing algorithm for a hypercube in which there are two virtual (unidirectional) channels for each physical unidirectional channel is deadlock-free.*

The FF algorithm requires only two virtual channels per physical channel regardless of the size of the hypercube, compared with the algorithm in [4] that requires 2^{n-1} virtual channels per physical channel.

6 Concluding Remarks

In this paper, we introduced the message flow model for the study of routing algorithms in wormhole-routed networks. This method may be used to develop deterministic, partially adaptive, and fully adaptive routing algorithms for wormhole-routed networks with arbitrary topologies. We conclude that the proposed message flow model is a practical model for routing in wormhole-routed networks.

References

1. W. J. Dally and C. L. Seitz, "Deadlock-free message routing in multiprocessor interconnection networks," *IEEE Transactions on Computers*, vol. C-36, pp. 547–553, May 1987.

2. L. M. Ni and P. K. McKinley, "A survey of wormhole routing techniques in direct networks," *IEEE Computer*, vol. 26, no. 2, pp. 62–76, Feb. 1993.

3. X. Lin, P. K. McKinley, and L. M. Ni, "The message flow model for routing in wormhole-routed networks," Tech. Rep. MSU-CPS-ACS-78, Michigan State University, January 1993. submitted for publication.

4. D. H. Linder and J. C. Harden, "An adaptive and fault tolerant wormhole routing strategy for k-ary n-cubes," *IEEE Transactions on Computers*, vol. 40, pp. 2–12, Jan. 1991.

5. X. Lin, P. K. McKinley, and L. M. Ni, "Deadlock-free multicast wormhole routing in 2D mesh multicomputers." accepted to appear in *IEEE Transactions on Parallel and Distributed Systems*.

SESSION 12A

GRAPH_THEORETIC INTERCONNECTION STRUCTURES (II)

Generalized Fibonacci Cubes

W. J. Hsu† and M. J. Chung ‡
† Division of Computer Technology, SAS
Nanyang Technological University, Singapore
‡ Department of Computer Science
Michigan State University, E. Lansing, Michigan

Abstract

We present a new class of interconnection topologies called the *generalized Fibonacci cubes* (GFCs, for short) that encompass a range of networks such as the popular Boolean cube (hypercube) and the recent *second-order Fibonacci cube* in [5]. We show that each GFC has a recursive and self-similar structure and hence exhibits fault tolerant features. We also show that each GFC admits simple embedding of other useful networks such as cycles, trees, and meshes. The GFCs may find applications in fault-tolerant computing.

1   Introduction

The study of *interconnection topologies* (alternatively, *network topologies*, interconnection networks) is an important subject in the parallel or distributed systems. To date, most research in this area has emphasized on the properties and applications of individual networks. This approach is time-consuming and offers little insight about the interrelations of different interconnection schemes. Here we present a large family of interconnection topologies called the *generalized Fibonacci cubes*, which covers a spectrum ranging from regular graphs such as the hypercubes to semi-regular graphs such as the second-order Fibonacci cube in [6]. Whenever possible, we will present results that are applicable to *all* members of this family of network topologies.

The generalized Fibonacci cubes are an infinite class of network topologies $\mathbf{F} = \{\Gamma_n^{[k]} : n \geq 0 \text{ and } k \geq 2\}$, where each member $\Gamma_n^{[k]}$ (called the k-th order **Fibonacci cube of dimension** n, see Section 2 for formal definitions) is a graph that can be associated with the n-th term of a k-th order recurrence: $f_{n+k} = f_{n+k-1} + f_{n+k-2} + \ldots f_n$, where $n \geq 0$; $f_0^{[k]} = f_1^{[k]} = f_2^{[k]} = \ldots f_{k-1}^{[k]} = 0$, and $f_k^{[k]} = 1$. In fact, the above recurrence defines the generalized Fibonacci numbers from which the name of the networks is derived. We will show that these recurrences suggest useful information about the structure of the generalized Fibonacci cubes.

The properties that will be considered for the new class of networks include the connectivity issues (e.g., shortest path, diameter, edge or node connectivity, etc.), recursive decompositions of a given GFC, relations among the GFCs, and relations with other common networks such as the linear array, mesh, tree, and the Boolean cube. Specifically, the following results will be reported in this paper:

1. We show that the GFCs all have recurrent structures, hence they admit very flexible decompositions. These decompositions are the basis of graph embeddings and recursive algorithms for the GFC (as will be

demonstrated in the paper).

2. We show that there is a natural hierarchy in **F** in terms of structural complexity. Networks that are higher in this hierarchy contains more nodes and have relatively denser interconnections than those that are lower in the hierarchy. This hierarchy forms the basis for a recursive-descent reconfiguration scheme for fault tolerance computing.

3. We show that the network $\Gamma_n^{[k]}$ is connected, and that the diameter, the node degrees, the number of independent paths, and the number of edge-independent spanning trees are all in the logarithmic order. These features are comparable to those of the hypercube and are superior to the Incomplete hypercube.

4. We show that *any* GFC can emulate common structures such as the linear array, certain types of trees and meshes (generalized Fibonacci trees and meshes). These features suggest that the GFCs can be used as an underlying networks for implementing reconfigurable arrays (e.g., a fault tolerant mesh).

We refer the readers to [8] or [3] for the properties of the generalized Fibonacci numbers and Fibonacci codes. Section 2 defines the generalized Fibonacci cubes in terms of the generalized Fibonacci code, it also examines certain basic properties of the networks. Section 3 discusses the relations of GFC and other common networks in terms of the embedding relations. Section 4 briefly summarizes the results. Because of the limited space, we will have to omit the proofs of the claims, which are available from the authors.

2 Fibonacci Codes and GFCs

It is natural to define the generalized Fibonacci cube based on the generalized Fibonacci numbers and Fibonacci codes [8]. For convenience, we need the following notations. Let S_1 and S_2 denote two strings. Then write $S_1 \circ S_2$ to denote the string obtained by concatenating S_2 after S_1. Let X denote a set of strings. Write $S \circ X$ to denote the set $\{S \circ x : x \in X\}$.

Definition 1 *The set of k-th order Fibonacci codes of degree n, $V_n^{[k]}$, is defined as follows: For $n \geq k \geq 2$, define $V_n^{[k]} = 0 \circ V_{n-1}^{[k]} \cup 10 \circ V_{n-2}^{[k]} \cup 110 \circ V_{n-3}^{[k]} \cup \ldots \cup (1)^{k-1} 0 \circ V_{n-k+1}^{[k]}$. Define $V_n^{[k]} = \phi$ for $n \leq k-2$, and $V_n^{[k]} = \{\lambda\}$ for $n = k-1$.*

Clearly, $|V_n^{[k]}| = f_n^{[k]}$ for all $n \geq 0$ and $k \geq 2$ (the *generalized Zeckendorf's theorem*[8]).

Definition 2 *Let the generalized Fibonacci codes $V_n^{[k]}$ be defined as in the preceding paragraph. Let $H(i,j)$ denote the Hamming distance between two binary numbers i and j. The k-th order Fibonacci cube of dimension n is a graph $\Gamma_n^{[k]} = (V_n^{[k]}, E_n^{[k]})$, where the vertices are labeled by the Fibonacci codes $V_n^{[k]}$ and $(i,j) \in E_n^{[k]}$ if and only if $H(i,j) = 1$.*

Let B_n denote the n-dimensional Boolean cube. It is well known that the set of nodes in B_n can be labelled exactly by the set of n-bit binary numbers (called the nodes' *addresses*); the edges by the Hamming distance. The following lemma states a basic relation between the hypercube and the GFCs.

Lemma 1 *1. $\Gamma_n^{[k]} \subseteq B_n$ for all $n \geq k \geq 2$.*

2. $\Gamma_n^{[k]} \subseteq \Gamma_{n+1}^{[k+1]}$ for all $n \geq k \geq 2$.

[**proof**] Since both networks are defined by using the binary numbers and the Hamming distance, the Fibonacci cubes are subgraphs of the

Boolean cube. In fact, $\Gamma_n^{[k]}$ can be obtained from B_n by deleting nodes (along with incident edges) that have k consecutive 1's in their addresses. (The proofs are omitted. See [7].) It is also straightforward to verify that there is a hierarchy among the family of generalized Fibonacci cubes. □

Intuitively, because of its asymmetric and relatively sparse interconnections, the Fibonacci cube has a weaker structure than the hypercube. However, it turns out that $\Gamma_n^{[k]}$ also has useful recursive structures. Given this recursive structure of the Fibonacci cubes, one can derive efficient recursive algorithms in a rather straightforward manner.

Lemma 2 *Let $\Gamma_n^{[k]}$ denote the k-th order n -dimensional GFC as defined before. Then, for all $n \geq k \geq 2$,*
$$\Gamma_{n+k}^{[k]} \subseteq (\Gamma_{n+k-1}^{[k]} \cup \Gamma_{n+k-2}^{[k]} \cup \ldots \Gamma_n^{[k]}).$$

[proof] (Omitted. [6]) □

3   Properties of GFCs

The following theorem lists some combinatorial properties of GFCs:

Theorem 1 *Let $\Gamma_n^{[k]}$ denote the k-th order n-dimensional generalized Fibonacci cube as defined before. Let $n \geq k \geq 2$.*

1. *Let $d_n^{[k]}$ denote the degree of an arbitrary node in $\Gamma_n^{[k]}$. Then $\lfloor \frac{n-k}{k+1} \rfloor \leq d_n^{[k]} \leq (n-k)$.*

2. *The number of edges is bounded by $\Theta(N \log N)$, where N is the number of nodes $\Gamma_n^{[k]}$.*

3. *The number of edge-disjoint spanning trees in $\Gamma_n^{[k]}$ is no less than $\lfloor \frac{n-k}{k+1} \rfloor$.*

[proof] (Omitted. [6]) □

The following theorem lists some important path-connectivity properties of GFCs:

Theorem 2 *Let $\Gamma_n^{[k]}$ denote the k-th order n-dimensional generalized Fibonacci cube as defined before.*

1. *$\Gamma_n^{[k]}$ is a connected graph; the diameter of $\Gamma_n^{[k]}$ is equal to $n-2$.*

2. *Let $d(i,j)$ denote the distance between Node i and Node j. Then for all $n \geq k \geq 2$, $d(i,j) = H(i,j)$.*

3. *Let $K_e(G)$ and $K_v(G)$ denote the edge connectivity and node connectivity of a graph G respectively. Then $K_e(\Gamma_n^{[k]}) = K_v(\Gamma_n^{[k]}) = \lfloor \frac{n-k}{k+1} \rfloor$.*

[proof] (Omitted. [6]) □

4   Embeddings

Definition 3   1. *The k-th order Fibonacci linear array of rank n, denoted $L_n^{[k]}$ is a linear array of size $f_n^{[k]}$. Define $L_n^{[k]}$ to be a null graph for $0 \leq n \leq k-2$.*

2. *Let $n_1, n_2 > 0$ denote integers. The k-th order Fibonacci mesh of rank (n_1, n_2), denoted by $M_{(n_1,n_2)}^{[k]}$, is a $f_{n_1}^{[k]} \times f_{n_2}^{[k]}$ mesh. Define $M_{(i,n)}^{[k]} = M_{(n,i)}^{[k]} = (\phi, \phi)$, for all $0 \leq i \leq k-2$ and $n \geq 0$.*

3. *The k-th order Fibonacci tree of rank n, denoted by $T_n^{[k]}$, is a subgraph of the binomial tree BT_n induced by the set of nodes in $V_n^{[k]}$.*

The *binomial tree* BT_n as mentioned in the definition of the Fibonacci tree is a rather well known notion in the context of hypercube applications (see, e.g., [1]). The following theorem lists a few important relations between GFCs and other common networks:

Theorem 3   1. *(Boolean Cube)*

(a) $\Gamma_n^{[k]} = B_{n-k}$ *for all* $k \le n < 2k$.

(b) $\Gamma_n^{[k]} \subseteq B_{n-k}$ *for all* $2k \le n$.

2. $\Gamma_n^{[k]} \supseteq L_{n-k}^{[k]}$ *for all* $n \ge k \ge 2$.

3. $\Gamma_n^{[k]} \supseteq T_{n-k}^{[k]}$ *for all* $n \ge k \ge 2$.

4. $\Gamma_n^{[k]} \supseteq (M_{(n,n)}^{[2]} \cup M_{(n+1,n+1)}^{[2]})$,

5. $\Gamma_n^{[k]} \supseteq (M_{(n,n+1)}^{[2]} \cup M_{(n,n-1)}^{[2]})$,

[proof] (Omitted. [6]) □

5 Conclusion

We have presented a family of interconnection topologies called the generalized Fibonacci cubes. The structural properties that were analyzed include the recursive decompositions, diameter, node or edge connectivities, and relations with other structures. We also showed that various types of networks can be directly embedded in the generalized Fibonacci cubes. To further investigate the applications of GFCs, [9] studied the Point-to-Point and Broadcasting types of communications. It is found that both operations can be done efficiently on *all* GFCs. A recent study shows that two common primitives *parallel prefixes* and *tree contraction* can be done in logarithmic time. These result show that these new topologies have attractive structures which are useful in developing either recursive-descent/recursive-doubling algorithms or fault-tolerant schemes.

Clearly there are many questions that are left unanswered. We call attentions to the following open problem that has applications in fault-tolerant computing:

Given a graph, how to quickly decide whether it is a GFC?

References

[1] D. P. Bertsekas and J. N. Tsitsiklis, *Parallel and Distributed Computation: Numerical Methods*, Prentice-Hall, Inc., 1989.

[2] *The Fibonacci Quarterly*: The Official Journal of the Fibonacci Association.

[3] R. L. Graham, D. E. Knuth, and O. Patashnik, *CONCRETE MATHEMATICS*, Addison-Wesley Publishing Co., 1989 (Chapter 6: Special Numbers).

[4] C. -T. Ho and S. L. Johnsson, Distributed Routing Algorithms for Broadcasting and Personalized Communication in Hypercubes, Proc. Int'l Conf. on Parallel Processing, 1986.

[5] W. -J. Hsu, Fibonacci Cubes-A New Interconnection Topology, Accepted for publication, IEEE Transactions on Parallel and Distributed Systems, to appear 1993.

[6] W. -J. Hsu and M. -J. Chung, On a Large Family of Interconnection Topologies, 1991.

[7] W. -J. Hsu and J. Liu, Fibonacci Codes as Formal Languages, Tech. Report CPS-91-05, Michigan State University, May 1991.

[8] D. E. Knuth, *THE ART OF COMPUTER PROGRAMMING*, Vol. 1-3, Addison-Wesley, 1973.

[9] J. -S. Liu and W. -J. Hsu, On Embedding Rings and Meshes in Fibonacci Cubes, Tech. Report CPS-91-01, Jan. 1991. Proc. Int'l Parallel Processing Symp., 1992. pp. 589-596.

HMIN : A New Method for Hierarchical Interconnection of Processors[1]

Yashovardhan R. Potlapalli and Dharma P. Agrawal
Department of Electrical and Computer Engineering,
Box 7911, North Carolina State University,
Raleigh, N.C. 27695-7911.

Abstract

Hierarchical interconnection schemes have been proposed in the literature to take advantage of locality of refernce. This paper presents a dynamic scheme for hierarchical interconnection called Hierarchical Multistage Interconnection Network (HMIN). We discuss the various design alternatives and evaluate each of them with respect to average delay and cost. HMINs are compared only with MINs because of a lack of a method of comparing the cost of a static HIN with a dynamic HMIN.

1   Introduction

It has been shown in the literature that the accesses in a multiprocessor system also follow the principle of locality of reference. Under this assumption, it has been shown that hierarchical networks have a lower cost and average delay. The cost of an interconnection network is proportional to the number of inputs and outputs, the number of switches and the number of links connecting the switches.

So far, the study of hierarchical networks has been restricted to the use of static interconnection networks as the building blocks. Here, we present a hierarchical network constructed using MINs called the Hierarchical Multistage Interconnection Network (HMIN). We define a generic routing algorithm for HMINs. We also present different structures for the HMIN and evaluate them based on cost and average delay. Dynamic MINs are a better choice because:

- self-routing property can be maintained in the network, and

- the average delay through the network could be smaller than $\log_2 N$,
  where N is the number of inputs and outputs.

- in hierarchical networks built with static interconnection networks, we have the following problems -

    - the communication delay between two processors depends on the distance between them and

[1]This work has been supported in part by the Army Research Office under contract DAAL03-89-K-0142

can vary greatly as compared to HMINs where the delay varies in a few fixed steps depending on the number of levels.

    - the communication between two processors more than one hop away uses the communication interface of all intermediate processors in the communication path and can interfere with the communication of those processors.

Section 2 presents the structure of and routing algorithm for HMINs. Section 3 presents the performance evaluation and discusses the choice of the structure for best performance.

2   Structure of HMINs

By definition, a HMIN has MINs at all the levels of hierarchy. At each level, the inputs the of the MIN are of two types: lower-level and higher-level. They are so called because the lower-level inputs are connected to the inputs of the MINs at the lower level and the higher-level inputs are connected to the inputs of the higher level MINs. The MINs at the lowest level have processors connected to their lower-level inputs while the highest level MINs have no higher-level inputs. Fig. 1 shows the structure of a HMIN. It must be noted here that a input and the corresponding output are connected to the same thing, only at different ports. This is made clearer by the following example: consider a HMIN with two levels. The lower-level inputs of the lower level MIN are processors. The outputs corresponding to these inputs, i.e., having the same port number, are also connected to the same processors. The higher-level inputs of the lower level MIN are connected to the the outputs of the higher-level MIN. The inputs of the higher level MIN corresponding to these outputs are the lower-level inputs. They are connected to the outputs of the lower level MIN corresponding to the higher-level inputs of the lower level MIN. The higher level MIN is the highest level as the HMIN has a two level hierarchy and so, it has no higher-level inputs.

Figure 2 shows a two level HMIN. We will be using this diagram for all subsequent discussions. The lower level MIN is made up of 2×2 switches and has d stages. The

MIN has only one higher-level input and thus has 2^d - 1 processors connected to it. The higher level MIN is also made up of 2 × 2 switches while it has D stages. All its inputs are lower-level inputs as the HMIN is two level. clearly, there are 2^D lower level MINs and so there are $2^D \times (2^d - 1)$ processors. Thus, the number of processors using this design is not a power of two.

As seen in the above example, the most obvious design procedure results in a HMIN with non-power-of-two inputs. This is not desirable in most cases. The reason for not having power-of-two inputs is that, at each level of hierarchy, we have chosen the number of MINs in the lower level to be not a power of two. This implies that we need to choose the number of lower- or higher-level inputs to be a power of two. As we are using power-of-two input MINs, we need to use only half the inputs as lower-level inputs. However, as most of the communication is going to be local, we do not need so many hierarchical links. Also, the use of a large number of hierarchical links results in a lower number of processors connected and so increases the cost of the network (a larger size network will be required to connect the same number of processors).

Clearly, the above solution is not a best one. A better method would be to choose non-power-of-two input MINs (generalized MINs) and use the extra inputs (inputs more than the nearest power of two) as higher-level inputs. It must be noted that the choice of generalized MIN does not depend only on the number of inputs of the HMIN. A suitable number of higher-level inputs must be used for performance reasons. This will be discussed in a later section.

The network obtained by the above method is just one of the many alternative structures possible in HMINs. A different structure would result if any of the redundant and fault-tolerant MINs were used as building blocks. The complexity of the MIN used determines the complexity of the routing algorithm. For this reason, the best choice is the use of generalized MINs.

2.1 Routing

In routing, we see the advantage of using MINs because MINs are self-routing. The routing strategy is to this property of MINs and recursively define the algorithm. Before we specify the routing algorithm, we need to determine the addressing scheme because a good addressing scheme can reduce the complexity of the routing algorithm.

As seen in Figure 3, the 0th input (output) of every MIN at every level is used as higher-level input (output) and the rest are used as lower-level inputs (outputs). The address of a processor has many fields: one for each level of hierarchy. Each message has to go up to the highest level in which the source and destination fields differ. To reach the level, the routing tag of the message must contain 0's. After reaching the highest level, the tag should contain the remaining destination address, i.e., the fields corresponding to the lower levels of hierarchy.

We use an example to demonstrate the method. For simplicity, let there be l levels and hence l fields. Let each field be b bits wide. Consider a case when the source and destination addresses differ in level i - 1 ($< l$). Then, the

tag has $(i - 1) \times b$ 0's. Then it has the lower $i \times b$ bits of the destination. Thus, the tag is of variable length with a maximum length of $(2 \times l - 1) \times b$ bits and a minimum length of b bits when the source and destination have only the highest level of hierarchy in common and when the source and destination have the lowest level hierarchy in common, respectively.

The routing algorithm is described below. The assumption is that the number of stages in each level of the hierarchy are kept in an array called size.

> *Routing Algorithm*
> *Compare hierarchy address fields of source and destination from highest level hierarchy to lowest level hierarchy*
> *if all fields are the same then*
> *   routing tag = lowest field of destination*
> *else /* tags differ in level i-1 but are same in level i */*
> *   routing tag = $\sum (size)|_0^{i-1}$ ## destination address$|_0^{i-1}$*
> *endif*
> *end*

Lemma 1: The algorithm will correctly route a packet from any source to any destination.

Proof: If both the source and destination are connected to the inputs of the same MIN, then, the packet has to be routed only across the MIN and can be done by using the local address field of the destination. If the source and destination are not connected to the same MIN, the routing has to be done using the hierarchical connections. The ascend the hierarchy, we need to use output "0" of every MIN. Thus, the routing tag has to be all zeros. We need to ascend the hierarchy till the fields corresponding to all the higher levels are identical. After we reach the level, we can get to the destination using the lower fields of the destination address. At the lowest level of hierarchy, we have a single MIN and the destination's local address is enough to route the packet across the local MIN.   □

3 Performance Evaluation

In this section, we discuss the different methods for measuring the performance of HMINs. We will also present a method for comparing the costs of an HMIN with a MIN. Since, both MINs and HMINs are built using the same type of switches (b×b crossbars), the number of switches used is a good measure of the cost of the networks (the number of links is proportional to the number of switches).

In a simple multistage network, the latency is fixed from any input-output connection. However, for HMINs, the latency depends on the *locality* of communication, i.e., if the communication is over the local (lowest level hierarchy) network, then the delay through the network is low; and the delay increases if the communication is across greater levels of hierarchy.

We consider a two-level HMIN with the size (number of stages) of the lower level MIN represented by d and the size of the higher level MIN represented by D. Thus, the number of processors in the system is $2^D \times (2^d - 1)$. Thus, the number of stages in MIN using 2×2 switches for these processors is $\log_2(2^D \times (2^d - 1))$.

The simplest method for determining the latency is to assume uniform communication and determine the average distance. The avearage distance ratio is the ratio of the average distance of the HMIN to the avearage distance of the MIN. The average distance ratio is calculated from the following equation :

Av. distance ratio =

$[(d \times (2^d - 1) + (D + 2d) \times 2^D \times (2^d - 1))/(2^D \times (2^d - 1))] / [\log_2(2^D \times (2^d - 1))]$

The results based on this model are shown in Fig. 4. This, however, is the worst case for the HMIN as communication is mostly localized. The second method is to multiply the distances used to measure the average distance with the probability that the path of that length will be used, i.e., a factor, α, is introduced to account for the *locality* in communication. The results are shown in Fig. 5. The equation for the average distance ration in this case is as below :

Av. distance ratio =

$[\alpha \times d + (1 - \alpha) \times (D + 2d)] / [\log_2(2^D \times (2^d - 1))]$

A deficiency with the above model is that it does not take into account the conflicts arising when more than one processor tries to use the hierarchical link. We model this as follows. We determine the number of hierarchical links based on the locality. This increases the size of the MIN at each level as the number of processors remains the same. Thus, the delay associated with each level is also increased. The increase in delay is a good measure of the actual increase in delay due to conflicts because it is determined by the number of extra hierarchical links required which itself depends on the number of conflicts. The results using this model are shown in Fig. 6. The avearage distance ratio is given by :

Av. distance ratio =

$[\alpha \times \log_2((2^d - 1)(1 - \alpha) + 2^d) + (1 - \alpha) \times (\log_2(2^{D+d}(1 - \alpha) + 2^D) + 2 \times \log_2((2^d - 1)(1 - \alpha) + 2^d))] / [\log_2(2^D \times (2^d - 1))]$

In both the above models, we have shown the effect of changing the structure of the HMIN on the average delay. However, we have not accounted for the fact that the *locality* is also dependent on the number of processors in the *local* network. There is an optimal number of processors required at the lowest level of hierarchy for the HMIN to perform most effectively. This implies that the performance will depend on the application to some extent. Thus, though the HMIN with only 4 processors in the local MIN and a higher level MIN of size 2^{14} seems to have the best average delay, it may not be the best structure. A more ideal structure would be the use of at least 16 processors at the local MIN and have a hierarchical MIN of size 2^{12}. The number 16 is chosen because it is large enough for a task to be scheduled on it and it is not too large for the effects of conflicts in the local MIN to be severe.

Fig. 7 shows the performance-to-cost ratio of the different structures.

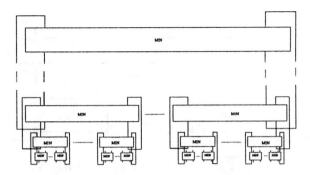

Figure 1: Generic structure of HMINs

4 Conclusions

In this work , we proposed a hierarchically structured MIN and presented its routing algorithm. We demonstrated that the HMIN is a cost-effective solution to the problem of interconnecting processors in a large multiprocessor environment by using some analytical models.

References

[1] J. Ding and L.N. Bhuyan,"Performance evaluation of multistage interconnection networks with finite buffers," 1991 Intl. Conf. on Parallel Processing, pp. 592-599, Aug. 1991.

[2] W.T. Chen and J.P. Sheu,"Performance analysis of multistage interconnection networks with hierarchical requesting models," IEEE Trans. Computers, Vol. C-37, No. 11, pp. 1438-1442, Nov. 1988.

[3] L.N. Bhuyan and D.P. Agrawal,"Design and performance of generalized interconnection networks," IEEE Trans. Computers, Vol. C-32, No. 12, pp. 1081-1090, Dec. 1983.

[4] S.P. Dandamudi and D.L. Eager,"Hierarchical interconnection networks for multicomputer systems," IEEE Trans. Computers, Vol. 39, No. 6, pp. 786-797, Jun. 1990.

[5] D.P. Agrawal and I.O. Mahgoub," Analysis of a class of cluster based multiprocessor systems," Information Sciences, Fall 1987, pp. 85-105.

[6] A. Kumar, D.P. Agrawal, I.O. Mahgoub, and, C.R. Green,"Impact of network failure on the performance degradation of a class of cluster based multiprocessors," IEEE Trans. Relaibility, pp. 39-44, April 1991.

[7] C. Chen, D.P. Agrawal, and, J.R. Burke," dbCube: A new class of hierarchical interconnection networks with area efficient layout," IEEE Trans. Parallel and Distributed Systems (to appear).

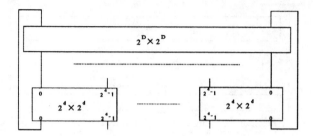

Figure 2: Structure of a two-level HMIN

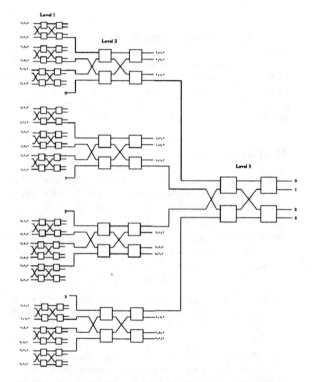

Figure 3: Detailed diagram of a 3-level HMIN

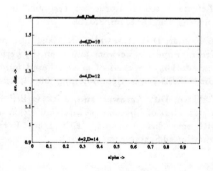

Figure 4: Average distances assuming uniform traffic

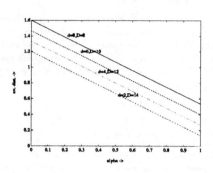

Figure 5: Average distances assuming locality in traffic

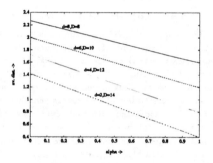

Figure 6: Average distances with locality in traffic (adjusting for conflicts)

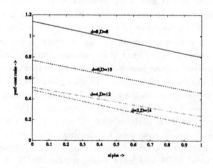

Figure 7: Performance-to-cost ratio for various structures

TIGHTLY CONNECTED HIERARCHICAL INTERCONNECTION NETWORKS FOR PARALLEL PROCESSORS

Peter Thomas Breznay   Mario Alberto Lopez

Department of Mathematics and Computer Science
University of Denver, Denver CO 80208.

Abstract – *A method for constructing hierarchical interconnection networks is presented. The method is based on connecting isomorphic clusters using a complete graph as the higher level network. Applying it to various classes of graphs, including hypercubes and meshes, results in networks with optimal connectivity, high bisection width, low degree, diameter and cost. With hypercube clusters, the degree, diameter and cost are approximately $\frac{1}{2}$, $\frac{1}{2}$ and $\frac{1}{4}$ of the same parameters in a comparable size hypercube. With mesh clusters, the performance parameters are polynomially better than those in a similar size mesh.*

1. Introduction

In the past two decades researchers have proposed numerous variations to the widely studied hypercube architecture aiming at better performance parameters (for example see [1, 5, 6, 7]). Recently an interest has emerged in hierarchical interconnection networks (HINs) proposed and studied by Dandamudi and Eager [3, 4], among others. We define a class of HINs which use a complete graph to connect lower level isomorphic clusters. In the singly linked version one link connects each pair of clusters; in the doubly linked network two links are used. We call the resulting structure *tightly connected network* or *TCN*. TCNs result in substantial improvements in various performance parameters, when compared to clusters of the same size. For example, a hypercube based TCN of size N has diameter $\frac{1}{2}\log N + 1$ and degree $\frac{1}{2}\log N + 2$ as opposed to $\log N$ for both degree and diameter in a hypercube of the same size. Similar improvements are obtained for other cluster structures, including meshes and cycles.

The paper is organized as follows. In Section 2 we define the singly and doubly linked TCN networks. In Section 3 we introduce the class of diagonal graphs and show that this class includes several known topologies, such as hypercubes, meshes, and cycles. In Section 4 we discuss the properties of doubly linked TCNs based on diagonal clusters, as well as routing and broadcasting algorithms. Singly linked TCNs are studied in more depth in [2]. Section 5 concludes.

2. Tightly Connected HINs

Let $G(V_G, E_G)$ be a connected simple graph with n nodes and m edges. Let $\deg(G)$ denote the maximum degree in V_G. In this paper we mostly consider regular graphs, in which case $\deg(G)$ is the degree of any node. Let the nodes of G be labeled $1, \ldots, n$. In the following all algebraic operations are performed (mod $n + 1$).

Definition 1 *Let ϕ and ψ be two permutations of the first n natural numbers. We say that ϕ and ψ are compatible if*
(i) ψ is idempotent, that is $\psi^2(x) = x$;
(ii) for any $x \in \{1, 2, \ldots, n\}$

$$\phi(x) + \phi(\psi(x)) = 0. \tag{1}$$

Note that the addition in (1) is taken (mod $n + 1$). We interpret ϕ and ψ as permutations of the *node set* V_G of G. Two properties of compatible permutations, which we use later, are discussed next.

Proposition 1 *Let ϕ and ψ be compatible permutations. If n is odd, ψ has exactly one fixed point. If n is even, ψ has no fixed point.*

Proof: Suppose $\psi(x) = x$. Then by (1), $\phi(x) + \phi(\psi(x)) = 2 \cdot \phi(x) = n + 1$. Therefore, if n is odd then the fixed point of ψ is $\phi^{-1}(\frac{n+1}{2})$. If n is even, $2 \cdot \phi(x) = n + 1$ is a contradiction. $\square$

In the following we assume that n is even. Thus ψ will be fixed-point free.

Proposition 2 *If ϕ and ψ are compatible permutations then $\phi \circ \psi$ and ψ are also compatible.*

Proof: Let $x \in \{1, \ldots, n\}$ and $y = \psi(x)$. By (1) $\phi(y) + \phi(\psi(y)) = 0$ but this means that $\phi(\psi(x)) + \phi(\psi^2(x)) = \phi \circ \psi(x) + \phi \circ \psi(\psi(x)) = 0$. $\square$

To establish the existence of compatible functions ϕ and ψ, we note that a multitude of fixed-point free idempotent permutations exist for a given n, because every *matching* of $1, \ldots, n$ generates such a permutation. Given such ψ, it can be shown that there are $(n/2)! \cdot 2^{n/2}$ permutations satisfying (1), see [2].

We now define the TCN structures.

Definition 2 *Let G be a simple connected graph, and let ϕ and ψ be compatible permutations of V_G. Consider $n + 1$ copies of G denoted by $G_0, \ldots, G_n$. For any $a \in \{0, \ldots, n\}$ and $x \in \{1, \ldots, n\}$ let (a, x) denote node x of graph G_a. We define the singly linked tightly connected network, $1\mathrm{TCN}(G, \phi, \psi)$, as the graph having nodes $\{(a, x) : 0 \leq a \leq n, 1 \leq x \leq n\}$ and edges defined as follows: (a, x) and (b, y) are connected if and only if*
(i) $a = b$ and (x, y) is an edge of G, or
(ii) $b = a + \phi(x)$ and $y = \psi(x)$.

The doubly linked tightly connected network, $2\mathrm{TCN}(G, \phi, \psi)$, is defined as the graph having the same links as $1TCN(G, \phi, \psi)$ with the additional links
(iii) $b = a + \phi(\psi(x))$ and $y = \psi(x)$.

Edges of type *(i)* are *internal* or *cluster links*; those of types *(ii)* and *(iii)* are *external* or *metalinks*. The metalinks of type (ii) are *simple*; those of type *(iii)* are *composed*. If ϕ and ψ are understood, we denote $1TCN(G, \phi, \psi)$ and $2TCN(G, \phi, \psi)$ simply by $1TCN(G)$ and $2TCN(G)$, respectively. Also, ϕ is the *cluster offset function*: node x of cluster G_a is connected to a node $\phi(x)$ clusters "ahead", i.e. to a node of $G_{a+\phi(x)}$. Similarly, ψ is the *node interface function*, which specifies which node of $G_{a+\phi(x)}$ node (a, x) is connected to, namely $\psi(x)$. The composed metalinks of $2TCN(G)$ follow the same pattern, except that the cluster offset function is $\phi \circ \psi$.

Compatibility condition (1) guarantees that each node of a TCN is incident to exactly one simple, and in case of a 2TCN, also one composed metalink. Thus the degree of each node in a 1TCN (resp. 2TCN) is one (resp. two) higher than the degree of the nodes of the cluster G. Illustrations of 2-dimensional hypercube-based 1TCN and 2TCN are given in Figures 1 and 2, respectively. In order to follow the notation of Definition 2, an n-dimensional cube has nodes labeled 1 through 2^n. Thus, ϕ and ψ can be handled as permutations of V_G. In the rest of the paper H_n and C_r^n denote the n-dimensional hypercube and wraparound mesh of radix r, respectively.

3. Diagonal Graphs

We introduce a class of graphs called *diagonal graphs* which includes hypercubes and even-radix meshes. Diagonal clusters enable us to to construct low cost TCNs.

Definition 3 *Let G be a simple connected graph with diameter d. G has the* diagonal node property *if for each $x \in V_G$ there is a unique $\overline{x} \in V_G$ such that $dist(x, \overline{x}) = d$.*

Clearly, if G has the diagonal property then $\overline{\overline{x}} = x$, i.e., $x \mapsto \overline{x}$ is an idempotent operator.

Definition 4 *Let G be a graph with diameter d that has the diagonal node property. We say that G is* diagonal *if for each $x, y \in V_G$, $d(x, y) + d(\overline{x}, y) = d$.*

Theorem 1 *(i) The direct product of finitely many diagonal graphs is diagonal.*
(ii) The direct product of finitely many even size cycles is diagonal.

Proof: See [2]. $\square$

Corollary 1 *Hypercubes, even radix meshes and multicubes are diagonal graphs.*

A *diagonal* TCN of G, denoted by $TCN(G, \phi)$, is defined by the node interface function $\psi(x) = \overline{x}$, and a compatible cluster offset function ϕ. In this case the simple and composed metalinks are

$$(a, x) \longleftrightarrow (a + \phi(x), \overline{x}) \qquad (2)$$

$$(a, x) \longleftrightarrow (a + \phi(\overline{x}), \overline{x}). \qquad (3)$$

respectively.
Additionally, the *standard diagonal* $TCN(G)$ is defined by using $\phi(x) = x$. Figure 2 illustrates $2TCN(H_2)$.

4. Doubly Linked TCNs

We present a routing algorithm and investigate various topological properties of 2TCNs, including diameter, cost, average distance, connectivity, cluster symmetry, bisection width, and recursivity. We also discuss broadcasting and spanning trees.

Suppose G has n vertices and m edges.

Theorem 2 *(i) The number of vertices of $2TCN(G, \phi, \psi)$ is $N = n^2 + n$*
(ii) The number of edges of $2TCN(G, \phi, \psi)$ is $M = m(n + 1) + 2\binom{n+1}{2}$
(iii) $deg(2TCN(G, \phi, \psi), (a, x)) = deg(G, x) + 2$

Proof: See [2]. $\square$

Corollary 2 *If G is a regular graph of degree k then $2TCN(G, \phi, \psi)$ is a regular graph with degree $k + 2$.*

The routing algorithm of $2TCN(G, \phi)$ takes advantage of the diagonality of G. We always use one of the metalinks from a cluster to another, choosing the one that gives a shorter distance from the source to the destination node. To route from (a, x) to (b, y) we first find the simple and composed interface nodes $z = \phi^{-1}(b - a)$ and $\overline{z}$ from G_a to G_b. The corresponding interface nodes in G_b are $\overline{z}$ for the simple, and z for the composed metalink. We compare d_{simple} with d_{comp}, where

$$d_{simple} = dist(x, z) + dist(y, \overline{z})$$

$$d_{comp} = dist(x, \overline{z}) + dist(y, z).$$

If $d_{simple} < d_{comp}$ we use the simple, if $d_{comp} < d_{simple}$ we use the composed metalink. In case of a draw, we can use either. Note that the draw occurs if and only if $d_{simple} = d_{comp} = diam(G)$. We will refer to this algorithm as the *diagonal routing* of $2TCN(G, \phi)$.

We now show that with diagonal routing the distance of any two vertices of $2TCN(G, \phi)$ is no more than $diam(G) + 1$. This means that while a 2TCN more than squares the number of nodes, it increases the degree by only two, and the diameter by only one.

Theorem 3 *If G is diagonal, $\psi(x) = \overline{x}$ and ϕ is compatible with ψ then the diameter of $2TCN(G, \phi)$ with diagonal routing is $diam(G) + 1$.*

Proof: Consider the nodes (a, x) and (b, y). Between the clusters G_a and G_b there are two metalinks: a simple metalink

$$(a, z) \longleftrightarrow (b, \overline{z}) \qquad (4)$$

by (2) and a composed metalink

$$(a, \overline{z}) \longleftrightarrow (b, z) \qquad (5)$$

by (3) where (a, z) is the simple interface node between G_a and G_b, that is $z = \phi^{-1}(b - a)$. By the diagonality of G, there is a path of length $d = diam(G)$ from (a, z) to $(a, \overline{z})$ through (a, x) and similarly, a path of length d from (b, z) to $(b, \overline{z})$ through (b, y). Thus, denoting

$$l_1 = dist(x, z), \qquad l_2 = dist(x, \overline{z})$$

$$m_1 = dist(y, \overline{z}), \qquad m_2 = dist(y, z)$$

we get $l_1 + l_2 = m_1 + m_2 = d$.
We now construct two paths from (a, x) to (b, y): path 1 goes from (a, x) to (a, z) within G_a, followed by metalink (4), and then by the path from $(b, \overline{z})$ to (b, y) within G_b.

Its length is $l_1 + m_1 + 1$. Similarly, path 2 has length $l_2 + m_2 + 1$ and goes from (a, x) to $(a, \overline{z})$ within G_a, followed by metalink (5), and then by the path from (b, z) to (b, y) within G_b. There are two cases. If $l_1 + m_1 \leq d$ then the length of path 1 is at most $d + 1$. Otherwise, $l_1 + m_1 > d$, and the length of path 2 is $l_2 + m_2 = (d - l_1) + (d - m_1) = 2d - (l_1 + m_1) < d$. In either case, $dist((a, x), (b, y)) \leq d + 1 = diam(G) + 1$, as claimed. $\square$
The following corollaries refer to diagonal routing.

Corollary 3 *(i)* $diam(2TCN(H_n, \phi)) = n + 1$
(ii) The cost of $2TCN(H_n, \phi)$ *is* $n^2 + 3n + 2$.

Corollary 4 *(i)* $diam(2TCN(C_{2n}^2, \phi)) = 2n + 1$
(ii) The cost of $2TCN(C_{2n}^2, \phi)$ *is* $12n + 6$.

Corollary 3 implies an improvement by a factor of 2 for the diameter, and 4 for the cost, as compared to a similar size hypercube. Corollary 4 represents improvements by factors of $2n$ and $\frac{4}{3}n$ for the diameter and cost of the mesh, respectively. Plots of the above results are shown in Figures 3, 4, 5 and 6.

Theorem 4 *The average distance* $2TCN(H_n, \phi)$ *with respect to diagonal routing is*
$$\overline{d} = \frac{S}{2^{2n} + 2^n - 1}$$
where
$$S = n2^{n-1} + \frac{(n+1)(2n)!}{(n!)^2} + 2\sum_{k=0}^{n}\sum_{r=0}^{k-1} k\binom{n}{r}\binom{n}{k-r-1}$$
is the total distance from a given point in $2TCN(H_n, \phi)$.
Proof: See [2]. $\square$

Theorem 5 $2TCN(G, \phi, \psi)$ *is cluster symmetric.*
Proof: See [2]. $\square$

We call a permutation ψ of V_G *fixed-edge free* if for any edge (x, y) of G, $(x, y) \neq (\psi(x), \psi(y))$ holds. One such permutation is $\psi(x) = \overline{x}$. Also, G is *optimally connected* if its connectivity is $\min_{x \in V_G} \deg(x)$.

Theorem 6 *If* G *is connected and* ψ *is fixed-edge free then* $2TCN(G, \phi, \psi)$ *is optimally connected.*
Proof: See [2]. $\square$

Corollary 5 $2TCN(H_n, \phi)$ *and* $2TCN(C_{2m} \times C_{2n}, \phi)$ *are optimally connected with connectivities* $n + 2$ *and* 6, *respectively.*

Theorem 7 *The bisection width of* $2TCN(H_n, \phi, \psi)$ *is* $\frac{N}{2} = 2^{2n-1} + 2^{n-1}$.
Proof: See [2]. $\square$
Now we prove a modularity property of hypercube-based 2TCNs. First we define an extension of the offset and a modified extension of the interface function.

Definition 5 *Let* ϕ *and* ψ *be compatible permutations of* $\{1, \ldots, 2^n\}$. *Define permutations* $\widetilde{\phi}$ *and* $\widetilde{\psi}$ *of* $\{1, \ldots, 2^{n+1}\}$ *as follows:*
$$\widetilde{\phi}(x) = \begin{cases} \phi(x) & \text{if } 1 \leq x \leq 2^n; \\ \phi(x - 2^n) + 2^n & \text{if } 2^n + 1 \leq x \leq 2^{n+1} \end{cases}$$
and
$$\widetilde{\psi}(x) = \begin{cases} \psi(x) + 2^n & \text{if } 1 \leq x \leq 2^n; \\ \psi(x - 2^n) & \text{if } 2^n + 1 \leq x \leq 2^{n+1}. \end{cases}$$

Definition 6 *Let* $E_n = 2TCN(H_n, \phi, \psi)$. *The* $(n+1)$-*dimensional extension of* E_n *is the* $2TCN$ $E_{n+1} = 2TCN(H_{n+1}, \widetilde{\phi}, \widetilde{\psi})$.

Theorem 8 *Let* $E_n = 2TCN(H_n, \phi, \psi)$ *and* $E_{n+1} = 2TCN(H_{n+1}, \widetilde{\phi}, \widetilde{\psi})$ *be given. In* E_{n+1}, $2^{n+1} + 1$ *copies of* E_n *can be constructed out of nodes and edges available in* E_{n+1} *such that each metalink of* E_n *has dilation 2.*

Proof: See [2]. $\square$
Finally, we consider the problem of broadcasting from a node (a, x). First we build a spanning tree T_a of G_a rooted at (a, x). At each node (a, y) of T_a we add both metalinks to T_a. Let $dist((a, x), (a, y)) = k$. We build subtrees of depth $diam(G) - k$ of the spanning trees of $G_{a+\phi(y)}$ and $G_{a+\phi\psi(y)}$ rooted at $(a + \phi(y), \psi(y))$ and $(a + \phi\psi(y), \psi(y))$, respectively. The rest of the nodes in each cluster are added to the tree when we process the other metalink connecting the cluster to G_a. By Theorem 3 the depth of the spanning tree is $diam(G) + 1$. As usual, broadcasting can be done using this tree.

5. Conclusions

We have presented a method for constructing hierarchical interconnection networks using a fully connected secondary network to connect isomorphic copies of a primary network. Each pair of clusters can be connected using only one link (1TCN) or two links (2TCN). When the clusters satisfy a diagonal property, substantial improvements in performance can be achieved, as compared to a cluster of similar size. This diagonal property is satisfied by many popular networks, including hypercubes, meshes, and even cycles. TCN structures are optimally connected and have very high bisection widths. Since the routing algorithms are not known to be optimal, further improvements can be expected with better routing algorithms.

References

[1] L. Bhuyan and D. Agrawal. Generalized hypercube and hyperbus structures for a computer network. *IEEE Trans. Comp.*, 33(4):323–333, 1984.

[2] P. Breznay and M.A. Lopez. A new class of hierarchical interconnection networks. Technical report, Dept. of Mathematics and Computer Science, University of Denver, 1993.

[3] S. Dandamudi and D. Eager. Hierarchical interconnection networks for multicomputer systems. *IEEE Trans. Comp.*, C-39(6):786–797, 1990.

[4] S. Dandamudi and D. Eager. On hypercube-based hierarchical interconnection network design. *J. Par. and Dist. Comp.*, 12:283–289, 1991.

[5] J. R. Goodman and C. H. Sequin. Hypertree: A multiprocessor interconnection topology. *IEEE Transactions on Computers*, 30(12), 1981.

[6] S. Latifi and A. El-Amawy. On folded hypercubes. In *Proc. Int'l Conf. on Par. Proc.*, 1989.

[7] F.P. Preparata and J. Vuillemin. The cube-connected cycles: A versatile network for parallel computation. *Comm. of the ACM*, 24(5):300–309, 1981.

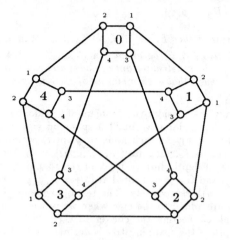

Figure 1: Two dimensional hypercube-based 1TCN

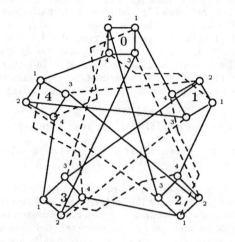

Figure 2: Two dimensional hypercube-based 2TCN. Composed metalinks are shown dashed.

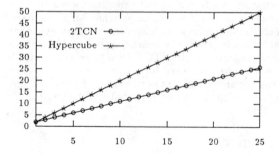

Figure 3: Diameter of 2TCN and hypercube as a function of cluster dimension.

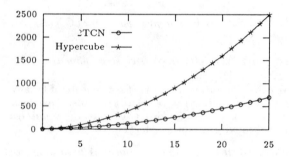

Figure 4: Cost of 2TCN and hypercube as a function of cluster dimension.

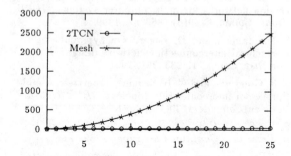

Figure 5: Diameter of 2TCN and mesh as a function of cluster dimension.

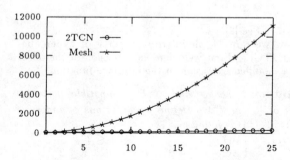

Figure 6: Cost of 2TCN and mesh as a function of cluster dimension.

The Folded Petersen Network : A New Communication–Efficient Multiprocessor Topology

Sabine Öhring
Department of Computer Science
University of Würzburg
8700 Würzburg, Germany

Sajal K. Das*
Department of Computer Science
University of North Texas
Denton, TX 76203-3886, USA

Abstract

We introduce and analyze a new interconnection topology, called the n–dimensional *folded Petersen network* (FP_n), which is constructed by iteratively applying the cartesian product operation on the well–known Petersen graph itself. The FP_n topology provides regularity, node– and edge-symmetry, optimal connectivity (maximal fault–tolerance), logarithmic diameter, and permits self–routing and broadcasting algorithms even in the presence of faults. With the same node–degree and connectivity, FP_n has smaller diameter and more nodes than the $3n$–dimensional binary hypercube and several other networks. This paper also embeds rings, meshes, hypercubes, and a variety of tree–related structures efficiently into FP_n.

Keywords: broadcasting, embedding, fault–tolerance, interconnection network, Petersen graph, routing.

1 Introduction

Existing static interconnection networks include complete binary trees, X-trees, meshes, hypercubes, cube–connected cycles, de Bruijn networks, Stirling networks, recursive combinatorial networks, and so on [DGD92, DM93, Lei92].

An interconnection network topology can be modeled by an undirected graph $G = (V, E)$, where the node–set V represents the processors and the edge–set E represents the communication links among them. The *distance*, $dist_G(u, v)$, between two nodes in G is the length of the shortest path between them. The *diameter* (d) is the maximum distance among all node–pairs. The *cost* of a regular network of degree δ is $C = d \cdot \delta$. The *node-connectivity* (κ) is the number of nodes whose removal results in a disconnected network. The *f–fault diameter* is the worst case diameter by removing f nodes.

The *Petersen graph*, P, with ten nodes has an outer 5-cycle, an inner 5-cycle and five spokes joining them. Fig. 1a) depicts P with decimal node–labeling. It is a 3–regular graph with $d = 2$ and $\kappa = 3$ [CW85].

An n-dimensional *binary hypercube*, Q_n, consists of the node–set $\mathbf{Z}_2^n = \{x_{n-1} \ldots x_0 | x_i = 0 \text{ or } 1\}$ and there is an edge between two nodes iff their binary labels differ in exactly one bit. A *generalized hypercube*, GQ_b^n, of base b and dimension

*This work is supported by TARP grant 003594003. E-mail : oehring@informatik.uni-wuerzburg.de and das@cs.unt.edu

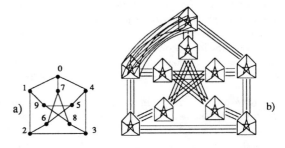

Figure 1: a) Petersen graph   b) Scheme of FP_2

n with nodes in $\mathbf{Z}_b^n$ is defined analogously [BA84]. The *de Bruijn graph* $DG(b, n)$ of base b and order n has node-set $V_{b,n} = \{x_{n-1} \ldots x_0 | x_i \in \mathbf{Z}_b, 0 \le i \le n - 1\}$ and edge-set $E_{b,n} = \{\{x_{n-1} \ldots x_0, x_{n-2} \ldots x_0 p\} \mid p, x_i \in \mathbf{Z}_b, 0 \le i \le n - 1\}$.

The *cartesian product* $G = G_1 \times G_2$ of two graphs $G_1 = (V_1, E_1)$ and $G_2 = (V_2, E_2)$ has the node–set $V = V_1 \times V_2$ and an edge $\{(u, x), (v, y)\} \in E$ iff either $u = v$ and $\{x, y\} \in E_2$, or $x = y$ and $\{u, v\} \in E_1$. This definition can be generalized to a product of n graphs.

In this paper, we propose a new interconnection topology, called the *folded Petersen network*. It is constructed from an iterative product on the Petersen graph, P.

2 Topological Properties and Performance

The n-dimensional folded Petersen graph is defined as $FP_n = P \times \ldots \times P = (V_P, E_P)$, where $V_P = \{(x_1, \ldots, x_n) \mid x_i \in \{0, 1, \ldots, 9\}, 1 \le i \le n\}$ and $E_P = \{\{(x_1, \ldots, x_n), (x_1, \ldots, x_{i-1}, y, x_{i+1}, \ldots, x_n)\}\} \mid y \in \{0, 1, \ldots, 9\}$ and $dist_P(y, x_i) = 1, 1 \le i \le n\}$. As depicted in Fig. 1b), FP_2 is obtained from P by replacing each of its nodes by P. As product of the node– and edge-symmetric Petersen graph [CW85], FP_n is also node– and edge-symmetric. Furthermore, FP_n is a regular graph with degree $\delta = 3n$, diameter $d = 2n$, and node-connectivity $\kappa = 3n$. It has 10^n nodes and cost $C = 6n^2$. As shown in Table 1, the folded Petersen network compares favorably with several others e.g. (generalized) hypercube, de Bruijn, Hyper Petersen $HP_n = Q_{n-3} \times P$, and hyper de Bruijn $HD(m, n) = Q_m \times DG(2, n)$ networks in terms of degree, diameter, connectivity or cost.

Network	Nodes	δ	d	κ	Cost
FP_n	10^n	$3n$	$2n$	$3n$	$6n^2$
Q_{3n}	8^n	$3n$	$3n$	$3n$	$9n^2$
GQ_{10}^n	10^n	$9n$	n	$9n$	$9n^2$
GQ_3^{2n}	9^n	$4n$	$2n$	$4n$	$8n^2$
HP_{3n}	$(5/4)\cdot 8^n$	$3n$	$3n-1$	$3n$	$9n^2-3n$
$DG(10,n)$	10^n	20	n	18	$20n$
$HD(2n,n)$	8^n	$2n+4$	$3n$	$2n+2$	$6n^2+12n$

Table 1: Comparison of topological properties

Multiple disjoint paths between two nodes provide a measure of fault–tolerance of a network. For two non–adjacent nodes of the Petersen graph P, there exist a path of length 2 and two node–disjoint paths of length 3, if there are no faulty nodes. Otherwise, a path of length 1 and two paths of length 4 exist.

The following theorem implies that the node–connectivity of FP_n is $3n$ and hence it is maximally fault–tolerant. Also the $(3n-1)$–fault diameter of this network is $2n+3$.

Theorem 1 : *Between two nodes x and y in a non–faulty FP_n, there exist $3n$ node–disjoint paths of length $\leq dist_{FP_n}(x,y)+3$. If there are at most $3n-1$ faulty nodes, then there exists at least one path of length $\leq 2n+3$.*

Proof : Let two nodes $x=(x_1,\ldots,x_n)$ and $y=(y_1,\ldots,y_n)$ in FP_n have identical values in k positions, denoted by λ_μ for $1\leq\mu\leq k$; and let i_j for $1\leq j\leq n-k$, be the positions in which they differ. Now in the Petersen graph P, suppose $(x_{i_j},x'_{i_j},y_{i_j})$ denotes the unique shortest path of length ≤ 2 between nodes x_{i_j} and y_{i_j}, and $x^l_{\lambda_\mu}$ denotes the lth, $1\leq l\leq 3$, neighbor of x_{λ_μ}. We construct the following node–disjoint paths between x and y. Fig. 2a) gives an example.

1. $n-k$ shortest paths $\mathcal{P}^1_m$, $1\leq m\leq n-k$, of length $dist_{FP_n(x,y)}$ by equalizing all differing elements in x and y in cyclic order $i_m,i_{m+1},\ldots,i_n,i_1,\ldots,i_{m-1}$.

2. $3k$ paths $\mathcal{P}^2_i$, $1\leq i\leq 3k$, of length $dist_{FP_n}(x,y)+2$: $\mathcal{P}^2_i=(x,(\ldots,x^l_{\lambda_\mu},\ldots),$ elements in i_j are equalized as in $\mathcal{P}^1_1,(\ldots,x^l_{\lambda_\mu},\ldots,y_{i_1},\ldots,y_{i_{n-k}},\ldots),(\ldots,x_{\lambda_\mu},\ldots),y)$, for $1\leq\mu\leq k$ and $l=1,2,3$.

3. $n-k$ paths $\mathcal{P}^3_j$, $1\leq j\leq n-k$, of length $dist_{FP_n}(x,y)+\gamma_{i_j}$, where $\gamma_{i_j}=3$ if $dist_P(x_{i_j},y_{i_j})=1$, otherwise $\gamma_{i_j}=1$ if $dist_P(x_{i_j},y_{i_j})=2$. Let $(x_{i_j},x^2_{i_j},\ldots,y_{i_j})$ denote the non–shortest path $p^1_{i_j}$ in P between x_{i_j} and y_{i_j} that routes over the neighbor $x^2_{i_j}$ of x_{i_j}. Then $\mathcal{P}^3_j=((x_1,\ldots,x_n),(\ldots,x^2_{i_j},\ldots),$ elements in $l\neq i_j$ are switched as in $\mathcal{P}^1_j,(\ldots,x^2_{i_j},\ldots,y_{i_s},\ldots)$ for $s\neq j$, switching i_jth elements as in path $p^2_{i_j},(\ldots,y_{i_1},\ldots,y_{i_{n-k}},\ldots))$.

4. $n-k$ paths $\mathcal{P}^4_j$, $1\leq j\leq n-k$, of length $dist_{FP_n}(x,y)+\gamma_{i_j}$. If $(x_{i_j},x^3_{i_j},\ldots,y_{i_j})$ denotes the non–shortest path $p^3_{i_j}$ in P between x_{i_j} and y_{i_j} that routes over the neighbor $x^3_{i_j}$ of x_{i_j}. Then $\mathcal{P}^4_j=((x_1,\ldots,x_n),(\ldots,x^3_{i_j},\ldots),$ elements in $l\neq i_j$ are switched as in $\mathcal{P}^1_j,(\ldots,x^3_{i_j},\ldots y_{i_s},\ldots)$ for $s\neq j$, switching i_jth element as in path $p^3_{i_j},(\ldots,y_{i_1},\ldots,y_{i_{n-k}},\ldots))$.

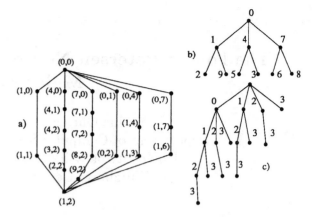

Figure 2: a) Disjoint paths in FP_2, Broadcasting b) in P, c) in Q_n

The only way to remove $3n$ nodes that dissconnects the FP_n network is to remove all neighbors of a particular node. Let E_{FP_n} and $E_{Q_{3n}}$ respectively denote the events that all $3n$ neighbors of a node in FP_n and Q_{3n} are faulty. The probabilities of such events are given by $P(E_{FP_n})=\frac{10^n}{\binom{10^n}{3n}}$ and $P(E_{Q_{3n}})=\frac{8^n}{\binom{8^n}{3n}}$ and therefore, $\frac{P(E_{FP_n})}{P(E_{Q_{3n}})}\leq(\frac{4}{5})^{3n^2-n}$. It implies that the probability of an FP_n network being disconnected is significantly less than that of Q_{3n}.

Fault–tolerance capabilities can be improved with the existence of a wide *container*, providing a large number of node–disjoint paths between any two nodes [EAL91]. The length of a container is the longest path length in it. The *container quality* between two nodes at distance r is given by $CQ(r)=\frac{K_1}{K_2}$, where K_1 is the number of node–disjoint paths and K_2 is the average length of all disjoint paths between those two nodes. A large value of $CQ(r)$ means a short container that includes many paths. Let $u=(u_1,\ldots,u_n)$ and $v=(v_1,\ldots,v_n)$ be two arbitrary nodes in FP_n with $dist_{FP_n}(u,v)=r$ and let i_j, $1\leq j\leq n-k$, be the positions in which they differ. Let α_1 and α_2 be the number of positions i_j with $dist_P(u_{i_j},v_{i_j})=1$ and $dist_P(u_{i_j},v_{i_j})=2$, respectively. Thus, $dist_{FP_n}(u,v)=\alpha_1+2\alpha_2=r$ and $\alpha_1+\alpha_2=n-k$. By Theorem 1, we derive $CQ_{FP_n}(r)\geq\frac{9n^2}{3nr-2n-4r+8k}$, and thus $\frac{9n^2}{3nr+6n-8r}\leq CQ_{FP_n}\leq\frac{9n^2}{3nr+6n-12r}$. This is an improvement over the hypercube Q_{3n}, since the term $\frac{CQ_{FP_n}-CQ_{Q_{3n}}}{CQ_{Q_{3n}}}\geq\frac{6r}{3nr+6n-2r}$ monotonically increases with r.

3 Routing and Broadcasting

Let $x=(x_1,\ldots,x_n)$ and $y=(y_1,\ldots,y_n)$ be the source and destination nodes in FP_n, respectively, such that $x_{i_j}\neq y_{i_j}$ for $1\leq j\leq n-k$ and $i_j\in\{1,\ldots,n\}$. A (shortest path) self–routing algorithm in a non–faulty FP_n is obtained by equalizing x_{i_j} and y_{i_j} for $1\leq j\leq n-k$ using self–routing in the Petersen graph P (see Fig. 2b)), which is completely defined by the routing from node 0 to all others. The following algorithm uses only local knowledge and tolerates upto $3n-1$ node failures in FP_n.

Routing–Algorithm

Begin $node_{int}:=$ source; $I := \{i_1, \ldots, i_{n-k}\}$;

While $node_{int} \neq$ destination **do**

Search the first index $i_l \in I$ such that there exists a path between $node_{int}$ and $node'_{int}$ that results from $node_{int}$ by replacing the i_l–th element by y_{i_l}.

If there exists no such path **then**

request from three neighbors of each element x_λ of $node_{int}$, $\lambda \neq i_j$, $1 \leq j \leq n-k$, such that there exists an $i_l \in I$ and a path from this neighbor to $node'_{int}$.

$node_{int} := node'_{int}$;

$I := I - \{i_l\}$. **End**

It can be shown that the path routed by this algorithm has a length $\leq 3(n-k) + 2 \cdot \min\{n-k, i'\}$, where i' is the largest value of i such that $\frac{3}{2} \cdot (i^2 + i) \leq 3n - 1$ [ÖD93b].

Next, we develop a broadcasting algorithm in FP_n. Because of the symmetry, let us restrict $(0, \ldots, 0)$ as the source node. The broadcasting is done by combining the schemes for binary hypercube Q_n and the Petersen graph (Fig. 2). The labels on the edges $\{x, y\}$ in the broadcast spanning tree in Q_n give the bit positions in which the nodes x and y differ.

Starting at node $(0, \ldots, 0)$ in FP_n, we go in each position to its three neighbors 1, 4 and 7 in P. Subsequently, from each node x at the current level of the broadcasting tree we go to all of its neighbors in P in the current dimension (due to the broadcasting in P) and to all the neighbors in the next dimension (due to the broadcasting in Q_n). An example is drawn in Fig. 3, where P^j represents the broadcasting–tree of Fig. 2b) with labels $(j, x), 0 \leq x \leq 9$.

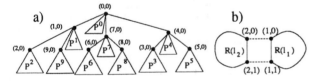

Figure 3: a) Broadcasting in FP_2 b) Join $R(l_1)$ and $R(l_2)$

Let us now calculate the average distance in FP_n.

Theorem 2 : *The number Z_i^n of nodes at distance i from any node in FP_n is $\sum_{j=0}^{\lfloor \frac{i}{2} \rfloor} 3^{i-j} 2^j \binom{i-j}{j} \binom{n}{i-j}$ for $0 \leq i \leq 2n$.*

Proof : FP_n is node–symmetric. Thus, w.l.o.g., let us compute the number of nodes Z_i^n at distance i, where $1 \leq i \leq 2n$, from the node $(0, \ldots, 0)$. Following the broadcasting scheme, the neighbors of each node x in FP_n are selected as follows. At level 1 of the broadcast–tree in P (Fig. 2b)) in the current dimension, either two neighbors of x in P or three neighbors in the next dimension are visited. At levels 0 and 2, we only go to the three neighbors of P in the next dimension. This implies : $Z_1^n = 3n$, $Z_2^n = 3 \cdot 2\binom{n}{1} + 3 \cdot 3 \cdot \binom{n}{2}$, and in general $Z_i^n = \sum_{\lceil j/2 \rceil}^{i} 3^j \cdot 2^{i-j} \cdot \binom{j}{i-j} \sum_{\lambda_{j-2}}^{n-2} \cdots \sum_{\lambda_1}^{n-2} \binom{n-\lambda_1}{2} = \sum_{j=0}^{\lfloor \frac{i}{2} \rfloor} 3^{i-j} 2^j \binom{i-j}{j} \binom{n}{i-j}$. $\square$

It turns out that the average distance $(\bar{d}_n)$ of FP_n, given by summation of distances to all nodes from a given node over the total number of nodes, is 25% less than the diameter.

Theorem 3 : $\bar{d}_n = \frac{3n}{2} \left(\frac{10^n}{10^n - 1} \right) \approx \frac{3n}{2}$.

Proof : Since $\sum_{i=1}^{2n} i \cdot Z_i^n = \frac{3n}{2} \cdot (10^n)$, $\bar{d}_n = \frac{3n}{2} \cdot \frac{10^n}{10^n - 1}$. $\square$

4 Network Embeddings

Graph embeddings describe the mapping of an interconnection network or task graph on another network.

An *embedding* of a *guest* graph G into a *host* graph H is a one-to-one mapping of their nodes along with an edge-mapping $\psi : E_G \rightarrow \{paths \text{ in } H\}$. The *dilation* ($\mathcal{D}$) of an embedding is the maximum distance in H between the images of adjacent nodes in G. The *expansion* is $E = \frac{|V_H|}{|V_G|}$. The *edge-congestion* ($\mathcal{E}$) is the maximum number of edges in G routed by the mapping ψ over a single edge of H.

Table 2 summarizes our results on embedding various networks into FP_n, where $\Delta = \frac{10^n}{(\frac{75}{64})2^{3n} - \sqrt{11}(2^{\frac{3n}{2}})}$ is the minimum expansion for a choice of $k = \lfloor n/2 \rfloor$. A brief discussion of the embeddings is given below and the details are in [ÖD93b].

Guest network	$\mathcal{D}$	$\mathcal{E}$	E
Ring $R(k), 4 \leq k \leq 10^n$	1	1	1
Mesh $M(2^{i_1} \cdot 5^{j_1}, \ldots, 2^{i_q} \cdot 5^{j_q})$ $\prod_{k=1}^{q} 2^{i_k} \cdot 5^{j_k} = 10^n$	2	2	1
Hypercube $\binom{n}{i} Q_{3(n-i)+i}, \forall 0 \leq i \leq n$	2	2	1
Complete binary tree $CBT(3n-1), 2\, CBT(3n-4)$	1	1	1
X-tree $\binom{n}{k} X(3n-2k-1), \forall 0 \leq k \leq n$	2	2	1
Tree machine $TM(3n-1)$	2	2	$\frac{2}{3} \cdot (\frac{5}{4})^n$
Mesh of trees $MT(3k-1, 3(n-k)-1)$, $2\, MT(3k-1, 3(n-k)-4)$, $2\, MT(3k-4, 3(n-k)-1)$, $4\, MT(3k-4, 3(n-k)-4), 0 \leq k \leq n$	1	1	Δ

Table 2: Embeddings in FP_n

Cycles of length 5, 6, 8 and 9 are subgraphs of P, while those of length 3, 4, 7 and 10 are not [CW85]. FP_2 contains subgraphs $\{i, i+1\} \times P \cong Q_1 \times P = HP_4$ for $i = 0, 2, 4, 6, 8$. We have shown in [DÖ92] that any cycle of length l, where $4 \leq l \leq 20$, is isomorphic to a subgraph of the hyper Petersen network HP_4 containing the edge $\{(0,0), (0,1)\}$. By combining rings $R(l_1)$ in $\{0, 1\} \times P$ and $R(l_2)$ in $\{2, 3\} \times P$ in the way depicted in Fig. 3b), we obtain a ring of length $4 \leq l_1 + l_2 \leq 40$ in $\{0, 1, 2, 3\} \times P$. Similarly we do in $\{6, 7\} \times P$ and $\{8, 9\} \times P$. Next, an edge–automorphism of P is used on the edges in $R(L_1)$, where $L_1 = l_1 + l_2$, such that the resulting cycle contains the edge $\{(3,0), (3,1)\}$. By connecting $R(L_1)$ in $\{0, 1, 2, 3\} \times P$ with $R(L_2)$ in $\{4, 5\} \times P$, we get a cycle $R(L_1 + L_2)$, where $4 \leq L_1 + L_2 \leq 60$. This cycle can then be connected with another in $\{6, 7, 8, 9\} \times P$ to form a cycle of length l, $4 \leq l \leq 100$. Recursively, we can construct all cycles of length l, $4 \leq l \leq 10^n$, in FP_n for $n \geq 2$.

One can easily construct an embedding of an 1×10 mesh (as subgraph) and an 2×5 mesh (dilation 2). All possible meshes M_{k+1} embedded in FP_{k+1} result from taking 10 copies $(f_k(M_k), j), 0 \leq j \leq 9$, of a mesh M_k which is embeddable in FP_k with expansion 1. These are then put together either one after another along one dimension (Case 1 of Figure 4), or putting two of these 10 copies in one dimension and five in another dimension (Case 2 of Figure 4).

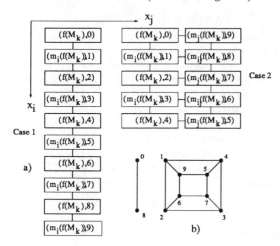

Figure 4: a) Embedding of M_{k+1} b) Q_1 and Q_3 in P

In Figure 4a), m_i and m_{ij} respectively denote the mirror images by reflection in the dimension x_i and in both x_i and x_j, which are two different directions.

As Figure 4b) shows, one can construct an embedding of Q_1 and Q_3 in P with dilation and edge–congestion 2. By using cartesian products of hypercubes embedded in FP_{n-1} and P, we can inductively show that $\binom{n}{i}$ instances of $Q_{3(n-i)+i}$, $0 \le i \le n$, are embeddable in FP_n.

For $FP_1 = P$, there are three possible subgraph embeddings f_1^i, $1 \le i \le 3$, of an $AddCBT(2)$, that is a complete binary tree (CBT) of height 2 with an additional tail of length two connected with the root. In f_1^1, the root is mapped on r^i, the first node in the tail adjacent to the root to u^i and its neighbor, the second node in the tail, to v^i (see Fig. 5a)). It holds $r^2 = u^1$, $u^2 = r^1$, and $u^3 = r^1$. In Fig. 5a) the superscript "1" for r, u and v is omitted. Suppose $AddCBT(3n-1)$ and two instances of $CBT(3n-4)$ can be embedded as subgraphs in FP_n by the mappings f_n^i, $i = 1, 2, 3$, and f_n'. The mapping f_n^i fulfills the properties that the root in f_n^2 is mapped to $r^2 = u^1$, and the neighbor of the root in f_n^2 (f_n^3) is mapped to $u^2 = r^1$ (or $u^3 = r^1$). Using edge–automorphisms in FP_{n+1}, we construct the mappings f_{n+1}^i, $1 \le i \le 3$, and f_{n+1}' for $AddCBT(3n+2)$ and $CBT(3n-1)$ as in Fig. 5.

The mesh of trees $MT(m,n)$ is a subgraph of $CBT(m) \times CBT(n)$. Using the above results, one instance of $MT(3k-1, 3(n-k)-1)$, two instances of $MT(3k-1, 3(n-k)-4)$, two instances of $MT(3k-4, 3(n-k)-4)$ and four instances of $MT(3k-4, 3(n-k)-4)$ are subgraphs of $FP_n = FP_k \times FP_{n-k}$.

5 Conclusion

In this paper, we have presented a new interconnection network topology, called the n-folded Petersen network. It is defined by the iterative product of the Petersen graph,

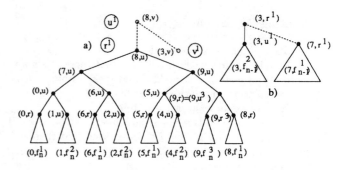

Figure 5: a) f_{n+1}^1 b) f_{n+1}'

and possesses many important properties such as regularity, symmetry, small diameter and degree, simple routing and broadcasting and maximal fault–tolerance. Furthermore, it is versatile since it can efficiently simulate rings, meshes, hypercubes, and several tree structures. We have also studied the dynamic embeddings of dynamic trees into FP_n and several types of dynamic grids that arise in dynamic programming or multigrid methods [ÖD93c]. Currently, we are working on a generalization of the FP_n network in the sense that the number of nodes is not restricted to a power of 10.

References

[BA84]   L. Bhuyan and D.P. Agrawal. Generalized hypercubes and hyperbus structures for a computer network. *IEEE Transactions on Computers*, C-33:323–333, 1984.

[CW85]   C Chartrand and R.J. Wilson. The Petersen Graph. In *Graphs and Applications* (Eds. F. Harary and J.S. Maybee), pp 69 –100, 1985.

[DGD92] S.K. Das, J. Ghosh, and N. Deo. Stirling networks : A versatile combinatorial topology for multiprocessor systems. *Discrete Applied Mathematics*, 37/38 :119–146, July 1992.

[DM93]   S.K. Das and A. Mao. Embeddings in recursive combinatorial networks. In *Proc. 18th Int. Workshop on Graph–Theoretic Concepts in Computer Science, Germany, June 1992.* LNCS, vol. 657, pp 184–204.

[DÖ92]   S.K. Das and S. Öhring. Embeddings of Tree–Related Topologies in Hyper Petersen Networks. Technical Report CRPDC-92-16, Univ. North Texas, Univ. Wuerzburg, Sept 1992.

[EAL91] A. El-Amawy and S. Latifi. Properties and performance of folded hypercubes. *IEEE Transactions on Parallel and Distributed Systems*, 2(1):31–42, Jan 1991.

[Lei92] F.T. Leighton. *Introduction to Parallel Algorithms and Architectures : Arrays - Trees - Hypercubes.* Morgan Kaufmann Publishers, San Mateo, CA, 1992.

[ÖD93a] S. Öhring and S.K. Das. Dynamic embeddings of trees and quasi–grids into hyper-de bruijn networks. to appear in the *Proc. 7th Int. Parallel Processing Symposium*, Newport Beach, California, April 1993.

[ÖD93b] S. Öhring and S.K. Das. The Folded Petersen Network : A New Versatile Multiprocessor Topology. Technical Report CRPDC-93-2, Univ. Wuerzburg, Univ. North Texas, Feb 1993.

[ÖD93c] S. Öhring and S.K. Das. Mapping Dynamic Data Structures on Product Networks. Technical Report CRPDC-93-5, Univ. North Texas, March 1993.

Hierarchical WK-Recursive Topologies for Multicomputer Systems

Ronald Fernandes and Arkady Kanevsky
Department of Computer Science
Texas A&M University, TX 77843
{ronaldf,arkady}@cs.tamu.edu

Abstract : *We present two Hierarchical networks that use the WKR network as a basic module. The new networks retain the recursive structure of the WKR network, at the same time have reduced diameter. Various properties of the new networks are discussed and are compared with those of other hierarchical networks.*

1 Introduction

In massively parallel point-to-point multicomputer networks, the topology of the network plays a crucial role in the performance and cost of systems. For systems with large number of processors, even structures such as the hypercube require prohibitively large number of links, and a very high node degree. To avoid this problem, use of hierarchical interconnection networks has been suggested in the literature [4, 5, 6, 7].

The WKR network, proposed in [1], is a hierarchical network that is recursively defined and is expandable to any level. Though the WKR network has a lot of useful properties [1], it is unsuitable for large multicomputer networks since its diameter is an exponential function of the expansion level. To reduce the diameter, the WKR was expanded to three dimensions in [2]. In this paper, we construct a hierarchy of WKR networks, to get a new network that retains the recursive structure but has reduced diameter. We call the new network the *Hierarchical WKR* network. We also present a Pyramid-like variant of the Hierarchical WKR network, called the *Pyramid WKR* network.

This paper is organized as follows. In Section 2, we describe the WKR network. In Section 3, we present the Hierarchical WKR network and its properties. In Section 4, we discuss implementation of parallel algorithms on the Hierarchical WKR network. Section 5 describes the Pyramid WKR network. The two new networks are compared with other hierarchical networks in Section 6. Section 7 has concluding remarks.

2 WK Recursive Network

A WK-Recursive network has two parameters : the amplitude W and the expansion level L, and is recursively described as:

$L = 1$: There are W *real* nodes, configured as a fully connected undirected graph K_W. Each node has W links, of which $(W-1)$ links are used for the connection and one is free. The whole network has W free links and forms one *virtual* node for $L = 2$.

$L > 1$: The network consists of W copies of virtual nodes, each comprising a WK-Recursive network of amplitude W and level $(L-1)$. The W virtual nodes are connected as a fully connected graph K_W. The resulting network also has W free links.

We denote a WKR network with amplitude W and expansion level L by (W,L)-WKR. Fig. 1(a) shows a $(4,2)$-WKR. A (W,L)-WKR has W^L nodes. Its diameter $D_{wk}(W,L)$ is independent of W. By induction on

L, $D_{wk}(W,L) = (2^L - 1)$. Since there are W links per node and W free links, the total number of bidirectional links, $I_{wk}(W,L)$ is given by

$$I_{wk}(W,L) = W(W^L - 1)/2. \qquad (1)$$

The addressing scheme for a (W,L)-WKR [1] is summarized here. After fixing an origin and an orientation, every real node within each $(W,1)$-WKR substructure is labeled by an index $n_1 \in \{0, 1, \ldots W-1\}$. Likewise, every $(W,1)$-WKR within each $(W,2)$-WKR substructure is labeled by an index n_2. We proceed in this fashion upto level L. Thus each of the W^L real nodes is labeled by an L-plet of indices $(n_L, \ldots, n_2, n_1)$, where each $n_i \in \{0 \ldots W-1\}$. We consider a clockwise orientation for the addressing scheme. Fig. 1(a) shows the addressing scheme for the $(4,2)$-WKR. A self routing message passing scheme is described in [1].

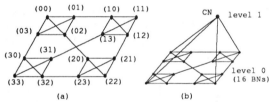

Fig. 1: A $(4,2)$-WKR and a $(4,2,1)$-HWK.

3 Hierarchical WKR Network

In the Hierarchical WK-Recursive network (henceforth called *HWK*), a basic module consists of a (W,L)-WKR, with the W free links connected to an additional *communication node* or *CN*. In a basic module, each node of the (W,L)-WKR is called a *base node* or *BN* and the CN is called the *parent CN* of every BN. While each BN has W links, the CN has $2W$ links, of which W are used within the basic module, the rest are used in connecting to other basic modules. Fig. 1(b) shows a basic module comprising a $(4,2)$-WKR and a CN.

A basic module trivially constitutes an HWK of unit height and is denoted by $(W,L,1)$-HWK. An HWK of height 2, denoted by $(W,L,2)$-HWK, is built from W^L basic modules by connecting the W^L CNs of the basic modules in a (W,L)-WKR configuration. The W corner nodes of the (W,L)-WKR network of CNs are connected to a new CN that forms an apex which is used for further expansion of the network. In general, a (W,L,h)-HWK is an HWK of height h, such that the lowest level (level 0) has $(W^L)^h$ BNs, and level h comprises the apex node. We use the term *node* to include both a BN and a CN. The CNs have the same processing power as the BNs, and are not used only for communication. Fig. 2 shows a $(4,2,2)$-HWK.

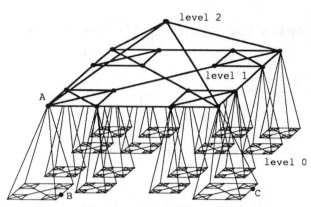

Fig. 2: A (4,2,2)-HWK.

3.1 Parameters of the HWK

Number of nodes: Let n be the total number of BNs in a basic module. Let N_1 be the total number of BNs in a (W, L, h)-HWK. Then $n = W^L$ and $N_1 = n^h$. The total number of CNs is given by $N_2 = \sum_{i=0}^{h-1} n^i$. Hence the total number of nodes in the network is

$$N = N_1 + N_2 = (n^{h+1} - 1)/(n - 1). \qquad (2)$$

Diameter: In the worst case, a message travels up to level $(h-1)$, then across the (W, L)-WKR of level $(h-1)$, and down the network. At each level except level $(h-1)$, the message has to be routed to a corner node of a (W, L)-WKR. At level $(h-1)$, the message travels the diameter of the (W, L)-WKR. Hence the diameter of a (W, L, h)-HWK is given by

$$D_{hw}(W, L, h) = h2^L - 1. \qquad (3)$$

Number of Links: The vertical links of the network forms a forest of W-ary trees of unit depth, each tree rooted at a CN. There are N_2 such trees, each having W edges. Also, there are N_2 different (W, L)-WKRs, each having $I_{wk}(W, L)$ horizontal links. Hence the total number of bidirectional links is given by

$$I_{hw}(W, L, h) = N_2 * (W + I_{wk}(W, L)). \qquad (4)$$

3.2 Addressing Scheme

In a HWK, a node address consist of two parts: a local address and a module address. The complete address of a node is formed by prefixing its module address to its local address. The local address of a node is obtained as per the address scheme of the (W, L)-WKR network it belongs to. The module address of every node (except the apex) is the complete address of the parent CN of the (W, L)-WKR it belongs to.

The apex node of the HWK is given a null address ϕ. For the rest of the nodes, the local address A_0 is first obtained as per the addressing scheme described in Section 2. Then, starting with level $(h-1)$ and proceeding level by level down to level 0, we obtain the complete address of every node in a level by prefixing its module address to its local address. Thus at level 0, a BN is labeled as $A = \langle A_{h-1} A_{h-2} .. A_1 A_0 \rangle$, where each A_i is an L-plet of symbols from the set $\{0 \ldots W-1\}$. The local address of the BN is A_0 and its module address is $\langle A_{h-1} A_{h-2} .. A_1 \rangle$. Thus in Fig. 2, node A has address (33), node B has address (33)(21) and node C has address (22)(12).

3.3 Message Routing

A message that has its source node and destination node in the same module is routed using the message passing scheme for the WKR [1]. Messages whose source and destination lie in different basic modules are routed via one or more CNs. We define the *Least Common Level* or LCL of two nodes as the lowest level that a message has to travel up the hierarchy in order to go from one node to the other. Message passing is done in three steps. The message first travels up to the LCL. It then gets routed within the (W, L)-WKR network in the LCL to the proper ancestor of the destination node. It finally travels down to the destination node. Whenever a message has to go to a higher or lower level, it does so via the nearest corner node of the basic module.

Consider a message that has to be routed from source node addressed $\langle A_{h-1}^s A_{h-2}^s \ldots A_x^s \rangle$ to destination node addressed $\langle A_{h-1}^d A_{h-2}^d \ldots A_y^d \rangle$. Thus the source node resides in level x and the destination node resides in level y. The LCL of the source and destination is the highest value j such that A_j^s is different from A_j^d. Note that for the W corner nodes of a (W, L)-WKR with local address $(n_L n_{L-1} \ldots n_1)$, all the n_i's correspond to the same symbol, say $a' \in \{0 \ldots W-1\}$. In the message routing procedure, we denote such a local address as $(a')^L$. The message routing procedure is as follows.

Procedure Msg-Routing

> *begin*
>> *if* $\langle A_{h-1}^s A_{h-2}^s \ldots A_x^s \rangle = \langle A_{h-1}^d A_{h-2}^d \ldots A_y^d \rangle$
>> *then destination is the source, exit.*
>> *else*
> 1.   *Set* $lcl = h - 1$
> 2.   *While* $(A_{lcl}^s = A_{lcl}^d)$ $lcl = lcl - 1$
> 3.   *For* $i = x$ *to* $lcl - 1$ *step 1*
>>   *begin*
> 3a.   *let* a' *be the leftmost symbol in* A_i^s
> 3b.   *route message to* $\langle A_{h-1}^s \ldots A_{i+1}^s (a')^L \rangle$
> 3c.   *route message to* $\langle A_{h-1}^s \ldots A_{i+1}^s \rangle$
>>   *end*
> 4.   *Route message from* $\langle A_{h-1}^s \ldots A_{lcl}^s \rangle$ *to* $\langle A_{h-1}^d \ldots A_{lcl}^d \rangle$
> 5.   *For* $i = lcl - 1$ *downto* y *step -1*
>>   *begin*
> 5a.   *let* a' *be the leftmost symbol in* A_i^d
> 5b.   *route message to* $\langle A_{h-1}^d \ldots A_{i+1}^d (a')^L \rangle$
> 5c.   *route message to* $\langle A_{h-1}^d \ldots A_i^d \rangle$
>>   *end*
> *end.*

Steps 1 and 2 determine the LCL. Step 3 routes the message up to the source's parent node in the LCL. Step 4 routes the message to the parent node of the destination in the LCL. Step 5 routes the message down to the destination node. Using this message routing procedure, a message from node B to node C in Fig. 2 is routed as follows: $(33)(21) \rightarrow (33)(22) \rightarrow (33) \rightarrow (32) \rightarrow (23) \rightarrow (22) \rightarrow (22)(11) \rightarrow (22)(12)$.

4 Parallel Algorithms

The (W, L, h)-HWK is similar to the the pyramid network, in that the number of nodes decrease as we go to higher levels, and at level h we have a single apex.

Hence many pyramid-based algorithms that use child-parent tree links can be implemented on the HWK. In this Section, we present an example of such an algorithm. Other algorithms for the HWK can be found in [3]. We descibe an algorithm for the *selection* problem which finds the kth smallest value out of N_1 elements, one element per BN. The algorithm is the modification of the one given for a pyramid computer [8]. We assume that in one unit of time, a node can receive and send messages simultaneously along any/all of its links.

The selection problem can be solved by solving the weighted selection problem for which we are given k and N_1 pairs (v_i, w_i), where v_i is the *value* of the pair and w_i is its positive integral weight. The v_is are all distinct and we need to find the v_I such that

$$\sum\{w_i : v_i \leq v_I\} \geq k \text{ and } \sum\{w_i : v_i < v_I\} < k \quad (5)$$

Each of the original items at level 0 of the HWK has a weight of 1, and intermediate calculations produce values that have greater weights. Initially every item is active, and may later become inactive. The algorithm is as follows:

Repeat the following 3 steps until the kth item is found

1. *Each node at level $(\log_n N_1)/2$ takes as its value, the median of active elements below it, and as its weight, the number of active elements below it.*

2. *The apex finds M_w, the weighted median of the items found in step 1 and transmits M_w to all the node of the base.*

3. *Each base node sends up 1 if its item is less than or equal to M_w. These 1's are summed up on the way to the apex. The apex determines that M_w is the kth item if exactly k 1's arrive. If M_w is too large, it sends down a message that deactivates all values as large as M_w. If M_w is too small, it sends down a message which deactivate all items as small as M_w.*

During each iteration of the algorithm, at least 1/4 of all active elements become inactive [8]. Hence at most $\log_{3/4} N_1$ iterations are required. If $T(h)$ is the time required for one iteration of the algorithm, then

$$T(h) = 2T(h/2) + O(h2^L). \quad (6)$$

The $O(h2^L)$ represents the time required for step 3 of the iteration. The solution of the equation gives $O(2^L h(\log h))$ time per iteration. Since there are a maximum of $O(\log N_1)$ iterations, the total time complexity for the algorithm is $O(2^L(\log^2 N_1)(\log \log N_1))$.

5 Pyramid WKR Network

The HWK has poor fault tolerance, since a single faulty CN disconnects the network. To increase the fault tolerance, we add links in the HWK in such a way that every level of the HWK becomes a WKR network. The resulting network is called a *Pyramid WKR network* or *PWK*. Let (W, L, h)-PWK denote a PWK of height h, such that the base level is a $(W, L * h)$-WKR and every (W, L)-WKR substructure of the network has a CN. Fig. 3 shows a $(4, 2, 2)$-PWK.

Since a PWK is a HWK with extra links, the PWK inherits many properties of the HWK. In particular,

a (W, L, h)-PWK has the same number of BNs and CNs and the same diameter as a (W, L, h)-HWK. The addressing and message passing schemes described in Section 3 can also be used for a (W, L, h)-PWK. Moreover, for messages to be routed along the same level, the message routing scheme in [1] can also be used whenever there is a shorter distance. Parallel algorithms for the HWK can also be implemented on the PWK. However, unlike the HWK, the BNs of a PWK do not have the same degree. Other differences are as follows.

Number of links: For a (W, L, h)-PWK, there are a total of $\sum_{i=1}^{h} I_{wk}(W, i * L)$ horizontal links. The number of vertical links is the same as that of the (W, L, h)-HWK. Hence the total number of links in a (W, L, h)-PWK is given by $I_{pw}(W, L, h) = (\sum_{i=1}^{h} I_{wk}(W, i * L)) + W N_2$.

Connectivity : A network of connectivity x remains connected inspite of upto $(x - 1)$ node/link failures. The connectivity of the PWK is as follows.

Theorem 1. A (W, L, h)-PWK has a connectivity W.
Proof: Given in [3].

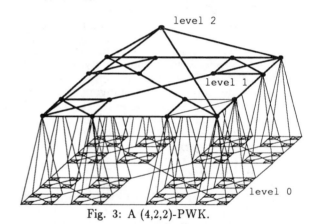

Fig. 3: A $(4,2,2)$-PWK.

6 Comparison of Parameters

We compare the parameters of the HWK and the PWK with those of other hierarchical topologies like the Enhanced Hypercube [7], the Hypernet [5], the HCN [6] and the pyramid. Parameters considered are the diameter, number of links and the product of diameter and number of links (called DI-product).

For an Extended Hypercube EH(k, l), the basic module is a k-dimension binary hypercube, and the hierarchy has l levels. An EH(k, l) has 2^{kl} PEs and $(2^{kl} - 1)/(2^k - 1)$ CNs. Its diameter is $k + 2(l - 1)$ for $l \geq 1$. The number of links satisfies equation $I_{EH}(k, l) = 2^k * I_{EH}(k, l - 1) + I_{EH}(k, 1)$, where $I_{EH}(k, 1) = k * 2^{(k-1)} + 2^k$.

For Hierarchical Cubic Networks or HCNs [6], there are two levels of hierarchy. A HCN(m,n) has m modules arranged in a binary hypercube of dimension m, each module being an n-dimensional binary hypercube. When $m = n$, the HCN is called complete. In such cases, there are 2^{2m} nodes in the network, the diameter is $(m + n)$ and every node has degree $(n + 1)$.

For a Hypernet with cubelets as the basic structure [5], a (d, h)-hypernet has a d-dimension binary hypercube as its basic module and h levels of hierarchy.

We consider the pyramid (a 2-PC[8]), with square meshes at each level. Every node of the pyramid (except the apex) is connected to one parent node. A pyramid of height h has $N_p = \sum_{i=0}^{h} 4^i$ nodes, $2 * \sum_{i=1}^{h} 2^i (2^i - 1))$ horizontal links and $N_p - 1$ vertical links.

Diameter: Fig. 4 shows the variation of the diameter with number of nodes for various networks. The HWK and the PWK have a lower diameter compared to the complete HCN, the pyramid and the Hypernet, but higher diameter than the Extended Hypercube.

Link Cost: The comparison of number of links is shown in Fig. 5. The PWK has marginally more number of links than the HWK. They both have a lower number of links than the other networks considered.

DI-product: This parameter is representative of the cost-performance ratio of a network. The comparison of the DI-product is shown in Fig. 6. The HWK and the PWK have a lower product compared to the complete HCN, the Hypernet, and the pyramid. The DI-product of the HWK, the PWK and the Extended Hypercube are of comparable value.

Connectivity: The HWK has unit connectivity. Connectivity is increased with the PWK. The EH suffers the same poor connectivity as the HWK. The rest of the networks have good connectivity.

Extensibility: The HWK is fully extensible by adding another level without changing the degree of existing nodes. The same is true for the EH and the Hypernet. For the PWK, the BH and the HCN, the degree of some/all existing node must be increased when extending the network.

7 Conclusions

Both, the HWK and the PWK retain the recursive structure of the WKR. HWKs can be built for any levels of hierarchy, using basic modules. PWK has improved connectivity over that of the HWK. Both the networks permit pyramid based algorithms, yet have a much better cost-performance ratio than the pyramid, and many other hierarchical networks. This makes the HWK and the PWK attractive networks for applications like image processing and computer vision.

References

[1] G. Della Vecchia and C. Sanges, "Recursively Scalable networks for message passing architectures", *Proc Int Conf. Parallel Processing and Applications*, Sept 1987, L'Aquila, Italy, pp 33-40.

[2] R. Fernandes, "Recursive Interconnection Networks for Multicomputer Networks", *Proc. 21st ICPP*, Aug. 1992, Vol I, pp 76-79.

[3] R. Fernandes and A. Kanevsky, "Hierarchical WKR Networks" Tech Report, TAMU 93-21.

[4] S. W. Wu and M. T. Liu, "A cluster structure as an interconnection network for large multicomputer systems", *IEEE T.C.*, C-30, pp 254-264.

[5] K. Hwang and J. Ghosh, "Hypernet: A Communication-Efficient Architecture for Constructing Massively Parallel Computers.", *IEEE T.C.* C-36, 1987, pp. 1450-1466.

[6] K.Ghose and K.R.Desai, "The HCN: A Versatile Interconnection Network Based on Cubes", *Proc. Supercomputing* 1989, Reno, pp 426-435.

[7] J.M. Kumar and L.M. Patnaik, "Extended Hypercube: A Hierarchical Interconnection Network of Hypercubes", *IEEE Trans Parallel Dist. Sys.* Vol 3, 1992, pp 45-57.

[8] Q. Stout "Sorting, Merging, Selecting and Filtering on Tree and Pyramid Machines", *Proc. 12th ICPP*, 1983, pp 214-221.

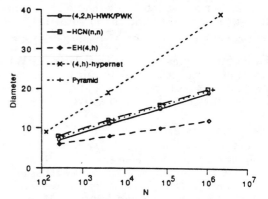

Figure 4: Diameter v/s Number of Nodes.

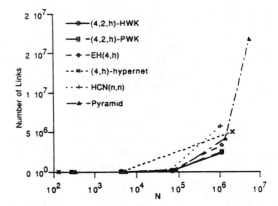

Figure 5: Number of Links v/s Number of Nodes.

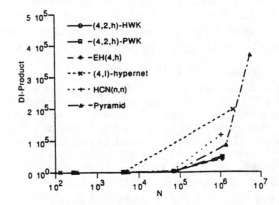

Figure 6: DI-Product v/s Number of Nodes.

Substructure Allocation in Recursive Interconnection Networks

Ronald Fernandes and Arkady Kanevsky
Department of Computer Science
Texas A&M University
College Station, TX 77843
{ronaldf,arkady}@cs.tamu.edu

Abstract : *In a multiuser message passing MIMD system, substructure allocation is an important aspect of system design. In this paper, we present four substructure allocation algorithms for a multiuser WK-Recursive network. Two algorithms are bit-map based and two are tree based. The algorithms are compared using simulation.*

1 Introduction

The WK-Recursive network or WKR network was proposed in [1]. It has excellent properties for massively parallel scalable message passing MIMD systems. A WKR network is recursively defined, being described by two parameters: the amplitude and the expansion level. In the multiuser WKR network, a user requests a WKR-substructure by specifying the amplitude and the expansion level of the substructure. On completion of the task, the allocated substructure is deallocated. Allocation algorithms must include good substructure recognition capabilities and low overhead. In this paper, we present four algorithms for substructure allocation for the WKR network. The first two algorithms are bit-mapped based, wherein each node of the network is associated with a bit whose value indicates whether or not the node is free. The third and fourth algorithms use trees to keep track of free substructures.

This paper is organized as follows. In Section 2, we describe the WKR network and its addressing scheme. In Section 3, the allocation algorithms are described. In Section 4, simulation results for the algorithms are presented. Section 5 has concluding remarks.

2 WK-Recursive Topology

A WK-Recursive network with amplitude W and expansion level L is denoted by (W, L)-WKR, and is recursively described as:

$L = 1$: There are W *real* nodes, configured as a fully connected undirected graph K_W. Each node has W links, of which one is free. The whole network has W free links and forms one *virtual* node for $L = 2$.

$L > 1$: The network consists of W copies of virtual nodes, each comprising a WK-Recursive network of amplitude W and level $(L - 1)$. The W virtual nodes are connected as a fully connected graph K_W. The resulting network also has W free links.

Fig.1(a) shows a (5,2)-WKR. A (W, L)-WKR has a total of W^L nodes. Its diameter equals $(2^L - 1)$. For any fixed amplitude W, the network is scalable

to any level without changing the number of links per processor [1]. The WKR is versatile because parallel algorithms for such networks can be designed irrespective of the amplitude or expansion level. The WKR is suitable for applications having high degree of message locality within virtual nodes [2].

The addressing scheme for a (W, L)-WKR is as follows [1]. After fixing an origin and an orientation, every real node within each $(W, 1)$-WKR substructure is labeled with an index $a_1 \in \{0, 1, \ldots W - 1\}$. Likewise, every $(W, 1)$-WKR within each $(W, 2)$-WKR substructure is labeled with an index a_2. We proceed in this fashion upto level L. Thus each of the W^L real nodes is labeled by an L-plet of indices $(a_L, \ldots, a_2, a_1)$, where each $a_i \in \{0 \ldots W - 1\}$. Also, an address $A' = \langle a_L, \ldots, a_{x+1} \rangle$ represents a (W, x)-WKR substructure of the original WKR. We consider a clockwise orientation for the indexing scheme. Fig.1(a) shows the addressing scheme for the (5,2)-WKR. In this paper, $X \cdot Y$ represents the concatenation of address X and Y.

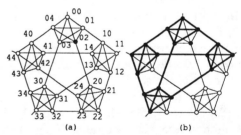

Figure 1: Embedding a (4,2)-WKR in a (5,2)-WKR

3 Substructure Allocation

We assume that the original network is a (W, L)-WKR and that the user makes a request for a (w, l)-WKR, where $w \leq W$ and $l \leq L$. Let such a request be called a *(w, l)-request*. Allocation algorithms are described with respect to the addressing scheme of Section 2. For the first two algorithms a string of W^L allocation bits keeps track of availability of nodes such that an allocation bit with value 0(1) indicates the availability (unavailability) of the corresponding node.

3.1 Set Product Method

Let S_W be the ordered set $\{0, 1 \ldots (W - 1)\}$. Let $S \subseteq S_W$. We have a total ordering among all subsets of S_W that are of the same cardinality. Thus among all subsets of S_4 that are of cardinality three, $\{0, 1, 2\} < \{0, 1, 3\} < \{0, 2, 3\} < \{1, 2, 3\}$. Let S^y denote the set obtained by taking the cartesian product of S, y times. Thus $\{0,2,3\}^2 = \{00,02,03,20,22,23,30,32,33\}$. A (W,L)-WKR can be described as $(S_W)^L$. This is because the address of every node of the (W,L)-WKR is contained in $(S_W)^L$. For $S \subseteq S_W$ such that $|S| = w$, S^L describes a (w,L)-WKR substructure embedded in the (W,L)-WKR.

In general, for $l \leq L$, every (w,l)-WKR substructure lies within a (W,l)-WKR substructure of the network, and there are $W^{(L-l)}$ such substructures. To satisfy a (w,l)-request, we find the smallest addressed (W,l)-WKR substructure that can embed a (w,l)-WKR which is described by S^l, where $S \subseteq S_W, |S| = w$. If $A' = \langle a_L a_{L-1} \ldots a_{l+1} \rangle$ is the address of the (W,l)-WKR, then the granted substructure can be completely described by the tuple (A', S^l). Since the address of the original (W,L)-WKR is denoted by ϕ, the $(4,2)$-WKR of Fig.1(b) (shown in bold) is completely described as $(\phi, \{0,1,3,4\}^2)$. The allocation and deallocation schemes are formally given below.

Processor Allocation :

Step 1 Determine the (W,l)-WKR of smallest address $A' = \langle a_L a_{L-1} \ldots a_{l+1} \rangle$, such that for the smallest $S \subseteq S_W, |S| = w$, all nodes in the set $\{A' \cdot X | X \in S^l\}$ are free. If no such A' exists, put the request in the waiting queue and exit.

Step 2 Set the allocation bit of every element of the set $\{A' \cdot X | X \in S^l\}$ to 1.

In Fig.1(a), only node (02) is allocated prior to a $(4,2)$-request. The Set Product algorithm allocates the $(4,2)$-WKR shown in bold in Fig.1(b) to the request.

Processor Deallocation : Reset the allocation bit of every node in the set $\{A' \cdot x | x \in S^l\}$ to 0.

Number of Recognizable Substructures : Since there are $\binom{W}{w}$ sets $S \subseteq S_W$ where $|S| = w$, the Set Product method recognizes a total of $N_{sp} = W^{(L-l)} * \binom{W}{w}$ different (w,l)-WKR substructures.

Time Complexity : Checking all bits of a (w,l)-WKR takes $O(w^l)$ time. Hence allocation time is $O(w^l * N_{sp})$. Deallocation time is $O(w^l)$.

3.2 Recursive Allocation Method

The Set Product method causes *virtual fragmentation*, since it does not recognize every available (w,l)-WKR substructure when $w < W$. For example, the $(3,2)$-WKR shown in bold lines in Fig.2(b) cannot be recognized from Fig.2(a) by the Set Product method. Described below is the Recursive Allocation method which exhaustively searches all available (w,l)-WKRs.

Processor Allocation :

Step 1 Determine the (W,l)-WKR substructure of smallest address $A' = \langle a_L a_{L-1} \ldots a_{l+1} \rangle$ for which we can select (the smallest) set $S \subseteq S_W, |S| = w$ that satisfies the following property:

- $\forall x \in S$, there exists a $(w,l-1)$-WKR embedded in the $(W,l-1)$-WKR of address $A' \cdot x$ with $(w-1)$ corner nodes labeled $A' \cdot x \cdot y^{(l-1)}$ where $y \in (S - \{x\})$.

If no such A' exists, put the request in the waiting queue and exit.

Step 2 Set the allocation bit of every node of the recognized (w,l)-WKR to 1.

The problem of finding a $(w,l-1)$-WKR with $(w-1)$ corner nodes specified is called the constrained recognition problem, and is formulated as follows. Given a $(W,l-1)$-WKR of address $\langle A'' = A' \cdot x \rangle$ and a set $S' = S - \{x\}$, embed a $(w,l-1)$-WKR such that $(w-1)$ corner nodes are specified. Each specified corner node is of the form $\langle A'' \cdot y^{(l-1)} \rangle$ where $y \in S'$. The $(w,l-1)$-WKR is recursively embedded as follows : Select the least $p \in (S_W - S')$ such that

1. $\forall u \in S'$ we can embed a $(w,l-2)$-WKR in the $(W,l-1)$-WKR with all w endpoints specified as $\langle A'' \cdot u \cdot v^{(l-2)} \rangle$, where $v \in S' \cup \{p\}$.

2. We can recursively embed a $(w,l-2)$-WKR in submodule $A'' \cdot p$ with $(w-1)$ corner nodes specified as $\langle A'' \cdot p \cdot v^{(l-2)} \rangle$ where $v \in S'$.

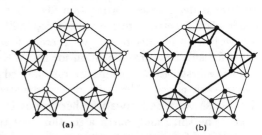

Figure 2: Embedding a $(3,2)$-WKR in a $(5,2)$-WKR

For the $(5,2)$-WKR of Fig.2(a), a request for a $(3,2)$-WKR is satisfied by $S = \{0,1,3\}$. The three constrained recognition problems are solved as follows. The $(3,1)$-WKR with specified corner nodes (01) and (03) is embedded in the $(5,1)$-WKR labeled 0 by selecting p as 2. The $(3,1)$-WKR with specified corner nodes (10) and (13) is embedded in the $(5,1)$-WKR labeled 1 by selecting p as 1. The $(3,1)$-WKR with specified corner nodes (30) and (31) is embedded in the $(5,1)$-WKR labeled 3 by selecting p as 4.

Processor Deallocation : Reset the allocation bit of every node of the allocated structure to 0.

Number of Recognizable Substructures : For an $S \in S_W$, $|S| = w$, there are w constrained recognition problems. For each constrained recognition problem, a choice for selecting the value p from $(W - w + 1)$ indexes is made at $(l-1)$ levels. S can be selected from S_W in $\binom{W}{w}$ ways. Also, there are $W^{(L-l)}$ different (W,l)-WKRs in the (W,L)-WKR. Thus the algorithm recognizes a total of N_{ra} different (w,l)-WKR substructures, where

$N_{ra} = W^{(L-l)} * \binom{W}{w} * w * (W-w+1)^{(l-1)}$.

N_{ra} also represents the total number of (w,l)-WKR substructures in a (W,L)-WKR.

Time Complexity : Checking all allocation bits of a (w,l)-WKR takes $O(w^l)$ time. Hence, allocation time is $O(w^l * N_{ra})$. Deallocation time is $O(w^l)$.

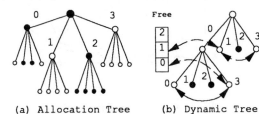

(a) Allocation Tree   (b) Dynamic Tree

Figure 3: Tree Methods for a (4,2)-WKR

3.3 Allocation Tree Method

This method grants a (W,l)-WKR substructure to a (w,l)-request, even when $w < W$. We maintain a W-ary tree of height L, called the *allocation tree*. The root of the tree is at the L-th level of the tree. Each node at the ith level of the tree is associated with a (W,i)-WKR. Each tree node is associated with an allocation bit that has value 0(1) if the substructure it represents is unallocated (partially/fully allocated). Fig.3(a) shows the allocation tree of a (4,2)-WKR for which the (4,1)-WKR labeled 2, and two (4,0)-WKRs labeled (01) and (02) are allocated. The allocation and deallocation procedures are as follows.

Processor Allocation :

Step 1 Search the tree at level l from left to right for a node with allocation bit of value 0. If unsuccessful, put the request in the waiting queue and exit.

Step 2 Set the allocation bits of the node and all its children to 1.

Step 3 Traverse the tree upto the root setting all bits along the path to 1.

Processor Deallocation :

Step 1 Reset the allocation bit of the tree node corresponding to the relinquished substructure and all of its children to 0.

Step 2 If all the siblings of the tree node have an allocation bit of 0, set the allocation bit of the parent to 0 and repeat Step 2 for the parent.

Number of Recognizable Substructures : For a (w,l)-request, the Allocation Tree method recognizes $W^{(L-l)}$ substructures. When $w < W$, unused processors in the granted (W,l)-WKR are not allocated to other incoming requests, causing *internal fragmentation.*

Time Complexity : Since the allocation tree is static, the entire tree can be represented as a bit-map. Thus each tree node is accessed in unit time. For processor allocation, Step 1 takes $O(W^{(L-l)})$ time, Step 2 takes

$(W^{l+1}-1)/(W-1)$ time units and Step 3 takes $(L-l)$ time units. Hence, allocation time is $O(W^L)$. For processor deallocation, Step 1 takes $(W^{l+1}-1)/(W-1)$ time units, and Step 2 requires a maximum of $W*(L-l)$ time units. Hence deallocation time is $O(W^L)$.

3.4 Dynamic Tree Method

Like the Allocation Tree method, this algorithm grants only (W,l)-WKR substructures and results in internal fragmentation when $w < W$. This method is similar to the one in [3]. It uses a dynamic W-ary tree and a 1-Dimensional array of free lists. A leaf node of the tree either represents a free substructure or a fully allocated substructure. A non-leaf node always represents a partially allocated substructure. An array *Free* of size $L+1$ keeps track of free substructures. *Free[i]* provides a pointer to the head of a doubly linked list that contains all free (W,i)-WKR substructures of the network. In Fig.3(b), two (4,0)-WKRs labeled (01) and (02), and one (4,1)-WKR labeled 2 are allocated. The allocation and deallocation procedures are:

Processor Allocation :

Step 1 If the list of *Free[l]* is not empty, allocate the head of the *Free[l]* to the request and remove it from the list. Exit.

Step 2 $i = l + 1$.

Step 3 If $i > L$, keep the request in the waiting list and exit. Else, if there is an available substructure in *Free[i]*, allocate the smallest indexed (W,l)-WKR to the request. Remove the (W,i)-WKR from the list of *Free[i]*. Put the remaining substructures in the list of *Free[i-1]*, *Free[i-2]* ... *Free[l]*. (There will be $(W-1)$ such free substructures at each of these levels). Exit.

Step 4 $i = i + 1$. Goto Step 3.

Processor Deallocation :

Step 1 If the deallocated substructure has expansion level L, put the node in the list of *Free[L]*. Exit.

Step 2 If all the siblings of the deallocated node are free, remove the node and its $W-1$ siblings from the free list, mark the parent as free and Goto 1.

Number of Recognizable Substructures : Same as that for the Allocation Tree method.

Time Complexity : For substructure Allocation, Steps 1 and 2 each take $O(1)$ time. Steps 3 and 4 run at most L times. If in Step 3 a free substructure is available, then granting the request takes $O(WL)$ steps. Hence the time complexity for processor allocation is $O(WL)$. Deallocation also takes $O(WL)$ time.

4 Performance Analysis

We present simulation results for the four algorithms on a multiuser $(6,4)$-WKR. The inter-arrival and the task completion times of requests are assumed to be exponentially distributed. The mean task completion time is assumed to be 10.0 seconds. Requests are granted in a FCFS manner. For each (w, l)-request, $3 \le w \le 6$ and $0 \le l \le 4$. Both w and l have uniform probability over their respective range. Simulations were carried out for 75000 seconds of real time.

Fig. 4 shows the variation of average allocation time with the expansion level l of incoming requests and Fig. 5 shows the variation of average deallocation time with l. Mean inter-arrival time is taken at 10 seconds. These graphs show that the tree based algorithms have a lower overhead than the bit-map based algorithms. The Dynamic Tree algorithm has the least overhead.

Fig. 6 shows how the average delay between arrival and granting of a request varies with mean inter-arrival time. The variation of processor utilization with mean inter-arrival time is shown in Fig. 7. The following observation can be made from these two graphs. For mean inter-arrival times greater than 500 seconds, the bit-map based algorithms have lower delay time and higher processor utilization compared to the tree based algorithms, This is because the tree based algorithms give rise to internal fragmentation. The Recursive Allocation method is marginally better than the Set Product method. This is due to virtual fragmentation that occurs in the latter. For mean inter-arrival times less than 500 seconds, the tree based algorithms have reduced average delay compared to the bit-map based algorithms. This this is because the tree based algorithms have lower overhead and that fragmentation does not cause a problem at such low arrival rates.

5 Conclusion

Two bit-mapped allocation algorithms and two tree-based allocation algorithms are presented for a multiuser WKR network. The tree based algorithms have lower overheads but give rise to internal fragmentation, making them suitable only for low task arrival rates. The Recursive Allocation algorithm recognizes all available substructures of the network and has improved performance over the Set Product algorithm.

Acknowledgements: The first author wishes to thank Debendra Das Sharma for useful discussions.

References

[1] G. Della Vecchia and C. Sanges, "Recursively Scalable networks for message passing architectures", *Proc Int Conf. Parallel Processing and Applications*, 1987, L'Aquila, Italy. pp 33-40.

[2] R. Fernandes, "Recursive Interconnection Networks for Multicomputer Networks", *Proceedings 21st ICPP*, Aug. 1992, Vol. 1, pp 76-79.

[3] D. Das Sharma and D.K.Pradhan, "A Novel Approach for Subcube Allocation in Hypercube Multiprocessors", *Proc. 4th Symposium on Parallel and Distributed Computing*, 1992, pp 336-345.

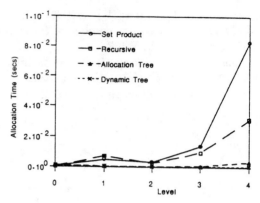

Figure 4: Allocation Time v/s Level.

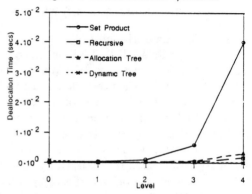

Figure 5: Deallocation Time v/s Level.

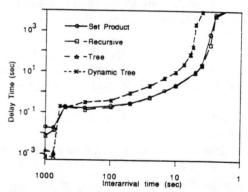

Figure 6: Delay Time v/s Mean Inter-arrival Time.

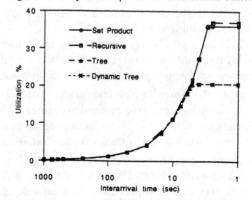

Figure 7: Processor Utilization v/s Inter-Arrival Time.

SESSION 13A

ARCHITECTURE (II)

Coherence, Synchronization and State-sharing in Distributed Shared-memory Applications[*]

R. Ananthanarayanan   Mustaque Ahamad   Richard J. LeBlanc

College of Computing
Georgia Institute of Technology
Atlanta, GA 30332-0280

Abstract

Distributed shared memory consistency protocols suffer from poor performance due their lack of application specific knowledge which can be exploited in message passing systems. Explicit synchronization can be used in memory coherence activities to realize the benefits of application specific information if the user is allowed to associate data with synchronization operations. In this paper, we present a refinement of synchronization and data sharing patterns and identify attributes that can be used to associate data with suitably modified synchronization primitives. We present modifications for semaphores, read-write locks, and barriers which can be used to implement a variety of interaction patterns including mutual exclusion, producers-consumers, readers-writers etc. We claim that the performance of applications programmed using shared memory can be comparable to their message passing implementation which use intermediary entities such as mailboxes. On the other hand, programming is easier with our system, since it allows for richer patterns of interaction than when directly using low-level messages.

1 Introduction

It is generally agreed that concurrent programming using shared memory is easier than using message passing. However, the price paid for the ease of programming in shared memory systems is the cost of maintaining memory coherence. Thus far, the overall performance of applications that use a shared memory abstraction in distributed systems is inferior to the implementations of these applications using message passing. There are two sources of this problem:

Cost of coherence maintenance. The excessive cost of coherence maintenance is due to the lack of application specific knowledge. In an earlier paper, we have analyzed some applications, and showed that traditional shared-memory protocols result in extraneous messages, messages which are used to ensure coherence at points in the application where no coherence is required by the programmer [4].

Granularity of coherence. The granularity problem arises due to a different kind of mismatch between the application and the shared-memory system. The unit of shar-

ing between different components of the application is dictated by the language in which the applications are written, e.g., built-in types such as integers, or programmer defined types such as records, etc. However, distributed shared memory protocols which depend on hardware mechanisms are inherently tied to the hardware data representation unit as the unit of memory coherence. For example, shared memory systems that are integrated into the virtual memory system use page faults to trigger potential violation of memory coherence [10]. Such a system is bound to the hardware page as the unit of memory coherence.

To alleviate the performance problems in shared memory systems, it has been suggested that synchronization information in the program be used. Such protocols ensure coherence only at synchronization points [7, 8, 5, 9]. In this paper, we investigate how the benefits of such protocols can be realized in distributed systems where synchronization is not implemented using memory operations. In our approach, we allow the programmer to specify an association between data and synchronization primitives.

Existing weak consistency protocols, with the exception of Entry Consistency, are geared towards multiprocessor systems. Particularly, they assume synchronization algorithms using memory operations. However, many distributed synchronization algorithms can be implemented more efficiently using message passing, without using memory operations. This explicit synchronization gives us the opportunity to adapt and modify the primitives for synchronization so that the programmer can convey application specific information to a combined synchronization and memory coherence system. Different applications use different synchronization mechanisms which capture the required patterns of interaction between the components of the application. In previous work, release consistency considered critical-section type of accesses and entry consistency generalized it to include readers-writers type of accesses. Our work extends this to include producer-consumer and single program multiple data (SPMD) style synchronization patterns. We distill the common characteristics of these patterns into a set of attributes. Extensions need to be incorporated into synchronization primitives and/or operations so that the programmer can specify values for these attributes. Memory coherence activities will be triggered based on the values for these attributes.

In solving the coherence problem, our scheme allows for

[*]This work was supported in part by NSF contracts CCR-91-06627 and CCR-86-19886.

avoiding hardware assistance in certain identified cases of memory access. In cases where the capability of dynamic access detection is needed, the extra information that our scheme requires is used to avoid problems of false-sharing and granularity.

The rest of the paper is organized as follows: Section 2 presents an overview of the relationship between shared data and synchronization. Section 3 describes the attributes of this association and section 4 discusses an implementation strategy which uses the association and its attributes. In section 5, we describe how the association between data and synchronization can be specified in various interaction patterns. In section 6, we argue that the cost of coherence maintenance can be compared with a certain variation of message passing. Finally, related work and conclusions are presented in sections 7 and 8.

2   Shared Data and Synchronization

In distributed systems, traditionally the coherence of shared data is maintained by a Distributed Shared Memory (DSM) subsystem and synchronization primitives are realized outside of the DSM system. The combined DSM and the synchronization system is called the Augmented Shared Memory System (ASM).

The proposed memory model, called the *synchronized shared memory model* can be represented by a five tuple, $M(D, P, S, F, O)$, where,

- D is the set of shared data items.

- P is the set of synchronization primitives (*e.g.* locks, semaphores).

- O is the set of operations allowed on elements of P.

- S is the set of synchronization variables, each of a type in P.

- $F(D, S)$ is a function which associates the data items in D with the synchronization variables.

An execution of a process is a sequence E of operations from O, along with read and write accesses to D. A concurrent program is a set of such executions, $E_1 \ldots E_n$. The sequence of O operations and read/write accesses to data items in D in each E_i follows *program order*. The O operations in each E_i affect the values read by accesses to D. Specifically, each O operation falls into either an *acquire* or *release* category [8]. Each acquire in an execution E_i has a *matching* release in some execution E_j in that they form a producer-consumer kind of relationship with respect to the data protected by the synchronization variable used in the acquire and release operations. An acquire operation on a synchronization variable S_k in E_i, together with $F(S_k, D_k)$ is an assertion by the programmer that data items in D_k will be accessed following the acquire of S_k, and that the values of data items in D_k are expected to be the values produced before the corresponding release operation in E_j. In such a system, memory coherence can be specified as the values of D that can be expected at each of the O operations in each of E_i's.

Since the memory coherence is valid only at synchronization operations, and for data for which the function $F(D, S)$ is specified, the model is called *synchronized*

shared memory (SSM) model. Note that an ASM system can be used to physically realize the SSM model. However, an ASM system can ensure memory coherence based on other models. For example, it can provide Release Consistent [8] shared memory. However, since SSM uses more information about the shared data, it can provide weaker coherence of data, resulting in fewer messages used in coherence activities, and hence in better performance. Our work focusses on (1) how the function F can be specified by the programmer, and (2) how it can be used to transfer data from where it is generated to where it will be used.

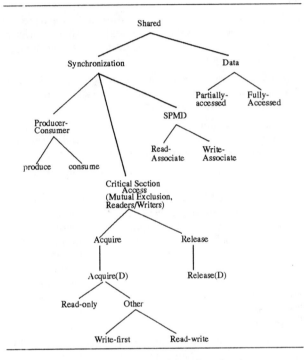

Figure 1: Attributes of Synchronization

3   Attributes of Association

In this section, we study the access patterns for shared data and their programming using various synchronization constructs. We identify a set of attributes which allows the specification of the function $F(D, S)$ for a range of synchronization patterns. Figure 1 gives an overall view of the refinement. All shared accesses can be divided into synchronization and data accesses. Much of the earlier work only explores the critical section subtree; we contend that since the other interaction patterns are significantly different, further performance improvements can be obtained when they are dealt with differently. Furthermore, the refined classification for synchronization also helps us in determining whether the data is dynamically or statically associated with synchronization. Each acquire operation can also be divided into different categories. We also refine data accesses and divide them according to the amount of access. The rest of this section discusses these issues.

The data sharing patterns using different types of synchronization primitives can be specified using a small number of attributes. These attributes can be classified into the following three classes.

3.1 Type of Access

Each synchronization operation can be classified into *acquire* or a *release* operation [8]. An acquire operation is an assertion that the associated data will be accessed by the program, and hence is expected to be recent. A release operation is an assertion that the program has finished access to the data, and will not use it until it executes another acquire on the data. Each acquire-type operation can be further classified as follows:

• *Write-first* acquire. These acquires are for purpose of only generating new values of the data, without considering old values of the data. Initialization of data is one type of write-first acquire.

• *Read-only* acquire. Such an acquire is an assertion that the data will only be read from; no new values are generated at the corresponding release.

• *Read-write* acquire. This type of acquire is needed if a new value of the data is to be generated and if this new value depends on the current value of the data.

Read-only acquires are compatible with each other. Any other combination of acquires are non-compatible; the 'second' processor which initiated the non-compatible acquire will be blocked until the 'first' processor performs a release corresponding to the acquire. For example, the operation read-lock on a read-write lock is compatible with read-lock by another processor. On the other hand, a P() on a mutual exclusion semaphore is a read-write acquire, and hence, not compatible with a P() by another processor which will be blocked. These notions of compatibility allow for programs using the synchronization operations to be data race free [1].

The type of acquire is inherently conveyed in certain types of synchronization operations. For example, the readlock() operation on a lock asserts that the programmer needs only to read the data. This approach was taken by the Midway system [5]. We contend that the acquire operations of other synchronization primitives also follow this basic assumption by the programmer: there is some data that will be generated, some that will be read from, and others both read and written into on the acquire. For example, we will discuss how barriers can be modified to capture this information in section 5.4.

Some synchronization operations fall in both the acquire and the release category. For example, a barrier is an assertion by the programmer that new values for a particular data partition have been generated, and hence acts as a release for that data partition (section 5.4). However, it also acts as an acquire for the data partitions generated by other processors (which will be used in the computation of the new value in the next iteration).

Associated with each synchronization primitive is a queue of blocked processes, each of which is waiting on a non-compatible acquire. A write-type acquire (write-first or read-write), generates a new value for the associated data. The value that is made accessible by an acquire of a blocked process is the value at the corresponding release.

The ordering of accesses due to different types of acquires is shown in figure 2. For write-first acquire, no data needs to be transferred into the processor being granted the acquire. For read-only or read-write acquires, data needs to be transferred if there has been any intervening write-type acquire since the last acquire by the processor.

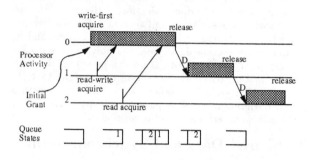

Figure 2: Data Flow on Synchronization

3.2 Time of Association

The time of association of data and synchronization can be divided into two categories — *static* and *dynamic*. With static association, the data that will be made accessible by an acquire is known at the time of the acquire. In case of dynamic association, the data is known only after the acquire is complete. For example, consider the example where a shared counter is protected by a mutual exclusion semaphore. The data that is made accessible by a P() on the semaphore is statically determined, and is the counter variable. On the other hand, at the time of a P() on a counting semaphore used to implement a bounded buffer (see section 5.3) the data that is made accessible by the P() is not known. However, on successful completion of the P() it is possible to determine which data is made accessible.

3.3 Amount of Access

Data that is associated to a synchronization operation may not always be accessed fully after the acquire operation. If the release corresponding to an acquire operation prepares the associated data, ahead of actual access by the acquiring processor, so as to reduce the latency of access to the data, and the processor actually accesses only part of it, then the data transferred on the release may be unnecessarily large. On the other hand, in situations, it would be beneficial to always pre-prepare the data, since the data will always be accessed in its entirety. It is not possible to determine these cases by program analysis alone, since these access characteristics may depend on input. Consequently, the association between an acquire operation and the data that it makes available can be divided into two categories: *fully accessed* and *partially accessed*. We contend that the programmer should be given the option to classify the data into one of these the categories. With this information, unnecessary data transfer can be avoided. It should be noted that if such information is not supplied by the programmer, data can be arbitrarily put into either category without affecting the correctness of the program;

obviously, however, performance may be compromised in certain cases.

The type and time of association, and amount of access are factors which determine the memory coherence activities. In other words, the tuples defined by

{write-first, read-only, read-write}
× {static, dynamic}
× {fully-accessed, partially-accessed}

define a space which is used to categorize each acquire and release in a program. Each of the points in this space corresponds to a different data sharing pattern and leads to different memory coherence activity (section 4). For example, write-first, dynamic and fully accessed attributes specify the data access pattern for a producer process. Classifying the acquire and release operations in a given program into these points is achieved using programmer provided annotations (section 5).

4 Implementation

An implementation of a synchronization variable S_i needs to maintain its state which consists of information such as a queue of blocked processes that are waiting to acquire S_i and the mode in which it has been granted. In our system we need to store additional state information so S_i can be associated with data. The additional information includes a resource queue of available data in case of synchronization variables with dynamically associated data and a dirty vector indicating which processors have clean copies of the associated data in their caches. The processor where the state information of a synchronization variable is stored is called its *owner*. The owner for a variable may not be fixed; it can change when the synchronization variable is granted in response to an acquire request. If multiple processors can be granted an acquire (as with read-only acquire), the last processor to be granted the acquire can be the owner.

When a processor needs to perform an acquire, it sends a request to the owner. This gives rise to the problem of locating the owner, which is the same as that of finding a mobile resource in a distributed setting (A fixed owner precludes supporting efficient runs, in which a processor repeatedly acquires and releases the lock, without being interrupted by the acquire of another processor). Numerous existing techniques can be used to solve this problem, including the ones which employ forwarding addresses, and edge-reversal [11] etc. The overall cost of state-sharing will depend on number of messages used by the location finding algorithm. We will use a fixed rendezvous node, called the *finder* for each synchronization variable. The finder node keeps track of the location of the owner. All requests for acquire are sent to the finder which will then forward the request to the current owner.

A generic algorithm that implements the acquire and release operations for a synchronization variable is shown in Figure 3. It makes use of the attributes that define the sharing pattern of the data which is protected using the synchronization variable. In this algorithm, we assume that data is fully accessed. In section 4.1 we show that actions needed to handle the amount of access attribute are orthogonal to the implementation in Figure 3.

When a processor executes an acquire, it sends a request to the owner of the synchronization variable. Once the owner gets the acquire, the owner blocks the request if it conflicts with the current state of the synchronization variable (lines 1-2). If the acquire can be granted then the following actions are performed. In case of statically associated data, the owner checks to see if the data at the requester is dirty, and if so the owner, which is guaranteed to have a consistent copy of the data, sends the data to the requester (lines 5-6). In case of dynamically associated data, the owner sends the address of the data and the process where the data is available, by removing an entry from the resource queue (lines 8-9). Thus the requester either gets a copy of the data and updates its local cache (line 12), or has consistent copy in its local cache, or must fetch a copy of the data from some other processor whose identity has been sent as part of the grant (lines 14-15).

A release operation checks to see if any processors are blocked on an acquire request. If so, the releasing processor de-queues all processors which can be granted an acquire from the wait queue, and sends the data to each processor that have a dirty cached copy (lines 16-19). If there are no processors waiting, then the data is added to the resource queue, in case of dynamic association (line 21), and is otherwise left in the local cache.

Processors always cache any data that is received on an acquire. On a subsequent acquire, the cached data may still be up-to-date. The algorithm in figure 3 sends update to a processor only if the processor being granted the acquire has a copy which is older than the most recent one. The following paragraph describes an implementation of the procedure dirty(s, d).

Let lock L protect data D. Associated with L is a boolean vector, called the *dirty vector*, denoted DV_L. There is an entry in DV_L for each processor, and has the following semantics:

$$DV_L[i] := 0 \text{ if and only if processor } i \text{ has a dirty copy of D.}$$
$$1 \text{ if and only if processor } i \text{ has a clean copy of D.}$$

DV_L is initially all 0's, except for the owner node. Each processor, i, on being granted the lock modifies the vector, as follows: If lock is granted in read-mode, then i sets $DV_L[i]$ to 1, indicating that at that point it has a clean copy of D. If lock is granted in write-type-mode (write-first or read-write), then i sets $DV_L[i]$ to 1 as above. In addition, it sets $DV_L[j]$, for all $j \neq i$, to 0 indicating that only i has the clean copy. If processor i is to be granted the lock, then if $DV_L[i] = 0$, the data D is transferred along with the lock and DV_L, otherwise only DV_L is sent. Thus, the owner of a lock can determine if a cached copy at the processor to be granted the lock is up-to-date and can avoid unnecessary data transfer.

4.1 Partially Accessed Data

The algorithm in figure 3 assumes that data made accessible by the acquire will indeed be entirely required by the processor being granted the acquire. That is, it assumes that the data is of fully-accessed type. In case of partially-accessed data, the amount of data transfer may be unnecessarily large. To circumvent this problem, access to data should be dynamically detected (by using the read/write to data rather than acquire of the synchronization variable). We use page faults to achieve this effect as follows: On the acquire the data that is protected by the acquire is invalidated locally from the cache of the processor. Access to any part of the data will result in a page-fault. The required page can be fetched at that time, and thus fetching only data that is needed.

```
// handle_acquire() is executed at the owner in response
// to an acquire request. acquire() and release() are
// executed at the processor where the application is
// being executed. State is the state associated
// with the synch. variable — wait_q, resource_q, etc.

handle_acquire(r, m)
// acquire request in mode m from processor r.
(1)   if (conflicting(m, current-mode))
(2)       enqueue(r, wait_q);
(3)   else
(4)       if (static_association)
(5)           if (dirty(r, d) Send(Grant, d, State) to r;
(6)           else Send(Grant, State) to r;
(7)       else // dynamic association
(8)           (addr, s) = dequeue(resource_q);
(9)           Send(Grant, {addr, s}, State) to r;

d = acquire(r, m)
// acquire by proc. r, d is the address of dynamically
// associated data, m is the mode of access.
(10) Send(acquire(r, m)) to owner;
(11) Receive(msg);
(12) if (msg.d) update local copy of d;
(13) if (dynamic_association)
(14)     Get d from processor(msg.s);
(15)     return addressof(d);

release(r, d)
// release by processor r, d is supplied by the
// programmer in case of dynamic association, other-
// wise is fixed for the synch. variable or processor.
(16) if (blocked_processors)
(17)     S = dequeue(wait_q);
// S is the set of processors to be granted the acquire.
(18)     For all s in S:
(19)         if (dirty(s, d)) send(Grant, d, State) to s;
(20)             else send(Grant, State) to s;
(21) else
(23)     if (dynamic_association)
(24)         enqueue(d, r) on resource queue;
(25)     else // nothing
```

Figure 3: Generic algorithm for synchronization-based memory coherence

On the other hand, we can avoid problems related to granularity and false-sharing associated with using hardware mechanisms. First, since the *size* of the data accessed is known, the request for the data can fetch only the required amount of data (for instance, a fraction of a page). Sec-

ond, since the data that is protected by different synchronization primitives is known, suitable *data allocation* can be performed (by the compiler) so that two different data items made accessible by different synchronization variables can be located in different pages. Hence, the problem of false sharing can be minimized.

5   Programming Common Data Sharing Patterns

In this section, we consider commonly used data sharing patterns and discuss how the operations on the synchronization variables that control access to the data can be classified using the attributes discussed in section 3. For each pattern , we (1) classify the pattern of interaction using the attributes discussed in section 3, (2) consider modifications to the operations of the synchronization primitives so that the programmer can specify these attributes, and (3) present optimizations, specific to the pattern, and of the generic algorithm in figure 3.

5.1   Critical Section Access

In critical section accesses, a mutual exclusion semaphore or an exclusive lock protects the critical resource. A P() on this semaphore is an acquire of the read-write type. Since read-write type acquires are incompatible with one another, only one process is granted the semaphore at a time. The association is static since the acquire on the semaphore always allows access to the same data. No modifications are necessary on the semaphore primitive; however, the programmer has to explicitly specify the data associated with a semaphore variable. This specification can be done in a manner similar to that in the example using read-write locks in section 5.2. Further, in the generic algorithm of figure 3, the procedure to determine whether the cached data at a processor is dirty need not be used, since all acquires are of read-write type. Hence, a previously cached copy of the data can never be current, unless that processor is the one that last modified it. Consequently, as an optimization, only the identity of the processor which was last granted the acquire needs to be maintained instead of the dirty vector.

5.2   Readers-Writers

Readers-writers pattern of interaction can be implemented using read-write locks, which already have two different types of acquire operations: read lock (read-only acquire) and write lock (read-write acquire). To accommodate write-first accesses, we introduce a third operation write-first lock. The association between the lock and the data is static, since it is the only 'resource' being guarded by the lock. In the algorithm of figure 3, the dirty vector needs to be maintained, since non-modifying acquires are possible, and hence, a previously cached copy may still be current. The following code fragment shows how a lock can be associated with a simple counter:

```
L : ReadWriteLock;
counter :  integer PROTECTED BY L;
// PROTECTED BY implies static association
```

5.3 Producers-Consumers

Another common interaction pattern between components of an application is that of producer-consumer. A shared buffer into which producers add items, and from which consumers delete items is one way of facilitating the producer-consumer interaction.

We introduce the *Queueing Semaphore* abstraction to aid the implementation. A P() on a queueing semaphore returns the address of the data which can be acquired. A V() on the semaphore takes the address of the released data as an extra parameter. Similarly, the initialization of a queueing semaphore can specify a positive number of initial items to be made available.

In Figure 4 the buffer elements are declared to be dynamically associated with two queueing semaphores, Full and Empty (line 1D). All buffer elements are enqueued in Empty (line 3D), and none in Full (line 2D). The semaphore Empty makes data available to be only written into, and hence classified as write-first. Consequently, when P(Emtpy) is executed (line 1A), no coherence activity is needed on the item returned by the P(). On the other hand, Full makes data available for reading, and hence classified as read-only. Thus, the P(Full) operation (line 1G) results in coherence activity of the data it returns. Either the data was directly sent as a result of the V(Full) (in line 3A), or was queued in the resource queue, in which case the data needs to be fetched from the processor at which it is available. The two V() operations (lines 3A and 3G) are supplied with the respective data they queue — buffer elements with produced data, and buffer elements with no data. The V(Full) may result in transfer of data if it finds a processor blocked. In contrast, V(Empty) does not result in any data transfer, since acquires on Empty are write-first.

The essential difference between a traditional semaphore and a queueing semaphore is that (1) traditional semaphore maintains only a count of available resources, whereas the other maintains a queue of the resources itself and (2) the queueing strategies (add/delete routines) are defined externally for traditional semaphores, whereas in the case of modified semaphores, the queueing strategy is built-in and is FIFO with respect to produced and consumed items. Due to the known queueing strategy, the P() on the modified semaphore can detect which data item will be accessed, and hence can facilitate coherence activity of that particular data item only. The fixed FIFO strategy of the modified semaphores works well in the programming of producer-consumer applications. Using a generalized version of the queueing semaphore, we show that even if the queueing strategy is programmer defined, but known to the semaphore implementation, then similar performance gains can be achieved [3].

5.4 Single Program Multiple Data

In SPMD-style interaction, the shared data is divided into non-overlapping partitions. A number of processes work in phases on different partitions of the data. In each phase, each processes generates new values for the data partition that it is responsible for. In generating the new values for a data partition, the process uses values of data parti-

Global
```
(1D)    Buf[N] : Item, QUEUED BY Full, Empty;
        // QUEUED BY implies dynamic association.
(2D)    Full :  Counting Semaphore (0), READ_ONLY;
(3D)    Empty : Counting Semaphore (N,
                {Buf[1] ... Buf[N]}), WRITE_FIRST;
```

AddItem(p) // p is an in parameter
```
(1A)    i = P(Empty);
(2A)    // put p into item denoted by address i
(3A)    V(Full, i);
```

GetItem(p) // p is an out parameter
```
(1G)    i = P(Full);
(2G)    copy item denoted by address i into p
(3G)    V(Empty, i);
```

Figure 4: Bounded buffer using queueing semaphore. Line numbers 1D-3D, 1A-3A and 1G-3G are used to identify statements in the declaration, AddItem() and Getitem(), respectively.

tions of other processes from the previous phase. The division of data into partitions, and allocation of partitions to processes can change between different execution of the program depending on available resources such as physical processors. Similarly, the data partitions required for computing a new partition value can differ from processor to processor.

Barriers are commonly used to implement SPMD-style interaction. However, barrier calls are not aware of the partitioning, and hence may result in extraneous messages to maintain coherence of the data partitions. We propose the following operations on barriers, which can be used to specify partitioning information. The operation write_associate(data) on a barrier s is used to convey the information that the processor executing the operation will be responsible for generating new values of data. The operation read_associate(s, data), similarly, is an assertion that data will be required for reading by the processor on a subsequent barrier() operation.

These calls correspond to the different types of acquires in section 3.1. We use an example to illustrate this. The code for a synchronous iterative algorithm that solves a set of linear equations is shown in Figure 5. The coordinator process tests for convergence and sets the termination condition; a work process Worker[i] computes the new value for variable $x[i]$. The computation of $x[i]$ reads the values of all $x[j], i \neq j$ which are produced in the preceding iteration. This is specified by the **read_associate** call in the worker code. Since Worker[i] computes a new value for $x[i]$ and writes it, $x[i]$ is the parameter to the **write_associate** call. Thus, only the data $x[i]$ is shipped out of processor i on the barrier at processor i. The dirty vector of the generic algorithm in figure 3 is not needed in the case of barriers, since a new value of the write-associated data is always generated on each barrier. Further, the owner can be one fixed process, like the coordinator. The owner collates the different data partitions as they are generated,

Global

```
a[n,n], b[n] : Real; // linear equation descriptors
s : Barrier;
x[n] : Real, PROTECTED BY s; // vector of unknowns
done : Boolean, PROTECTED BY s;
```

Coordinator

```
    read_associate(s, ALL);
    write_associate(s, done);
    while (!done)
        barrier(s);
        if (converged) done := T;
```

Worker[i]

```
    read_associate(s, ALL);
    write_associate(s, X[i]);
    while (!done)
        barrier(s);
        x[i] := expression(x's);
```

Figure 5: Linear Solver using Barriers

and sends them to the required processors. Notice that even though the coordinator performs the convergence test in parallel with the generation of new $x[i]$, it makes use of values produced from the previous iteration because coherence activities will take place only when the barrier call is made.

6 Performance

In this section, we briefly discuss the cost of state-sharing in applications which are suitably annotated as described in the previous section. The total cost is expressed in terms of messages, some of which are control messages, (e.g. request for data), and others data transfer messages.

In the examples we discussed in the previous section, our system allows for optimal cost of data-sharing with respect to both latency and number of messages. As shown in figure 2, transfer of information from one component of the application to another can be achieved with a single message. This is because,

- the request for acquire can be hidden,

- data to be transferred is always known to the processor performing the release, even in case of dynamically associated data

- and data is transferred only if necessary, and the releasing processor can determine if the granted processor has a dirty copy.

We feel that such performance gains can be achieved for a variety of applications because they often use synchronization patterns that were discussed in the previous section. In fact, the performance of applications implemented using our system is equivalent to the performance of the same applications implemented using certain variations of message passing. In the following, we sketch an informal argument

to support this; a more elaborate discussion is beyond the scope of this paper, due to lack of space, and can be found in [3]. In particular, it can be shown to be equivalent to message passing with intermediary globally named entities such as mailboxes. An acquire is equivalent to a receive(), and release() is equivalent to a send(). The synchronization variables are the mailboxes. The number of messages used to match a send-receive pair of a mailbox is also the same as the number of messages used to match a release-acquire pair in our system : a request needs to be sent to the owner of the mailbox or the synchronization variable, which will then send a reply with the required data. Since both mailboxes and synchronization variables are globally named entities, the problem of locating either one is the same. Further, as shown above, once the acquire request has been queued with the owner of the synchronization variable, only one message is needed to transfer data to the requester, which is the same as with a mailbox.

While in this section, we have considered performance of our solution in qualitative manner supported by informal arguments, actual implementation is needed to measure the run-time costs. Such effort is underway, and the protocol is being implemented on the *clouds* operating system.

7 Related work

Investigating ways to improve performance of shared memory systems, both in tightly coupled parallel systems and in loosely coupled distributed systems, has been an area of active research. In this section, we discuss some of the work related to the results in this paper.

Early work on distributed shared memory is characterized by systems such as Ivy [10]. In these systems, coherence activities are triggered by reads and writes to memory and the sequential consistency model is implemented. To reduce the cost of coherence maintenance, Ramachandran et. al. suggested combining synchronization and data transfer by requiring the data to be locked before its access [12]. However, access patterns for shared data or its association with synchronization operations was not investigated.

The idea of maintaining coherence on synchronization operations, rather than on reads and writes was suggested by Dubois et. al. [7]. Extending on this observation, the release consistency model offered better performance by further refining a synchronization operation into acquire and release [8]. The idea of Data-Race-Free (DRF) programs, which is similar to the properly labeled (PL) programs expected by release consistency, was proposed by Adve and Hill [1]. These papers focus on hardware optimizations that are possible with the weakened consistency and assume that synchronization operations are made known to the hardware. On the other hand, the Midway system which supports entry consistency [5], presents software implementations by exploiting association between data and synchronization variables. Entry consistency also refines the acquire operation by allowing shared or exclusive operations. Lazy release consistency [9] provides further optimizations for release consistency by making data coherent only at the node that acquires the synchronization variable. Most of these systems explore the critical section type synchronization, except Midway and Lazy Release Consistency, which also investigate barriers.

In contrast to using synchronization, Munin [6] explored data sharing patterns of programs, and proposed the use of multiple memory consistency protocols which are triggered by annotations classifying shared data according to these patterns. Causal Memory [2] also does not use synchronization to relax memory consistency; rather it uses the causal relationships established by ordinary reads and writes and provides a weaker memory model.

All the above systems, except Munin, do not make wide use of programmer annotations other than requiring the programs to be properly labeled [8]. In [4], we proposed the association of particular data items with specific synchronization operations in case of critical section accesses. The main differences between the results of this paper and the other systems are as follows:

- Our solution further refines access patterns of shared data. Particularly, only we handle dynamically associated data.

- The specification of various patterns using a small number of attributes, and the generic algorithms that implement synchronization in conjunction with memory coherence are also unique to our system. Further, we present possible ways to obtain attribute information from programs by considering modifications to existing synchronization primitives.

- With respect to implementation, in our system a releasing processor can determine whether data at the acquiring processor is consistent without requiring any compiler support for tracking of memory accesses as in Midway. Instead, only a single dirty vector needs to be maintained with the synchronization variable.

8 Conclusions

We have explored how the characteristics of synchronization in applications programmed with a distributed shared memory abstraction can be used to reduce the cost of coherence maintenance for shared data. Our system is unique in considering multiple patterns of sharing, characterizing these patterns using a set of attributes, and proposing modifications to synchronization primitives so that the attributes for a program can be determined from programmer provided annotations. We contend that the performance of applications programmed using our system can be comparable to their message passing implementation which use intermediary entities such as mailboxes. On the other hand, our system allows for more patterns of interactions between application components, since it supports a variety of synchronization primitives. Therefore, the shared memory system is better suited for expressing what the programmer wishes to convey, and hence should be easier to program in.

In our future work, we want to implement the synchronization operations proposed in the paper and will extend a language to facilitate the annotations. That will allow us to quantify the performance gains made possible by the techniques developed in this paper.

References

[1] Sarita V. Adve and Mark D. Hill. A unified formalization of four shared-memory models. Technical Report CS-1051, University of Wisconsin, Madison, September 1991.

[2] Mustaque Ahamad, Phillip W. Hutto, and Ranjit John. Implementing and programming causal distributed shared memory. In *11th International Conference on Dist. Comput.*, May 1991.

[3] R. Ananthanarayanan, M. Ahamad, and R. J. LeBlanc. Coherence, synchronization and state-sharing in distributed shared-memory applications. Technical Report GIT-CC-93/02, Georgia Institute of Technology, January 1993.

[4] R. Ananthanarayanan, M. Ahamad, and R.J. LeBlanc. Application specific coherence control for high performance distributed shared memory. In *Proceedings of the Third Symposium on Experimental Distributed and Multiprocessor Systems*, March 1992.

[5] Brian N. Bershad and Matthew J. Zekauskas. Midway: Shared memory parallel programming with entry consistency for distributed memory multiprocessors. Technical Report CMU-CS-91-170, Carnegie-Mellon University, September 1991.

[6] J. B. Carter, J. K. Bennett, and W. Zwaenepoel. Implementation and performance of munin. In *Proceedings of the 13th ACM Symposium on Operating Systems Principles*, pages 152–164, October 1991.

[7] Michel Dubois, Christoph Scheurich, and Faye Briggs. Memory access buffering in multiprocessors. In *Proceedings of the 13th Annual International Symposium on Computer Architecture*, pages 434–442, 1986.

[8] Kourosh Gharachorloo, Daniel Lenoski, James Laudon, Phillip Gibbons, Anoop Gupta, and John Hennessy. Memory consistency and event ordering in scalable shared-memory multiprocessors. In *Proceedings of the 17th Annual International Symposium on Computer Architecture*, pages 15–26, 1990.

[9] P. Keleher, A.L. Cox, and W. Zwaenepoel. Lazy release consistency for software distributed shared memory. *SIGARCH Computer Architecture News*, 20(2), May 1992.

[10] Kai Li and Paul Hudak. Memory Coherence in Shared Virtual Memory Systems. In *Proc. 5th ACM Symp. Principles of Distributed Computing*, pages 229–239. ACM, August 1986.

[11] Mitchell L. Neilsen and Masaaki Mizuno. A dag-based algorithm for distributed mutual exclusion. In *In Proceedings of 11th International Conference on Distributed Computing Systems*, 1991.

[12] Umakishore Ramachandran, Mustaque Ahamad, and M. Yousef Khalidi. Coherence of distributed shared memory: Unifying synchronization and data transfer. In *Proceedings of the 18th International Conference on Parallel Processing*, pages 160–169, August 1989.

A Characterization of Scalable Shared Memories*

Prince Kohli   Gil Neiger   Mustaque Ahamad

College of Computing
Georgia Institute of Technology
Atlanta, GA 30332-0280

Abstract

The traditional consistency requirements of shared memory are expensive to provide both in large scale multiprocessor systems and in distributed systems that implement a shared memory abstraction. As a result, several memory systems have been proposed that enhance performance and scalability by providing weaker consistency. The differing models used to describe such memories make it difficult to relate and compare them. We develop a simple non-operational model and identify parameters that can be varied to describe existing memories and to identify new ones. We show how a uniform framework makes it easy to compare and relate various memories.

1 Introduction

The programming and performance of parallel and distributed applications depends on the mechanisms used for sharing state across processors in such systems. Shared memory is attractive because it simplifies programming since processors can access both local and remote state using the standard read and write operations. However, the strong consistency guarantees provided by traditional memories limit the scalability of shared memory systems. This has been recognized in multiprocessors and distributed systems. A number of new memory models have been proposed [1,3,5,6,13,14,16] that seek to enhance performance and scalability by weakening the consistency guarantees. One approach, taken in the systems described in [1,5,6], provide strong consistency only for a subset of the operations and other operations can be executed more efficiently due to their weakened consistency. In these systems, programs that meet certain requirements (e.g., data-race-free) can be programmed as if the system provides strong consistency. The second approach is taken in distributed systems [3,14], where the application programmer must directly program with the weakly consistent memory.

Many systems advocate shared memories that provide weaker consistency. These memories are defined using different models. For example, *Processor Consistency* (PC) as defined by Gharachorloo et al. provides an operational definition by stating how read and write operations are executed. On the other hand, the *Total Store Ordering* (TSO) memory model implemented by the SPARC architecture is defined using an axiomatic approach. The different models used for defining memories make it hard

*This work was supported in part by the National Science Foundation under grant CCR-9106627.

to relate and compare them. We present a simple non-operational model and identify key parameters that can be varied to systematically define and relate the memories that have been proposed. Memories in our model are characterized by the *execution histories* that are allowed by a given memory model. Informally, weaker consistency places fewer demands on the execution histories and, as a result, permits a larger set of histories. Thus, we are able to use a set-based approach to relate and compare the various memories.

In the full paper, we use the model to demonstrate that Lamport's Bakery algorithm for solving the n-processor mutual exclusion problem executes correctly with release consistency when labeled operations are sequentially consistent (RC_{sc}) but does not when the labeled operations are processor consistent (RC_{pc}). This shows that the RC_{sc} and RC_{pc} models differ for applications that may use read and write operations to achieve mutual exclusion in a shared memory system.

2 The Model

This section describes the model that underlies our definitions and results. It is similar to those used by Misra [15] and Herlihy and Wing [9]. A system is a finite set of *processors* and a finite set of *locations*. Processors execute read and *write* operations. Each operation acts on a named location and has an associated value. For example, a write operation executed by processor p, denoted by $w_p(x)v$, stores value v in location x; similarly, a read operation, $r_p(x)v$, reports that v is stored in location x. The *execution* of a processor is a sequence of read and write operations. The execution history of processor p, denoted by H_p, is the sequence $o_{p,1}, o_{p,2}, \ldots, o_{p,i}, \ldots$, where $o_{p,i}$ is the ith operation issued by processor p. The set of processor execution histories is the *system execution history*.

We characterize various memories by the set of system execution histories that can be produced when processors execute with a certain type of memory. In particular, we need to develop rules to determine if a certain system execution history H is possible with a given type of memory. Our general approach consists of showing that each processor can assume that the memory has performed some set of operations in a *sequential order*. This set of operations must include the processor's operations and can also include operations of other processors to shared locations. Since a processor view is sequential, there is a unique "most recent" write preceding each read operation and it is required that each read operation return the value written by that write. As far as a processor is concerned, it can assume that the shared memory executed only the operations included in its view, one at a time, in the order defined by the view. This defines the state of the memory when an operation is executed by the processor.

For each processor p, there exists a *sequential* execution history $S_{p+\delta_p}$ that includes all operations in H_p as well as a subset δ_p of operations of other processors. Furthermore, the value v returned by a read operation to x is written by a write operation to x that precedes it in $S_{p+\delta_p}$. We say that such a sequential history is *legal*.

The consistency guarantees of a memory model place restrictions on the sequential histories for the processors. The following parameters characterize these restrictions:

1. Set of Operations: The membership of δ_p needs to be specified for a given memory model. Two natural choices for the set δ_p are, first, all operations of other processors, and second, all write operations of other processors. In the first case, we use a to denote δ_p for any processor p in the system and hence refer to $S_{p+\delta_p}$ simply as S_{p+a}. In the second case, we denote δ_p by w and hence use S_{p+w} instead of $S_{p+\delta_p}$.

There exist models that further distinguish memory operations. Examples of these include labeled operations in *release consistency* [6] and strong and weak operations in *hybrid consistency* [4].

2. Mutual Consistency: Although processors can define their own views of memory, there may need to be mutual consistency requirements as the views result from accessing a shared memory. For example, a memory model may require that all writes to a given location appear in the same order in the sequential histories for all processors. This particular form of consistency is equivalent to *coherence*, which is provided in several memory models [2,6].

3. Ordering: The order of the operations in the processor views must reflect somehow the actual ordering of these operations. *Program order* is one such commonly used order which states that $o_{p,i}$ is ordered before $o_{p,j}$ when $i < j$ ($o_{p,i}$ and $o_{p,j}$ are operations of processor p). A memory model may require that this or some other order derived from H be preserved between operations when they are included in the sequential execution history or view of a processor. The following orders are used in defining many of the memories.

- **Program order**: When $o_{p,i}$ precedes $o_{p,j}$ in the program ($i < j$), we say $o_{p,i}$ *is ordered before* $o_{p,j}$ *by the program order*. This orders all operations of a given processor.

  Some memory definitions consider *non-blocking operations* [7]. In this case, all orderings defined by program order may not be maintained and operations of a processor are only *partially ordered*.

- **Writes-before order**: Suppose that o_1 is a write to some location and o_2 is a read of the same location by a processor[1] such that o_2 reads the value written by o_1. We call this the *writes-before* order and it captures the natural requirement that, if a read operation returns

the value written by a certain write operation, then the write operation must be ordered before the read.

- **Causal order**: The *happens-before* relation defined by Lamport [11] can also be adapted to a shared memory system and captures the causal relationship between the read and write operations. This is the transitive closure of the union of program order and writes-before order.

3 Memories Definitions

The model developed in the previous section can be used to define a variety of memories. Basically, a memory can be characterized by specifying the three parameters we identified: set of operations, mutual consistency, and ordering.

3.1 Sequential Consistency

We first consider *sequential consistency* (SC) [12], which is a widely accepted correctness condition for shared memory. In SC, the results of a system execution history H should be equivalent to some sequential execution of the operations of all processors in which the program order between operations is maintained. This can be captured in our model by requiring that the view of processor p is a sequential history S_{p+a}. The ordering requirement for operations is the one defined by the program order relation. In the case of SC, there is no need for a mutual consistency requirement.

To see how memories with weaker consistency require all three kinds of requirements, we consider the *total store ordering* (TSO) [16] and the *processor consistency* (PC) [6] memory models that have been implemented in the SPARC and DASH architectures, respectively. We then consider other weak memories, including *pipelined RAM* (PRAM) [14] and *Causal Memory* [3].

3.2 Processor Consistency

We consider processor consistency as defined by Gharachorloo et al. [6]. This definition explicitly requires coherence that, for each memory location, there is a unique ordering of the writes to that location. If a processor executes a write followed by a read of a different location, PC allows the read to "bypass" the write. PC also allows writes by different processors to appear in different orders in processor views.

We define PC as follows. The view of a processor p includes not only the operations of p but also the write operations of all other processors. Thus, S_{p+w} denotes p's view. The mutual consistency condition is the following. If we denote by $S_{p+w}|_{w,x}$ the sequential history obtained from S_{p+w} after all read operations (to any location) and write operations to locations other than x have been deleted, then mutual consistency for PC requires that, for all processors p and q, $S_{p+w}|_{w,x} = S_{q+w}|_{w,x}$. It is easy to see that this mutual consistency condition implies coherence.

For ordering operations within a processor's view, PC uses a "semi-causality" relation which is defined by augmenting the weaker program order with weakened forms of the "writes-before" and "reads-before" orders[2] [2].

---

[1] We will use single subscripts when the extra information is irrelevant.

[2] *Reads-before* relates a read of an old value to a write of a new value.

3.3 Release Consistency

In the DASH architecture, processor consistency is provided only for memory operations that implement "synchronization" between processors. Such operations are called *labeled* and others are *ordinary*. *Release consistency* (RC) is designed to be used with programs that are *properly labeled*: all ordinary operations are "bracketed" between labeled operations that correspond to *acquire* (read) and *release* (write) operations on a synchronization variable. RC ensures that an ordinary operation completes before the following release operation is performed. This allows for added efficiency for ordinary operations. For a program to execute correctly, stronger consistency needs to be provided for synchronization operations. Two consistency requirements are identified: RC_{sc} guarantees that the labeled operations are SC, and RC_{pc} guarantees that they are PC.

In a system execution with a memory that provides release consistency, there are labeled operations as well as ordinary read and write operations. As in the case of most of the memories considered above, processor p's view must consist of all of its own operations and all write operations of other processors. The mutual consistency requirement is that of coherence. The ordering requirements for RC are somewhat complex. In the case of RC_{sc}, if $S_p|_l$ is the subsequence of S_p containing only labeled operations, then the sequences $S_p|_l$ meet the requirements of SC. A similar condition holds for RC_{pc}. The following conditions control the ordering of ordinary operations with respect to labeled operations:

- Let o be an ordinary operation of p that follows a labeled read operation (acquire) operation o_r of p. Let o_w be the write operation (possibly by another processor) that is read by operation o_r. Then, o follows o_w in all histories in which they both appear.

- If o is an ordinary operation of p that precedes a labeled write operation (release) operation o_w of p, then o precedes o_w in all histories in which they both appear.

3.4 Other Memories

We can also use our model to define other memories that have been presented in the literature. In particular, we consider TSO, PRAM and Causal Memory. TSO [16] can be characterized within our model, but space constraints do not permit its inclusion here. We discuss it in our full paper. PRAM [14] "pipelines" writes to memory, allowing the effects of writes (as perceived by other processors) to be arbitrarily delayed. Since a processor only executes its operations and the received writes of others, δ_p consists of the write operations of other processors. PRAM has no mutual consistency requirement. The ordering requirements are specified by program order. PRAM allows executions that are not allowed by TSO.

Causal Memory [3] is similar to PRAM in the sense that processor views include local operations and write operations of other processors. However, Causal Memory requires that the causal order be preserved between operations in processor views. Because the causal order is stronger than program order, there exist execution histories that are allowed by PRAM but not allowed by Causal Memory. There are also executions allowed by Causal Memory but not by TSO.

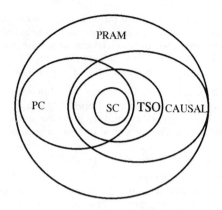

Figure 1: Relationship Between Memories

4 Relating Memories

Our model provides a natural framework for relating and comparing memories. A memory model is characterized by the set of system execution histories that are allowed by the model. Thus, to show that one memory model A is strictly stronger than B, we need to show that each system execution history allowed by A is also a system execution history of B. Furthermore, there must exist histories allowed by B that are not histories of A. We can also use a Venn-diagram representation for the set of histories allowed by various memories. In this representation, if A is strictly stronger than B, then A is contained in B. Figure 1 shows how the memories we discussed in the previous section are related. SC is the strongest and hence the set representing it is contained in the sets of all other memories. PRAM is the weakest memory. TSO is weaker than SC, incomparable to PC and is strictly stronger than Causal Memory. Causal Memory and PC are not comparable. This is because, while causal memory's ordering condition requires the stronger causal order, it lacks PC's mutual consistency requirement.

5 Distinguishing RC_{sc} and RC_{pc}

It has been claimed that the RC_{sc} and RC_{pc} memory models should be equivalent for most practical applications [6]. In this section, we show that a well-known algorithm that uses read and write operations to implement access to a critical section distinguishes RC_{sc} and RC_{pc}. The Bakery algorithm was proposed by Lamport [10] and assumes that the data used by it is stored in sequentially consistent shared memory. To execute the algorithm on an RC_{sc} memory, all memory operations executed to enter or exit the critical section are labeled. It is not hard to see that the program will now be *properly labeled* [2]. Gibbons et al. [8] showed that any program proved correct with SC will remain correct if properly labeled and run with RC_{sc}. Thus, the Bakery algorithm can correctly control access to a critical section when run with RC_{sc}. In the full paper, we show that the Bakery algorithm fails to execute correctly when the memory model is changed to RC_{pc} so that the labeled operations are now guaranteed only to be PC. Thus, the Bakery algorithm shows that RC_{sc} and RC_{pc} differ in power for applications that implement processor coordination with read and write operations.

6 Comparison with Related Work

In this paper, we developed a uniform framework that can be used to define a variety of memories. Other papers that have had similar goals are [15] and [16]. In [15], only atomic memory is considered, and this is stronger than sequential consistency. In [16], the TSO memory model is introduced using an axiom-based specification which can be used to capture several memories. The axiom-based model is also formal and precise and can be used to define and relate the memories as we have done in this paper.

We prefer the model used in this paper for several reasons. First, it is implementation independent but captures the essence of how memories can be implemented. For example, the per-processor view can be thought of as the behavior of a local cache; the set of operations and the ordering requirements for the view specify how the cache should be accessed and updated. Also, processor views are a natural extension of how sequential consistency is defined and understood. SC requires all processors to agree on their views, whereas weaker memories allow them to differ in the set of operations they include and also in the ordering between the operations. We believe that the parameters of the model identified by us make it easy to relate and compare the various memories.

7 Concluding Remarks

It is difficult to relate and compare the numerous high performance memories that have been proposed because they are often defined using different models. We developed a framework that allows the characterization of many of the existing memories. It identifies the parameters that can be varied to get the various memories and provides us a simple technique to relate them. We used our model to characterize memories that include sequential consistency, processor consistency, release consistency, total store ordering, PRAM and causal. The model makes it easy to understand the relationships between the memories. We also used it to show that the RC_{sc} and RC_{pc} models are not identical when read and write operations are used to implement processor coordination.

We have focused on the characterization of and the relationships between different existing memories in this paper. However, the model also helps us in identifying new memories. For example, a mutual consistency condition that requires coherence can be added to causal memory or perhaps such coherence can only be required for labeled operations. The model can also help us to precisely characterize the program model for a given memory which can be used to identify the programs that will execute correctly on a certain weak memory when they are correct on sequentially consistent memory. We will address these issues in our future work.

References

[1] Sarita V. Adve and Mark D. Hill. Weak ordering — a new definition. In *Proceedings of the 17th Annual International Symposium on Computer Architecture*, 1990.

[2] Mustaque Ahamad, Rida Bazzi, Ranjit John, Prince Kohli, and Gil Neiger. The power of processor consistency (extended abstract). In *Proceedings of the Fifth Symposium on Parallel Algorithms and Architectures.* ACM Press. June 1993.

[3] Mustaque Ahamad, James E. Burns, Phillip W. Hutto, and Gil Neiger. Causal memory. *Proceedings of the Fifth International Workshop on Distributed Algorithms*, volume 579 of *Lecture Notes on Computer Science*, Springer-Verlag, October 1991.

[4] Hagit Attiya and Roy Friedman. A correctness condition for high performance multiprocessors. In *Proceedings of the Twenty-Fourth ACM Symposium on Theory of Computing*, May 1992.

[5] Michel Dubois, Christoph Scheurich, and Faye Briggs. Synchronization, coherence, and event ordering in multiprocessors. *IEEE Computer*, Feb 1988.

[6] Kourosh Gharachorloo, Daniel Lenoski, James Laudon, Phillip Gibbons, Anoop Gupta, and John Hennessy. Memory consistency and event ordering in scalable shared-memory multiprocessors. In *Proceedings of the Seventeenth International Symposium on Computer Architecture*, May 1990.

[7] Phillip B. Gibbons and Michael Merritt. Specifying nonblocking shared memories (extended abstract). In *Proceedings of the Fourth Symposium on Parallel Algorithms and Architectures*, June 1992.

[8] Phillip B. Gibbons, Michael Merritt, and Kourosh Gharachorloo. Proving sequential consistency of high-performance shared memories (extended abstract). In *Proceedings of the Third Symposium on Parallel Algorithms and Architectures*, July 1991.

[9] Maurice P. Herlihy and Jeannette M. Wing. Linearizability: A correctness condition for concurrent objects. *ACM Transactions on Programming Languages and Systems*, July 1990.

[10] Leslie Lamport. A new solution of Dijkstra's concurrent programming problem. *Com. ACM*, Aug 1974.

[11] Leslie Lamport. Time, clocks, and the ordering of events in a distributed system. *Com. ACM*, July 1978.

[12] Leslie Lamport. How to make a multiprocessor computer that correct executes multiprocess programs. *IEEE Transactions on Computers*, Sept 1979.

[13] J. Lee and U. Ramachandran. Synchronization with multiprocessor caches. In *Proceedings of the 17th Annual International Symposium on Computer Architecture*, 1990.

[14] Richard J. Lipton and Jonathan S. Sandberg. PRAM: A scalable shared memory. Technical Report 180-88, Department of Computer Science, Princeton University, September 1988.

[15] Jayadev Misra. Axioms for memory access in asynchronous hardware systems. *ACM Transactions on Programming Languages and Systems*, Jan 1986.

[16] Pradeep S. Sindhu, Jean-Marc Frailong, and Michel Cekleov. Formal specification of memory models. Technical Report CSL-91-11, Xerox Corporation, Palo Alto Research Center, December 1991.

Real-Time Control of a Pipelined Multicomputer for the Relational Database Join Operation.

Yoshikuni Okawa, Yasukazu Toteno, and Bi Kai

Faculty of Engineering, Osaka University

Yamadaoka, Suita, Osaka 565, Japan

Abstract—We propose a database system which consists of a personal computer, disks and a pipeline of microcomputers. The relational join is a target. We find that there exist two different algorithms: one contributes the speed up of the pipeline cycle time, and the other reduces the number of the necessary disk accesses. The controller watches the state of the pipeline, and if the pipeline is critical, then it switches to the high speed algorithms. Likewise, the controller selects the most appropriate strategy for the given situation. We have built an experimental parallel processing system. The results show the feasibility of the proposed control algorithm.

I. INTRODUCTION

With the advent of personal computers and workstations the cost of the small scale computers and the peripheral equipments (e. g., disks) drops rapidly. While a special purpose database machine could improve the performance, it fails to compete with a no-specific database machine which is merely a combination of an existing workstation, disks, and networks. This, surely, is another outcome by the principle of mass-production. As DeWitt[1][2] has already pointed out, this ordinary type of databases is, and will be the main stream of the present and future database machines, at least, in the commercial sector.

Our database machine consists of a high performance personal computer (controller), disks (data storage), and microcomputers (processing element). We take a relational database as a sample database, and the join[3] as the target operation. Though the band width of data stream in this system is not so high as that of a special purpose database machine, the performance to cost ratio is incredibly high.

The join operation of a relational database gives the heaviest load to the processor. To overcome the comparably low speed of microcomputers we adopt a pipeline architecture[4] in which multiple microcomputers are connected by bank-switchable memory blocks in a line. There are a few reports which execute the join operation in various parallel processors, under the condition of all necessary data being in RAMs[5]. This condition is not realistic, since we are always able to consider a larger relation (or a table) than the memory limit. By this reason we strictly define the processing time as a time interval from the start of the first table read from disks till the end of the last table write.

There is no urgent need to speed up processing elements, since, if speeded up, we have only to wait the disk operation. The crucial issue is not the problem of speed up, but the balance between the disk access operation and the processing capability. We call an algorithm to execute the join operation in the processing pipeline as a "strategy." First, we propose three different strategies. Next, we have investigated the characteristics of their processing speed. We find that the first strategy contributes to the speed up of the pipeline cycle time, the second shortens the disk accessing time, and the third is a mix of the two strategies.

The controller watches the status of the processing pipeline and the disk access operations. If it finds that the disk access is the bottleneck, then it switches to the lighter disk access strategy. Conversely, if the pipeline becomes the bottleneck, the controller adopts the higher cycle time strategy. Thus, by changing strategies the controller keeps the balance between the disk access operation and the processing speed. This adaptive control scheme is the central idea of our database processing system.

We actually have built an experimental database system with a parallel processing pipeline. Using a database kept in the library of our school several benchmark tests have been executed. The results show that our control scheme works nicely.

II. PARALLEL PROCESSING SYSTEM

We adopt a pipeline as a base architecture. It consists of microcomputers and bank switchable memory blocks. A microcomputer has four ports, and a bank switchable memory block has two. A microcomputer has its own memory, which is called a local RAM.

A pipeline is a linear array of microcomputers, where two dual-ported bank switchable memory blocks are inserted in between two consecutive microcomputers. Fig. 1 shows the basic configuration of our database system. The controller is a complete computer. A pipeline is plugged into its extended bus. Tables are stored in the disks. The controller can access the first two and last two of the bank switchable memory blocks, but not others. A microcomputer has four bank memory blocks (two in its front and two in its rear), and its own local RAM.

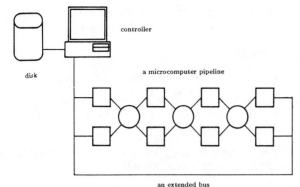

Fig. 1   The basic structure of our system.

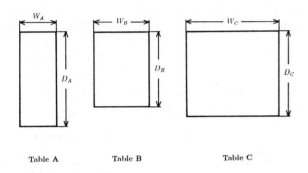

Fig. 2   The related three tables and their symbols.

III. THE RELATIONAL JOIN OPERATION

Basically, the relational join is an operation between two independent relations (use the word 'table' instead of 'relation' hereafter.) As indicated in Fig. 2, we call two given source tables as Table A, and Table B. The resulted table is Table C. The number of tuples (or rows) in each table is denoted by D_A, D_B, and D_C. The number of bytes (or width) of each table is w_A, w_B, and w_C. Each source table is divided into equal sized subtables. Call this subtable as a page. The number of pages in each table is n_A, n_B, and n_C. By the definition of the join operation a page of Table C is generated from a pair of two source tables. Thus, we have

$$n_C = n_A n_B. \tag{1}$$

r is the averaged number of tuples in a page of Table C, which is defined by

$$r = \frac{D_C}{D_A D_B}. \tag{2}$$

This factor r plays the most important role in the relational join, since r determines the substaintial part of the disk access (Table C) and besides we have no means to estimate it before an operation begins.

IV. THE BASIC MODEL OF THE PIPELINE JOIN

We propose the basic model to execute the relational join operation in this type of pipelines. For simplicity we adopt the nested loop algorithm[3]. Two source

tabels (Table A and B) are stored in a disk, where A is the inner table and B is the outer. The resulted Table C has also to be stored in a disk. The processing time is defined as a time interval from the first fetch of Table A to the last store of Table C. We assume the total volume of the storage is large enough to accommodate all necessary tables. No overflow in the storage devices is considered.

The basic model of the pipeline join operation

The controller reads one or more pages of Table A from the disk, and feeds them to the pipeline one at a time. Each microcomputer in the pipeline takes a designated page, and stored it in its local RAM. We call this process 'A transfer.' After A transfer, each microcomputer in the pipeline has one and strictly one page of Table A.

Next, the controller feeds each page of Table B into pipeline from the first page to the last. Each microcomputer does the join using a page of Table A in its local RAM and a page of Table B in a bank memory. "What parts of two tables are joined, and how, etc." is given by the controller as a program, which is stored in the local RAM. We call this process as 'B transfer.'

'A transfer' plus 'B transfer' is called a pipeline stage or simply a stage. After a stage, if no page of Table A is left, then the whole process ends. If some pages of Table A are left to be processed, then the next stage is started.

end of the model description

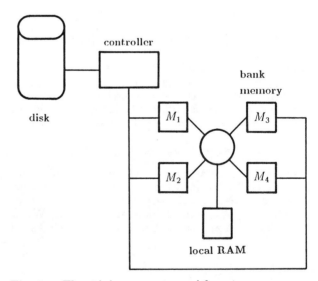

Fig. 3   The minimum system with one processor.

First, we will consider the simplest case of single microcomputer in a pipeline, and derive the equation to estimate the total processing time. Next, we try to extend the result to the general case of p microcomputers in a pipeline.

The configuration of the simplest case is shown in Fig. 3. Only one microcomputer and four bank memory blocks are implemented in the pipeline. The controller takes the first page of Table A (write as A_1, similar with other pages) from the disk, and sends it to the bank memory M_1. After that transfer the controller switches the memory to the microcomputer. The microcomputer moves A_1 from M_1 into its local RAM. After this move operation the microcmputer switches the memory to the controller. While this move is being executed in the pipeline, the controller begins to read B_1 from the disk and write it into M_2. This is a 'A transfer.' After that, B transfer is initiated.

Now, we derive the equation to estimate the total processing time. Assume n_A, and n_B is large, since we are dealing with databas. Under this assumption we could concentrate onto the most frequently repeated part of the pipeline join processing, and discard all other miscellaneous trivial parts. The critical path is, clearly, the parallel execution part. In this interval each microcomputer combines two pages of Table A and B, and constructs one page of Table C. The controller gets one page of Table C and writes it into the disk, and also takes one page of Table B from the disk and sends it into a bank memory. We call this critical section as the base cycle. Since the base cycle is determined by the larger of the two, we have

$$t_{cycle} = max(t_{AB}, t_B + t_C) \qquad (3)$$

where t_{AB} is the expected time a microcomputer joins two pages, t_B is the expected time of reading one page of Table B from a disk, and t_C is the expected time of storing one page of Table C into a disk.

In each stage one page of Table A and whole pages of Table B are sent to the pipeline. The time for one stage cycle is approximately estimated by

$$t_{stage} = t_A + n_B max(t_{AB}, t_B + t_C) \qquad (4)$$

where t_A is the time to fetch one page of Table A and store it into a bank memory. Eq.(4) does not hold strictly. It must be interpreted as the approximation when n_B is sufficiently large. Assume that Eq.(4) holds. Then, the total processing time is given by

$$T = n_A(t_A + n_B max(t_{AB}, t_B + t_C)). \qquad (5)$$

Now we turn to a general case of the pipeline join. The difference is that we have p microcomputers in a pipeline instead of a single processor in the above example. We propose three different algorithms. We call this algorithm a "pipeline strategy", or simply "strategy."

α strategy

- All microcomputers have the same page of Table A in their local RAMs

- Each microcomputer joins full page of Table A and $\frac{1}{p}$ of Table B.

Clearly, the α strategy decreases the process time of a base cycle by $\frac{1}{p}$. Then, Eq.(5) becomes

$$T\alpha = n_A(t_A + n_B max(\frac{t_{AB}}{p}, t_B + t_C)). \qquad (6)$$

Intuitively speaking, the α strategy speeds up the apparent processing ability of a pipeline by p.

β strategy

- Every microcomputer has a different page of Table A in its local RAM.

- Each microcomputer joins one page of Table A and one full page of Table B.

Clearly, the β strategy has a virtual effect to enlarge the local RAM of a processor by p. Eq.(5) becomes

$$T_\beta = \frac{n_A}{p}(pt_A + n_B max(t_{AB}, t_B + pt_C)). \qquad (7)$$

Intuitively speaking, the β strategy has an effect to decrease the number of accesses of Table B by increasing the capacity of a local RAM in a processor.

$\alpha\beta$ strategy

- Divide p into p_α and p_β, such that $p = p_\alpha p_\beta$.

- p_α microcomputers execute the α strategy.

- p_β microcomputers execute the β strategy.

Clearly, the $\alpha\beta$ strategy speeds up t_{AB} by p_α, and also decreases the number of the disk accesses to $\frac{1}{p_\beta}$. The total processing time becomes

$$T = \frac{n_A}{p_\beta}(p_\beta t_A + n_B max(\frac{t_{AB}}{p_\alpha}, t_B + p_\beta t_C)). \qquad (8)$$

The $\alpha\beta$ algorithm shows an intermediate characteristics between the α strategy and the β strategy.

V. ANALYSIS

We have analyzed the proposed computing model, which is summarized as follows:

- If r is large, the system tends to become disk-bound.

- If r is small, the system is likely to be in the pipeline-bound state.

- If the number of tuples to be joined γ increases, a page of Table C can not be stored in a single bank memory, which we call a pipeline jam. This jam greatly deteriorates the processing capability of the pipeline.

- If γ is small, the number of disk accesses increases. Thus, there exists an optimum γ.

VI. ADAPTIVE CONTROL OF A PIPELINE

n_A and n_B, and the control strategy have to be optimized to shorten the total processing time. This is the real-time control problem of this system. r is the most influential variable in the total processing time, but the controller has no means to measure it. Also we must anticipate that r varies while a processing proceeds. Different pages have possibly different outcomes. Thus, r becomes the major disturbance to the control system.

Our basic idea is that by selecting a suitable startegy in a given situation, we are able to keep an operation of a pipeline in an optimum state in that given situation.

The control algorithm for the pipeline join

- In the first one stage select an arbitrary strategy.

- Record the result.

- If the controller had been fully busy in that stage, that is, the system is in the disk-bounded state, then, select the β strategy in the next stage.

- If the controller had been idle, that is, it is in the pipeline-bound, then, select the α strategy in the next pipeline stage.

- In both cases if the pipeline is in the jam, then decrease γ.

- If the pipeline is not in the jam, then increase γ.

- Repeat a stage, until the whole operations are finished.

end of the algorithm

This adaptive control algorithm has one delay of a stage, but it can follow the change of r in real-time. The algorithm we are proposing consists entirely of changes in its control softwares, and does not need any physical change like changing cable connections of networks by switches, etc. The necessary cost to execute this strategy is almost zero. This is the core of our proposals.

VII. Experiments

We have designed and built an experimental system from a commercially available personal computer and its disks. The six microcomputers are connected in a pipeline. Thus, the following four strategies are available:

- α strategy $\qquad$ ($p_\alpha = 6$, and $p_\beta = 1$)

- $\alpha\beta 1$ strategy $\qquad$ ($p_\alpha = 3$, and $p_\beta = 2$)

- $\alpha\beta 2$ strategy $\qquad$ ($p_\alpha = 2$, and $p_\beta = 3$)

- β strategy $\qquad$ ($p_\alpha = 1$, and $p_\beta = 6$)

We use the databases in our library as the test tables. First, the controller processes the database using a sigle strategy. Six different pairs of tables are tested. In each of these experiments the controller keeps a single strategy from the beginning to the end. That is, there is no strategy change in this experiment.

Next, the adaptive control scheme is implemented in the controller. Starting from each strategy, the controller changes the strategy according to the algorithm shown in the previous section. All the computing time converge to the optimum values. We concluded that the feasibility of our proposed method is verified.

VIII. Conclusion

We try to build a relational database system combining a high performance personal computer and its disks. The processing system consists of microcomputers linearly connected by bank switchable memory blocks. The target is the join operation.

We have found that there exist two completely different strategies to execute the relational join in a pipeline. The difference comes from the different usages of CPU and memory resources. One strategy utilizes the CPU resource to speed up the pipeline stage, while the other uses the memory resource to reduce the number of necessary disk accesses. The intermediate strategies between the two are also possible. To implement these strategies only changes in software are required. No extra hardware is necessary. This is the greatest merit of this algorithm.

The most important factor in database operation is the variation of the join ratio r. This variation affects directly the efficiency of the parallel processing system. Generally speaking, if r increases, then, the system is apt to become the disk bound state. On the contrary, if r decreases, the system tends to be in the pipeline bound.

To overcome this difficult situation we adopt the concept of the feedback control. The basic idea is that if the controller is busy, then the disk access is the bottle neck, and if the controller is idle then the pipeline is critical. If the disk access is critical, we can shift the strategy to a slower speed and less frequently disk access startegy, and vice versa.

We have built an experimental parallel processing system. Test data are taken from the library. Several benchmark tests are executed. By the inspection we conclude the proposed algorithm contributes to the increase of the overall system efficiency. There are many exciting problems to be attacked in this new type of databases, which is practical and also fruitful.

IX. Reference

1. H. Boral, and D. DeWitt, Database Machine: An idea whose time has passed? A critique of the future of database machines, Proc. of the 1983 Workshop on Database Machine, pp. 166-187, 1983.

2. D. DeWitt, and J. Gray, Parallel Database Systems: The future of high performance database systems, Comm. of the ACM, Vol. 35, No. 6, pp. 85-98, 1992.

3. P. Mishra, and M. H. Eich, Join processing in relational database, ACM Computing Survey, Vol. 24, No. 1, pp. 63-113, 1992.

4. W. Kim, D. Gajski, and D. J. Kuck, A Parallel Pipelined Relational Query Processor, ACM Trans. on Database Systems, Vol. 9, No. 2, pp. 214-242, 1984.

5. O. Frieder, Multiprosessor Algorithms for Relational-Database Operations on Hypercube Systems, IEEE Computer, Vol. 23, No. 11, pp. 13-28, 1990.

P³M: A Virtual Machine Approach to Massively Parallel Computing
(Extended Abstract)

Fabrizio Baiardi
Dipartimento di Informatica
Università di Pisa
Pisa, Italy
baiardi@dipisa.di.unipi.it

Mehdi Jazayeri
Hewlett Packard Laboratories
1501 Page Mill Road
Palo Alto, California 94304
jazayeri@hpl.hp.com

While technology improvements in VLSI favor the building of massively parallel computers, the key obstacle to the widespread adoption of such computers is the lack of a methodology for developing software that achieves both portability and high performance. This paper presents a virtual machine intended to serve as the cornerstone of such a methodology. The P³M virtual machine can be mapped efficiently to different physical machines and enables the development of software tools and applications that can be ported across different parallel machines.

INTRODUCTION

Technology improvements in VLSI favor the building of massively parallel computers as viable alternative to traditional supercomputer systems. A massively parallel computer includes a large number of nodes, each including a processor and a local memory; nodes are connected by a sparse interconnection network that supports message exchange among the nodes. The interconnection network must be sparse so that it may scale to massively parallel systems with up to 10^3-10^6 nodes.

The key obstacle to the widespread adoption of such parallel computers is the difficulty in program development. Indeed, because the main power of a parallel computer is derived from its ability to execute parallel threads of computation, an application has to be decomposed into a large number of parallel components–referred to as processes, actors, or objects depending on the computational model underlying the adopted programming language–and each component has to be assigned, or *mapped*, to a node of the system.

The complexity of software development may be reduced through the introduction of a programming environment for the development of parallel applications. The key components of such an environment are a methodology, a programming language supported by appropriate tools, and a virtual (or abstract) machine for massively parallel systems.

After discussing the main problems posed by the adoption of a massively parallel system, we discuss the advantages offered by the adoption of a virtual machine and introduce the virtual machine P³M

along with the rationale behind its definition. We demonstrate the machine's generality by outlining the implementations of run time systems for CSP and OR-Parallel Prolog, two languages which represent two very different styles of concurrent programming.

P³M is being developed in the context of the Pisa Parallel Processing Project (P4) whose goal is a complete methodology and environment for developing portable software tools and applications for massively parallel machines. Our target machines are characterized as MIMD with direct networks. P4 is a collaborative research effort of the Dipartimento di Informatica, Università di Pisa and the Pisa Science Center, Hewlett-Packard Laboratories.

PROGRAM MAPPING AND MASSIVELY PARALLEL SYSTEMS

To be executed on a massively parallel system, an application must first be decomposed into a large number of parallel, communicating components and then each component must be mapped to the system nodes. Mapping has a major influence on both the execution time and on efficient resource utilization. This is mostly due to the sparse interconnection structure among system nodes.

The mapping of an application consists of trying to superimpose the program graph, i.e. the graph that describes the communications among program components, onto the system graph, i.e. the one describing the system interconnection structure. In general, the two graphs may be structurally very different and, as a consequence, two communicating components may be mapped onto two nodes that are not directly connected; to allow two components to communicate independently from their allocation, some *message routing* functions must be supported by each node of the system. A *local* communication is a communication that is implemented through a message that crosses at most one link; any other communication is *nonlocal* and it requires the execution of some routing functions. The goals of any mapping strategy are:

i) to balance the computational load among the system nodes;

ii) to balance the load among the communication resources of the system;
iii) to exploit any locality in the program.

P³M: A VIRTUAL MACHINE FOR MASSIVELY PARALLEL SYSTEMS

Before discussing P³M and introducing its instruction set, we first present the rationale behind its design. P³M stands for *Parallel Processor, Parallel Memory*.

Rationale for P³M

The main idea behind the definition of P³M is to enable the partitioning of physical resources of a physical machine among computation, local communication, and nonlocal communication, at the virtual machine level. This ability is fundamental when efficient resource utilization is a goal or if a set of applications with broadly different communication and computation loads has to be supported by the same system. P³M assumes that a software configuration of the interconnection network is possible with satisfactory performance. Since each node of the system includes a processor, then it may be specialized and devoted to the support of functionalities such as message routing or emulation of a shared memory and the performance of this solution may be close to that of a system that exploits special-purpose components. This assumption is in line with the development of technology and it is the main starting point of P³M. Under this assumption, the amount of resources devoted to the support of process communication can be chosen on the basis of the application to be executed.

Partitioning the Resources of a Physical Machine

The specialization of a physical architecture into a P³M virtual machine takes place through a *configuration* step that partitions system nodes into two sets, the *processing set* and the *data set*. Nodes in the processing set are devoted to the execution of processes of the program while those in the data set support nonlocal communication and the sharing of information. Both the processing and the data set can be structured, i.e. partitioned into subsets. In this case, nodes in distinct subsets of the data, or processing set are connected only by paths crossing nodes belonging to the processing or data set. An example is shown in Fig. 1. This structuring may be useful if information has to be shared among a subset of the processes only or nonlocal communications involve distinct subsets of the processes. Given a massively parallel system, a configuration, and a program, processes can be mapped onto the nodes in the processing set. Communications between processes mapped onto nodes of the processing set that are directly connected are

implemented through the link between the nodes; other communications are implemented through the nodes in the data set. To guarantee that any process communication can be implemented, the mapping has to satisfy the **nonlocal communication constraint**:

> two processes that communicate or share information have to be mapped onto the same node or onto two nodes that are either directly connected or both connected to the data set (to the same subset of the data set if several subsets exist).

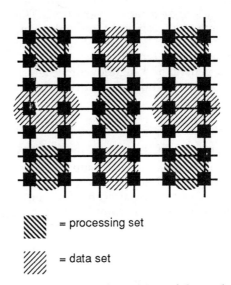

⬛⬛⬛ = processing set

▨▨▨ = data set

Fig. 1 A structured partition of the nodes

This constraint guarantees that either a communication is local, and it will be implemented through just one link, or that it is not local, and it will be supported by nodes in the data set.

The partitioning of nodes into two sets corresponds to the introduction of two interconnection networks:

i) a direct one, including all the links between nodes in the processing set, used for local communications,
ii) an indirect one, including all the remaining links and the nodes of the data sets, used for nonlocal communications.

This solution guarantees that

i) program locality is exploited since local and nonlocal communications are implemented through distinct sets of resources and no interference between the two kinds of communication is possible;
ii) the routing of nonlocal communication and the management of shared information do not interfere with the computation of the processes.

The P³M virtual machine leaves open the binding of a

node to the processing or to the data set. This allows a physical machine to be configured according to the needs of a specific application. At one extreme, if the application requires no global communication, the data set is empty. The modeling step may be eventually done automatically, by a modeler programming tool.

Shared Tuple Space

The partitioning of nodes into two sets implies that the virtual machine includes two kinds of instructions for local and nonlocal communication: the instructions for local communication are fairly standard message exchange instructions. The instructions to implement nonlocal communication, instead, are read/write operations on a shared memory organized in the style of Linda tuples [Carriero89]. Tuples provide a general and high-level mechanism for information sharing among the nodes in the processing set.

The Instruction Set

For the purposes of this paper, the important point about the P^3M machine is the support of both local communication and access to a shared tuple space; due to space restrictions, we cannot present the complete instruction set. The complete set of operations supported by our current prototype is given in [Mackey92].

Any P^3M machine must support:

  i) traditional sequential operations,
  ii) local communication primitives
    - send(ConNo, ProcessId, ChannelNo, Var),
    - receive(ConNo, ProcessId, ChannelNo, Var),
    - test(ConNo, ProcessId, ChannelNo),
    - test&receive(ConNo, ProcessId, ChannelNo, Var),
  iii) tuple primitives
    - out(RelationName, <tuple>),
    - in(RelationName, <tuple>),
    - read(RelationName, <tuple>)
    - test(RelationName,<tuple>)& update(<op_tuple>),
    - inp(RelationName, <tuple>),
    - readp(RelationName, <tuple>),
    - testp(RelationName,<tuple>)& update(<op_tuple>).

USING P^3M TO SUPPORT HIGH LEVEL PROGRAMMING LANGUAGES

An important requirement of a virtual machine for massively parallel systems is that it must support diverse programming languages efficiently. This means that different programming models, or runtime systems for different classes of programming languages, can be implemented in a highly parallel fashion. This section shows that this is indeed the case for P^3M.

We show how two representative programming languages, CSP and Prolog, may be implemented on a P^3M machine. CSP and Prolog span a wide range of programming languages. The CSP implementation represents the class of static applications with explicit parallelism where the programmer explicitly creates the threads of execution and where new threads may be activated but not created at run time. Prolog represents the class of search-tree applications with implicit dynamic parallelism where the run time system implicitly generates and controls the threads according to the structure of the tree considered in a given execution.

We stress that the role of P^3M is to support local communication and information sharing in a language-independent way. The virtual machine to support the execution of a given programming language has to be built on top of P^3M through the functionalities supplied by P^3M

Explicit Parallelism

A CSP program may be described as a network of processes where processes are connected by unidirectional channels. To execute a CSP program, first of all the physical machine has to be configured as a P^3M machine, i.e. the nodes have to be partitioned into processing and data sets. The main parameter affecting this configuration is the "distance" between the interconnection structure of the physical architecture and the logical interconnection among the CSP processes. This distance is proportional to parameters such as the difference between the degree of each node and the average number of channels used by each process. If the distance is large, then a large number of nodes is included in the data set, otherwise a small number, or even none, is included.

After choosing a configuration, the processes of the program are mapped onto nodes of the processing set. The mapping has to satisfy the nonlocal communication constraint, and its goals are load balancing and exploitation of locality, i.e. two heavily communicating processes should be mapped onto directly connected elements of the processing set. Given a configuration and a process mapping satisfying the nonlocal communication constraint, the logical channels of the program can be partitioned into two sets: those that can be implemented through P^3M channels and those to be implemented through the data set. A channel belongs to the first set if it connects two processes mapped onto directly connected nodes of the processing set. All other channels belong to the second set.

Assuming that a configuration has been chosen and a mapping satisfying the local communication constraint has been determined, we next consider the implementation of CSP communication channels on P^3M.

The implementation of CSP channels and I/O commands through P^3M channels and corresponding operations is fairly straightforward. Therefore, here we only consider the implementation of a CSP channel through the tuples of the data set. A CSP channel includes two distinct pieces of information: the value to be transmitted and the status of the communication, i.e. whether one or both of the processes connected by the channel are ready to communicate. Furthermore, the status of processes has to be recorded since communication with a terminated process leads to a failure. These two pieces of information can be represented by two separate tuples belonging, respectively, to two relations, Status and Msg. Both tuples are identified through a key that identifies the channel. The status of a communication is represented by the status of the channel being used and both the status of the sending and the receiving process. The possible states of a channel are "empty" (initial state), "receiverReady" and "senderReady" denoting, respectively, that neither the sender nor the receiver have committed to a communication, the receiver has committed and the sender has committed to a communication. The status of a process is "active" or "terminated."

The Status relation is of the form:

(Status, <ChanId, ChanStatus, ProcStatus, MsgNo>).

ChanId is an integer that identifies the CSP channel and it is used as a key of the tuple. MsgNo is an integer used to tag and match the messages transmitted along the channel.

The Msg relation is of the form:

(Msg, <ChanId, MsgValue, MsgNo>)

In both relations, the MsgNo field records the number of currently in-progress communication.

Given these two relations, it is relatively straightforward to implement the i/o commands of CSP. The main task of the CSP runtime support executed by each node of the processing set is to hide the implementation of a logical channel so that the code produced by the CSP compiler sees a uniform interface. The ability of operating on CSP channels independently from their implementation is important especially for nondeterministic constructs that involve several CSP channels, each of which can be implemented through either a P^3M channel or through tuple operations.

Implicit Parallelism

By illustrating the problems posed by the implementation of a declarative language and the range of solutions offered by P^3M, this section shows how P^3M helps in the high-level design of a run-time system by allowing the stepwise refinement of a series of increasingly more parallel versions of the system.

An OR-parallel implementation of Prolog carries out many possible executions in parallel [Warren87]. These possible executions can be described by a search tree where a path from the root of the tree to a leaf node describes a possible execution. Each node of the tree is associated with a goal; the root is associated with the initial query to be refuted. Each node of the tree also has an associated execution environment including the bindings produced for the program variables so far in this execution. Each node produces its environment by unifying one atom of the goal associated with its parent with one clause of the program and by adding the bindings returned by the unification to those of its parent. The children of a node are all the possible executions that may follow from the current node, and they may be performed in parallel. A leaf is associated either with the empty goal, corresponding to a successful refutation, or to a goal whose atoms cannot be unified with any clause, i.e. a failure.

An OR-parallel implementation includes several parallel instances of a Prolog interpreter connected to a shared memory and exploring distinct paths. The environments produced and used by each instance represent a large amount of shared data. As a matter of fact, two instances visiting two distinct nodes of the tree share all the bindings associated with the two nodes' common ancestors in the tree. The management of this shared data is key to efficient execution of OR-parallel Prolog.

Each instance of the interpreter visits the tree in a left-to-right order. When reaching a choice point, i.e. a node with more than one child, it chooses one path and records the others in its stack. The stack is stored in the shared memory together with the bindings produced by the unifications along the path currently considered. When reaching a leaf node, an instance I1 can either backtrack to one of the choice points stored in its stack or, if the stack is empty, steal a choice point C from the stack of an instance I2 and switch to C. When such a switching takes place, I1 needs to access all the bindings produced on the path from the root to C. This is why bindings are stored in the shared memory.

A first, naive implementation of an OR-Prolog system on P^3M would use one process in the processing set for each interpreter instance and store all the environments and instance stacks in the data set to be shared by all instances. As an optimizing refinement, a complex scheduling policy could be used to assign choice points when an interpreter is ready to switch. For example, the overhead of switching to a subtree that is too small dominates any benefit of a parallel search. To prevent such a scheduling policy from becoming a further source of overhead, it should be implemented by a separate scheduling process S(Ij), one for each instance Ij. S(Ij) knows the node currently considered by Ij and it searches the stack of other instances, stored in the data set, to determine the choicepoint for S(Ij) to follow when it reaches a leaf. S(Ij) can inform other scheduling processes when Ij inserts a choicepoint in its stack. S(Ij) is mapped onto a node of the processing set connected both to the data set and to the node where Ij has been mapped. According to the scheduling policy, the cooperation among the scheduling processes can be implemented through P^3M channels only or both through channels and the data set.

As a further optimization, each instance Ij can be decomposed into two processes MUj and UPj. While UPj attempts the unification of an atom of a goal with the head of a program clause, MUj retrieves from the data set the tuples recording the bindings required by UPj. In other words, MUj loads the local cache of UPj in parallel with the computation of UPj. Obviously, MUj has to be connected to the data set, while UPj has to be connected only to MUj and S(Ij).

CURRENT STATUS

The P^3M machine described in this paper has been developed in the context of a larger project at the HP Pisa Science Center to develop a complete environment and methodology for the building of massively paralle programs. We have developed a prototype of the complete environment. Here we list a sample of the tools that have been implemented:

Parallel Tuple Space: To experiment with, and to evaluate, distribution strategies for tuples in a virtual machine environment and, in particular, their relation with message routing strategies, a prototype of the data set of P^3M was implemented on the Meiko Computing Surface consisting of 32 nodes; the system was partitioned into processing and data sets, 16 nodes each [Caselli92].

Virtual Machine Emulator: To gain experience with the use of P^3M in modeling different physical machines, a software emulator of P^3M has been designed and built[Mackey92]. This emulator runs on a Unix workstation.

Tuple Space Animator: To visualize the behavior of allocation and routing strategies used in the implementation of the tuple space, as well as to visualize the access patterns to the tuples, a Tuple Space Animator has been implemented. The animator plots various data about the use of the tuple space as an application is executing.

Machine Configurator: To investigate the feasibility of an automatic configuration of an architecture into a P^3M machine, we have developed a programming tool that given an application and an architecture, configures the architecture into a P^3M machine by applying a genetic algorithm to implement a stepwise refinement of the configuration.

RELATED WORK

The definition of general-purpose massively parallel systems is currently pursued by several research projects [Dally91, Ranade87, May90, Johnsson90] The research issues of this area are reviewed in [Baiardi91]. The tuple space model of P^3M is based directly on Linda [Carriero89]. The full version of this paper, [Baiardi92] compares the different approaches.

REFERENCES

[Baiardi91] F. Baiardi, M. Danelutto, M. Jazayeri, S. Pelagatti and M.Vanneschi. Architectural models and design methodologies for general purpose highly-parallel computers. IEEE CompEuro'91, Bologna (Italy), May 1991, pp. 18-25.

[Baiardi92] F. Baiardi and M.Jazayeri, An abstract machine for highly parallel architectures. Report No. HPL-PSC-92-37, Hewlett-Packard Laboratories, Pisa Science Center (Italy), June 1992.

[Carriero89] N. Carriero and D. Gelernter. How to write parallel programs: a guide to the perplexed. ACM Computing Surveys, 21(3) 323-58, Sept. 1989

[Caselli92] F. Caselli. Modello di cooperazione a comunicazione generativa: analisi delle caratteristiche e possibili implementazioni. Laurea Thesis, Dip. di Informatica, Univ. di Pisa, March 1992.

[Dally91] W. J. Dally, D. S. Wills, and R. Lethin. Mechanisms for parallel computers. NATO Advanced Study Institute on Parallel Computing and Distributed Memory Multiprocessors, Springer-Verlag, 1991.

[Hoare78] C. A. R. Hoare. Communicating Sequential Processes. Commun. of the ACM, 8(21): 666-667, August 1978.

[Johnsson90] S. Johnsson. Communication in Network Architectures. In VLSI and Parallel Computers, R. Suaya and G. Birtwistle eds, Morgan Kaufmann, 1990.

[Mackey92] M. Mackey and T. Sullivan. P3M Machine Interface Definition. Technical Report. Hewlett-Packard Labs, Pisa Science Center, 1992.

[May90] D. May and P. Thompson. Transputers and routers: components for concurrent machines. INMOS Tech. Rep. 1990.

[Ranade87] A. G. Ranade. How to Emulate Shared Memory. In Proc. of 28th IEEE Symposium on Foundations of Computer Science, pp. 185-194, 1987.

[Warren87] D. H. D. Warren. OR-parallel Execution Models of Prolog. In Proc. of the Int. Joint Conference on Theory and Practice of Software Development, Pisa, 1987.

PIPELINE PROCESSING OF MULTI-WAY JOIN QUERIES IN

SHARED-MEMORY SYSTEMS

Kian-Lee TAN   Hongjun LU

Department of Information Systems and Computer Science

National University of Singapore

10 Kent Ridge Crescent, Singapore 0511

Internet: {tankl,luhj}@iscs.nus.sg

ABSTRACT – *This paper explores the pipeline processing of multi-way join queries using hash-based join algorithm. The basic approach that has been adopted in the literature is: "split" the join tree into segments and for each segment, split each base relation into buckets and pipeline the segment using the buckets. The effectiveness of the approach depends on two closely related factors – the number of buckets and the number of segments. We investigate how varying these factors may affect the system performance. Two greedy heuristics, proposed to generate query evaluation plans with "optimal" pipeline length, are shown to perform best in most cases.*

INTRODUCTION

Pipelining is the technique whereby several operations are executed in an *interleaved* fashion by streaming the output of one operator, before the operator finishes processing, to the input of another operator. Pipelining may provide a lower elapsed time by avoiding the materialization and possibly disk I/Os of some intermediate results. It will also provide a faster *initial* response time of a query. This is especially important for interactive systems. As such, it has been exploited in relational database systems such as System R.

In relational database systems, *join* is the most costly operation. With novel applications, such as graphics, geometric modeling and object-oriented database systems, it will not be uncommon to have queries involving many relations. Most of the previous work on multi-way join query optimization have reduced the problem to that of finding an optimal join order. The development of effective schemes to exploit pipelining to optimize multi-way join queries has received lesser attention. This is so since incorporating pipelining increases the search space drastically.

For multi-way join queries, pipelining can be performed using hash-join algorithm in two phases [1,2].

In the *splitting phase*, the join processing tree is split into *segments*, each of which is a right-deep tree that comprises several relations. We call the inner relations the *building* relations and the outer relation the *probing* relation. For each segment, partition the relations into buckets. In the *pipelining phase*, each segment is processed in a pipeline fashion: hash tables for buckets from the building relations are built and buckets from the probing relation is used to find matching tuples in the hash tables.

The approaches adopted in the literature concentrate on two extremes:

- several *short* pipelines [1,2]. In this approach, the basic strategy is to split the join processing tree into segments such that the *entire* of all the building relations within each segment fit in memory. The result of an earlier segment becomes the input of the next segment.

- a single *long* pipeline [2]. For this method, all the relations are processed in a single pipeline. Since the memory may not be able to contain all the building relations, each relation may be split into buckets and the buckets are processed in a pipeline fashion each time.

These work prompted several questions: 1) what is the relative performance of the two approaches under different circumstances? 2) would varying the length of the pipeline improve the performance? 3) what heuristics can be applied to incorporate pipelining without increasing the optimization cost drastically? In this paper, we address the above issues for the space of *right-deep processing trees*. It has been shown that such trees have the greatest potential for pipelining [2].

MULTI-WAY JOIN ALGORITHMS

We propose two heuristics to study the impact of varying the pipeline length on system performance.

Both techniques comprise the two phases as described above. The difference between the two heuristics is on how the tree is split.

Segment-based Pipeline Processing (SPP)

Algorithm SPP splits the processing tree based on the number of segments that is desired. First, the maximum number of segments, max_seg, to split the processing tree is determined. For $n+1$ relations, the maximum number of segments is at most n. Next, it iterates through all the possible number of segments, from 1 to max_seg. For iteration k, it generates the pipeline plan for k segments. The relations within each segment are selected such that the sum of the sizes of the building relations within each segment is approximately the same. It then computes the cost of this plan, which is the sum of the cost of each segment. The plan with the minimum cost is then the optimal solution.

Bucket-based Pipeline Processing (BPP)

For algorithm SPP, since the sum of the sizes of the building relations may vary, the number of buckets for the various segments of a tree may be different. Algorithm BPP "fixes" the number of buckets for each segment. Varying the number of buckets gives us a family of plans with different pipeline length. By setting the number of buckets to a large value, we can get a long pipeline and a small number of buckets gives a short pipeline. Note that it is possible that a relation is so large that it not only do not fit in memory but requires the number of buckets to exceed the value of k in iteration k. In such cases, the relation participates only in a single join.

While the proposed heuristics have the advantage, as in all hash-based algorithms, that each base relation is read no more than 3 times, its main problem is the extra I/O cost incurred for writing and reading the intermediate results.

PIPELINING IN PARALLEL SYSTEMS

The parallel architecture considered is *shared memory (SM)* system. In SM systems, all the processors share a common pool of memory and the disks. Each relation is horizontally partitioned across the disks in the system. Initially, all relations are evenly distributed among the disks to facilitate full concurrent access to the relation. Access to the memory is regulated by a locking mechanism. Shared-memory systems have the advantage that load-balancing can be easily achieved. To explore pipelining in such systems, all processors will read in portion of the probing relation and all of them can probe for matches in the hash tables in memory. Thus all processors will contribute to the final result of the pipeline.

Based on the description of the proposed heuristics, it suffices for us to show how each segment is processed in shared-memory systems. We assume a worse-case scenario, that is all the joins have different joining attributes. Thus, after each join, the intermediate results must be rehashed. We use R_i to denote the relation that corresponds to the ith join in the pipeline.

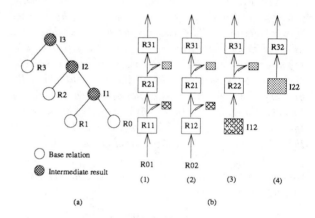

Figure 1: Pipelining a 4-relation segment.

The processing of each segment comprises three phases. In the first phase, for each of the building relations, the following is done: all the processors will read portion of the relation and partition it into $b+1$ buckets. As in the hybrid join algorithm, the first bucket is kept in memory while the other partitions are written to the disk. In the second phase, all the processors will read portion of the probing relation and partition it using the same hashing function and join attribute as relation R_1. At the same time, the first bucket is used to join with the other buckets from the other relations. Since each join may be on different join attributes from the previous join, the intermediate result must be re-hashed such that portion of it will be used for the next join in the pipeline while the rest are partitioned and copied to the appropriate buffer for the bucket to be written out to disk for later processing. When phase 2 completes, we have a partial result of the multi-way join. Thus all processors contribute to the result of the pipeline. At this time, the first bucket of R_1 is no longer needed and the space may be used to build the hash table for the second bucket of R_1. The second bucket of R_0 is used as the probing bucket. This begins phase 3. This process is repeated for all the buckets of R_1. When there is no more buckets from R_1 (and R_0), the process is repeated for all buckets from R_2 with the corresponding buckets from the intermediate results I_1 as the probing buckets. All this will continue until all the buckets are processed. Figure 1 illustrates how a 4-relation segment is pipelined. The right-deep join tree is shown in Figure 1a. Suppose $\sum_{i=1}^{3} |R_i| > M$ where M is the memory size available for the join but $1/2 \cdot \sum_{i=1}^{3} |R_i| = M$. Four iterations of processing this segment are shown in Figure 1b.

OPTIMIZATION WITH PIPELINING

The algorithms described above can be added into existing optimizer to generate an "optimal" pipelined plan easily. Using a *two-phase optimization* approach, the optimization process is split into two phases – *generating an optimal join order* followed by *improving the pipeline processing of the selected tree*. Any existing optimization algorithms can be employed for the first phase. The second phase may be achieved by the algorithms proposed in this work. While the two-phase approach may be simple to implement without affecting existing optimizer, it may prune away a better plan. The *single-phase optimization* approach integrates the two phases into one. Thus, for all possible plans, as each plan is generated, its pipelining is improved using the proposed algorithms. The best plan is then kept as the optimal plan. While this approach generates better plans, it incurs higher optimization cost.

A PERFORMANCE STUDY

To study the performance of the algorithms, we developed an analytical model [3]. The model captures the I/O and CPU cost of each phase in a memory-resident pipeline segment and computes the elapsed time of a k-segment tree as:

$$\sum_{i=1}^{k} \sum_{j=1}^{3} \max(I/O_{ij}, CPU_{ij})$$

where I/O_{ij} and CPU_{ij} are the I/O and CPU cost of phase j in segment i respectively.

For the experiments, we vary the number of processors, the memory size and the number of relations. We can only present here some interesting results. In particular, we present the case when the number of processors is 4 and each query has 20 relations. A total of 200 20-relation queries are tested. Two memory sizes are used for illustration purposes. Small and large memory sizes are represented respectively by 4 MBytes and 64 MBytes. For small memory size, no more than 1 entire relation fit in memory. For large memory size, several entire relations fit in memory.

The other experiments show similar results. We use the *CostMultiplier* as a metric to measure the relative performance of the algorithms which is defined as follows:

$$CostMultiplier(H, opt) = \frac{Elapsed(H)}{Elapsed(opt)}$$

where H is any algorithm, *opt* is the best algorithm (in terms of elapsed time) among all the algorithms studied and $Elapsed(A)$ is the elapsed time for algorithm A. Thus a value of 1 implies that the algorithm is the best and a value of 1.2 means that the algorithm is 20% worse than the best algorithm.

Experiment 1. This experiment studies each iteration of algorithm SPP. The purpose is to see how the number of segments affect the performance of SPP and whether we can restrict to a small number of segments rather than trying out all the number of segments. Table 1 shows the results. For Table 1 and subsequent tables, the columns of each row are in percentages, that is they reflect the percentages of the queries tested that fall into the range of the *CostMultiplier* value indicated by the column.

Table 1: Vary number of segments.

#segments	Cost multiplier (large memory)				
k	1.0	< 1.1	< 1.2	< 1.3	≥ 1.3
1	14	78	6	1	1
2	15	78	5	2	0
3	8	83	4	4	1
6	5	81	7	5	2
8	17	69	6	5	3
10	44	42	6	4	4
#segments	Cost multiplier (small memory)				
k	1.0	< 1.1	< 1.2	< 1.3	≥ 1.3
1	84	16	0	0	0
2	14	86	0	0	0
3	3	97	0	0	0
6	1	97	2	0	0
8	0	85	15	0	0
10	0	85	15	0	0

There are several interesting observations that we made. First there is no single pipeline length that is superior in all cases, that is different number of segments performs well in different situations. This is seen from the fact that no single column in Table 1 has 100% for the CostMultiplier value. Second, for large amount of memory, using a large number of segments performs best for a large number of queries. When the memory is large, a large number of segments implies that fewer relations are processed as a pipe each time. More portions of each relation is used to built the hash table in memory. This means that more intermediate results at each level flow through the pipe and less are written out. On the other hand, by decreasing the number of segments, more intermediate results need to be written out. Since the I/Os incurred are mainly random I/Os, a small number of segments becomes more expensive. However, when memory is scarce, a small number of segments becomes superior. A short pipeline implies the need to write out the entire of several intermediate results. By exploiting a longer pipeline (i.e. small number of segments), this may be avoided by allowing the intermediate result to flow through the pipe. Moreover, most of the I/Os are random for small memory size.

Experiment 2. This experiment is similar to that of Experiment 1 except that we study each iteration of the algorithm BPP instead, that is how the number of buckets affect the performance of BPP. The purpose is to see whether we can restrict to just some values of k rather than a large number of k values. Table 2 shows the results. The result of this study is similar to that in Experiment 1, that is to effectively exploit pipelining in system with large (small) memory, each of the relations should be split into a small (large) number of buckets.

Table 2: Vary number of buckets.

#buckets	Cost multiplier (large memory)				
k	1.0	< 1.1	< 1.2	< 1.3	≥ 1.3
0	63	33	1	3	0
1	30	66	4	0	0
2	4	71	16	6	3
4	0	52	29	9	10
6	0	47	28	10	15
8	1	43	28	12	16
max	1	40	28	13	18
#buckets	Cost multiplier (small memory)				
k	1.0	< 1.1	< 1.2	< 1.3	≥ 1.3
0	2	65	6	6	21
1	2	65	6	6	21
2	2	65	6	6	21
4	2	65	6	6	21
6	6	61	6	8	19
8	49	19	6	7	19
max	47	14	8	7	24

Experiment 3. In this experiment, we compare the performance of the two heuristics with that of the two conventional approaches – algorithm RDHS that

processes all relations in a single pipeline and algorithm staticRD that keeps entire relations in memory [2]. Table 3 shows the relative performance of the algorithms for a total of 400 queries. Both small and large memory are studied. It is clear from the results that while RDHS and staticRD may perform well in some cases, better performance are given by the two proposed heuristics. This is so since both RDHS and staticRD are considered in the two heuristics as the number of iterations varies. We also computed a 95% confidence interval of the relative performance of the algorithms and shows that RDHS and staticRD range from 5% – 20% worse than the two heuristics, that is when algorithm RDHS and staticRD performs worse, they perform badly. Thus by restricting to either one of these two algorithms will lead to poor performance.

Table 3: Relative performance of the algorithms.

Algorithm	Cost multiplier				
	1.0	< 1.1	< 1.2	< 1.3	≥ 1.3
SPP	80	15	3	1	1
BPP	88	9	1	1	0
RDHS	50	16	8	11	10
staticRD	30	24	13	5	25

CONCLUSION

In this paper, we have studied how to effectively pipeline a multi-way join query. The length of the pipeline may affect the system performance significantly. Our study shows that previous work that restrict to a single long pipeline or several short pipelines are not effective. There is no single pipeline length that is optimal for all cases. While pipelining a large number of joins is favorable when memory is scarce, a short pipeline performs better for large memory. The two proposed heuristics not only provide better performance but can be easily incorporated into existing optimizer. Moreover, enlarging the search space to explore pipelining using the heuristics do not increase the optimization overhead drastically.

REFERENCE

[1] Chen, M.S., *et. al.*, Using Segmented Right-Deep Trees for the Execution of Pipelined Hash Joins, *Proc. VLDB 92*, Aug 1992, 15-26.

[2] Schneider, D. and DeWitt, D., Tradeoffs in Processing Complex Join Queries via Hashing in Multiprocessor Database Machines, *Proc. VLDB 91*, Aug 1991, 469-480.

[3] Tan, K.L. and Lu, H., Incorporating Pipelining into Query Optimizer, *Tech. Rep. TR B4/93, National University of Singapore*, Ap 1993.

PANEL: IN SEARCH OF A UNIVERSAL (BUT USEFUL) MODEL OF PARALLEL COMPUTATION

Howard Jay Siegel, Panel Moderator
Parallel Processing Laboratory
School of Electrical Engineering
Purdue University
West Lafayette, IN 47907-1285, USA
hj@ecn.purdue.edu

Is it possible to create one idealized parallel computer model to provide an interface between hardware and software performance that supports a machine-independent framework for specifying various computational costs? Is it necessary to have more than one? What attributes should this model(s) have? Does such a model(s) exist today? How can we create such a model(s)? Is it important to have such a model(s)?

The following people have agreed to participate as panelists to discuss these issues:

H. J. Siegel (Moderator), Purdue University
James C. Browne, University of Texas at Austin
Tom DeBoni, Lawrence Livermore Nat'l Lab.
Jack B. Dennis, MIT
Dennis Gannon, Indiana University
Joseph JaJa, University of Maryland
Leah H. Jamieson, Purdue University

The topic for this panel was motivated by the outcome of a workshop sponsored by the National Science Foundation entitled ''The Purdue Workshop on Grand Challenges in Computer Architecture for the Support of High Performance Computing.'' This workshop involved 20 computer architects from across the country. The design of such a universal model (or small set of models) was identified as a ''grand challenge'' for computer architects, as discussed in [1].

Each member of the panel was asked to prepare a paragraph position statement for these proceedings. Their position statements are reproduced unedited below.

James C. Browne:

Plausibility arguments that there exists a ''universal'' model of parallel computation are easy to put forward. Given sufficiently restrictive definitions and acceptance of a sufficiently high level of abstraction rigorous proofs can probably be given of ''universality.'' The real interest attaches to the constructive power of the model and to its ability to support translations across levels of abstraction to realizable systems. That is, the models of interest should not only support analyses for properties from representations but also realizations of concrete systems.

The model of parallel computation I will propose can, in principle, be translated to either hardware or software. The model, in its most abstract form, contains two elements, a set of operations and a set of composition rules. The substance of the presentation will be a demonstration that the common models of parallel computation for hardware and software systems can be cast in this model with appropriate choices of operations and composition rules.

Tom DeBoni:

The von Neumann model of computation has succeeded because it explicitly concerns itself with straightforward, sequential, one-thing-at-a-time-ness, a mode of operation humans can use efficiently. A general model of parallel computation must focus on getting many things done at the same time, a mode of operation humans are not good at. The difference, then, seems to be the deemphasis of time and the introduction of simultaneity in the desired new formalism. This can be accomplished by using mathematics as the basis of the new model, reintroducing the time dimension as sparsely as possible and only when needed, and relying on automatic methods to handle the ''dirty details'' of mapping programs to architectures. This avoids the questions of how to accomplish such broad ends, much as the von Neumann model did. The success of any new model will require the invention of ''implementation details,'' similar to the mechanisms that grew out of operating system and language research to support and refine the von Neumann model.

Jack B. Dennis:

The ability to quantify costs of program execution is in conflict with the power of composing programs from independently written modules. In the long run there will be greater benefit from computer architectures that support modular program construction than from those that permit precise performance prediction. The path to parallel computers that can support general purpose modular program construction is known and should be pursued with vigor. The semantic model such advanced parallel computers implement will be based on dataflow and functional programming principles.

Dennis Gannon:

Two factors are essential for a useful model of parallel computation. First, the model must allow a way to reason about and predict performance for scalable algorithm design. Second, algorithms designed with the model should translate directly into a real programming language that runs on contemporary parallel systems. In other words, analytic predictions made with the model should be valid for the corresponding real software. Consequently, models like the PRAM are relatively useless, but the Bulk Synchronous Parallel Model (BSP) is relatively good. In particular, BSP algorithm design is relatively consistent with the data parallel and object parallel programming styles. On the other hand, BSP is not very good with systolic or data flow programming models, nor is it very good for describing computation that involves nested concurrency.

Joseph JaJa:

The success of massively parallel computers depends on performance, cost, and programming effort. A balance between the three factors, architecture, software, and algorithms, under a unified logical model is required before massively parallel processing becomes a reality. These factors have been developing in different directions with competing models. I believe that the shared-memory model, where a collection of processes communicate asynchronously by accessing shared variables, can form the basis of a unified logical framework in which fundamental issues in hardware, software and algorithms can be addressed. The important data parallel model should be easy to incorporate efficiently into the shared memory model.

Leah H. Jamieson:

A useful universal model of parallel computing should consist of three parts: (1) An algorithm representation that supports analysis of computation, communication, and synchronization complexity. Overall complexity must be expressible in terms of these separate components: without the ability to identify the components, it may not be possible to relate asymptotic complexity in the general model to asymptotic complexity on even idealized machines. (2) Mappings that allow (possibly stochastic) translation of complexity in terms of theoretical communication and synchronization operations to expected or worst-case communication and synchronization times under specific topologies. These mappings provide the link between the machine-independent complexity model and the more realistic machine- and data-dependent performance. (3) A growth model that governs what grows and what doesn't. An algorithm in which the number of processors is allowed to grow with the problem size may have radically different behavior than one where the problem size grows but the machine size remains fixed, or the problem size remains fixed and the machine size grows.

Models of sequential computing provided a theory of scalability at a time when the dimensions of scalability were few: how big is the problem, how fast is the processor and, possibly, how much memory is there? Models of parallel computing should serve the same function, but at a time when the dimensions of scalability must be expanded to include questions about how many processors there are, how they are connected, how they might delay one another, and how many of the dimensions might be permitted to grow simultaneously. In effect, a model of parallel computing is a model of algorithm scalability in a multidimensional space.

REFERENCE

[1] Howard Jay Siegel, Seth Abraham, William L. Bain, Kenneth E. Batcher, Thomas L. Casavant, Doug DeGroot, Jack B. Dennis, David C. Douglas, Tse-yun Feng, James R. Goodman, Alan Huang, Harry F. Jordan, J. Robert Jump, Yale N. Patt, Alan Jay Smith, James E. Smith, Lawrence Snyder, Harold S. Stone, Russ Tuck, and Benjamin W. Wah, "Report of the Purdue Workshop on Grand Challenges in Computer Architecture for the Support of High Performance Computing," *Journal of Parallel and Distributed Computing,* Vol. 16, No. 3, pp. 199-211, Nov. 1992.

TABLE OF CONTENTS - FULL PROCEEDINGS

Volume I = Architecture
Volume II = Software
Volume III = Algorithms & Applications

A1

A2

A3

A7